GUIDED NOTEBOOK WITH INTEGRATED REVIEW WORKSHEETS

INTERMEDIATE ALGEBRA
THIRTEENTH EDITION

Margaret L. Lial
American River College

John Hornsby
University of New Orleans

Terry McGinnis

Pearson

Reproduced by Pearson from electronic files supplied by the author.

ISBN-13: 978-0-13-489638-0
ISBN-10: 0-13-489638-6

1 2019

CONTENTS for INTEGRATED REVIEW WORKSHEETS

CONTENTS for GUIDED LECTURE NOTEBOOK

Chapter 1 LINEAR EQUATIONS, INEQUALITITES, AND APPLICATIONS

Learning Objectives
Simplify, and then use the addition property of equality.
Simplify, and then use the multiplication property of equality.
Solve applications involving the sums of quantities.
Solve linear inequalities using both properties of inequalities.

Key Terms

Use the vocabulary terms listed below to complete each statement in exercises 1–3.

linear equation   **solution set**   **equivalent equations**

1.   Equations that have exactly the same solutions sets are called
     ________________________________.

2.   An equation that can be written in the form $Ax + B = C$, where A, B, and C are real numbers and $A \neq 0$, is called a ________________________________.

3.   The set of all numbers that satisfy an equation is called its ________________.

Integrated Review Worksheets for *Intermediate Algebra*, 13th edition

Objective 1   Simplify, and then use the addition property of equality.

Review these examples for Objective 1:

1.   Solve $5t-16+t+4=9+5t+6$.

$$5t-16+t+4=9+5t+6$$
$$6t-12=15+5t$$
$$6t-12-5t=15+5t-5t$$
$$t-12=15$$
$$t-12+12=15+12$$
$$t=27$$

Check by substituting 27 in the original equation. The solution set is $\{27\}$.

2.   Solve $4(3+6x)-(5+23x)=19$.

$$4(3+6x)-(5+23x)=19$$
$$4(3)+4(6x)-1(5)-1(23x)=19$$
$$12+24x-5-23x=19$$
$$x+7=19$$
$$x+7-7=19-7$$
$$x=12$$

Check by substituting 12 in the original equation. The solution set is $\{12\}$.

Now Try:

1.   Solve
$$8t-9+t+7=12+8t+15.$$

____

2.   Solve
$$5(7+8x)-(29+39x)=14.$$

____

Objective 1 Practice Exercises

Solve each equation. First simplify each side of the equation as much as possible. Check each solution.

1. $3(t+3)-(2t+7)=9$

1. _______________

2. $-4(5g-7)+3(8g-3)=15-4+3g$

2. _______________

3. $3.6p+4.8+4.0p=8.6p-3.1+0.7$

3. _______________

Objective 2   Simplify, and then use the multiplication property of equality.

Review these examples for Objective 2:	**Now Try:**
3. Solve $9m+4m=39$.	**3.** Solve $12m+8m=80$.

$$9m+4m=39$$
$$13m=39$$
$$\frac{13m}{13}=\frac{39}{13}$$
$$m=3$$

Check by substituting 3 in the original equation. The solution set is $\{3\}$.

4. Solve $3(2x-7)+21=-18$.   **4.** Solve $4(2x-5)+20=-32$.

$$3(2x-7)+21=-18$$
$$3(2x)+3(-7)+21=-18$$
$$6x-21+21=-18$$
$$6x=-18$$
$$\frac{6x}{6}=\frac{-18}{6}$$
$$x=-3$$

Check by substituting –3 in the original equation. The solution set is $\{-3\}$.

Objective 2 Practice Exercises

Solve each equation and check your solution.

4. $-7b+12b=125$ **4.** _______________

5. $3w-7w=20$ **5.** _______________

6. $-11h-6h+14h=-21$ **6.** _______________

Objective 3   Solve applications involving sums of quantities.

Review these examples for Objective 3:

5. George and Al were opposing candidates in the school board election. George received 21 more votes than Al, with 439 votes cast. How many votes did Al receive?

Step 1 Read the problem carefully. We are given total votes and asked to find the number of votes Al received.

Step 2 Assign a variable.
Let x = the number of votes Al received.
Then $x + 21$ = the number of votes George received.

Step 3 Write an equation.

The total is for Al plus votes for George

$$439 = x + (x+21)$$

Step 4 Solve the equation.

$$439 = x + (x+21)$$
$$439 = 2x + 21$$
$$439 - 21 = 2x + 21 - 21$$
$$418 = 2x$$
$$\frac{418}{2} = \frac{2x}{2}$$
$$209 = x \quad \text{or} \quad x = 209$$

Step 5 State the answer. Al received 209 votes.

Step 6 Check. George won $209 + 21 = 230$ votes. The total number of votes is $209 + 230 = 439$. The answer checks.

Now Try:

5. On a psychology test, the highest grade was 38 points more than the lowest grade. The sum of the two grades was 142. Find the lowest grade.

____

6. Penny is making punch for a party. The recipe requires twice as much orange juice as cranberry juice and 8 times as much ginger ale as cranberry juice. If she plans to make 176 ounces of punch, how much of each ingredient should she use?

Step 1 Read the problem. The three amounts of ingredients must be found.

Step 2 Assign a variable.
   Let x = the number of ounces of cranberry juice.
Then $2x$ = the number of ounces of orange juice, and $8x$ = the number of ounces of ginger ale.

Step 3 Write an equation.
Cranberry plus orange plus ginger ale is total

$$x + 2x + 8x = 176$$

Step 4 Solve the equation.
$$x + 2x + 8x = 176$$
$$11x = 176$$
$$\frac{11x}{11} = \frac{176}{176}$$
$$x = 16$$

Step 5 State the answer. There are 16 ounces of cranberry juice, $2(16) = 32$ ounces of orange juice, and $8(16) = 128$ ounces of ginger ale.

Step 6 Check. The sum is 176. All conditions of the problem are satisfied.

6. Linda wishes to build a rectangular dog pen using 52 feet of fence and the back of her house, which is 36 feet long to enclose the pen. How wide will the dog pen be if the pen is 36 feet long?

____

Objective 3 Practice Exercises

Write an equation for each of the following and then solve the problem. Use x as the variable.

7. Denali in Alaska is 5910 feet higher than Mount Rainier in Washington. Together, their heights total 34,730 feet. How high is each mountain?

7. _________________

Mt. Rainier ____________

Denali _______________

8. Charles bought five general admission tickets and four student tickets for a movie. He paid $35.25. If each student ticket cost $3.50, how much did each general admission ticket cost?

8. _________________

____

9. Pablo, Faustino, and Mark swim at a public pool each day for exercise. One day Pablo swam five more than three times as many laps as Mark, and Faustino swam four times as many laps as Mark. If the men swam 29 laps altogether, how many laps did each one swim?

9. _________________

Mark _________________

Pablo_________________

Faustino _________________

Objective 4   Solve linear inequalities using both properties of inequality.

Review this example for Objective 4:

7. Solve $4x+3-7>-2x+8+3x$. Graph the solution set.

Step 1 Combine like terms and simplify.
$$4x+3-7>-2x+8+3x$$
$$4x-4>x+8$$

Step 2 Use the addition property of inequality.
$$4x-4-x>x+8-x$$
$$3x-4>8$$
$$3x-4+4>8+4$$
$$3x>12$$

Step 3 Use the multiplication property of inequality.
$$\frac{3x}{3}>\frac{12}{3}$$
$$x>4$$

The solution set is $(4,\ \infty)$. The graph is shown below.

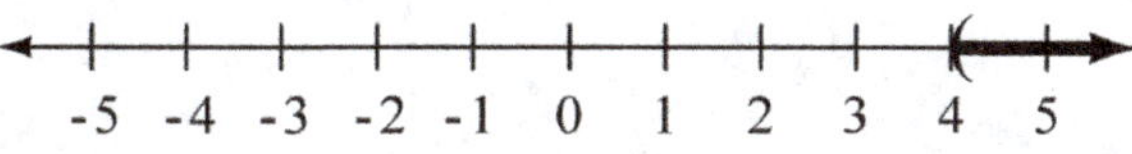

Now Try:

7. Solve $8x-5+4\geq 6x-3x+9$. Graph the solution set.

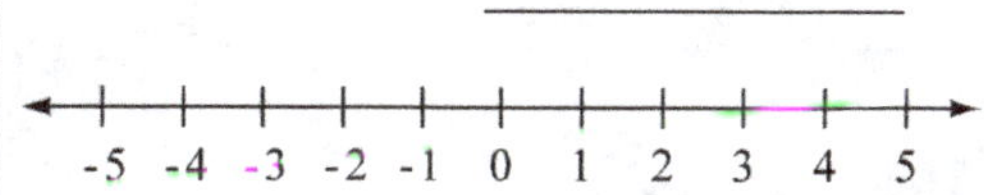

Objective 4 Practice Exercises

Solve each inequality. Write the solution set in interval notation and then graph it.

10. $4(y-3)+2>3(y-2)$

10. _____________________

11. $-3(m+2)+3\leq -4(m-2)-6$

11. _____________________

12. $7(2-x)\leq -2(x-3)-x$

12. _____________________

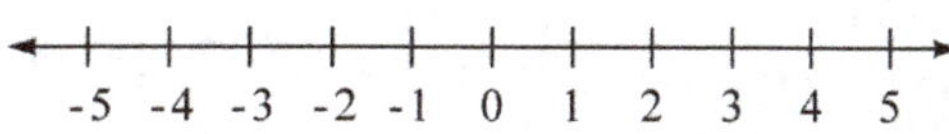

Chapter 2 LINEAR EQUATIONS, GRAPHS, AND FUNCTIONS

Learning Objectives
Graph linear equations in the form $Ax + By = C$
Determine if the slope of a line is positive, negative, zero, or undefined.
Use slopes to determine whether two lines are parallel, perpendicular, or neither.
Write an equation of a line using its slope and any point on the line.
Graph linear inequalities in two variables.

Key Terms

Use the vocabulary terms listed below to complete each statement in exercises 1−7.

graph　　　　**graphing**　　　　***y*-intercept**　　　　***x*-intercept**

perpendicular lines　　　　**parallel lines**

linear inequality in two variables

1. If a graph intersects the y-axis at k, then the _________________ is $(0, k)$.

2. If a graph intersects the x-axis at k, then the _________________ is $(k, 0)$.

3. Two lines in a plane that never intersect are called _________________.

4. An inequality that can be written in the form $Ax + By < C$, $Ax + By > C$, $Ax + By \leq C$, or $Ax + By \geq C$ is called a _________________.

5. The process of plotting the ordered pairs that satisfy a linear equation and drawing a line through them is called _________________.

6. The set of all points that correspond to the ordered pairs that satisfy the equation is called the _________________ of the equation.

7. Two lines that intersect in a 90° angle are called _________________.

Objective 1   Graph linear equations of the form $Ax + By = C$.

Review this example for Objective 1:

1.   Graph $x + 5y = 0$.

To find the y-intercept, let $x = 0$.
To find the x-intercept, let $y = 0$.

$$\begin{array}{c|c} 0 + 5y = 0 & x + 5(0) = 0 \\ 5y = 0 & x + 0 = 0 \\ y = 0 & x = 0 \end{array}$$

The x- and y-intercepts are the same point (0, 0). We must select two other values for x or y to find two other points. We choose $y = 1$ and $y = -1$.

$$\begin{array}{c|c} x + 5(1) = 0 & x + 5(-1) = 0 \\ x + 5 = 0 & x - 5 = 0 \\ x = -5 & x = 5 \end{array}$$

We use (–5, 1), (0, 0), and (5, –1) to draw the graph.

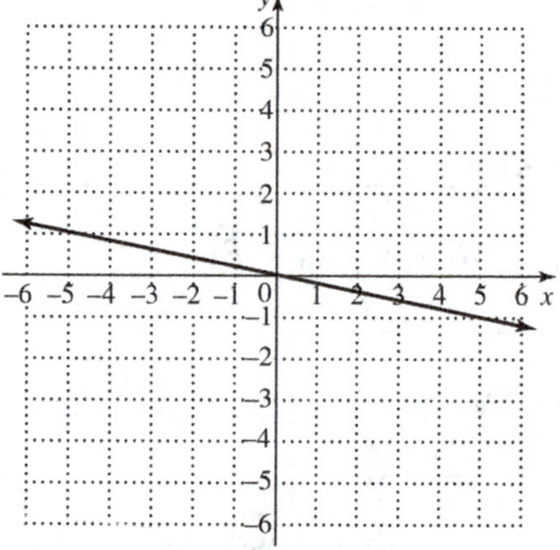

Now Try:

1.   Graph $3x - y = 0$.

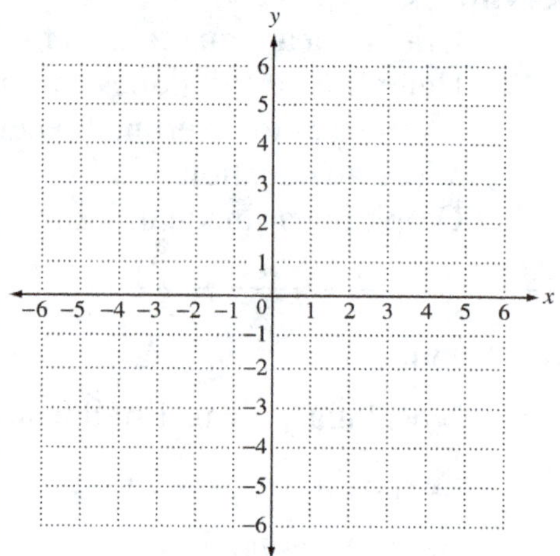

Objective 1 Practice Exercises

Graph each equation.

1. $-3x - 2y = 0$

1.

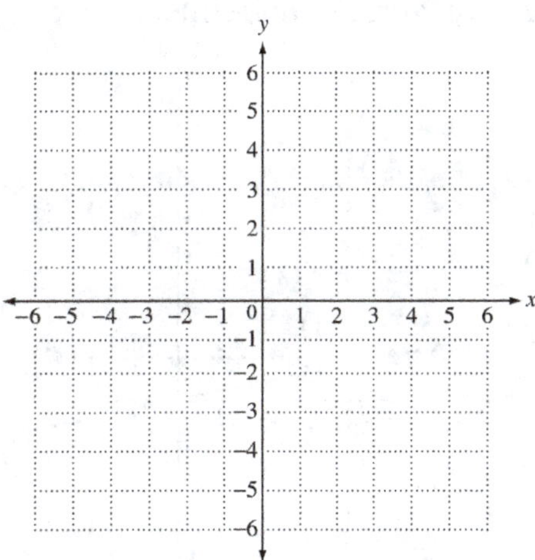

2. $x + y = 0$

2.

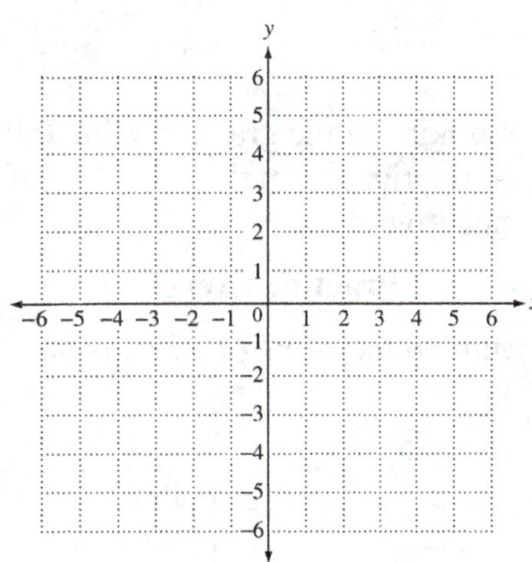

3. $y = 2x$

3.

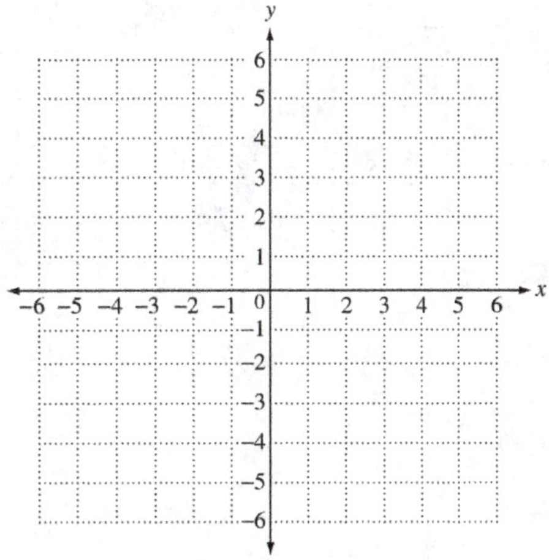

Objective 2   Determine if the slope of a line is positive, negative, zero, or undefined.

Review these examples for Objective 2:

2.   Determine if the slope of a line is positive, negative, zero, or undefined.

Now Try:

2.   Determine if the slope of a line is positive, negative, zero, or undefined.

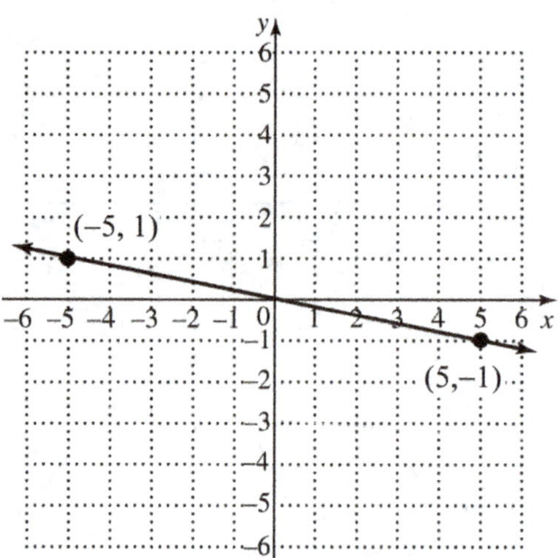

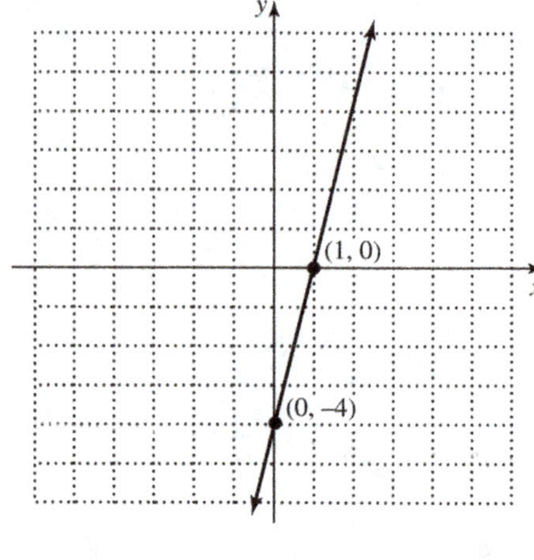

Looking at the graph the line falls as we move to the right from the left. Therefore it has a negative slope.

Objective 2 Practice Exercises

Determine if the slope of a line is positive, negative, zero, or undefined.

4.

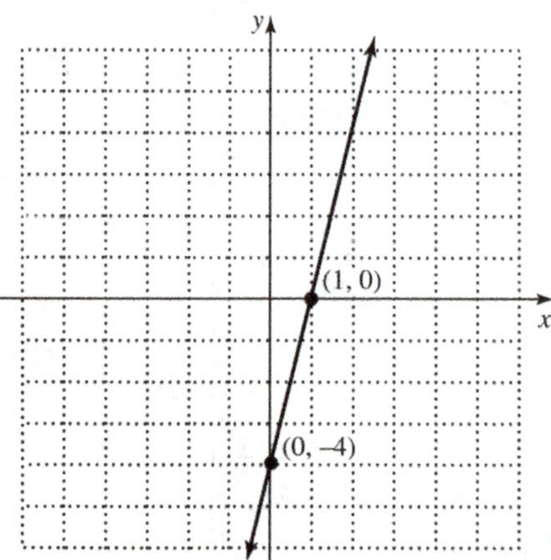

4. _______________

    Copyright © 2020 Pearson Education, Inc.

5.

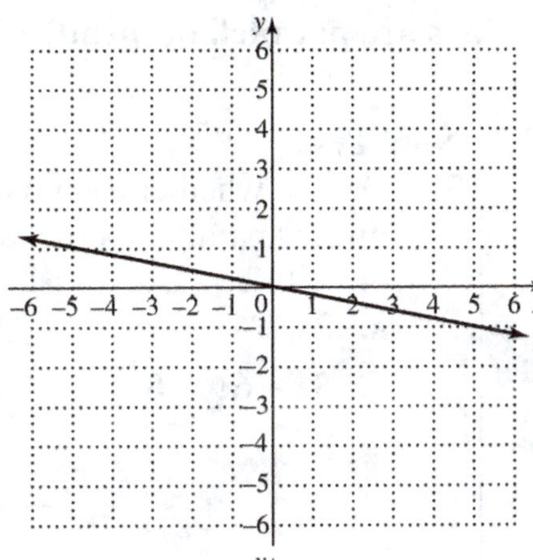

5. _______________

6.

6. _______________

Objective 3 Use slopes to determine whether two lines are parallel, perpendicular, or neither.

Review these examples for Objective 3:

3. Decide whether each pair of lines is parallel, perpendicular, or neither.

   a. $5x - y = 3$

   $15x - 3y = 12$

   Solve each equation for y.
   $$y = 5x - 3$$
   $$y = 5x - 4$$
   Both lines have slope 5, so the lines are parallel.

   b. $x + 3y = 8$ and $-3x + y = 5$

   Find the slope of each line by first solving each equation for y.

   $$3y = -x + 8 \qquad \qquad y = 3x + 5$$
   $$y = -\frac{1}{3}x + \frac{8}{3}$$

   The slope is $-\frac{1}{3}$. The slope is 3.

   Because the slopes are not equal, the lines are not parallel.

   Check the product of the slopes: $-\frac{1}{3}(3) = -1$.

   The two lines are perpendicular because the product of their slopes is -1.

Now Try:

3. Decide whether each pair of lines is parallel, perpendicular, or neither.

   a. $2x - 4y = 7$

   $3x - 6y = 8$

   ____

   b. $9x - y = 7$ and $x + 9y = 11$

   ____

Objective 3 Practice Exercises

In each pair of equations, give the slope of each line, and then determine whether the two lines are **parallel,** **perpendicular,** *or* **neither.**

7. $-x + y = -7$

   $x - y = -3$

7. ________________

8. $4x + 2y = 8$
      $x + 4y = -3$

8. _______________

9. $9x + 3y = 2$
      $x - 3y = 5$

9. _______________

Objective 4   Write an equation of a line using its slope and any point on the line.

Review these examples for Objective 4:

4. Write an equation in slope-intercept form of the line passing through the given point and having the given slope.

    a. $(0,\ 2)$, $m = -3$

    Because the point $(0, 2)$ is the y-intercept, $b = 2$. Substitute $b = 2$ and $m = -3$ directly in the slope-intercept form.
    $$y = mx + b$$
    $$y = -3x + 2$$

    b. $(2,\ 9)$, $m = 5$

    Since the line passes through the point $(2, 9)$, we can substitute $x = 2$, $y = 9$, and slope $m = 5$ into $y = mx + b$ and solve for b.
    $$y = mx + b$$
    $$9 = 5(2) + b$$
    $$-1 = b$$
    Now substitute the values of m and b into slope-intercept form.
    $$y = mx + b$$
    $$y = 5x - 1$$

Now Try:

4. Write an equation in slope-intercept form of the line passing through the given point and having the given slope.

    a. $(0, -5)$, $m = \dfrac{3}{4}$

    ____

    b. $(-1,\ 4)$, $m = 6$

    ____

Objective 4 Practice Exercises

Write an equation in slope-intercept form of the line passing through the given point and having the given slope.

10. $(0, -4)$, $m = \dfrac{2}{3}$

10. ________________

11. $(3,\ 6)$, $m = -2$

11. ________________

12. $(-2,\ 0)$, $m = 1$

12. ________________

Objective 5   Graph linear inequalities in two variables.

Review these examples for Objective 5:

5. Graph $3x - 2y \le 6$.

The inequality $3x - 2y \le 6$ means that
$$3x - 2y < 6 \text{ or } 3x - 2y = 6.$$
We begin by graphing the line $3x - 2y = 6$ with intercepts $(0, -3)$ and $(2, 0)$. This boundary line divides the plane into two regions, one of which satisfies the inequality. We use the test point $(0, 0)$ to see whether the resulting statement is true or false, thereby determining whether the point is in the shaded region or not.

$$3x - 2y \le 6$$
$$3(0) - 2(0) \overset{?}{\le} 6$$
$$0 - 0 \overset{?}{\le} 6$$
$$0 \le 6 \quad \text{True}$$

Since the last statement is true, we shade the region that includes the test point $(0, 0)$. The shaded region, along with the boundary line, is the desired graph.

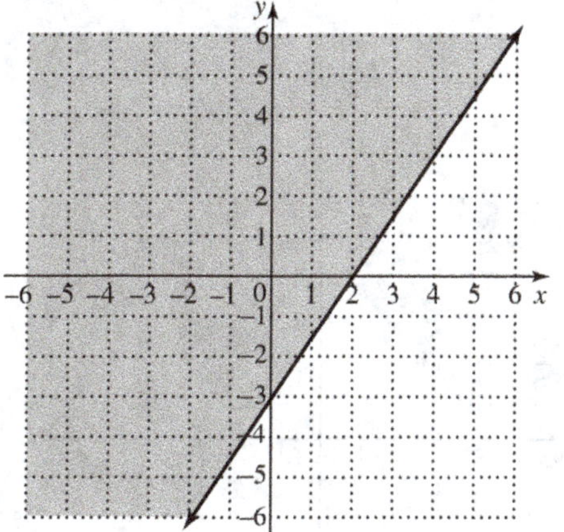

Now Try:

5. Graph $2x + 5y \le -8$.

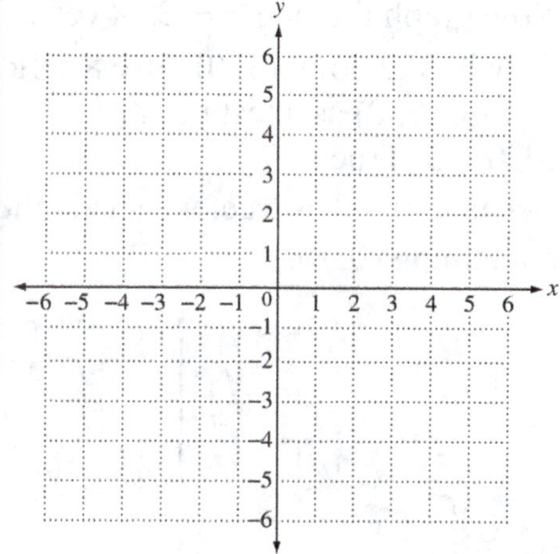

6. Graph $x - 4 \leq -1$.

First, solve the inequality for x.

$x \leq 3$

Now graph the line $x = 3$, a vertical line through the point $(3, 0)$. Use a solid line, and choose $(0, 0)$ as a test point.

$0 \leq 3$ True

Because $0 \leq 3$ is true, we shade the region containing $(0, 0)$.

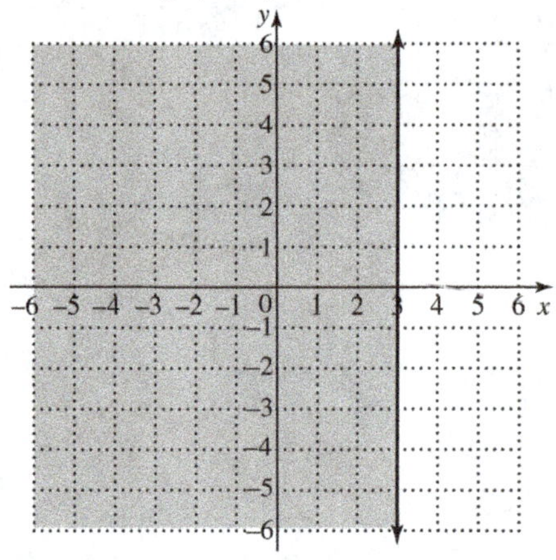

6. Graph $y \geq -1$.

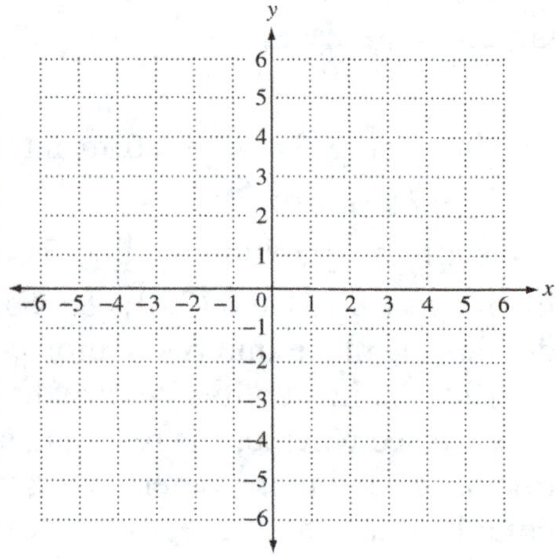

Objective 5 Practice Exercises

Graph each linear inequality.

13. $y \geq x - 1$

13.

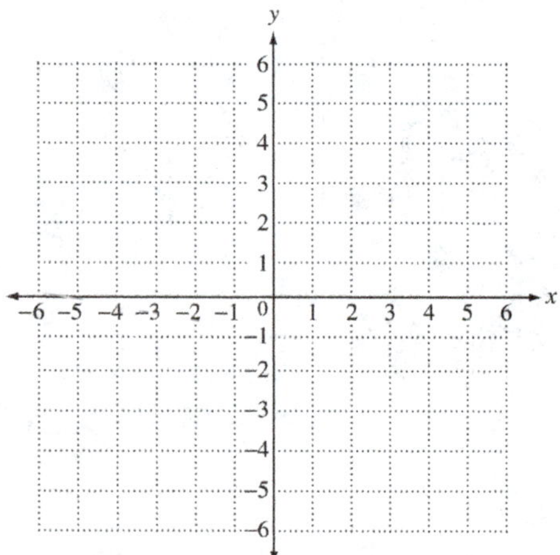

14. $y > -x + 2$

14.

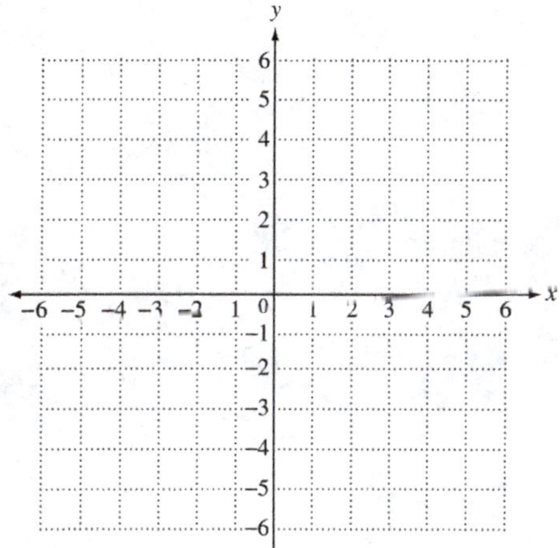

15. $3x - 4y - 12 > 0$

15.

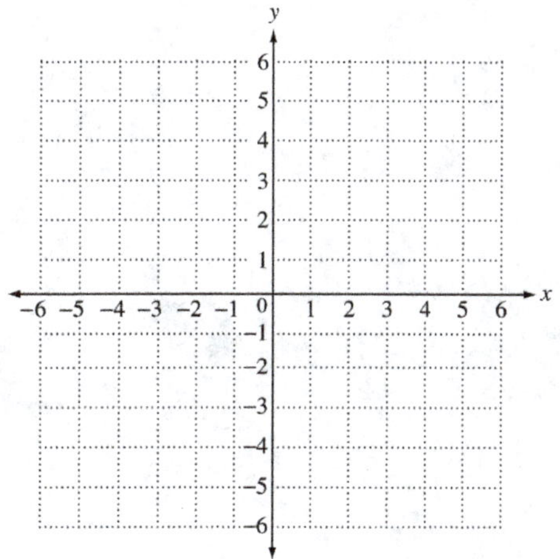

Chapter 3 SYSTEMS OF LINEAR EQUATIONS

Learning Objectives
Determine whether a given ordered pair is a solution of a given equation.
Determine whether a given ordered pair is a solution of a system.
Solve linear systems by substitution.
Multiply when using the elimination method.
Solve problems about quantities and their costs.

Key Terms

Use the vocabulary terms listed below to complete each statement in exercises 1−9.

linear equation in two variables   **solution set of the system**

consistent system   **inconsistent system**   **elimination method**

system of linear equations   **independent equations**   **dependent equations**

substitution

1. An equation that can be written in the form $Ax + By = C$, where A, B, and C are real numbers and A, $B \neq 0$, is called a ________________________________.

2. Equations of a system that have different graphs are called ________________________________.

3. A system of equations with at least one solution is a ________________________________.

4. When one expression is replaced by another, ________________________ is being used.

5. The ________________________________ of linear equations is an ordered pair that makes all the equations of the system true at the same time.

6. Using the addition property to solve a system of equations is called the ________________________________.

7. Equations of a system that have the same graph (because they are different forms of the same equation) are called ________________________________.

8. A system with no solution is called a(n) ________________________________.

9. A(n) ________________________________ consists of two or more linear equations with the same variables.

Objective 1   Determine whether a given ordered pair is a solution of a given equation.

Review these examples for Objective 1:

1. Decide whether each ordered pair is a solution of the equation $4x + 5y = 40$.

   a. $(5,\ 4)$

   Substitute 5 for x and 4 for y in the given equation.
   $$4x + 5y = 40$$
   $$4(5) + 5(4) \overset{?}{=} 40$$
   $$20 + 20 \overset{?}{=} 40$$
   $$40 = 40 \quad \text{True}$$
   This result is true, so $(5,\ 4)$ is a solution of $4x + 5y = 40$.

   b. $(-3,\ 6)$

   Substitute -3 for x and 6 for y in the given equation.
   $$4x + 5y = 40$$
   $$4(-3) + 5(6) \overset{?}{=} 40$$
   $$-12 + 30 \overset{?}{=} 40$$
   $$18 = 40 \quad \text{False}$$
   This result is false, so $(-3,\ 6)$ is not a solution of $4x + 5y = 40$.

Now Try:

1. Decide whether each ordered pair is a solution of the equation $3x - 4y = 12$.

   a. $(8,\ 3)$

   ____

   b. $(5, -4)$

   ____

Objective 1 Practice Exercises

Decide whether the given ordered pair is a solution of the given equation.

1. $4x - 3y = 10;\ (1, 2)$

   1. _______________

2. $2x - 3y = 1;\ \left(0, \frac{1}{3}\right)$

   2. _______________

3. $x = -7;\ (-7, 9)$

   3. _______________

Objective 2   Determine whether a given ordered pair is a solution of a system.

Review this example for Objective 2:

2. Determine whether the ordered pair $(5, -2)$ is a solution of the system.

$$4x + 5y = 10$$
$$3x + 8y = 6$$

Again, substitute 5 for x and -2 for y in each equation.

$$\begin{array}{c|c} 4x + 5y = 10 & 3x + 8y = 6 \\ 4(5) + 5(-2) \overset{?}{=} 10 & 3(5) + 8(-2) \overset{?}{=} 6 \\ 20 - 10 \overset{?}{=} 10 & 15 - 16 \overset{?}{=} 6 \\ 10 = 10 \ \text{True} & \text{False} \ \ -1 = 6 \end{array}$$

The ordered pair $(5, -2)$ is not a solution of this system because it does not satisfy the second equation.

Now Try:

2. Determine whether the ordered pair $(6, 5)$ is a solution of the system.

$$5x - 6y = 0$$
$$6x + 5y = 50$$

____

Objective 2 Practice Exercises

Decide whether the given ordered pair is a solution of the given system.

4. $(2, -4)$

$$2x + 3y = 6$$
$$3x - 2y = 14$$

4. _______________

5. $(-3, -1)$

$$5x - 3y = -12$$
$$2x + 3y = -9$$

5. _______________

6. $(4, 0)$

$$4x + 3y = 16$$
$$x - 4y = -4$$

6. _______________

Objective 3   Solve linear systems by substitution.

Review these examples for Objective 3:

3. Solve the system by the substitution method.
$$2x + 5y = 22 \quad (1)$$
$$y = 4x \quad (2)$$

Equation (2) is already solved for y. We substitute $4x$ for y in equation (1).
$$2x + 5y = 22$$
$$2x + 5(4x) = 22$$
$$2x + 20x = 22$$
$$22x = 22$$
$$x = 1$$

Find the value of y by substituting 1 for x in either equation. We use equation (2).
$$y = 4x$$
$$y = 4(1) = 4$$

We check the solution $(1, 4)$ by substituting 1 for x and 4 for y in both equations.

$$2x + 5y = 22 \qquad\qquad y = 4x$$
$$2(1) + 5(4) \overset{?}{=} 22 \qquad\qquad 4 \overset{?}{=} 4(1)$$
$$2 + 20 \overset{?}{=} 22 \qquad\qquad \text{True } 4 = 4$$
$$22 = 22 \ \text{True}$$

Since $(1, 4)$ satisfies both equations, the solution set of the system is $\{(1, 4)\}$.

4. Solve the system by the substitution method.
$$4x + 5y = 13 \qquad (1)$$
$$x = -y + 2 \quad (2)$$

Equation (2) gives x in terms of y. We substitute $-y + 2$ for x in equation (1).
$$4x + 5y = 13$$
$$4(-y + 2) + 5y = 13$$
$$-4y + 8 + 5y = 13$$
$$y + 8 = 13$$
$$y = 5$$

Find the value of x by substituting 5 for y in either equation. We use equation (2).
$$x = -y + 2$$
$$x = -5 + 2 = -3$$

Now Try:

3. Solve the system by the substitution method.
$$x + y = 7$$
$$y = 6x$$

____

4. Solve the system by the substitution method.
$$2x + 3y = 6$$
$$x = 5 - y$$

____

We check the solution $(-3, 5)$ by substituting -3 for x and 5 for y in both equations.

$$4x + 5y = 13 \qquad\qquad x = -y + 2$$
$$4(-3) + 5(5) \overset{?}{=} 13 \qquad\qquad -3 \overset{?}{=} -5 + 2$$
$$-12 + 25 \overset{?}{=} 13 \qquad\qquad \text{True} \quad -3 = -3$$
$$13 = 13 \ \text{True}$$

Both results are true, so the solution set of the system is $\{(-3, 5)\}$.

Objective 3 Practice Exercises

Solve each system by the substitution method. Check each solution.

7. $3x + 2y = 14$

$\qquad y = x + 2$

7. _______________

8. $x + y = 9$

$\qquad 5x - 2y = -4$

8. _______________

9. $3x - 21 = y$

$\qquad y + 2x = -1$

9. _______________

Objective 4   Multiply when using the elimination method.

Review these examples for Objective 4:

Now Try:

5. Solve the system.
$$2x - 4y = 10 \quad (1)$$
$$x + 5y = -2 \quad (2)$$

5. Solve the system.
$$5x - 6y = 8$$
$$x + 2y = -8$$

____

Step 1 The equations are already in $Ax + By = C$ form.

Step 2 Adding the two equations gives $3x + y = 8$, which does not eliminate either variable. However, multiplying equation (2) by -2 and then adding will eliminate x.

Step 3 Add the two equations.
$$2x - 4y = 10 \quad (1)$$
$$\underline{-2x - 10y = 4 \quad (2)}$$
$$-14y = 14$$

Step 4 Solve.
$$y = -1$$

Step 5 Find the value of x by substituting -1 for y in either of the original equations.
$$x + 5y = -2$$
$$x + 5(-1) = -2$$
$$x - 5 = -2$$
$$x = 3$$

Step 6 Check the ordered-pair solution $(3, -1)$ in both of the original equations.

$$2x - 4y = 10 \qquad\qquad x + 5y = -2$$
$$2(3) - 4(-1) \overset{?}{=} 10 \qquad 3 + 5(-1) \overset{?}{=} -2$$
$$10 = 10 \text{ True} \qquad \text{True } -2 = -2$$

Since $(3, -1)$ is a solution of both equations, the solution set is $\{(3, -1)\}$.

6. Solve the system.
$$3x + 8y = -2 \quad (1)$$
$$2x + 7y = 2 \quad (2)$$

6. Solve the system.
$$3x + 4y = 24$$
$$4x + 3y = 11$$

____

To eliminate x, multiply equation (1) by 2 and multiply equation (2) by -3. Then add.

$$6x + 16y = -4$$
$$-6x - 21y = -6$$
$$-5y = -10$$
$$y = 2$$

Find the value of x by substituting 2 for y in either equation (1) or (2).

$$2x + 7y = 2$$
$$2x + 7(2) = 2$$
$$2x + 14 = 2$$
$$2x = -12$$
$$x = -6$$

Check that the solution set of the system is $\{(-6, 2)\}$.

Objective 4 Practice Exercises

Solve each system by the elimination method. Check your answers.

10. $6x + 7y = 10$ 10. _______________

          $2x - 3y = 14$

11. $8x + 6y = 10$ 11. _______________

          $4x - y = 1$

12. $6x + y = 1$ 12. _______________

          $3x - 4y = 23$

Objective 5   Solve problems about quantities and their costs.

Review this example for Objective 5:

7.   The total receipts for a basketball game were $4690.50. There were 723 tickets sold, some for children and some for adults. If the adult tickets cost $9.50 and the children's tickets cost $4, how many of each type were there?

Step1 Read the problem.

Step 2 Assign variables. Let x = the number of adult tickets and y = the number of children's tickets.

Step 3 Write two equations. The total number of tickets is 723. The total value of the tickets is $4690.50.
$$x + y = 723 \qquad (1)$$
$$9.50x + 4y = 4690.5 \quad (2)$$

Step 4 Solve the system. Solve using elimination. Multiply equation (1) by –4.
$$-4x - 4y = -2892$$
$$\underline{9.50x + 4y = 4690.5}$$
$$5.50x \qquad = 1798.50$$
$$x = 327$$
Substitute 327 for x in equation (1) to find y.
$$x + y = 723$$
$$327 + y = 723$$
$$y = 396$$

Step 5 State the answer. The number of adult tickets is 327 and the number of children's tickets is 396.

Step 6 Check. The sum of the tickets is 723. The value of the tickets is
$$9.50(327) + 4(396) = 4690.50$$
This checks.

Now Try:

7.   Twice as many general admission tickets to a basketball game were sold as reserved seat tickets. General admission tickets cost $10 and reserved seat tickets cost $15. If the total value of both kinds of tickets was $26,250, how many tickets of each kind were sold?

____

Objective 5 Practice Exercises

Write a system of equations for each problem, then solve the problem.

13. There were 411 tickets sold for a soccer game, some for students and some for nonstudents. Student tickets cost $4.25 and nonstudent tickets cost $8.50 each. The total receipts were $3021.75. How many of each type were sold?

13.
student tix _______________

nonstudent tix _____________

14. A cashier has some $5 bills and some $10 bills. The total value of the money is $750. If the number of tens is equal to twice the number of fives, how many of each type are there?

14.
$5 bills _______________

$10 bills _______________

15. Luke plans to buy 10 ties with exactly $162. If some ties cost $14, and the others cost $25, how many ties of each price should he buy?

15.
$14 ties _______________

$25 ties _______________

Chapter 4 EXPONENTS, POLYNOMIALS, AND POLYNOMIAL FUNCTIONS

Learning Objectives
Use the power rules for exponents to simplify expressions.
Use combinations of the power rules and the product rule to simplify expressions.
Simplify polynomials by combining like terms.
Add and subtract polynomials.
Multiply a monomial and a polynomial.

Key Terms

Use the vocabulary terms listed below to complete each statement in exercises 1−11.

exponential expression	**base**	**power**	**term**
like terms **polynomial**	**descending powers**		**degree of a term**
degree of a polynomial	**monomial** **binomial**		**trinomial**

1. 2^5 is read "2 to the fifth _____________________".

2. A number written with an exponent is called a(n) _____________________.

3. The _______________ is the number being multiplied repeatedly.

4. A polynomial in x is written in _____________________ if the exponents on x in its terms are decreasing order.

5. A _______________ is a number, a variable, or a product or quotient of a number and one or more variables raised to powers.

6. A polynomial with exactly three terms is called a _____________________.

7. A _____________________ is a term, or the sum of a finite number of terms with whole number exponents.

8. A polynomial with exactly one term is called a _____________________.

9. The _____________________ is the greatest degree of any term of the polynomial.

10. A _____________________ is a polynomial with exactly two terms.

11. Terms with exactly the same variables (including the same exponents) are called _____________________.

Objective 1   Use the power rules for exponents to simplify expressions.

Review these examples for Objective 1:

1. Use power rule (a) for exponents to simplify.

   a. $\left(5^6\right)^3$

   $$\left(5^6\right)^3 = 5^{6\cdot3}$$
   $$= 5^{18}$$

   b. $\left(x^3\right)^4$

   $$\left(x^3\right)^4 = x^{3\cdot4}$$
   $$= x^{12}$$

Now Try:

1. Use power rule (a) for exponents to simplify.

   a. $\left(7^2\right)^4$

   ____

   b. $\left(x^5\right)^6$

   ____

Objective 1 Practice Exercises

Simplify each expression. Write all answers in exponential form.

1. $\left(7^3\right)^4$

2. $-\left(v^4\right)^9$

3. $(-3x^4)^3$

1. _______________

2. _______________

3. _______________

Objective 2   Use combinations of the power rules and the product rule to simplify expressions.

Review these examples for Objective 2:	**Now Try:**

Review these examples for Objective 2:

2.   Simplify each expression.

a. $\left(\dfrac{3}{4}\right)^3 \cdot 3^2$

$$\left(\dfrac{3}{4}\right)^3 \cdot 3^2 = \dfrac{3^3}{4^3} \cdot \dfrac{3^2}{1}$$

$$= \dfrac{3^3 \cdot 3^2}{4^3 \cdot 1}$$

$$= \dfrac{3^{3+2}}{4^3}$$

$$= \dfrac{3^5}{4^3}, \quad \text{or} \quad \dfrac{243}{64}$$

b. $\left(-x^5 y\right)^4 \left(-x^6 y^5\right)^3$

$$\left(-x^5 y\right)^4 \left(-x^6 y^5\right)^3$$

$$= \left(-1 x^5 y\right)^4 \left(-1 x^6 y^5\right)^3$$

$$= (-1)^4 \left(x^5\right)^4 \left(y^4\right) \cdot (-1)^3 \left(x^6\right)^3 \left(y^5\right)^3$$

$$= (-1)^4 \left(x^{20}\right)\left(y^4\right) \cdot (-1)^3 \left(x^{18}\right)\left(y^{15}\right)$$

$$= (-1)^7 x^{20+18} y^{4+15}$$

$$= -1 x^{38} y^{19}$$

$$= -x^{38} y^{19}$$

Now Try:

2.   Simplify each expression.

a. $\left(\dfrac{5}{2}\right)^3 \cdot 5^2$

____

b. $\left(-x^5 y\right)^3 \left(-x^6 y^5\right)^2$

____

Objective 2 Practice Exercises

Simplify. Write all answers in exponential form.

4. $\left(-x^3\right)^2 \left(-x^5\right)^4$

4. ______________

5. $\left(2ab^2 c\right)^5 (ab)^4$

5. ______________

6. $\left(5x^2 y^3\right)^7 \left(5xy^4\right)^4$

6. ______________

Objective 3   Combine like terms.

Review this example for Objective 3:	**Now Try:**
3. Simplify the expression by combining like terms.	**3.** Simplify the expression by combining like terms.

$$19m^3 + 6m + 5m^3$$

$$19m^3 + 6m + 5m^3 = (19+5)m^3 + 6m$$

$$= 24m^3 + 6m$$

Now Try — **3.** $22m^2 + 15m^3 + 7m^2$

____

Objective 3 Practice Exercises

In each polynomial, combine like terms whenever possible. Write the result with descending powers.

7. $7z^3 - 4z^3 + 5z^3 - 11z^3$

7. _________________

8. $-1.3z^7 + 0.4z^7 + 2.6z^8$

8. _________________

9. $6c^3 - 9c^2 - 2c^2 + 14 + 3c^2 - 6c - 8 + 2c^3$

9. _________________

Objective 4   Add and subtract polynomials.

Review these examples for Objective :

4.   Find each sum.

a. Add $5x^4 - 7x^3 + 9$ and $-3x^4 + 8x^3 - 7$

$$\left(5x^4 - 7x^3 + 9\right) + \left(-3x^4 + 8x^3 - 7\right)$$
$$= 5x^4 - 3x^4 - 7x^3 + 8x^3 + 9 - 7$$
$$= 2x^4 + x^3 + 2$$

b. $\left(5x^4 - 7x^2 + 6x\right) + \left(-3x^3 + 4x^2 - 7\right)$

$$\left(5x^4 - 7x^2 + 6x\right) + \left(-3x^3 + 4x^2 - 7\right)$$
$$= 5x^4 - 3x^3 - 7x^2 + 4x^2 + 6x - 7$$
$$= 5x^4 - 3x^3 - 3x^2 + 6x - 7$$

5.   Perform the subtraction.

Subtract $8x^3 - 5x^2 + 8$ from $9x^3 + 6x^2 - 7$.

$$\left(9x^3 + 6x^2 - 7\right) - \left(8x^3 - 5x^2 + 8\right)$$
$$= \left(9x^3 + 6x^2 - 7\right) + \left(-8x^3 + 5x^2 - 8\right)$$
$$= x^3 + 11x^2 - 15$$

6.   Perform the indicated operations to simplify the expression

$$\left(5 - 2x + 9x^2\right) - \left(7 - 5x + 8x^2\right) + \left(6 + 3x - 5x^2\right)$$

Rewrite, changing the subtraction to adding the opposite.

$$\left(5 - 2x + 9x^2\right) - \left(7 - 5x + 8x^2\right) + \left(6 + 3x - 5x^2\right)$$
$$= \left(5 - 2x + 9x^2\right) + \left(-7 + 5x - 8x^2\right) + \left(6 + 3x - 5x^2\right)$$
$$= \left(-2 + 3x + x^2\right) + \left(6 + 3x - 5x^2\right)$$
$$= 4 + 6x - 4x^2$$

Now Try:

4.   Find each sum.

a. Add $15x^3 - 5x + 3$ and $-11x^3 + 6x + 9$

____

b. $\left(8x^2 - 6x + 4\right) + \left(7x^3 - 8x - 5\right)$

____

5.   Perform the subtraction.

Subtract $18x^3 + 4x - 6$ from $7x^3 - 3x - 5$.

____

6.   Perform the indicated operations to simplify the expression

$$\left(10 - 7x + 6x^2\right) - \left(5 - 11x + 3x^2\right)$$
$$+ \left(2 + 4x - 7x^2\right)$$

____

Objective 4 Practice Exercises

Add or subtract as indicated.

10. $\left(3r^3 + 5r^2 - 6\right) + \left(2r^2 - 5r + 4\right)$ 10. _____________

11. $\left(-8w^3 + 11w^2 - 12\right) - \left(-10w^2 + 3\right)$ 11. _____________

12. $\left(2x^2y + 2xy - 4xy^2\right) + \left(6xy + 9xy^2\right) - \left(9x^2y + 5xy\right)$ 12. _____________

Objective 5   Multiply a monomial and a polynomial.

Review this example for Objective 5:

7.   Find the product.

$$5x^2(7x+3)$$

Use the distributive property.
$$5x^2(7x+3) = 5x^2(7x) + 5x^2(3)$$
$$= 35x^3 + 15x^2$$

Now Try:

7.   Find the product.

$$8x^3(4x+8)$$

____

Objective 5 Practice Exercises

Find each product.

13. $7z\left(5z^3+2\right)$

13.   ________________

14. $2m\left(3+7m^2+3m^3\right)$

14.   ________________

15. $-3y^2\left(2y^3+3y^2-4y+11\right)$

15.   ________________

Chapter 5 FACTORING

Learning Objectives
>   Identify factors of integers.
>   Rewrite expressions using the distributive property.
>   Multiply two polynomials.
>   Square binomials.
>   Find the greatest common factor of a list of terms.

Key Terms

Use the vocabulary terms listed below to complete each statement in exercises 1−5.

factor   **binomial**   **factored form**

greatest common factor (GCF)   **factoring**

1. A polynomial with two terms is called a _________________________.

2. The process of writing a polynomial as a product is called _________________.

3. An expression is in _______________________ when it is written as a product.

4. The _______________________ is the largest quantity that is a factor of each
    of a group of quantities.

5. An expression A is a _____________________ of an expression B if B can be divided
    by A with 0 remainder.

Objective 1   Identify factors of integers.

Review this example for Objective 1:

1. Find all integer factors of the number 52.

$$1 \times 52 = 52$$
$$2 \times 26 = 52$$
$$4 \times 13 = 52$$
$$(-1) \times (-52) = 52$$
$$(-2) \times (-26) = 52$$
$$(-4) \times (-13) = 52$$

The integer factors of 52 are –52, –26, –13, –4, –2, –1, 1, 2, 4, 13, 26, and 52.

Now Try:

1. Find all integer factors of the number 12.

____

Objective 1 Practice Exercises

Find all integer factors of each number.

1. 8

2. 38

3. 42

1. _______________

2. _______________

3. _______________

      Copyright © 2020 Pearson Education, Inc.

Objective 2   Rewrite expressions using the distributive property.

Review these examples for Objective 5:	**Now Try:**

2. Use the distributive property to rewrite each expression.

   a. $4(8+7)$

$$4(8+7)=4\cdot8+4\cdot7$$
$$=32+28$$
$$=60$$

   b. $5(3+x+m)$

$$5\cdot3+5x+5m=5(3+x+m)$$

   c. $7(p-6)$

$$7(p-6)=7[p+(-6)]$$
$$=7p+7(-6)$$
$$=7p-42$$

   d. $-3(5x-2)$

$$-3(5x-2)=-3[5x+(-2)]$$
$$=-3(5x)+(-3)(-2)$$
$$=(-3\cdot5)x+(-3)(-2)$$
$$=-15x+6$$

3. Rewrite each expression.

   a. $-(5x+7)$

$$-(5x+7)=-1\cdot(5x+7)$$
$$=-1\cdot5x+(-1)\cdot7$$
$$=-5x-7$$

   b. $-(-6c-7)$

$$-(-6c-7)=-1(-6c-7)$$
$$=-1(-6c)-1(-7)$$
$$=6c+7$$

2. Use the distributive property to rewrite each expression.

   a. $3(11+7)$

____

   b. $12(y+6+x)$

____

   c. $17(x-6)$

____

   d. $-4(2x-5)$

____

3. Rewrite each expression.

   a. $-(3x+4)$

____

   b. $-(-8k-9)$

____

c. $-(-p-5r+9x)$

$-(-p-5r+9x)$
$=-1\cdot(-1p-5r+9x)$
$=-1\cdot(-1p)-1\cdot(-5r)-1\cdot(9x)$
$=p+5r-9x$

c. $-(-4x-5y+z)$

____

Objective 2 Practice Exercises

Use the distributive property to rewrite each expression. Simplify if possible.

4. $n(2a-4b+6c)$

4. _______________

5. $-2(5y-9z)$

5. _______________

6. $-(-2k+7)$

6. _______________

Objective 3   Multiply two polynomials.

Review these examples for Objective 4:

Now Try:

4. Multiply. $(x + 3)(2x^2 - 1)$

4. Multiply. $(3x - 2)(5x + 3)$

Multiply each term of the second polynomial by each term of the first.

$$\begin{aligned}
(x + 3)(2x^2 - 1) &= x(2x^2) + x(-1) \\
&\quad + 3(2x^2) + 3(-1) \\
&= 2x^3 - x + 6x^2 - 3 \\
&= x^3 + 6x^2 - x - 3
\end{aligned}$$

____

5. Multiply $(x^2 + 6)(5x^3 - 4x^2 + 3x)$.

5. Multiply
$(x^3 + 9)(4x^4 - 2x^2 + x)$

Multiply each term of the second polynomial by each term of the first.

$$\begin{aligned}
&(x^2 + 6)(5x^3 - 4x^2 + 3x) \\
&= x^2(5x^3) + x^2(-4x^2) + x^2(3x) \\
&\quad + 6(5x^3) + 6(-4x^2) + 6(3x) \\
&= 5x^5 - 4x^4 + 3x^3 + 30x^3 - 24x^2 + 18x \\
&= 5x^5 - 4x^4 + 33x^3 - 24x^2 + 18x
\end{aligned}$$

____

Objective 3 Practice Exercises

Find each product.

7. $(2m^2 + 1)(3m^3 - 4m)$

7. _______________

8. $(x + 3)(x^2 - 3x + 9)$

8. _______________

9. $(3x^2 + x)(2x^2 + 3x - 4)$

9. _______________

Objective 4   Square binomials.

Review these examples for Objective 5:

6. Square each binomial.

   a. $(b-5)^2$

   $(b-5)^2 = b^2 - 2(b)(5) + 5^2$

   $\quad\quad\quad = b^2 - 10b + 25$

   b. $(6x+3y)^2$

   $(6x+3y)^2 = (6x)^2 + 2(6x)(3y) + (3y)^2$

   $\quad\quad\quad\quad = 36x^2 + 36xy + 9y^2$

Now Try:

6. Square each binomial.

   a. $(c-4)^2$

   ____

   b. $(2a+9k)^2$

   ____

Objective 4 Practice Exercises

Find each square by using the pattern for the square of a binomial.

10. $(7+x)^2$

10. _______________

11. $(2m-3p)^2$

11. _______________

12. $(4y-0.7)^2$

12. _______________

Objective 5   Find the greatest common factor of a list of terms.

Review this example for Objective 6:	**Now Try:**
7. Find the greatest common factor for the list of terms.	7. Find the greatest common factor for the list of terms.

7.   Find the greatest common factor for the list of terms.

$$28m^4,\ 35m^6,\ 49m^9,\ 70m^5$$

$$28m^4 = 2\cdot 2\cdot 7\cdot m^4$$

$$35m^6 = 5\cdot 7\cdot m^6$$

$$49m^9 = 7\cdot 7\cdot m^9$$

$$70m^5 = 2\cdot 5\cdot 7\cdot m^5$$

Then, $\text{GCF} = 7m^4$.

Now Try:

7.   Find the greatest common factor for the list of terms.

$$54x^5,\ 48x^7,\ 42x^9,\ 30x^4$$

____

Objective 5 Practice Exercises

Find the greatest common factor for each list.

13. $xy^2,\ x^2y,\ x^2y^3$

13.   _______________

14. $6k^2m^4n^5,\ 8k^3m^7n^4,\ k^4m^8n^7$

14.   _______________

15. $9xy^4,\ 72x^4y^7,\ 27xy^2,\ 108x^2y^5$

15.   _______________

Chapter 6 RATIONAL EXPRESSSIONS AND FUNCTIONS

Learning Objectives
Add and subtract fractions.
Multiply and divide fractions.
Solve equations with fractions as coefficients.
Solve problems involving unknown numbers.
Write rational expressions in lowest terms.
Simplify a complex fraction by multiplying numerator and denominator by the LCD (Method 2).

Key Terms

Use the vocabulary terms listed below to complete each statement in exercises 1−12.

numerator	**denominator**	**proper fraction**
improper fraction	**equivalent fractions**	**lowest terms**
prime number	**composite number**	**prime factorization**
rational expression	**lowest terms**	**complex fraction**
LCD		

1. Two fractions are _________________________________ when they represent the same portion of a whole.

2. A fraction whose numerator is larger than its denominator is called an
    _____________________.

3. In the fraction $\frac{2}{9}$, the 2 is the _______________________________.

4. A fraction whose denominator is larger than its numerator is called a
    _______________________________.

5. A _______________________________ is a rational expression with one or more fractions in the numerator, denominator, or both.

6. The _______________________________ of a fraction shows the number of equal parts in a whole.

7. A _______________________________ has at least one factor other than itself and 1.

8. In a _______________________________ every factor is a prime number.

9. The factors of a _______________________________ are itself and 1.

10. A fraction is written in _______________________________ when its numerator and denominator have no common factor other than 1.

11. To simplify a complex fraction, multiply the numerator and denominator by the
____________________________ of all the fractions within the complex fraction.

12. The quotient of two polynomials with denominator not 0 is called a
____________________________.

Objective 1   Add and subtract fractions.

Review these examples for Objective 1:	**Now Try:**
1. Add. Write the sum in lowest terms.	1. Add. Write the sum in lowest terms.

$$\frac{5}{24}+\frac{7}{24}$$

Add numerators. Keep the same denominator.

$$\frac{5}{24}+\frac{7}{24}=\frac{5+7}{24}=\frac{12}{24},\text{ or } \frac{1}{2}$$

Write in lowest terms.

Now Try 1. Add. Write the sum in lowest terms.

$$\frac{5}{16}+\frac{7}{16}$$

2. Add. Write the sum in lowest terms.

$$\frac{5}{21}+\frac{3}{14}$$

Now Try 2. Add. Write the sum in lowest terms.

$$\frac{7}{12}+\frac{3}{8}$$

Step 1 To find the LCD, factor the denominators to prime factored form.
$$21=3\cdot 7 \quad \text{and} \quad 14=2\cdot 7$$
7 is a factor of both denominators.

$$\begin{array}{cc} 21 & 14 \\ \wedge & \wedge \end{array}$$

Step 2 LCD $=3\cdot 7\cdot 2=42$
In this example, the LCD needs one factor of 3, one factor of 7 and one factor of 2.

Step 3 Now we can use the second property of 1 to write each fraction with 42 as the denominator.

$$\frac{5}{21}=\frac{5}{21}\cdot\frac{2}{2}=\frac{10}{42} \quad \text{and} \quad \frac{3}{14}=\frac{3}{14}\cdot\frac{3}{3}=\frac{9}{42}$$

Now add the two equivalent fractions to get the sum.

$$\frac{5}{21}+\frac{3}{14}=\frac{10}{42}+\frac{9}{42}$$
$$=\frac{19}{42}$$

3. Subtract. Write difference in lowest terms.

$$\frac{26}{9}-\frac{5}{9}$$

Now Try 3. Subtract. Write difference in lowest terms.

$$\frac{11}{18}-\frac{7}{18}$$

Subtract numerators. Keep the same denominator.

$$\frac{26}{9} - \frac{5}{9} = \frac{26 - 5}{9}$$

$$= \frac{21}{9}$$

$$= \frac{7}{3}, \text{ or } 2\frac{1}{3}$$

Objective 1 Practice Exercises

Find each sum or difference, and write it in lowest terms.

1. $\dfrac{23}{45} + \dfrac{47}{75}$

1. ______________

2. $2\dfrac{3}{4} + 7\dfrac{2}{3}$

2. ______________

3. $12\dfrac{5}{6} - 7\dfrac{7}{8}$

3. ______________

Objective 2   Multiply and divide fractions.

Review these examples for Objective 2:

4. Find the product, and write it in lowest terms.

$$\frac{5}{12}\cdot\frac{3}{10}$$

$$\frac{5}{12}\cdot\frac{3}{10}=\frac{5\cdot3}{12\cdot10}$$ Multiply numerators.
Multiply denominators.

$$=\frac{5\cdot3}{4\cdot3\cdot2\cdot5}$$ Factor the denominator.

$$=\frac{1}{4\cdot2}$$ $\frac{3}{3}=1$ and $\frac{5}{5}=1$

$$=\frac{1}{8}$$ Write in lowest terms.

5. Find the quotient, and write it in lowest terms.

$$\frac{2}{5}\div\frac{8}{7}$$

$$\frac{2}{5}\div\frac{8}{7}=\frac{2}{5}\cdot\frac{7}{8}$$ Multiply by the reciprocal.

$$=\frac{2\cdot7}{5\cdot4\cdot2}$$ Multiply and factor.

$$=\frac{7}{20}$$

Now Try:

4. Find the product, and write it in lowest terms.

$$\frac{7}{15}\cdot\frac{3}{14}$$

____

5. Find the quotient, and write it in lowest terms.

$$\frac{6}{7}\div\frac{9}{8}$$

____

Objective 2 Practice Exercises

Find each product or quotient, and write it in lowest terms.

4. $\dfrac{25}{11}\cdot\dfrac{33}{10}$

4. _______________

5. $\dfrac{5}{4}\div\dfrac{25}{28}$

5. _______________

6. $4\dfrac{3}{8}\cdot2\dfrac{4}{7}$

6. _______________

Objective 3 Solve equations with fractions as coefficients.

Review these examples for Objective 3:

6. Solve $\dfrac{x}{3} - 8 = -\dfrac{x}{5}$.

$$\frac{x}{3} - 8 = -\frac{x}{5}$$

Step 1 $\quad 15\left(\dfrac{x}{3} - 8\right) = 15\left(-\dfrac{x}{5}\right)$

$\quad 15\left(\dfrac{x}{3}\right) + 15(-8) = 15\left(-\dfrac{x}{5}\right)$

$$5x - 120 = -3x$$

Step 2 $\quad 5x - 120 - 5x = -3x - 5x$

$$-120 = -8x$$

Step 3 $\quad \dfrac{-120}{-8} = \dfrac{-8x}{-8}$

$$x = 15$$

Step 4 Check by substituting 15 for x in the original equation.

$$\frac{x}{3} - 8 = -\frac{x}{5}$$

$$\frac{15}{3} - 8 \overset{?}{=} -\frac{15}{5}$$

$$5 - 8 \overset{?}{=} -3$$

$$-3 = -3 \quad \text{True}$$

The check confirms that the solution set is $\{15\}$.

Now Try:

6. Solve $\dfrac{x}{2} - 5 = -\dfrac{x}{3}$.

____

7. Solve $\frac{1}{4}(x+3)-\frac{2}{5}(x+1)=2$.

To clear fractions, multiply by 20, the LCD.

$$\frac{1}{4}(x+3)-\frac{2}{5}(x+1)=2$$

Step 1 $\quad 20\left[\frac{1}{4}(x+3)-\frac{2}{5}(x+1)\right]=20(2)$

$$20\left[\frac{1}{4}(x+3)\right]+20\left[-\frac{2}{5}(x+1)\right]=20(2)$$

$$5(x+3)-8(x+1)=40$$

$$5x+15-8x-8=40$$

$$-3x+7=40$$

Step 2 $\qquad -3x+7-7=40-7$

$$-3x=33$$

Step 3 $\qquad \frac{-3x}{-3}=\frac{33}{-3}$

$$x=-11$$

Step 4 Check to confirm that $\{-11\}$ is the solution set.

7. Solve

$$\frac{1}{7}(x+5)-\frac{1}{2}(x+4)=-2.$$

____

Objective 3 Practice Exercises

Solve each equation and check your solution.

7. $\quad \dfrac{2w}{3}+\dfrac{w}{4}=\dfrac{11}{2}$

7. _______________

8. $\quad \dfrac{2}{9}x-\dfrac{1}{6}x=\dfrac{2}{3}x+11$

8. _______________

9. $\quad \dfrac{1}{3}(2m-1)-\dfrac{3}{4}m=\dfrac{5}{6}$

9. _______________

Objective 4   Solve problems involving unknown numbers.

Review this example for Objective 4:

8. The product of 5, and a number decreased by 8, is 150. What is the number?

Step 1 Read the problem carefully. We are asked to find a number.

Step 2 Assign a variable to represent the unknown quantity.
Let x = the number.

Step 3 Write an equation.

$$5 \cdot (x - 8) = 150$$

Step 4 Solve the equation.
$$5(x - 8) = 150$$
$$5x - 40 = 150$$
$$5x - 40 + 40 = 150 + 40$$
$$5x = 190$$
$$\frac{5x}{5} = \frac{190}{5}$$
$$x = 38$$

Step 5 State the answer. The number is 38.

Step 6 Check. The number 38 decreased by 8 is 30. The product of 5 and 30 is 150. The answer, 38, is correct.

Now Try:

8. The product of 8, and a number decreased by 11, is 40. What is the number?

____

Objective 4 Practice Exercises

Write an equation for each of the following and then solve the problem. Use x as the variable.

10.   If 4 is added to 3 times a number, the result is 7. Find the number.

10. ______________

____

11. If -2 is multiplied by the difference between 4 and a number, the result is 24. Find the number.

11. ___________________

____

12. If four times a number is added to 7, the result is five less than six times the number. Find the number.

12. ___________________

____

Objective 5   Write rational expressions in lowest terms.

Review these examples for Objective 5:

9. Write the expression in lowest terms.

$$\frac{15k^3}{3k^4}$$

Write k^3 as $k \cdot k \cdot k$ and k^4 as $k \cdot k \cdot k \cdot k$.

$$\frac{15k^3}{3k^4} = \frac{3 \cdot 5 \cdot k \cdot k \cdot k}{3 \cdot k \cdot k \cdot k \cdot k}$$

$$= \frac{5 \cdot (3 \cdot k \cdot k \cdot k)}{k \cdot (3 \cdot k \cdot k \cdot k)}$$

$$= \frac{5}{k}$$

10. Write each rational expression in lowest terms.

a. $\dfrac{6x-18}{5x-15}$

$$\frac{6x-18}{5x-15} = \frac{6(x-3)}{5(x-3)} = \frac{6}{5}$$

b. $\dfrac{m^2 + 5m - 24}{3m^2 - 5m - 12}$

$$\frac{m^2 + 5m - 24}{3m^2 - 5m - 12} = \frac{(m+8)(m-3)}{(3m+4)(m-3)}$$

$$= \frac{m+8}{3m+4}$$

11. Write $\dfrac{3x-2y}{2y-3x}$ in lowest terms.

Factor -1 from the denominator.

$$\frac{3x-2y}{2y-3x} = \frac{3x-2y}{-1(-2y+3x)}$$

$$= \frac{3x-2y}{-1(3x-2y)}$$

$$= -1$$

Now Try:

9. Write the expression in lowest terms.

$$\frac{12k^5}{4k^8}$$

____

10. Write each rational expression in lowest terms.

a. $\dfrac{7x-35}{9x-45}$

____

b. $\dfrac{m^2 - 3m - 54}{2m^2 - 15m - 27}$

____

11. Write $\dfrac{4y-5x}{5x-4y}$ in lowest terms.

____

Objective 5 Practice Exercises

Write each rational expression in lowest terms. Assume that no values of any variable make any denominator zero.

13. $\dfrac{15ab^3c^9}{-24ab^2c^{10}}$

13. _________________

14. $\dfrac{16-x^2}{2x-8}$

14. _________________

15. $\dfrac{9x^2-9x-108}{2x-8}$

15. _________________

Objective 6   Simplify a complex fraction by multiplying numerator and denominator by the LCD (Method 2).

Review these examples for Objective 6:

12. Simplify the complex fraction.

$$\frac{12+\dfrac{4}{x}}{\dfrac{x}{5}+\dfrac{1}{15}}$$

Step 1 Find the LCD for all the denominators.

The LCD for x, 5, and 15 is $15x$.

Step 2 Multiply the numerator and denominator of the complex fraction by the LCD.

$$\frac{12+\dfrac{4}{x}}{\dfrac{x}{5}+\dfrac{1}{15}}=\frac{15x\left(12+\dfrac{4}{x}\right)}{15x\left(\dfrac{x}{5}+\dfrac{1}{15}\right)}$$

$$=\frac{180x+60}{3x^2+x}$$

$$=\frac{60(3x+1)}{x(3x+1)}$$

$$=\frac{60}{x}$$

13. Simplify the complex fraction.

$$\frac{\dfrac{9}{7n}-\dfrac{3}{n^2}}{\dfrac{8}{3n}+\dfrac{5}{6n^2}}$$

The LCD is $42n^2$.

$$=\frac{42n^2\left(\dfrac{9}{7n}-\dfrac{3}{n^2}\right)}{42n^2\left(\dfrac{8}{3n}+\dfrac{5}{6n^2}\right)}$$

$$=\frac{42n^2\left(\dfrac{9}{7n}\right)-42n^2\left(\dfrac{3}{n^2}\right)}{42n^2\left(\dfrac{8}{3n}\right)+42n^2\left(\dfrac{5}{6n^2}\right)}$$

$$=\frac{54n-126}{112n+35}, \text{ or } \frac{18(3n-7)}{7(16n+5)}$$

Now Try:

12. Simplify the complex fraction.

$$\frac{4+\dfrac{2}{x}}{\dfrac{x}{3}+\dfrac{1}{6}}$$

____

13. Simplify the complex fraction.

$$\frac{\dfrac{2}{9n}-\dfrac{2}{5n^2}}{\dfrac{4}{5n}+\dfrac{2}{3n^2}}$$

____

Objective 6 Practice Exercises

Simplify each complex fraction by multiplying numerator and denominator by the least common denominator.

16. $\dfrac{\dfrac{9}{x^2}-1}{\dfrac{3}{x}-1}$

16. ______________

17. $\dfrac{\dfrac{x-2}{x+2}}{\dfrac{x}{x-2}}$

17. ______________

18. $\dfrac{\dfrac{6}{k+1}-\dfrac{5}{k-3}}{\dfrac{3}{k-3}+\dfrac{2}{k+2}}$

18. ______________

Chapter 7 ROOTS, RADICALS, AND ROOT FUNCTIONS

Learning Objectives

Find square roots.
Find the cube, fourth, and other roots.
Simplify square root radicals using the product rule.
Simplify radical sums and differences.
Rationalize denominators with square roots.

Key Terms

Use the vocabulary terms listed below to complete each statement in exercises 1−12.

square root	principal square root	radicand
radical	radical expression	perfect square
irrational number	cube root	
like radicals	index	unlike radicals

1. The number or expression inside a radical sign is called the ____________________.

2. A number with a rational square root is called a ____________________.

3. In a radical of the form $\sqrt[n]{a}$, the number n is the ____________________.

4. The number b is a ____________________ of a if $b^2 = a$.

5. The expression $\sqrt[n]{a}$ is called a ____________________.

6. The expressions $2\sqrt{2}$ and $6\sqrt[3]{2}$ are ____________________.

7. The expressions $2\sqrt{2}$ and $7\sqrt{2}$ are ____________________.

8. The positive square root of a number is its ____________________.

9. A real number that is not rational is called an ____________________.

10. A ____________________ is a radical sign and the number or expression in it.

11. The number b is a ____________________ of a if $b^3 = a$.

Objective 1   Find square roots.

Review these examples for Objective 1:

1. Find the square roots of 64.

   What number multiplied by itself equals 64?
   $8^2 = 64$ and $(-8)^2 = 64$.
   Thus, 64 has two square roots: 8 and –8.

2. Find each square root.

   a. $\sqrt{121}$

   $11^2 = 121$, so $\sqrt{121} = 11$.

   b. $-\sqrt{\dfrac{16}{25}}$

   $-\sqrt{\dfrac{16}{25}} = -\dfrac{4}{5}$

3. Find the square of each radical expression.

   a. $\sqrt{17}$

   The square of $\sqrt{17}$ is $\left(\sqrt{17}\right)^2 = 17$.

   b. $\sqrt{w^2 + 3}$

   $\left(\sqrt{w^2 + 3}\right)^2 = w^2 + 3$

Now Try:

1. Find the square roots of 81.

   ____

2. Find each square root.

   a. $\sqrt{169}$

   ____

   b. $-\sqrt{\dfrac{9}{49}}$

   ____

3. Find the square of each radical expression.

   a. $\sqrt{19}$

   ____

   b. $\sqrt{n^2 + 5}$

   ____

Objective 1 Practice Exercises

Find the square root.

1. 625

   1. ______________

2. $\dfrac{121}{196}$

   2. ______________

Find the square of the radical expression.

3. $\sqrt{x^2 - 5}$

   3. ______________

Objective 2   Find the cube, fourth, and other roots.

Review these examples for Objective 2:

4. Find each cube root.

a. $\sqrt[3]{729}$

$\sqrt[3]{729} = 9$, because $9^3 = 729$.

b. $\sqrt[3]{-64}$

$\sqrt[3]{-64} = -4$, because $(-4)^3 = -64$.

c. $\sqrt[3]{\dfrac{125}{8}}$

$\sqrt[3]{\dfrac{125}{8}} = \dfrac{5}{2}$, because $\left(\dfrac{5}{2}\right)^3 = \dfrac{125}{8}$.

5. Find each root.

a. $\sqrt[4]{81}$

$\sqrt[4]{81} = 3$, because 3 is positive and $3^4 = 81$.

b. $\sqrt[5]{-1024}$

$\sqrt[5]{-1024} = -4$, because $(-4)^5 = -1024$.

Now Try:

4. Find each cube root.

a. $\sqrt[3]{343}$

____

b. $\sqrt[3]{-125}$

____

c. $\sqrt[3]{\dfrac{64}{27}}$

____

5. Find each root.

a. $\sqrt[4]{1296}$

____

b. $\sqrt[5]{-3125}$

____

Objective 2 Practice Exercises

Find each root.

4. $\sqrt[3]{-64}$

5. $\sqrt[4]{256}$

6. $\sqrt[7]{-1}$

4. _______________

5. _______________

6. _______________

Objective 3   Simplify square root radicals by using the product rule.

Review these examples for Objective 3:	**Now Try:**

Review these examples for Objective 3:

6.   Simplify each radical.

   a. $\sqrt{40}$

$$\sqrt{40} = \sqrt{4 \cdot 10}$$
$$= \sqrt{4} \cdot \sqrt{10}$$
$$= 2\sqrt{10}$$

   b. $\sqrt{75}$

$$\sqrt{75} = \sqrt{25 \cdot 2}$$
$$= \sqrt{25} \cdot \sqrt{2}$$
$$= 5\sqrt{2}$$

   c. $\sqrt{48}$

$$\sqrt{48} = \sqrt{16 \cdot 3} = \sqrt{16} \cdot \sqrt{3} = 4\sqrt{3}$$

7.   Find each product and simplify.

   a. $\sqrt{16} \cdot \sqrt{20}$

$$\sqrt{16} \cdot \sqrt{20} = 4\sqrt{20}$$
$$= 4\sqrt{4 \cdot 5}$$
$$= 4\sqrt{4} \cdot \sqrt{5}$$
$$= 4 \cdot 2 \cdot \sqrt{5}$$
$$= 8\sqrt{5}$$

   b. $\sqrt{27} \cdot \sqrt{50}$

$$\sqrt{27} \cdot \sqrt{50} = \sqrt{27 \cdot 50}$$
$$= \sqrt{9 \cdot 3 \cdot 25 \cdot 2}$$
$$= \sqrt{9} \cdot \sqrt{25} \cdot \sqrt{3 \cdot 2}$$
$$= 3 \cdot 5 \cdot \sqrt{6}$$
$$= 15\sqrt{6}$$

Now Try:

6.   Simplify each radical.

   a. $\sqrt{12}$

   b. $\sqrt{98}$

   c. $\sqrt{80}$

7.   Find each product and simplify.

   a. $\sqrt{25} \cdot \sqrt{18}$

   b. $\sqrt{6} \cdot \sqrt{12}$

Objective 3 Practice Exercises

Simplify the radical.

7. $\sqrt{405}$

7. _______________

Find each product and simplify.

8. $\sqrt{11} \cdot \sqrt{33}$

8. _______________

9. $\sqrt{18} \cdot \sqrt{24}$

9. _______________

Objective 4   Simplify radical sums and differences.

Review these examples for Objective 4:

8.   Add or subtract, as indicated.

a. $4\sqrt{5}+\sqrt{20}$

$$4\sqrt{5}+\sqrt{20}=4\sqrt{5}+\sqrt{4\cdot 5}$$
$$=4\sqrt{5}+\sqrt{4}\cdot\sqrt{5}$$
$$=4\sqrt{5}+2\sqrt{5}$$
$$=6\sqrt{5}$$

b. $2\sqrt{28}+8\sqrt{63}$

$$2\sqrt{28}+8\sqrt{63}=2\left(\sqrt{4}\cdot\sqrt{7}\right)+8\left(\sqrt{9}\cdot\sqrt{7}\right)$$
$$=2\left(2\sqrt{7}\right)+8\left(3\sqrt{7}\right)$$
$$=4\sqrt{7}+24\sqrt{7}$$
$$=28\sqrt{7}$$

c. $6\sqrt[3]{54}+2\sqrt[3]{3}$

$$6\sqrt[3]{54}+2\sqrt[3]{3}=6\left(\sqrt[3]{27}\cdot\sqrt[3]{3}\right)+2\sqrt[3]{3}$$
$$=6\left(3\sqrt[3]{3}\right)+2\sqrt[3]{3}$$
$$=18\sqrt[3]{3}+2\sqrt[3]{3}$$
$$=20\sqrt[3]{3}$$

Now Try:

8.   Add or subtract, as indicated.

a. $3\sqrt{6}+\sqrt{150}$

____

b. $3\sqrt{24}+7\sqrt{54}$

____

c. $4\sqrt[3]{128}+7\sqrt[3]{2}$

____

Objective 4 Practice Exercises

Simplify and add or subtract wherever possible.

10. $4\sqrt{128}+2\sqrt{32}$

10. _______________

11. $2\sqrt[3]{16}-5\sqrt[3]{2}$

11. _______________

12. $5\sqrt{32}-8\sqrt{18}+2\sqrt{20}$

12. _______________

Objective 5   Rationalize denominators with square roots.

Review these examples for Objective 5:

9.   Rationalize each denominator.

    a. $\dfrac{10}{\sqrt{5}}$

$$\dfrac{10}{\sqrt{5}} = \dfrac{10 \cdot \sqrt{5}}{\sqrt{5} \cdot \sqrt{5}} \quad \text{Multiply by } \dfrac{\sqrt{5}}{\sqrt{5}} = 1.$$

$$= \dfrac{10\sqrt{5}}{5}$$

$$= 2\sqrt{5} \qquad \text{Write in lowest terms.}$$

    b. $\dfrac{18}{\sqrt{27}}$

$$\dfrac{18}{\sqrt{27}} = \dfrac{18}{3\sqrt{3}}$$

$$= \dfrac{18 \cdot \sqrt{3}}{3\sqrt{3} \cdot \sqrt{3}} \quad \text{Multiply by } \dfrac{\sqrt{3}}{\sqrt{3}} = 1.$$

$$= \dfrac{18\sqrt{3}}{3 \cdot 3}$$

$$= 2\sqrt{3}$$

Now Try:

9.   Rationalize each denominator.

    a. $\dfrac{14}{\sqrt{7}}$

    ____

    b. $\dfrac{5}{\sqrt{75}}$

    ____

Objective 5 Practice Exercises

Rationalize each denominator.

13. $\dfrac{15}{\sqrt{10}}$

13. ______________

14. $\dfrac{6}{\sqrt{28}}$

14. ______________

15. $\dfrac{3\sqrt{5}}{\sqrt{125}}$

15. ______________

Chapter 8 QUADRATIC EQUATIONS, INEQUALITIES, AND FUNCTIONS

Learning Objectives
> Evaluate exponential expressions.
> Use the four steps for solving a linear equation to solve equations.
> Factor trinomials with a coefficient of 1 for the second-degree term.
> Factor trinomials using the FOIL method.
> Solve quadratic equations using the zero-factor property.

Key Terms

Use the vocabulary terms listed below to complete each statement in exercises 1−11.

linear equation	**solution set**	**equivalent equations**
prime polynomial	**factoring**	**greatest common factor**
quadratic equation	**standard form**	**double solution**
coefficient	**trinomial**	

1. Equations that have exactly the same solutions sets are called

   _________________________________.

2. An equation that can be written in the form $Ax + B = C$, where A, B, and C are real numbers and $A \neq 0$, is called a _________________________________.

3. In the term $6x^2 y$, 6 is the _________________________.

4. An equation that can written in the form $ax^2 + bx + c = 0$, with $a \neq 0$, is a

   _________________________.

5. The _________________________________ of a polynomial is the greatest term that is a factor of all the terms in the polynomial.

6. A _________________________________ is a polynomial that cannot be factored using only integers.

7. A polynomial with three terms is a _________________________.

8. An equation written in the form $ax^2 + bx + c = 0$ is written in the _________________________ of a quadratic equation.

9. Two factors are identical and both lead to the same solution, called a

   _________________________.

10. The set of all numbers that satisfy an equation is called its _________________________.

11. _________________________________ is the process of writing a polynomial as a product.

Objective 1   Evaluate exponential expressions.

Review this example for Objective 1:

1. Find the value of the exponential expression.

$$6^2$$

6^2 means $6 \cdot 6$, which equals 36.

Now Try:

1. Find the value of the exponential expression.

$$7^2$$

____

Objective 1 Practice Exercises

Find the value of each exponential expression.

1. 3^3

1. __________

2. $\left(\dfrac{2}{3}\right)^4$

2. __________

3. $(0.4)^2$

3. __________

Objective 2   Use the four steps for solving a linear equation.

Review these examples for Objective 2:

2. Solve $-5x + 8 = 23$.

Step 1 There are no parentheses, fractions, or decimals in this equation, so this step is not necessary.

$$-5x + 8 = 23$$

Step 2 $\quad -5x + 8 - 8 = 23 - 8$

$$-5x = 15$$

Step 3 $\quad \dfrac{-5x}{-5} = \dfrac{15}{-5}$

$$x = -3$$

Step 4 Check by substituting -3 for x in the original equation.

$$-5x + 8 = 23$$
$$-5(-3) + 8 \overset{?}{=} 23$$
$$15 + 8 \overset{?}{=} 23$$
$$23 = 23 \quad \text{True}$$

The solution, -3, checks, so the solution set is $\{-3\}$.

3. Solve $4x + 3 = 6x - 11$.

Step 1 There are no parentheses, fractions, or decimals in this equation, so begin with Step 2.

$$4x + 3 = 6x - 11$$

Step 2 $\quad 4x + 3 - 4x = 6x - 11 - 4x$

$$3 = 2x - 11$$
$$3 + 11 = 2x - 11 + 11$$
$$14 = 2x$$

Step 3 $\quad \dfrac{14}{2} = \dfrac{2x}{2}$

$$7 = x$$

Step 4 Check by substituting 7 for x in the original equation.

$$4x + 3 = 6x - 11$$
$$4(7) + 3 \overset{?}{=} 6(7) - 11$$
$$28 + 3 \overset{?}{=} 42 - 11$$
$$31 = 31 \quad \text{True}$$

The solution, 7, checks, so the solution set is $\{7\}$.

Now Try:

2. Solve $-8x + 11 = 59$.

____

3. Solve $5x + 4 = 8x - 20$.

____

4. Solve $9a - (4 + 3a) = 2a + 5$.

$$9a - (4 + 3a) = 2a + 5$$

Step 1 $\quad 9a - 4 - 3a = 2a + 5$

$$6a - 4 = 2a + 5$$

Step 2 $\quad 6a - 4 - 2a = 2a + 5 - 2a$

$$4a - 4 = 5$$

$$4a - 4 + 4 = 5 + 4$$

$$4a = 9$$

Step 3 $\quad \dfrac{4a}{4} = \dfrac{9}{4}$

$$a = \dfrac{9}{4}$$

Step 4 Check that the solution set is $\left\{\dfrac{9}{4}\right\}$.

4. Solve $10a - (11 + 3a) = 5a + 4$.

____

Objective 2 Practice Exercises

Solve each equation and check your solution.

4. $\quad 7t + 6 = 11t - 4$

4. _______________

5. $\quad 3a - 6a + 4(a - 4) = -2(a + 2)$

5. _______________

6. $\quad 3(t + 5) = 6 - 2(t - 4)$

6. _______________

Objective 3   Factor trinomials with coefficient 1 for the second-degree term.

Review these examples for Objective 3: | **Now Try:**

5. Factor $m^2 + 8m + 15$.

Look for integers whose product is 15 and whose sum is 8. Only positive signs are needed.

Factors of 15	Sums of Factors
15, 1	$15 + 1 = 16$
5, 3	$5 + 3 = 8$

From the table, 5 and 3 are the required integers.

$m^2 + 8m + 15$ factors as $(m + 5)(m + 3)$

Check   Use the FOIL method.

$$(m + 5)(m + 3) = m^2 + 3m + 5m + 15$$
$$= m^2 + 8m + 15$$

5. Factor $x^2 + 11x + 24$.

____

6. Factor the trinomial.

$x^2 - 7x + 18$

Look for integers whose product is 18 and whose sum is –7. Since the numbers have a positive product and a negative sum, we consider only pairs of negative integers.

Factors of 18	Sums of Factors
$-18, -1$	$-18 + (-1) = -19$
$-9, -2$	$-9 + (-2) = -11$
$-6, -3$	$-6 + (-3) = -9$

None of the pairs of integers has a sum of –7.

$x^2 - 7x + 18$ cannot be factored.
It is a prime polynomial.

6. Factor the trinomial.

$m^2 - 7m + 5$

____

7. Factor $x^2 - 6xy - 7y^2$.

Here, the coefficient of x in the middle term is –6y, so we need to find two expressions whose product is $-7y^2$ and whose sum is –6y.

Factors of $-7y^2$	Sums of Factors
$7y, -y$	$7y + (-y) = 6y$
$-7y, y$	$-7y + y = -6y$

$x^2 - 6xy - 7y^2$ factors as $(x - 7y)(x + y)$

7. Factor $p^2 - 5pq - 14q^2$.

____

　　　　

Check Use the FOIL method.

$$(x-7y)(x+y)=x^2+xy-7xy-7y^2$$
$$=x^2-6xy-7y^2$$

Objective 3 Practice Exercises

Factor completely. If a polynomial cannot be factored, write prime.

7. r^2+r+3

7. ___________________

8. $x^2-11x+28$

8. ___________________

9. $x^2-8x-33$

9. ___________________

   Copyright © 2020 Pearson Education, Inc.

Objective 4   Factor trinomials using the FOIL method.

Review these examples for Objective 4:

8. Factor $6x^2 + 13x + 7$.

The number 6 has several possible pairs of factors, but 7 has only 1 and 7, or -1 and -7. We choose positive factors since all coefficients in the trinomial are positive.

$$(\_\_+7)(\_\_+1)$$

The possible pairs of $6x^2$ are $6x$ and x, or $3x$ and $2x$.

$$(3x+7)(2x+1)$$
gives middle term $3x + 14x = 17x$. Incorrect.
$$(2x+7)(3x+1)$$
gives middle term $2x + 21x = 23x$. Incorrect.
$$(6x+7)(x+1)$$
gives middle term $6x + 7x = 13x$. Correct.

$6x^2 + 13x + 7$ factors as $(6x+7)(x+1)$.

Check. Multiply $(6x+7)(x+1)$ to obtain $6x^2 + 13x + 7$.

9. Factor $6x^2 - x - 15$.

The integer 6 has several possible pairs of factors, as does -15. Since the constant term is negative, one positive factor and one negative factor of -15 are needed. Since the coefficient of the middle term is relatively small, it is wise to avoid large factors. We try $3x$ and $2x$ as factors of $6x^2$ and 5 and -3 as factors of -15.

$$(3x+5)(2x-3)$$
has middle term $-9x + 10x = x$. Incorrect.
$$(3x-5)(2x+3)$$
has middle term $9x - 10x = -x$. Correct.

$6x^2 - x - 15$ factors as $(3x-5)(2x+3)$.

Now Try:

8. Factor $15x^2 + 26x + 7$.

———————

9. Factor $8x^2 + 2x - 21$.

———————

10. Factor $18x^2 - 3xy - 28y^2$.

There are several factors of $18x^2$, including
   $18x$ and x, $9x$ and $2x$, and $6x$ and $3x$.
There are many possible pairs of factors of
$-28y^2$, including
   $28y$ and $-y$, $-28y$ and y, $14y$ and $-2y$,
   $-14y$ and $2y$, $7y$ and $-4y$, $-7y$ and $4y$.

Once again, since the coefficient of the middle
term is relatively small, avoid the larger factors.
Try the factors of $6x$ and $3x$, and $4y$ and $-7y$.
   $(6x + 4y)(3x - 7y)$ Incorrect
The first binomial has a common factor of 2.
   $(6x - 7y)(3x + 4y)$
has middle term $24xy - 21xy = 3xy$. Incorrect.
Interchange the signs of the last two terms.
   $(6x + 7y)(3x - 4y)$
has middle term $-24xy + 21xy = -3xy$. Correct.

Thus, $18x^2 - 3xy - 28y^2$ factors as
$(6x + 7y)(3x - 4y)$

11. Factor the trinomial.

$$-105a^3 + 65a^2 - 10a$$

The common factor is $-5a$. Then use trial
and error.

$$-105a^3 + 65a^2 - 10a = -5a(21a^2 - 13a + 2)$$
$$= -5a(3a - 1)(7a - 2)$$

Check.

$$-5a(3a - 1)(7a - 2) = -5a(21a^2 - 13a + 2)$$
$$= -105a^3 + 65a^2 - 10a$$

---

10. Factor $24x^2 - 2xy - 15y^2$.

   ____

11. Factor the trinomial.

$$-18a^3 + 66a^2 - 60a$$

   ____

Objective 4 Practice Exercises

Factor each trinomial completely.

10. $8q^2 + 10q + 3$ 10. _______________

11. $3a^2 + 8ab + 4b^2$ 11. _______________

12. $4c^2 + 14cd - 8d^2$ 12. _______________

Objective 5   Solve quadratic equations using the zero-factor property.

Review these examples for Objective 5:	**Now Try:**

Review these examples for Objective 5:

12. Solve each equation.

   a. $(x+9)(5x-6)=0$

By the zero-factor property, either $x+9=0$ or $5x-6=0$.

$$x+9=0 \quad \text{or} \quad 5x-6=0$$
$$x=-9 \quad \text{or} \quad 5x=6$$
$$x=\frac{6}{5}$$

Check:

Let $x=-9$.
$$(x+9)(5x-6)=0$$
$$(-9+9)[5(-9)-6]\overset{?}{=}0$$
$$0(-51)\overset{?}{=}0$$
$$0=0 \quad \text{True}$$

Let $x=\frac{6}{5}$.
$$(x+9)(5x-6)=0$$
$$\left(\frac{6}{5}+9\right)\left[5\left(\frac{6}{5}\right)-6\right]\overset{?}{=}0$$
$$\left(\frac{51}{5}\right)(6-6)\overset{?}{=}0$$
$$0=0 \quad \text{True}$$

Both values check, so the solution set is
$$\left\{-9, \ \frac{6}{5}\right\}.$$

   b. $x(8x-11)=0$

Use the zero-factor property.
$$x=0 \quad \text{or} \quad 8x-11=0$$
$$8x=11$$
$$x=\frac{11}{8}$$

Check these solutions by substituting each in the original equation. The solution set is $\left\{0, \ \frac{11}{8}\right\}$.

Now Try:

12. Solve each equation.

   a. $(x+12)(4x-7)=0$

____

   b. $x(6x-11)=0$

____

13. Solve $8p^2 + 30 = 46p$.

$$8p^2 + 30 = 46p$$
$$8p^2 - 46p + 30 = 0 \quad \text{Standard form}$$
$$2(4p^2 - 23p + 15) = 0 \quad \text{Factor out 2.}$$
$$4p^2 - 23p + 15 = 0 \quad \text{Divide each side by 2}$$
$$(4p - 3)(p - 5) = 0 \quad \text{Factor.}$$
$$4p - 3 = 0 \quad \text{or} \quad p - 5 = 0 \quad \text{Zero-factor}$$
$$p = \frac{3}{4} \quad \text{or} \quad p = 5 \quad \text{property}$$

The solution set is $\left\{\frac{3}{4},\, 5\right\}$.

13. Solve $15p^2 + 36 = 57p$.

____

Objective 5 Practice Exercises

Solve each equation and check your solutions.

13. $2x^2 - 3x - 20 = 0$

13. _______________

14. $25x^2 = 20x$

14. _______________

15. $c(5c + 17) = 12$

15. _______________

Chapter 9 INVERSE, EXPONENTIAL, AND LOGARITHMIC FUNCTIONS

Learning Objectives
Solve equations of the form $x^2 = k$, where $k > 0$.
Solve radical equations having square root radicals.
Determine whether a graph or an equation represents a function.
Find domain and range of given functions.

Key Terms

Use the vocabulary terms listed below to complete each statement in exercises 1–9.

quadratic equation	zero-factor property	range
radical equation	extraneous solution	domain
components	relation	function

1. In an ordered pair (x, y), x and y are the _______________________.

2. An equation that can be written in the form $ax^2 + bx + c = 0$ is a

   _______________________.

3. An _______________________ is a potential solution to an equation that does not satisfy the equation.

4. An equation with a variable in the radicand is a _______________________.

5. The _______________________ states that if a product equals 0, then at least one of the factors of the product also equals zero.

6. Any set of ordered pairs is called a _______________________.

7. The set of all second components in the ordered pairs of a relation is the _______________________ of the relation.

8. A _______________________ is a set of ordered pairs in which each first component corresponds to exactly one second component.

9. The set of all first components in the ordered pairs of a relation is the _______________________ of the relation.

Objective 1   Solve equations of the form $x^2 = k$, where $k > 0$.

Review these examples for Objective 1:

1. Solve each equation. Write radicals in simplified form.

a. $x^2 = 36$

By the square root property, if $x^2 = 36$, then
$x = \sqrt{36} = 6$ or $x = -\sqrt{36} = -6$
The solution set is $\{-6, 6\}$, or $\{\pm 6\}$.

b. $x^2 = 13$

The solutions are $x = \sqrt{13}$ or $x = -\sqrt{13}$
The solution set is $\left\{-\sqrt{13},\ \sqrt{13}\right\}$, or $\left\{\pm\sqrt{13}\right\}$.

c. $5p^2 - 100 = 0$

$$5p^2 - 100 = 0$$
$$5p^2 = 100$$
$$p^2 = 20$$
$$p = \sqrt{20} \quad \text{or} \quad p = -\sqrt{20}$$
$$p = 2\sqrt{5} \quad \text{or} \quad p = -2\sqrt{5}$$

Check

$$
\begin{array}{c|c}
5p^2 - 100 = 0 & 5p^2 - 100 = 0 \\
5(2\sqrt{5})^2 - 100 \stackrel{?}{=} 0 & 5(-2\sqrt{5})^2 - 100 \stackrel{?}{=} 0 \\
5(20) - 100 \stackrel{?}{=} 0 & 5(20) - 100 \stackrel{?}{=} 0 \\
0 = 0 \ \text{True} & \text{True} \ 0 = 0
\end{array}
$$

The solution set is $\left\{2\sqrt{5}, -2\sqrt{5}\right\}$, or $\left\{\pm 2\sqrt{5}\right\}$.

d. $5x^2 + 16 = 31$

$$5x^2 + 16 = 31$$
$$5x^2 = 15$$
$$x^2 = 3$$
$$x = \sqrt{3} \quad \text{or} \quad x = -\sqrt{3}$$

Now Try:

1. Solve each equation. Write radicals in simplified form.

a. $x^2 = 81$

____

b. $x^2 = 23$

____

c. $3x^2 - 54 = 0$

____

d. $4x^2 + 14 = 22$

____

2. Solve $p^2 = -64$.

Because –64 is a negative number and because the square of a real number cannot be negative, there is no real number solution of this equation. The solution set is $\varnothing$.

2. Solve $n^2 = -25$.

____

Objective 1 Practice Exercises

Solve each equation by using the square root property. Express all radicals in simplest form.

1. $r^2 = 900$

1. ________________

2. $3x^2 - 5 = 22$

2 ________________

3. $p^2 = -144$

3. ________________

Objective 2   Solve radical equations having square root radicals.

Review these examples for Objective 2:	**Now Try:**
3. Solve $\sqrt{p+2}=5$.	**3.** Solve $\sqrt{p+5}=4$.

$$\left(\sqrt{p+2}\right)^2 = 5^2$$
$$p+2 = 25$$
$$p = 23$$

Check $\sqrt{p+2}=5$
$$\sqrt{23+2}\overset{?}{=}5$$
$$\sqrt{25}\overset{?}{=}5$$
$$5=5 \quad \text{True}$$

The solution set is $\{23\}$.

4. Solve $4\sqrt{x}=\sqrt{x+30}$.

4. Solve $5\sqrt{x}=\sqrt{x+72}$.

$$\left(4\sqrt{x}\right)^2 = \left(\sqrt{x+30}\right)^2$$
$$4^2\left(\sqrt{x}\right)^2 = \left(\sqrt{x+30}\right)^2$$
$$16x = x+30$$
$$15x = 30$$
$$x = 2$$

Check $4\sqrt{x}=\sqrt{x+30}$
$$4\sqrt{2}\overset{?}{=}\sqrt{2+30}$$
$$4\sqrt{2}\overset{?}{=}\sqrt{32}$$
$$4\sqrt{2}=4\sqrt{2} \quad \text{True}$$

The solution set is $\{2\}$.

Objective 2 Practice Exercises

Solve each equation.

4. $\sqrt{3x+1}=3$

4. _____________

5. $\sqrt{2+4k} = 3\sqrt{k}$

5. _______________

6. $\sqrt{4x+3} = \sqrt{3x+5}$

6. _______________

Objective 3   Determine whether a graph or an equation represents a function.

Review these examples for Objective 3:

5.   Determine whether each relation represented by a graph or an equation is a function.

a.

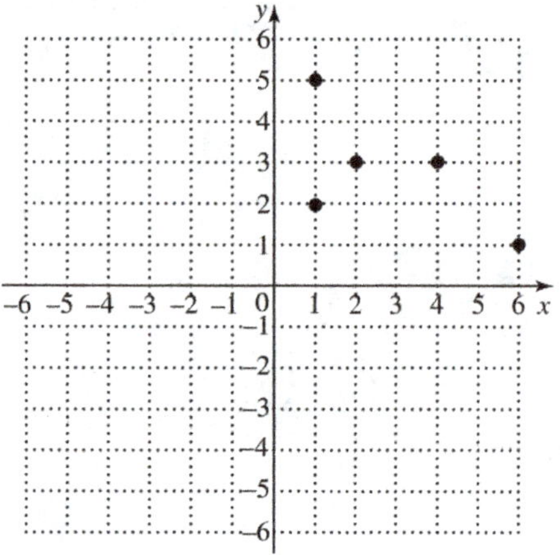

Because there are two ordered pairs with first component 1, this is not the graph of a function.

b.

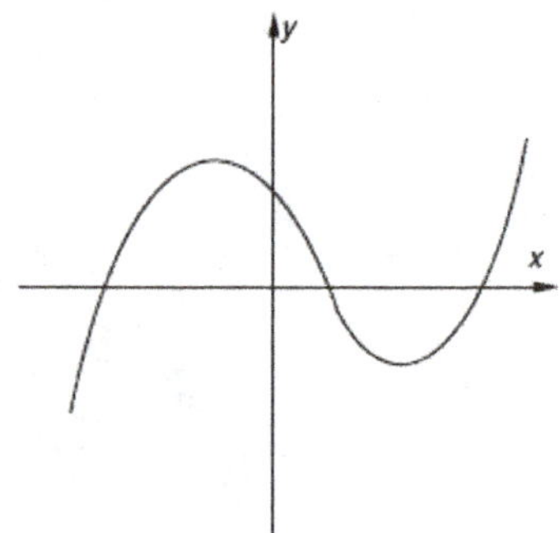

Use the vertical line test. Any vertical line intersects the graph just once, so this is the graph of a function.

c. $y = 4x - 2$

This linear equation is in the form $y = mx + b$. Since the graph of this equation is a line that is not vertical, the equation defines a function.

Now Try:

5.   Determine whether each relation represented by a graph or an equation is a function.

a.

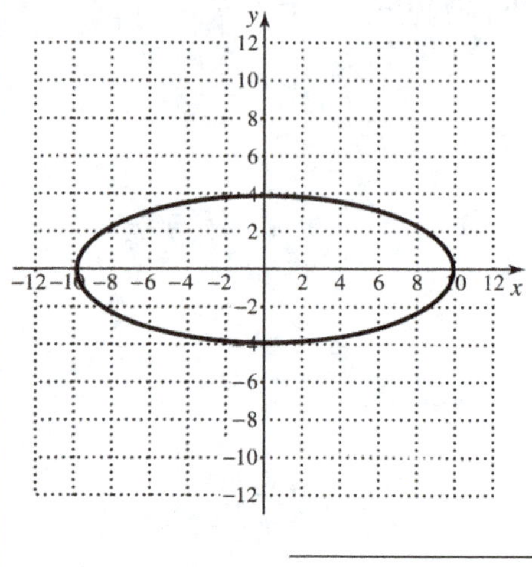

____

b.

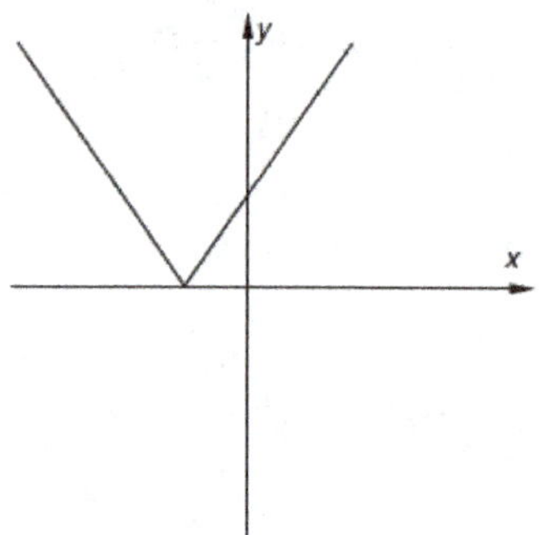

____

c. $y = -3x + 5$

____

Objective 3 Practice Exercises

Use the vertical line test to determine whether each relation graphed is a function.

7.

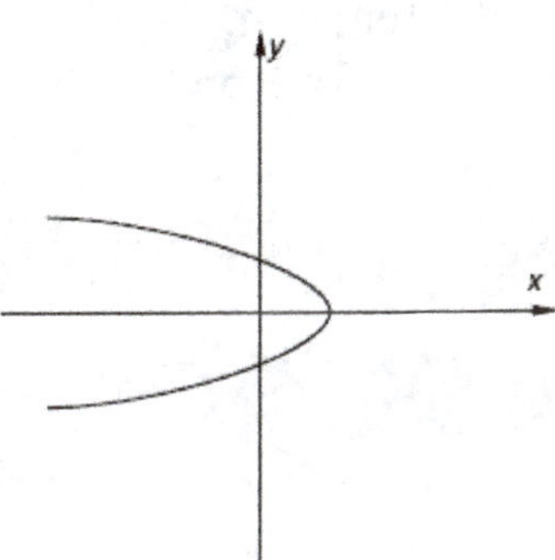

7. ______________________

8.

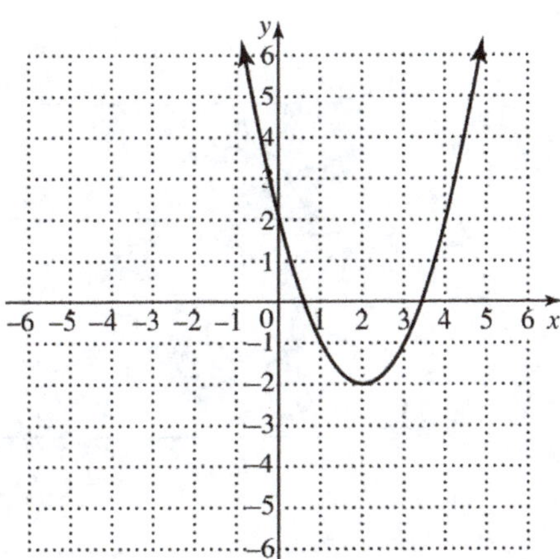

8. ______________________

Decide whether the equation defines y as a function of x.

9. $y = 7$

9. ______________________

Objective 4   Find domain and range of given functions.

Review this example for Objective 4:

6. Find the domain and range of the function.

$$y = 4x - 5$$

Any number may be input for x, so the domain is the set of all real numbers, or $(-\infty, \infty)$.
Any number may be the output for y, so the range is also the set of all real numbers, or $(-\infty, \infty)$.

Now Try:

6. Find the domain and range of the function.
$$y = -5x + 6$$

____

____

Objective 4 Practice Exercises

Find the domain and range of each function.

10. $y = -2x + 3$

10. __________

____

11. $y = -x - 1$

11. __________

____

12. $y = 2x^2$

12. __________

____

Chapter 10 NONLINEAR FUNCTIONS, CONIC SECTIONS, AND NONLINEAR SYSTEMS

Learning Objectives

Midpoint and Distance in the Coordinate Plane.
Solve quadratic equations by completing the square when the leading coefficient is 1.
Graph a line using its slope and a point on the line.
Graph quadratic equations.
Solve systems of linear inequalities by graphing.

Key Terms

Use the vocabulary terms listed below to complete each statement in exercises 1−12.

completing the square	perfect square trinomial	square root property
slope-intercept form	point-slope form	standard form
parabola vertex	axis of symmetry	line of symmetry
system of linear inequalities	solution set of a system of linear inequalities	

1. A _________________________ can be written in the form $x^2 + 2kx + k^2$ or $x^2 - 2kx + k^2$

2. The _________________________ says that, if k is positive and $a^2 = k$, then $a = \pm\sqrt{k}$.

3. Use the process called _________________________ in order to rewrite an equation so it can be solved using the square root property.

4. A linear equation in the form $y - y_1 = m(x - x_1)$ is written in _________________________.

5. A linear equation in the form $Ax + By = C$ is written in _________________________.

6. A linear equation in the form $y = mx + b$ is written in _________________________.

7. If a graph is folded on its _________________________, the two sides coincide.

8. The _________________________________ of a parabola that opens upward or downward is the lowest or highest point on the graph.

9. The _________________________________ of a parabola that opens upward or downward is a vertical line through the vertex.

10. The graph of the quadratic equation $y = ax^2 + bx + c$ is called a _________________________.

11. All ordered pairs that make all inequalities of the system true at the same time is called the ____.

12. A _________________________________ contains two or more linear inequalities (and no other kinds of inequalities).

Objective 1   Midpoint and Distance in the Coordinate Plane.

Review these examples for Objective 1:	**Now Try:**

Review these examples for Objective 1:

1. Find the midpoint of the line segment with endpoints of coordinates: $(3, 4)$ and $(-1, 2)$

   To find the midpoint use the midpoint formula

   $$\left(\frac{x_1 + x_2}{2}, \frac{y_1 + y_2}{2}\right)$$

   $$\left(\frac{3 + -1}{2}, \frac{4 + 2}{2}\right) = \left(\frac{2}{2}, \frac{6}{2}\right) = (1, 3)$$

   So, the midpoint of the line segment is $(1, 3)$.

2. Find the distance between the coordinates: $(3, 4)$ and $(-1, 2)$

   To find the distance use the distance formula:

   $$d = \sqrt{(x_2 - x_1)^2 + (y_2 - y_1)^2}.$$

   $$d = \sqrt{(-1 - 3)^2 + (2 - 4)^2}$$

   $$= \sqrt{(-4)^2 + (-2)^2}$$

   $$= \sqrt{16 + 4} = \sqrt{20} = 2\sqrt{5}.$$

Now Try:

1. Find the midpoint of the line segment with endpoints of coordinates: $(3, -4)$ and $(-2, 8)$

   ____

2. Find the distance between the coordinates: $(3, -4)$ and $(-2, 8)$

   ____

Objective 1 Practice Exercises

Given the following coordinates a) find the midpoint of the line segment b) find the distance between the given coordinates.

1. $(0, 3)$ and $(-4, 7)$

   1. Midpoint: _______________
      Distance: _______________

2. $(-2, 3)$ and $(1, 7)$

   2. Midpoint: _______________
      Distance: _______________

3. $(1, 1)$ and $(2, -4)$

3. Midpoint: _________________
Distance: _________________

Objective 2 Solve quadratic equations by completing the square when the coefficient of the second-degree term is 1.

Review these examples for Objective 1:

3. Complete each trinomial so that it is a perfect square. Then factor the trinomial.

a. $x^2 + 24x + $ _______

The perfect square trinomial will have the form $x^2 + 2kx + k^2$. Thus, the middle term $24x$, must equal $2kx$.

$$24x = 2kx$$
$$12 = k$$

Therefore, $k = 12$ and $k^2 = 12^2 = 144.$ The perfect square trinomial is $x^2 + 24x + 144,$

which factors as $(x+12)^2$.

b. $x^2 - 36x + $ _______

The perfect square trinomial will have the form $x^2 - 2kx + k^2$. Thus, the middle term $-36x$, must equal $-2kx$.

$$-36x = -2kx$$
$$18 = k$$

Therefore, $k = 18$ and $k^2 = 18^2 = 324.$ The required perfect square trinomial is

$x^2 - 36x + 324,$ which factors as $(x-18)^2$.

4. Solve $x^2 + 8x + 3 = 0.$

$$x^2 + 8x + 3 = 0$$

$$x^2 + 8x = -3$$

The expression on the left must be written as a perfect square trinomial in the form $x^2 + 2kx + k^2$.

$$x^2 + 8x + \underline{}$$

Here, $2kx = 8x,$ so $k = 4$ and $k^2 = 16.$ The required perfect square trinomial is $x^2 + 8x + 16$

which factors as $(x+4)^2$.

Therefore, if we add 16 to each side of

Now Try:

3. Complete each trinomial so that it is a perfect square. Then factor the trinomial.

a. $x^2 + 10x + $ _______

____

b. $x^2 - 22x + $ _______

____

4. Solve $x^2 + 10x - 7 = 0.$

____

$x^2 + 8x = -3$, the equation will have a perfect square trinomial on the left side, as needed.

$$x^2 + 8x + 16 = -3 + 16$$

$$(x+4)^2 = 13$$

Use the square root property.

$$x + 4 = \sqrt{13} \quad \text{or} \quad x + 4 = -\sqrt{13}$$

$$x = -4 + \sqrt{13} \quad \text{or} \quad x = -4 - \sqrt{13}$$

Check by substituting $-4 + \sqrt{13}$ and then $-4 - \sqrt{13}$ for x in the original equation. The solution set is $\{-4 + \sqrt{13},\ -4 - \sqrt{13}\}$.

Objective 2 Practice Exercises

Solve each equation by completing the square.

4. $r^2 + 8r = -4$

4. _______________

5. $x^2 - 4x = 2$

5. _______________

6. $x^2 + 2x = 63$

6. _______________

Objective 3   Graph a line by using its slope and a point on the line.

Review this example for Objective 3:

5. Graph the equation by using the slope and y-intercept.

$$2x - 3y = 6$$

Step 1 Solve for y to write the equation in slope-intercept form.

$$2x - 3y = 6$$
$$-3y = -2x + 6$$
$$y = \frac{2}{3}x - 2$$

Step 2 The y-intercept is $(0, -2)$. Graph this point.

Step 3 The slope is $\frac{2}{3}$. By definition,

$$\text{slope } m = \frac{\text{change in } y \text{ (rise)}}{\text{change in } x \text{ (run)}} = \frac{2}{3}$$

From the y-intercept, count up 2 units and to the right 3 units to obtain the point $(3, 0)$.

Step 4 Draw the line through the points $(0, -2)$ and $(3, 0)$ to obtain the graph.

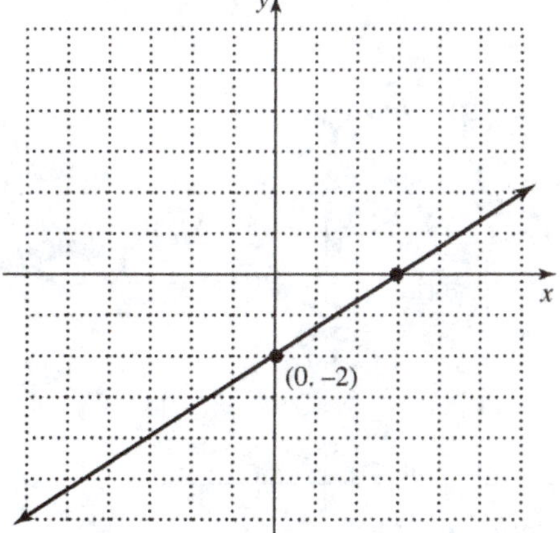

Now Try:

5. Graph the equation by using the slope and y-intercept.

$$y = \frac{2}{3}x$$

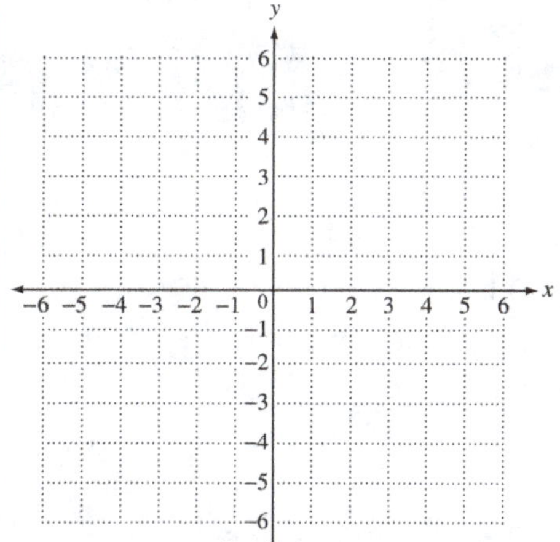

Objective 3 Practice Exercises

Graph each equation by using the slope and y-intercept.

7. $4x - y = 4$ **7.**

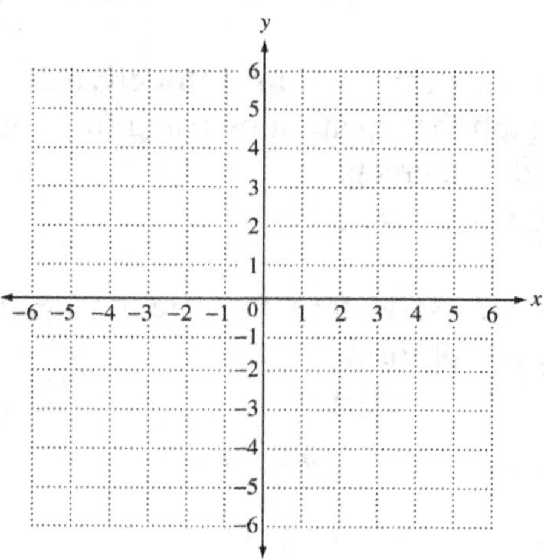

8. $y = -3x + 6$ **8.**

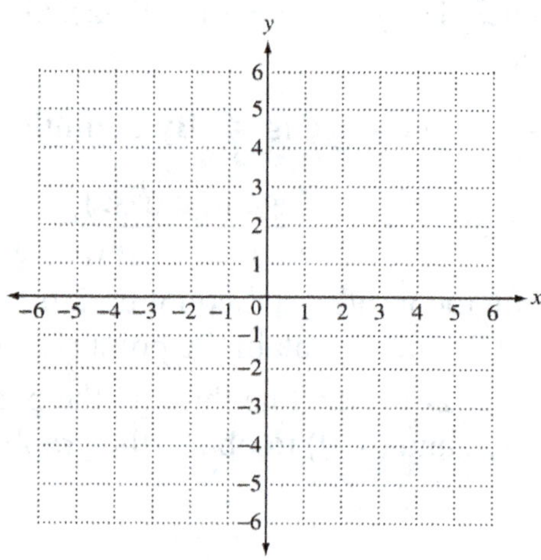

Graph the line passing through the given point and having the given slope.

9. $(4, -2)$; $m = -1$ **9.**

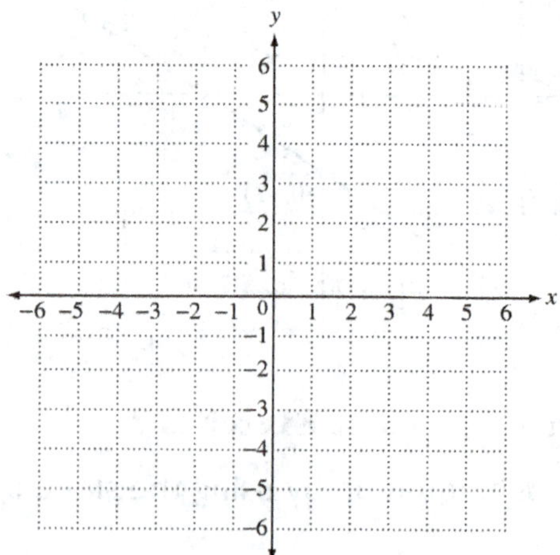

Objective 4 Graph quadratic equations.

Review this example for Objective 4:

6. Graph $y = 9 - x^2$. Identify the vertex.

Let $x = 3$.

$$y = 9 - 3^2$$
$$y = 9 - 9$$
$$y = 0$$

Ordered pair: (3, 0)

Let $x = 0$.

$$y = 9 - 0^2$$
$$y = 9 - 0$$
$$y = 9$$

Ordered pair: (0, 9)

Let $x = -3$.

$$y = 9 - (-3)^2$$
$$y = 9 - 9$$
$$y = 0$$

Ordered pair: (−3, 0)

Select several values for x and find the corresponding values for y. Plot the ordered pairs and join them with a smooth curve.

x	y
−4	−7
−3	0
−2	5
0	9
3	0
4	−7

The vertex is the highest point on the graph, (0, 9).

Now Try:

6. Graph $y = -x^2 - 1$. Identify the vertex.

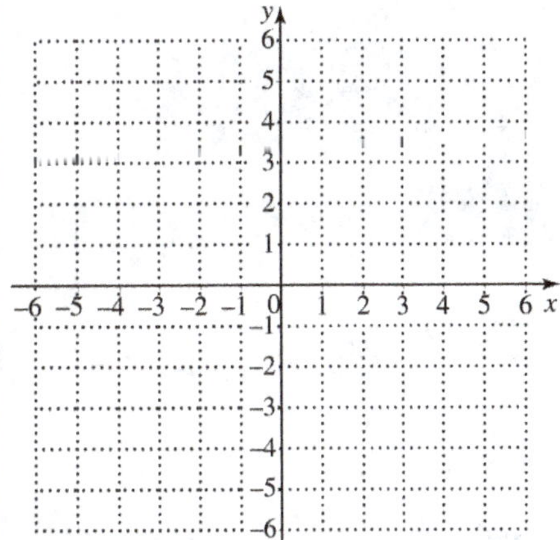

Objective 4 Practice Exercises

Graph each equation. Give the coordinates of the vertex in each case.

10. $y = -x^2$

10. vertex: _______________

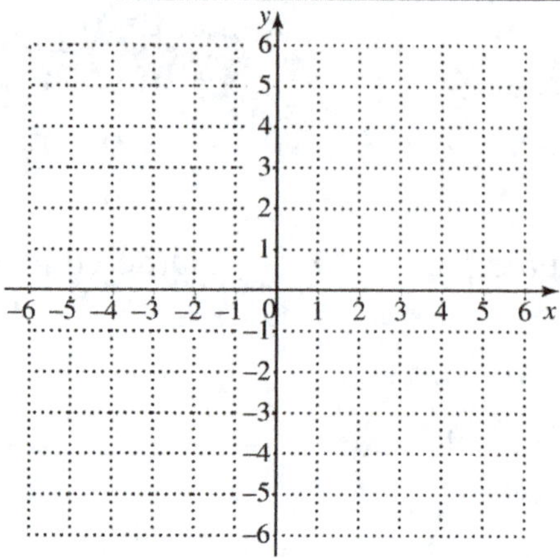

11. $y = x^2 - 1$

11. vertex: _______________

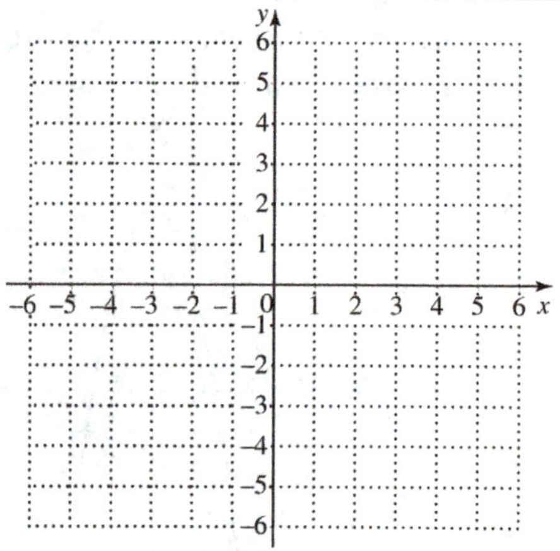

Objective 5   Solve systems of linear inequalities by graphing.

Review these examples for Objective 5:

7. Graph the solution set of the system.

$$x + y \leq 3$$

$$5x - y \geq 5$$

To graph $x + y \leq 3$, graph the solid boundary line $x + y = 3$ using the intercepts (0, 3) and (3, 0). Determine the region to shade using (0, 0) as a test point.

$$x + y \overset{?}{\leq} 3$$

$$0 + 0 \overset{?}{\leq} 3$$

$$0 \leq 3 \quad \text{True}$$

Shade the region containing (0, 0).

To graph $5x - y \geq 5$, graph the solid boundary line $5x - y = 5$ using the intercepts (0, –5) and (1, 0). Determine the region to shade using (0, 0) as a test point.

$$5x - y \geq 5$$

$$5(0) - 0 \overset{?}{\geq} 5$$

$$0 \geq 5 \quad \text{False}$$

Shade the region that does not contain (0, 0).

The solution set of this system includes all points in the intersection (overlap) of the graph of the two inequalities. This intersection is the gray shaded region and portions of the two boundary lines that surround it.

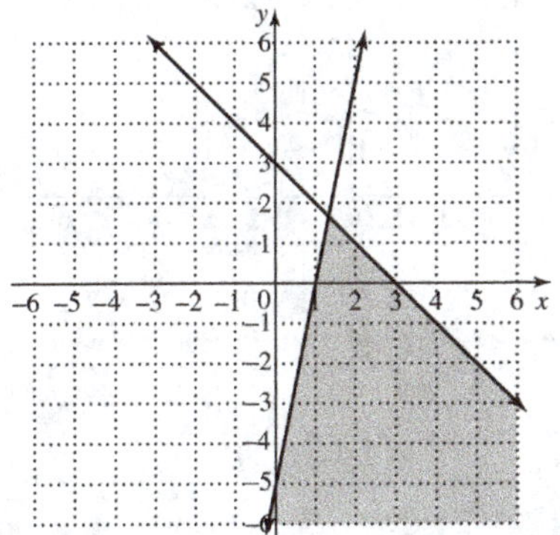

8. Graph the solution set of the system.

$$4x - y > 2$$

$$x + y > -2$$

Now Try:

7. Graph the solution set of the system.

$$3x - y \leq 3$$

$$x + y \leq 0$$

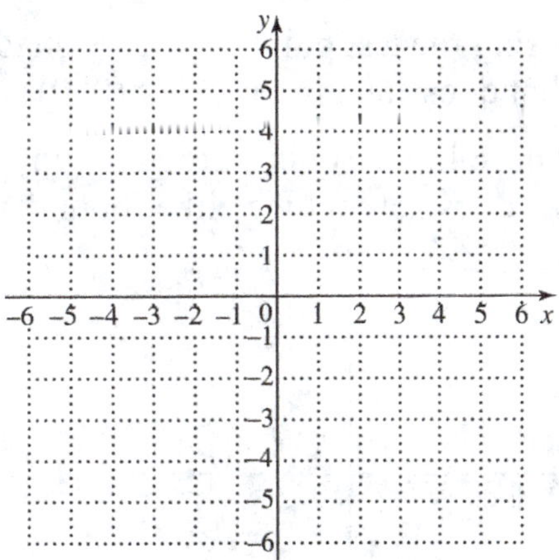

8. Graph the solution set of the system.

$$6x - \ y > 6$$

$$2x + 5y < 10$$

Graph $4x - y > 2$ using a dashed line for

$4x - y = 2$ and intercepts $(0, -2)$ and $\left(\frac{1}{2}, 0\right)$.

Using the test point $(0, 0)$, we shade the region that does not contain the point $(0, 0)$.

Graph $x + y > -2$ using a dashed line for $x + y = -2$ and intercepts $(-2, 0)$ and $(0, -2)$. Using the test point $(0, 0)$, we shade the region that does contain the point $(0, 0)$.

The solution set is marked in gray. The solution set does not include either boundary line.

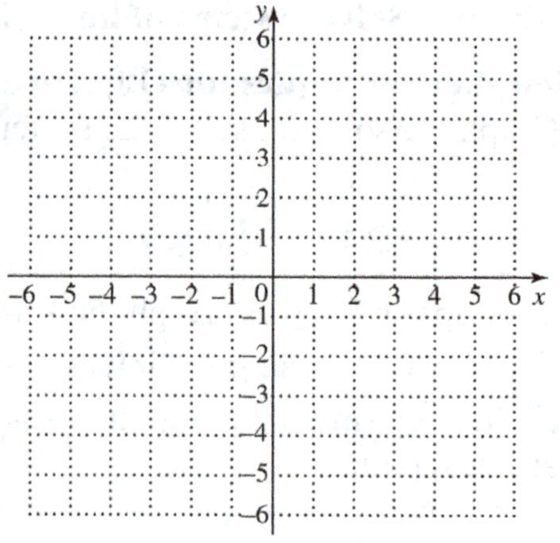

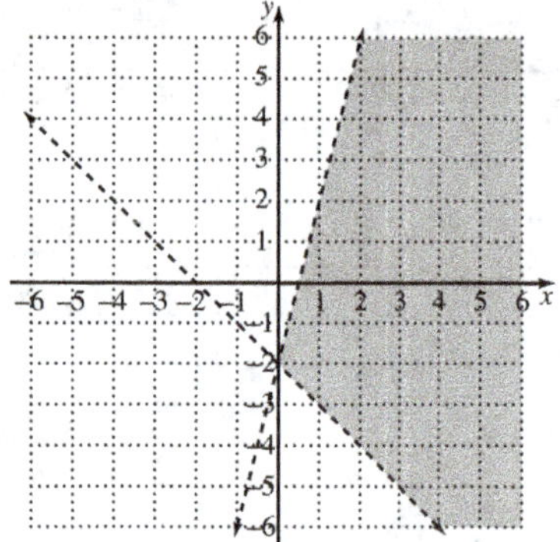

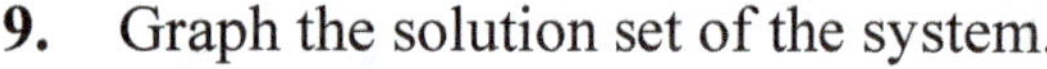

9. Graph the solution set of the system.
$$y \geq -1$$
$$2x - y > -1$$

Recall that $y = -1$ is a horizontal line through the point $(0, -1)$. Use a solid line, and shade the region with the point $(0, 0)$.

The graph of $2x - y > -1$ is created using a dashed line through the intercepts $(0, 1)$ and $\left(-\frac{1}{2}, 0\right)$. Shade the region with the point $(0, 0)$.

9. Graph the solution set of the system.
$$x - 3y \leq -7$$
$$x < 2$$

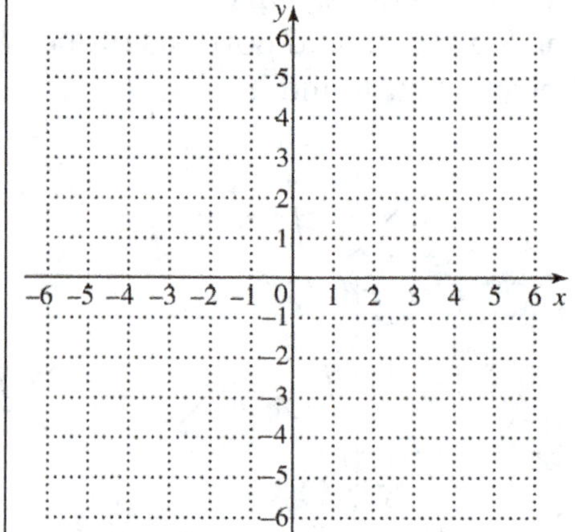

The solution set is marked in gray. The solution set includes the boundary line $y \geq -1$.

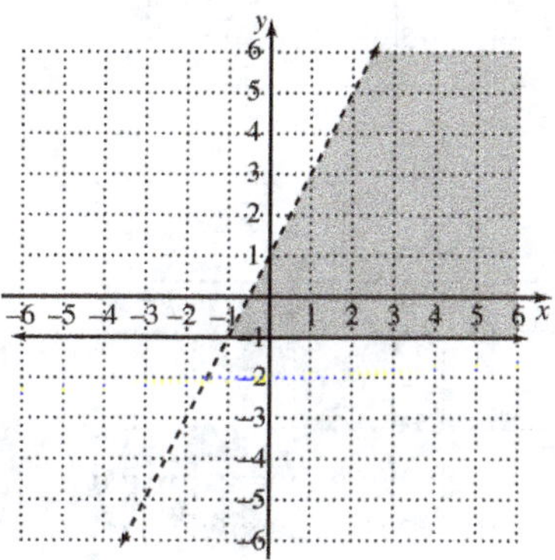

Objective 5 Practice Exercises

Graph the solution of each system of linear inequalities.

12. $4x + 5y \leq 20$

$\qquad y \leq x + 3$

12.

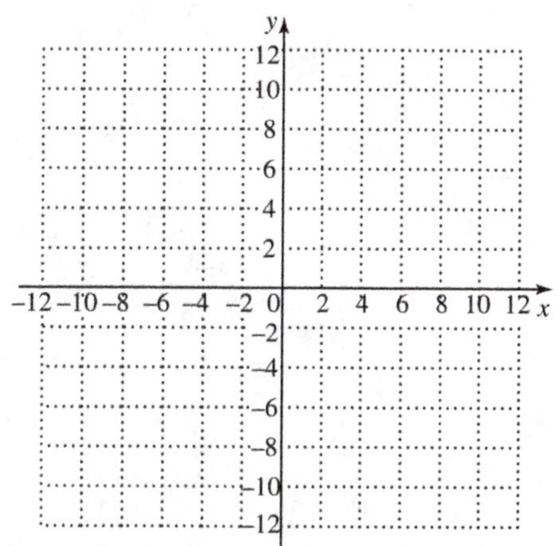

13. $x < 2y + 3$

$\qquad 0 < x + y$

13.

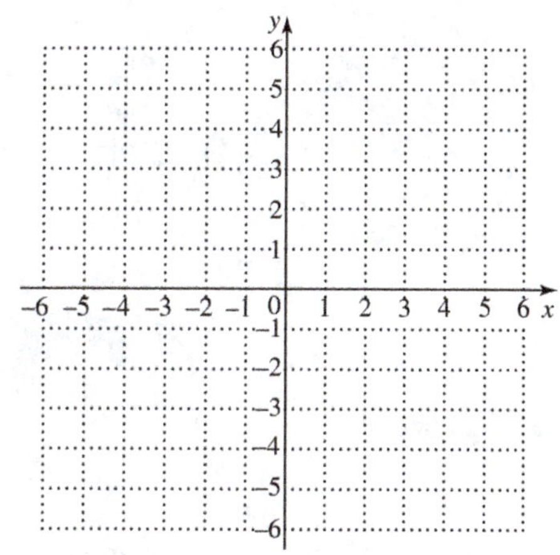

14. $y < 4$

$x \geq -3$

14.

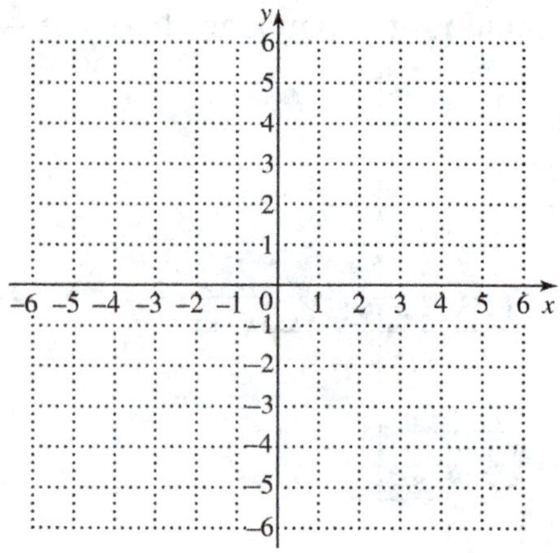

Chapter 11 FURTHER TOPICS IN ALGEBRA

Learning Objectives
Evaluate algebraic expressions, given values for the variables.
Use point-slope form to write an equation of a line.
Find greater powers of binomials.

Key Terms

Use the vocabulary terms listed below to complete each statement in exercises 1−7.

variable	**constant**	**algebraic expression**
equation	**solution**	**set** **elements**

1. A(n) _________________________ is a statement that says two expressions are equal.

2. The objects that belong to a set are its _______________________________ .

3. A _____________________________ is a symbol, usually a letter, used to represent an unknown number.

4. A collection of numbers, variables, operation symbols, and grouping symbols is an_______________________________________.

5. A _________________________________ is collection of objects.

6. Any value of a variable that makes an equation true is a(n) _________________ of the equation.

7. A _________________________________ is a fixed, unchanging number.

Objective 1   Evaluate algebraic expressions, given values for the variables.

Review these examples for Objective 2:

1. Evaluate each expression for $x = 4$.

a. $3x - 5$

$$3x - 5 = 3(4) - 5 \quad \text{Let } x = 4.$$
$$= 12 - 5 \quad \text{Multiply.}$$
$$= 7 \quad \text{Subtract.}$$

b. $5x^2$

$$5x^2 = 5 \cdot 4^2 \quad \text{Let } x = 4.$$
$$= 5 \cdot 16 \quad \text{Square 4.}$$
$$= 80 \quad \text{Multiply.}$$

3. Find the value of each expression for $x = 7$ and $y = 6$.

$$3x + 4y + 2$$

Replace x with 7 and y with 6.
$$3x + 4y + 2 = 3 \cdot 7 + 4 \cdot 6 + 2$$
$$= 21 + 24 + 2 \quad \text{Multiply.}$$
$$= 47 \quad \text{Add.}$$

Now Try:

1. Evaluate each expression for $x = 6$.

a. $4x - 7$

____

b. $7x^2$

____

2. Find the value of each expression for $x = 8$ and $y = 4$.
$$5x + 6y + 1$$

____

Objective 1 Practice Exercises

Find the value of each expression if $x = 2$ and $y = 4$.

1. $9x - 3y + 2$

1. ______________

2. $\dfrac{2x + 3y}{3x - y + 2}$

2. ______________

3. $\dfrac{3y^2 + 2x^2}{5x + y^2}$

3. ______________

Objective 2   Use point-slope form to write an equation of a line.

Review this example for Objective 3:	**Now Try:**

Review this example for Objective 3:

3. Write an equation of each line. Give the final answer in slope-intercept form.

The line passing through $(5, -3)$ with slope $\dfrac{5}{4}$.

$$y - y_1 = m(x - x_1)$$
$$y - (-3) = \frac{5}{4}(x - 5)$$
$$y + 3 = \frac{5}{4}x - \frac{25}{4}$$
$$y = \frac{5}{4}x - \frac{37}{4}$$

Now Try:

3. Write an equation of each line. Give the final answer in slope-intercept form.

The line passing through $(6, 11)$ with slope $-\dfrac{2}{3}$.

____

Objective 2 Practice Exercises

Write an equation for the line passing through the given point and having the given slope. Write the equations in slope-intercept form, if possible.

4. $(-3, 4); \ m = -\dfrac{3}{5}$

4. ___________________

5. $(-4, -3); \ m = -2$

5. ___________________

6. $(2, 2); \ m = -\dfrac{3}{2}$

6. ___________________

Objective 3   Find greater powers of binomials.
Review these examples for Objective 5:

Now Try:

4.   Find each product.

4.   Find each product.

a. $(x+4)^3$

a. $(x+6)^3$

$(x+4)^3$
$= (x+4)^2(x+4)$
$= (x^2+8x+16)(x+4)$
$= x^3+8x^2+16x+4x^2+32x+64$
$= x^3+12x^2+48x+64$

---

b. $(5y-4)^4$

b. $(3x-5)^4$

$(5y-4)^4$
$= (5y-4)^2(5y-4)^2$
$= (25y^2-40y+16)(25y^2-40y+16)$
$= 625y^4-1000y^3+400y^2$
$\quad\ -1000y^3+1600y^2-640y$
$\quad\ +400y^2-640y+256$
$= 625y^4-2000y^3+2400y^2-1280y+256$

Objective 3 Practice Exercises

Find each product.

7. $(a-3)^3$

7. ___________

8. $(j+3)^4$

8. ___________

9. $(4s+3t)^4$

9. ___________

Intermediate Algebra Integrated Review Answers

Chapter 1 Linear Equations, Inequalities, and Applications

Key Terms
1. equivalent equations   2. linear equation   3. Solution set

Objective 1
Now Try
1. $t = 29$ 2. $x = 8$

Practice Exercises
1. $t = -7$ 3. $p = 7.2$

Objective 2
Now Try
3. $m = 4$ 4. $x = -4$

Practice Exercises

5. $w = -5$

Objective 3
Now Try
5. 52 was the lowest grade.   6. The pen is 8 feet wide.

Practice Exercises
7. Rainier: 14410 ft, Denali: 20320 ft.

9. Mark: 3 laps, Pablo: 11 laps, Faustino: 12 laps

Objective 4
Now Try
7. $x \geq 2$,

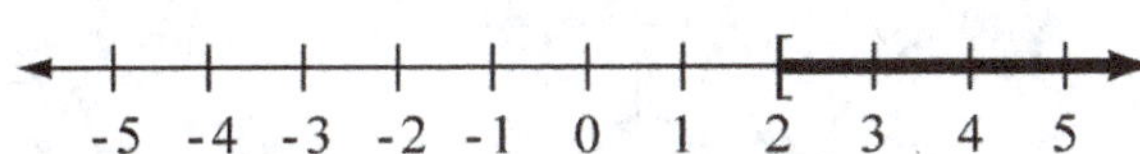

Practice Exercises
11. $m \leq 5$,

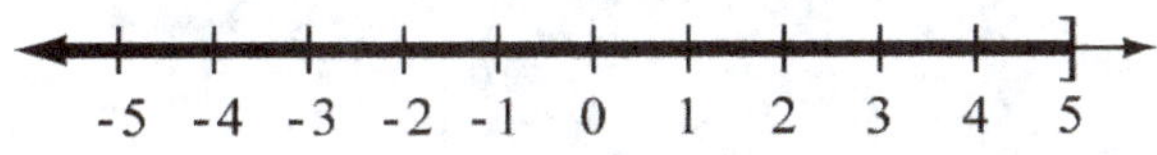

Chapter 2 Linear Equations and Inequalities in One Variable

Key Terms

1. y-intercept 2. x-intercept 3. parallel lines

4. linear inequality in two variables 5. graphing

6. graph 7. perpendicular lines

Objective 1

Now Try

1.

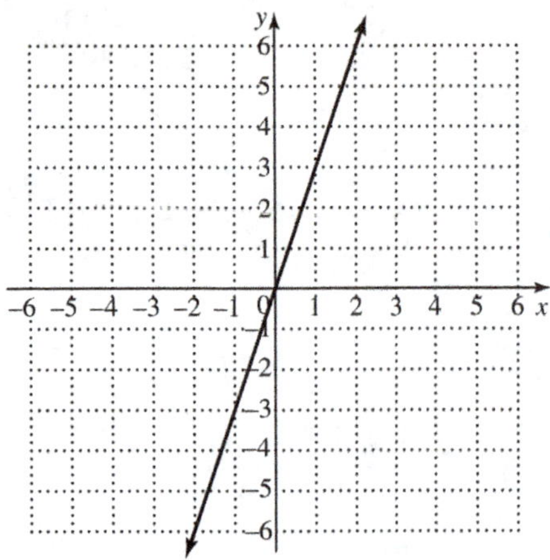

Practice Exercises

1.

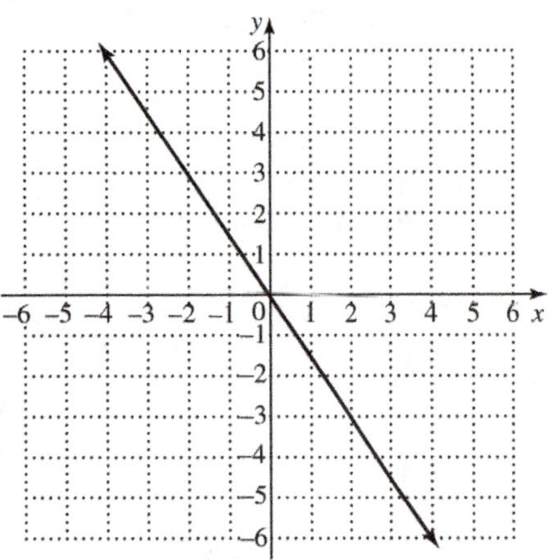

3.

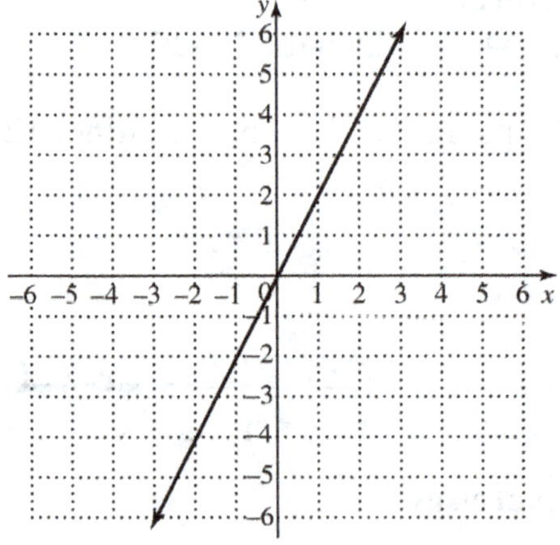

Objective 2

Now Try

1. positive

Practice Exercises

5. negative

Objective 3
Now Try
3a. parallel   3b. perpendicular

Practice Exercises
7. parallel   9. perpendicular

Objective 4
Now Try
4a. $y = \frac{3}{4}x - 5$ 4b. $y = 6x + 10$

Practice Exercises
11. $y = 2x$

Objective 5
Now Try
5.   6.

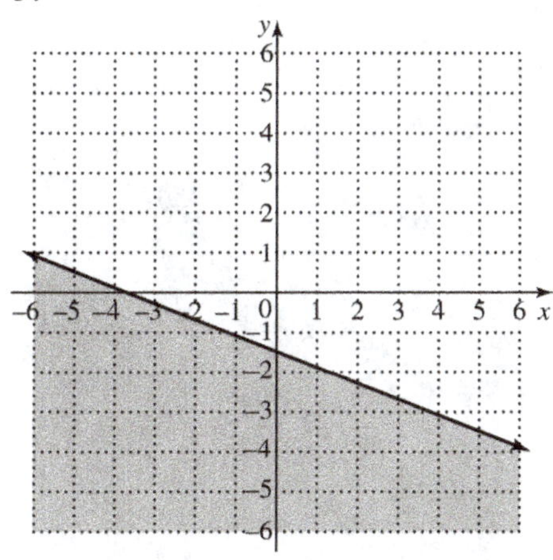   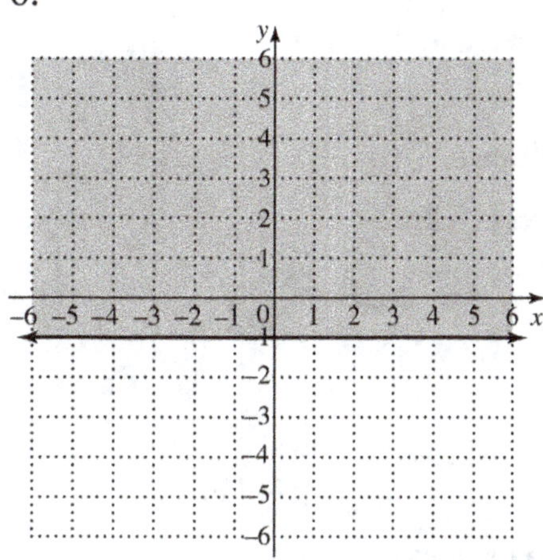

Practice Exercises
13.   15.

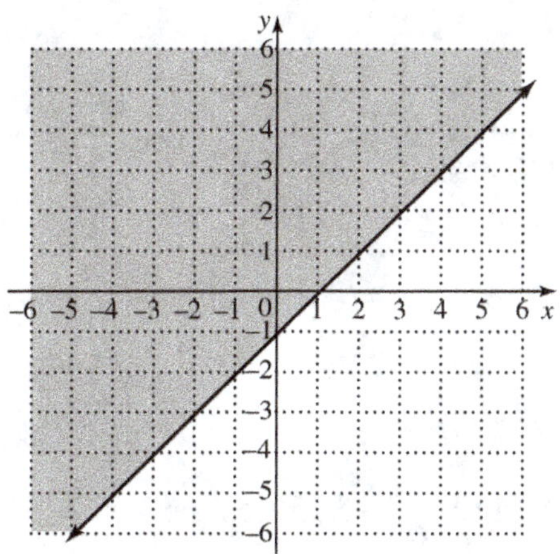   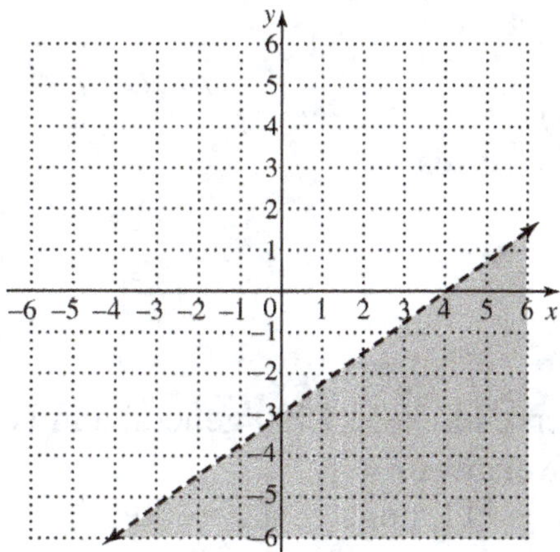

                    A-3

Chapter 3 Systems of Linear Equations

Key Terms

1. linear equation in two variables
2. independent equations
3. consistent system
4. Substitution
5. solution of the system
6. elimination method
7. dependent equations
8. inconsistent system
9. system of linear equations

Objective 1
Now Try
1a. solution
1b. not a solution

Practice Exercises
1. not a solution
3. solution

Objective 2
Now Try
2. not a solution

Practice Exercises
5. solution

Objective 3
Now Try
3. $(1, 6)$
4. $(9, 4)$

Practice Exercises
7. $(2, 4)$
9. $(4, -9)$

Objective 4
Now Try
5. $(-2, -3)$
6. $(-4, 9)$
Practice Exercises
11. $(0, 1)$

Objective 5
Now Try
7. 750 reserved tickets, 1500 general admission tickets
Practice Exercises
13. student $= 111$, nonstudent $= 300$
15. 8 - $14 ties, 2 - $25 ties

Chapter 4 Exponents, Polynomials, and Polynomial Functions

Key Terms

1. power
2. exponential expression
3. base
4. decreasing powers
5. term
6. trinomial
7. polynomial
8. monomial
9. degree of the polynomial
10. binomial
11. like terms

Objective 1

Now Try

1a. 7^8

1b. x^{30}

Practice Exercises

1. 7^{12}

3. $-27x^{12}$

Objective 2

Now Try

2a. $\frac{3125}{8}$

2b. $x^{27}y^{13}$

Practice Exercises

5. $5^{11}x^{18}y^{37}$

Objective 3

Now Try

3. $15m^3 + 29m^2$

Practice Exercises

7. $-3z^3$

9. $8c^3 - 8c^2 - 6c + 6$

Objective 4

Now Try

4a. $4x^3 + x + 12$

4b. $7x^3 + 8x^2 - 14x - 1$

5. $-11x^3 - 7x + 1$

6. $-4x^2 + 8x + 7$

Practice Exercises

11. $-8w^2 + 21w^2 - 15$

Objective 5

Now Try

7. $32x^4 + 64x^3$

Practice Exercises

13. $35z^4 + 14z$

15. $-6y^5 - 9y^4 + 12y^3 - 33y^2$

Chapter 5 Factoring

Key Terms
1. binomial   2. factoring   3. factored form   4. greatest common factor

5. factor

Objective 1
Now Try
1.1, 2, 3, 4, 6, and 12

Practice Exercise
1. 1, 2, 4 and 8   3. 1, 2, 3, 6, 7, 14, 24, and 48

Objective 2
Now Try
2a. 54   2b. $12y + 72 + 12x$ 2c. $17x - 102$

2d. $-8x + 20$ 3a. $-3x - 4$ 3b. $8k + 9$ 3c. $4x + 5y - z$

Practice Exercises
5. $-10y + 18z$

Objective 3
Now Try
5. $14x^2 - x - 6$ 6. $4x^7 - 6x^5 + 37x^4 - 18x^2 + 9x$

Practice Exercises
7. $6m^5 + 4m^4 - 5m^3 + 2m^2 - 4m$ 9. $6x^4 + 11x^3 - 9x^2 - 4x$

Objective 4
Now Try
7a. $c^2 - 8c + 16$ 7b. $4a^2 + 36ak + 81k^2$

Practice Exercises
11. $4m^2 - 12mp + 9p^2$

Objective 5
Now Try
8. $6x^4$

Practice Exercises
13. xy 15. $9xy^2$

Chapter 6 Rational Expressions and Functions

Key Terms

1. equivalent fractions   2. improper fraction   3. numerator

4. proper fraction   5. complex fraction   6. denominator

7. composite number   8. prime factorization   9. prime number

10. lowest terms   11. LCD   12. rational expression

Objective 1

Now Try

1. $\dfrac{3}{4}$ 2. $\dfrac{23}{24}$ 3. $\dfrac{2}{9}$

Practice Exercises

1. $1\dfrac{31}{225}$ 3. $4\dfrac{23}{24}$

Objective 2

Now Try

4. $\dfrac{1}{10}$ 5. $\dfrac{16}{21}$

Practice Exercises

5. $1\dfrac{2}{5}$

Objective 3

Now Try

6. $x = 6$ 7. $x = 2$

Practice Exercises

7. $w = 11$ 9. $m = -14$

Objective 4

Now Try

8. $x = \dfrac{51}{8}$

Practice Exercises

11. $x = 16$

Objective 5
Now Try

9. $\dfrac{3}{k^3}$ 10a. $\dfrac{7}{9}$ 10b. $\dfrac{m+6}{2m+3}$ 11. -1

Practice Exercises

13. $-\dfrac{5b}{8c}$ 15. $\dfrac{9(x+3)}{2m+3}$

Objective 6
Now Try

12. 2 13. $\dfrac{n-9}{3(2n+5)}$

Practice Exercises

17. $\dfrac{x^2-4n+4}{x^2+2x}$

Chapter 7 Roots, Radicals, and Root Functions

Key Terms

1. radicand
2. perfect square
3. index
4. square root
5. radical expression
6. unlike radicals
7. like radicals
8. Principal square root
9. irrational number
10. radical
11. cube root

Objective 1

Now Try

1. 9 and -9

2a. 13

2b. $-\dfrac{3}{7}$

3a. 19

3b. $n^2 + 5$

Practice Exercises

1. 25

3. $x^2 - 5$

Objective 2

Now Try

4a. 7

4b. -5

4c. $\dfrac{4}{3}$

5a. 6

5b. -5

Practice Exercises

5. 4

Objective 3

Now Try

6a. $2\sqrt{3}$

6b. $2\sqrt{7}$

6c. $4\sqrt{5}$

7a. $15\sqrt{2}$

7b. $6\sqrt{2}$

Practice Exercises

7. $9\sqrt{5}$

9. $12\sqrt{3}$

Objective 4

Now Try

8a. $8\sqrt{6}$

8b. $33\sqrt{6}$

8c. $23\sqrt[3]{2}$

Practice Exercises

11. $-\sqrt[3]{2}$

Objective 5

Now Try

9a. $2\sqrt{7}$

9b. $\dfrac{\sqrt{3}}{3}$

Practice Exercises

13. $\dfrac{3\sqrt{10}}{2}$

15. $\dfrac{3}{5}$

Chapter 8 Quadratic Equations, Inequalities, and Functions

Key Terms

1. equivalent equations
2. linear equation
3. coefficient
4. quadratic equation
5. greatest common factor
6. prime polynomial
7. trinomial
8. Standard form
9. Double solution
10. solution set
11. factoring

Objective 1

Now Try

1. 49

Practice Exercises

1. 27   3. 0.16

Objective 2

Now Try

2. $x = -6$ 3. $x = 8$ 4. $a = \frac{15}{2}$

Practice Exercises

5. $a = 4$

Objective 3

Now Try

5. $(x + 8)(x + 3)$ 6. prime   7. $(p - 7q)(p + 2q)$

Practice Exercises

7. prime   9. $(x - 11)(x + 3)$

Objective 4

Now Try

8. $(3x + 1)(5x + 7)$ 9. $(4x + 7)(2x - 3)$ 10. $(6x - 5y)(4x + 3y)$

11. $-6a(3a - 5)(a - 2)$

Practice Exercises

11. $(3a + 2b)(a + 2b)$

Objective 5

Now Try

12a. $x = -12, \frac{7}{4}$ 12b. $x = 0, \frac{11}{6}$ 13. $p = 3, \frac{4}{5}$

Practice Exercises

13. $x = 4, \frac{-5}{2}$ 15. $c = -4, \frac{3}{5}$

Chapter 9 Exponential and Logarithmic Functions
Key Terms

1. components
2. quadratic equation
3. extraneous solution

4. radical equation
5. zero-factor property
6. relation

7. range
8. function
9. domain

Objective 1
Now Try
1a. $x = 9 \text{ or } -9$
1b. $x = \pm\sqrt{23}$
1c. $x = \pm 3\sqrt{3}$

1d. $x = \pm\sqrt{2}$
2. no real solution

Practice Exercises
1. $r = 300 \text{ or } -300$
3. no real solution

Objective 2
Now Try
2. $p = 11$
4. $x = 3$

Practice Exercises
5. $k = \dfrac{2}{5}$

Objective 3
Now Try
5a. no
5b. yes
5c. yes

Practice Exercises
7. no
9. yes

Objective 4
Now Try
6. $D: (-\infty, \infty), R: (-\infty, \infty)$

Practice Exercises
11. $D: (-\infty, \infty), R: (-\infty, \infty)$

Chapter 10 Nonlinear Functions, Conic Sections, and Nonlinear Systems
Key Terms

1. perfect square trinomial 2. square root property 3. completing the square

4. point-slope form 5. standard form 6. slope-intercept form

7. line of symmetry 8. vertex 9. axis of symmetry

10. parabola 11. solution set of a system of linear inequalities

12. system of linear inequalities

Objective 1
Now Try

1. $\left(\frac{1}{2}, 2\right)$ 2. 13

Practice Exercises

1. Midpoint: $(1, 3)$, Distance: $4\sqrt{2}$ 3. Midpoint: $\left(\frac{3}{2}, -\frac{3}{2}\right)$, Distance: $\sqrt{26}$

Objective 2
Now Try

3a. $+ 25, (x + 5)^2$ 3b. $+ 121, (x - 11)^2$ 4. $x = -5 \pm 4\sqrt{2}$

Practice Exercises

5. $x = 2 \pm 2\sqrt{6}$

Objective 3
Now Try

5.

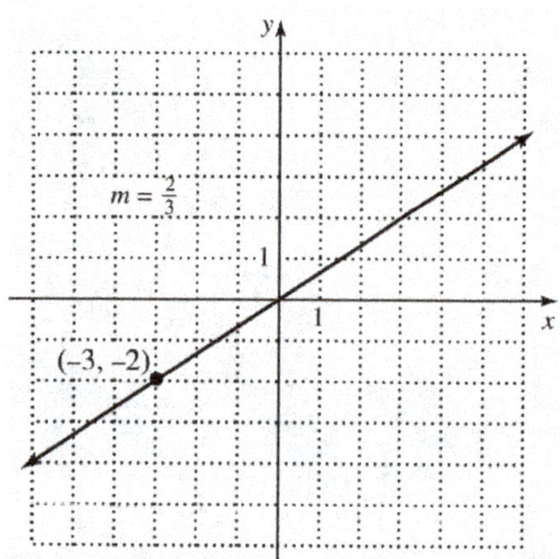

Practice Exercises

7.

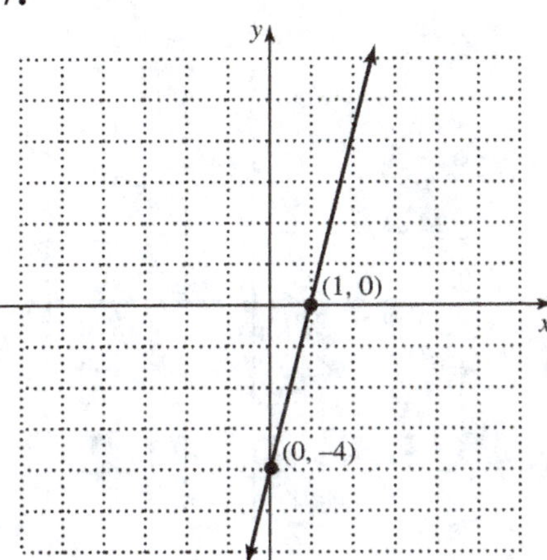

9.

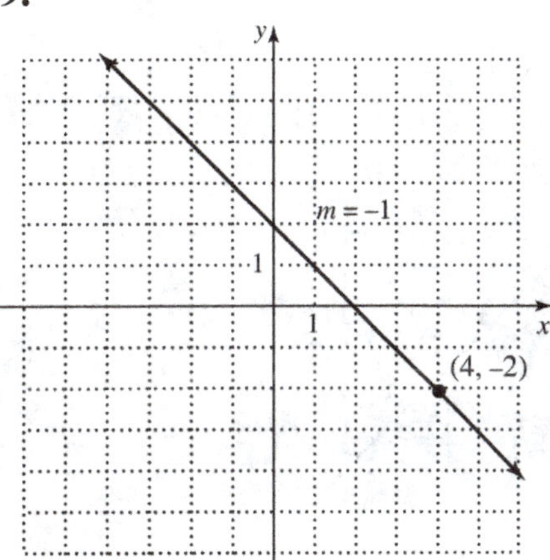

Objective 4
Now Try
6. vertex: $(0, -1)$

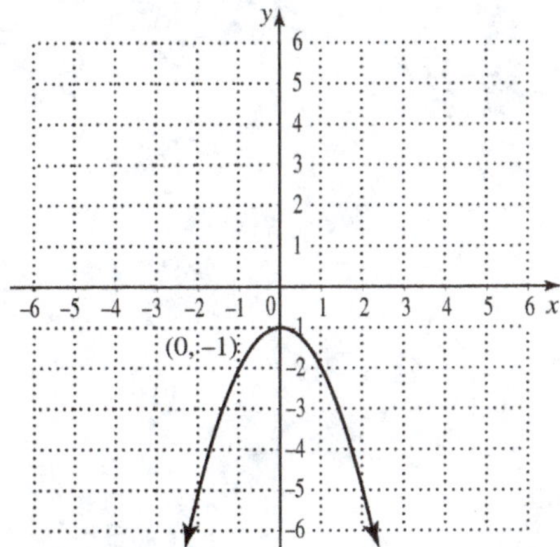

Practice Exercises
11. vertex: $(0, -1)$

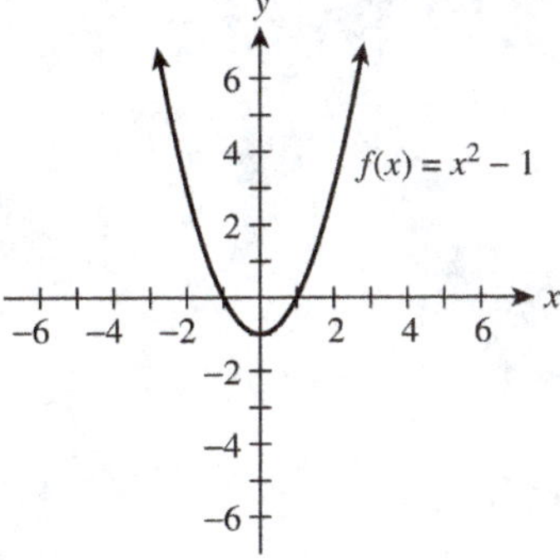

Objective 5
Now Try

7.

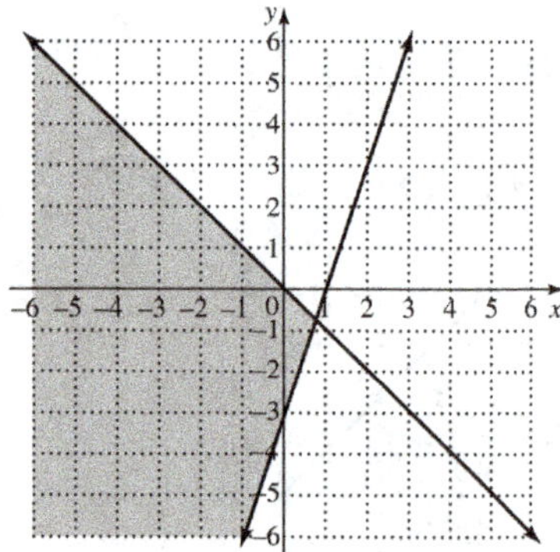

8.

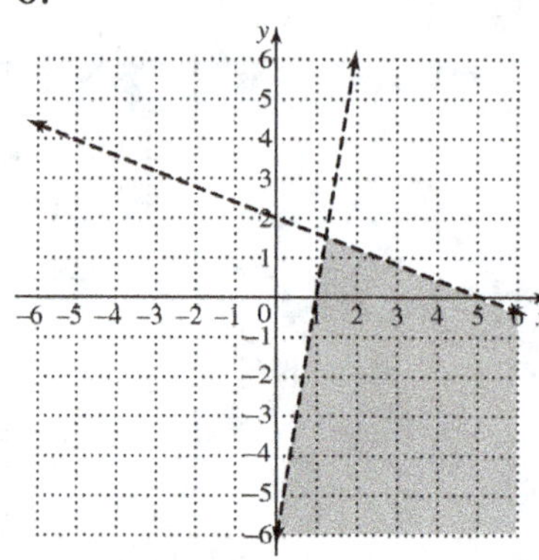

9.

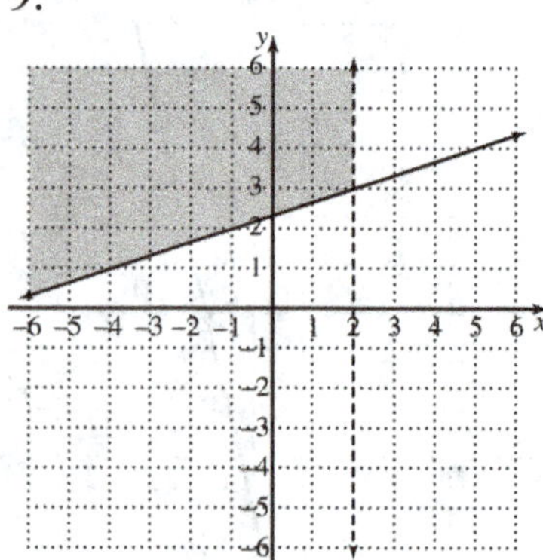

Practice Exercises

13.

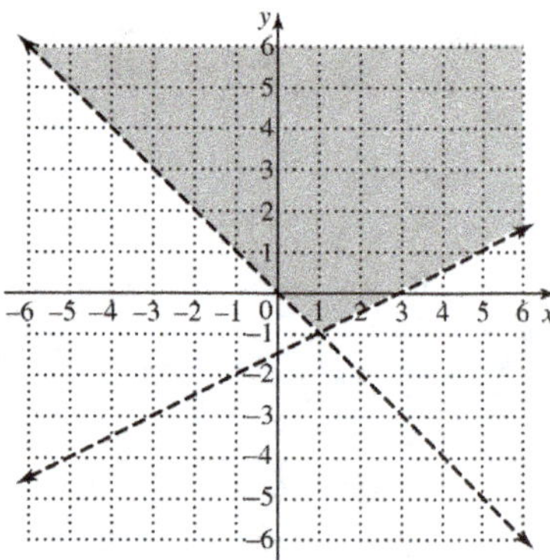

Chapter 11 Further Topics in Algebra

Key Terms

1. equation   2. elements   3. variable   4. algebraic expression

5. set   6. solution   7. constant

Objective 1
Now Try

1a. -49 1b. $256a^4$ 1c. $-96y^5$

Practice Exercises

1. -125 3. $16z^2$

Objective 2
Now Try

4. $y = -\frac{2}{3}x + 15$

Practice Exercises

5. $y = -2x - 11$

Objective 3
Now Try

6a. $x^3 + 18x^2 + 108x + 216$ 6b. $81x^4 - 540x^3 + 1050x^2 + 1500x + 625$

Practice Exercises

7. $a^3 - 9a^2 + 27a - 27$ 9. $256s^4 + 768s^3t + 864s^2t^2 + 432st^3 + 81t^4$

Chapter R REVIEW OF THE REAL NUMBER SYSTEM

R.1   Fractions, Decimals, and Percents

Learning Objectives	
1	Write fractions in lowest terms.
2	Convert between improper fractions and mixed numbers.
3	Perform operations with fractions.
4	Write decimals as fractions.
5	Perform operations with decimals.
6	Write fractions as decimals.
7	Write percents as decimals and decimals as percents.
8	Write percents as fractions and fractions as percents.

Key Terms

Use the vocabulary terms listed below to complete each statement in exercises 1−7.

numerator	**denominator**	**proper fraction**
improper fraction	**lowest terms**	**decimals**
percent		

1.   We use ________________________ to show parts of a whole.

2.   ____________________ means per one hundred.

3.   A fraction whose numerator is larger than its denominator is called an
     ____________________.

4.   In the fraction $\frac{2}{9}$, the 2 is the ____________________________.

5.   A fraction whose denominator is larger than its numerator is called a
     ____________________.

6.   The ____________________________ of a fraction shows the number of equal
     parts in a whole.

7.   A fraction is written in ____________________________________ when its numerator
     and denominator have no common factor other than 1.

                                    1

Objective 1   Write fractions in lowest terms.

Review this example for Objective 1:

1. Write the fraction in lowest terms.

  a. $\dfrac{15}{25}$

  $$\dfrac{15}{25} = \dfrac{3 \cdot 5}{5 \cdot 5} = \dfrac{3}{5} \cdot \dfrac{5}{5} = \dfrac{3}{5} \cdot 1 = \dfrac{3}{5}$$

  b. $\dfrac{28}{168}$

  $$\dfrac{28}{168} = \dfrac{28}{28 \cdot 6} = 1 \cdot \dfrac{1}{6} = \dfrac{1}{6}$$

  c. $\dfrac{120}{150}$

  $$\dfrac{120}{150} = \dfrac{4 \cdot 30}{5 \cdot 30} = \dfrac{4}{5} \cdot 1 = \dfrac{4}{5}$$

Now Try:

1. Write the fraction in lowest terms.

  a. $\dfrac{9}{15}$

  ____

  b. $\dfrac{9}{36}$

  ____

  c. $\dfrac{48}{150}$

  ____

Objective 1 Practice Exercises

For extra help, see Example 1 on page 2 of your text.

Write each fraction in lowest terms.

1. $\dfrac{42}{150}$

  1. _______________________

2. $\dfrac{180}{216}$

  2. _______________________

3. $\dfrac{132}{292}$

  3. _______________________

Objective 2   Convert between improper fractions and mixed numbers.

Review these examples for Objective 2:

2. Write $\dfrac{53}{6}$ as a mixed number.

  We divide the numerator of the improper fraction by the denominator.

  $$6\overline{)53} \quad \begin{array}{r} 8 \\ \underline{48} \\ 5 \end{array} \qquad \dfrac{53}{6} = 8\dfrac{5}{6}$$

Now Try:

2. Write $\dfrac{74}{5}$ as a mixed number.

  ____

3. Write $5\frac{3}{8}$ as an improper fraction.

We multiply the denominator of the fraction by the whole number and add the numerator to get the numerator of the improper fraction.
$$8 \cdot 5 + 3 = 40 + 3 = 43$$
The denominator of the improper fraction is the same as the denominator in the mixed number, which is 8 here. Thus, $5\frac{3}{8} = \frac{43}{8}$

3. Write $12\frac{2}{7}$ as an improper fraction.

____

Objective 2 Practice Exercises

For extra help, see Examples 2–3 on page 2 of your text.

Write the improper fraction as a mixed number.

4. $\dfrac{321}{15}$

4. _______________

Write each mixed number as an improper fraction.

5. $13\frac{5}{9}$

5. _______________

6. $22\frac{2}{11}$

6. _______________

Objective 3   Perform operations with fractions.

Review these examples for Objective 3:

4. Find the product, and write it in lowest terms.

$$\frac{5}{12} \cdot \frac{3}{10}$$

$$\frac{5}{12} \cdot \frac{3}{10} = \frac{5 \cdot 3}{12 \cdot 10} \qquad \text{Multiply numerators.}$$
$$\text{Multiply denominators.}$$

$$= \frac{5 \cdot 3}{4 \cdot 3 \cdot 2 \cdot 5} \qquad \text{Factor the denominator.}$$

$$= \frac{1}{4 \cdot 2} \qquad \frac{3}{3} = 1 \text{ and } \frac{5}{5} = 1$$

$$= \frac{1}{8} \qquad \text{Write in lowest terms.}$$

Now Try:

4. Find the product, and write it in lowest terms.

$$\frac{7}{15} \cdot \frac{3}{14}$$

____

5. Find the quotient, and write it in lowest terms.

a. $\dfrac{2}{5} \div \dfrac{8}{7}$

$\dfrac{2}{5} \div \dfrac{8}{7} = \dfrac{2}{5} \cdot \dfrac{7}{8}$ Multiply by the reciprocal.

$\qquad = \dfrac{2 \cdot 7}{5 \cdot 4 \cdot 2}$ Multiply and factor.

$\qquad = \dfrac{7}{20}$

c. $2\dfrac{5}{8} \div \dfrac{3}{4}$

$2\dfrac{5}{8} \div \dfrac{3}{4} = \dfrac{21}{8} \div \dfrac{3}{4}$ Write the mixed number as an improper fraction.

$\qquad = \dfrac{21}{8} \cdot \dfrac{4}{3}$ Multiply by the reciprocal of the second fraction.

$\qquad = \dfrac{21 \cdot 4}{8 \cdot 3}$ Multiply numerators. Multipy denominators.

$\qquad = \dfrac{3 \cdot 7 \cdot 2 \cdot 2}{2 \cdot 2 \cdot 2 \cdot 3}$ Factor the numerator and the denominator.

$\qquad = \dfrac{7}{2}$ Lowest terms

6. Add or subtract as indicated. Write the answers in lowest terms as needed.

a. $\dfrac{5}{24} + \dfrac{7}{24}$

Add numerators. Keep the same denominator.

$\dfrac{5}{24} + \dfrac{7}{24} = \dfrac{5+7}{24} = \dfrac{12}{24}$, or $\dfrac{1}{2}$

Write in lowest terms.

b. $\dfrac{5}{21} + \dfrac{3}{14}$

$21 = 3 \cdot 7$ and $14 = 2 \cdot 7$, so LCD $= 3 \cdot 7 \cdot 2 = 42$
Write equivalent fractions with the common denominator.

$\dfrac{5}{21} + \dfrac{3}{14} = \dfrac{5}{21} \cdot \dfrac{2}{2} + \dfrac{3}{14} \cdot \dfrac{3}{3}$

$\qquad = \dfrac{10}{42} + \dfrac{9}{42}$

$\qquad = \dfrac{19}{42}$

5. Find the quotient, and write it in lowest terms.

a. $\dfrac{6}{7} \div \dfrac{9}{8}$

____

c. $\dfrac{5}{4} \div 2\dfrac{7}{24}$

____

6. Add or subtract as indicated. Write the answers in lowest terms as needed.

a. $\dfrac{5}{16} + \dfrac{7}{16}$

____

b. $\dfrac{7}{12} + \dfrac{3}{8}$

____

c. $\dfrac{26}{9} - \dfrac{5}{9}$

Subtract numerators. Keep the same denominator.

$$\dfrac{26}{9} - \dfrac{5}{9} = \dfrac{26-5}{9}$$

$$= \dfrac{21}{9}$$

$$= \dfrac{7}{3}, \text{ or } 2\dfrac{1}{3}$$

d. $4\dfrac{5}{12} - 1\dfrac{11}{16}$

$4\dfrac{5}{12} - 1\dfrac{11}{16} = \dfrac{53}{12} - \dfrac{27}{16}$ Write each mixed number as an improper fraction.

$= \dfrac{53 \cdot 4}{48} - \dfrac{27 \cdot 3}{48}$ Find a common denominator.

$= \dfrac{212}{48} - \dfrac{81}{48}$

$= \dfrac{131}{48}$ or $2\dfrac{35}{48}$ Subtract. Write as a mixed number.

c. $\dfrac{11}{18} - \dfrac{7}{18}$

____

d. $2\dfrac{5}{8} - 1\dfrac{15}{32}$

____

Objective 3 Practice Exercises

For extra help, see Examples 4–6 on pages 3–5 of your text.

Find each product or quotient, and write it in lowest terms.

7. $4\dfrac{3}{8} \cdot 2\dfrac{4}{7}$

7. ________________

Find each sum or difference, and write it in lowest terms.

8. $\dfrac{23}{45} + \dfrac{47}{75}$

8. ________________

9. $12\dfrac{5}{6} - 7\dfrac{7}{8}$

9. ________________

Name: Date:
Instructor: Section:

Objective 4   Write decimals as fractions.

| **Review these examples for Objective 4:** | **Now Try:** |

Review these examples for Objective 4:

7. Write each decimal as a fraction. Do not write in lowest terms.

   a. 0.87

We read 0.87 as "eighty-seven hundredths," so the fraction form is $\frac{87}{100}$. Using the shortcut method, since there are two places to the right of the decimal point, there will be two zeros in the denominator.

$$0.87 = \frac{87}{100}$$

2 places   2 zeros

   b. 0.043

We read 0.043 as "forty-three thousandths."

$$0.043 = \frac{43}{1000}$$

3 places   3 zeros

   c. 5.3084

Here we have 4 places.
$$5.3084 = 5 + 0.3084$$
$$= \frac{50,000}{10,000} + \frac{3084}{10,000} \quad \text{The LCD is 10,000.}$$
$$= \frac{53,084}{10,000}$$

4 zeros

Now Try:

7. Write each decimal as a fraction. Do not write in lowest terms.

   a. 0.72

____

   b. 0.053

____

   c. 3.7058

____

Objective 4 Practice Exercises

For extra help, see Example 7 on page 6 of your text.

Write each decimal as a fraction. Do not write in lowest terms.

10.   0.007   10. _______________

11.   18.03   11. _______________

12.   30.0005   12. _______________

Objective 5   Perform operations with decimals.

Review these examples for Objective 5:

8. Add or subtract as indicated.

  a. $7.34 + 15.6 + 2.419$

  Place the digits of the numbers in columns, so that tenths are in one column, hundredths in another column, and so on.

$$\begin{array}{r} 7.34 \\ 15.6 \\ +\ 2.419 \\ \hline 25.359 \end{array} \quad \text{Align decimal points.}$$

  To avoid errors, attach zeros to make all the numbers the same length.

$$\begin{array}{r} 7.34 \\ 15.6 \\ +\ 2.419 \\ \hline \end{array} \quad \text{becomes} \quad \begin{array}{r} 7.340 \\ 15.600 \\ +\ 2.419 \\ \hline 25.359 \end{array}$$

  b. $56.8 - 42.307$

$$\begin{array}{r} 56.8 \\ -\ 42.307 \\ \hline \end{array} \quad \text{becomes} \quad \begin{array}{r} 56.800 \\ -\ 42.307 \\ \hline 14.493 \end{array}$$

9. Multiply or divide as indicated.

  a. 37.4×5.26

  There is 1 decimal place in the first number and 2 decimal places in the second number. Therefore, there are $1 + 2 = 3$ decimal places in the answer.

$$\begin{array}{r} 37.4 \\ \times\ 5.26 \\ \hline 2244 \\ 748\ \\ 1870\ \ \\ \hline 196.724 \end{array}$$

  b. 0.08×0.6

  Here $8 \times 6 = 48$. There are 2 decimal places in the first number and 1 decimal place in the second number. Therefore, there are $2 + 1 = 3$ decimal places in the answer.

$$0.08 \times 0.6 = 0.048$$

Now Try:

8. Add or subtract as indicated.

  a. $5.23 + 28.9 + 3.741$

  ____

  b. $64.5 - 37.218$

  ____

9. Multiply or divide as indicated.

  a. 26.8×9.37

  ____

  b. 0.04×0.7

  ____

c. $9.581 \div 5.78$ (Round the answer to two decimal places.)

Move the decimal point two places to the right in 5.78, to get 578. Do the same thing with 9.581 to get 958.1.

$$578\overline{)958.1}$$

Move the decimal point straight up and divide as with whole numbers.

$$
\begin{array}{r}
1.657 \\
578\overline{)958.100} \\
\underline{578} \\
3801 \\
\underline{3468} \\
3330 \\
\underline{2890} \\
4400 \\
\underline{4046} \\
354
\end{array}
$$

We carried out the division to three decimal places so that we could round to two decimal places, obtaining the quotient 1.66.

10. Multiply or divide as indicated.

 a. 24.927×10

  Move the decimal point one place to the *right*.
  $24.927 \times 10 = 249.27$

 b. 24.927×1000

  Move the decimal point three places to the *right*.
  $24.927 \times 1000 = 24{,}927$

 c. $24.927 \div 10$

  Move the decimal point one place to the *left*.
  $24.927 \div 10 = 2.4927$

 d. $24.927 \div 1000$

  Move the decimal point three places to the *left*.
  $24.927 \div 1000 = 0.024927$

---

c. $7.648 \div 5.36$ (Round the answer to two decimal places.)

____

10. Multiply or divide as indicated.

 a. 51.372×10

  ____

 b. 51.372×1000

  ____

 c. $51.372 \div 10$

  ____

 d. $51.372 \div 1000$

  ____

Objective 5 Practice Exercises

For extra help, see Examples 8–10 on pages 6–8 of your text.

Add or subtract as indicated.

13. $45.83 + 20.923 - 5.7$ **13.** _______________

Multiply or divide as indicated.

14. 14.64×0.16 **14.** _______________

15. $429.2 \div 1000$ **15.** _______________

Objective 6   Write fractions as decimals.

Review these examples for Objective 6:

11. Write each fraction as a decimal.

     a. $\dfrac{27}{8}$

Divide 27 by 8. Add a decimal point and as many 0s as necessary.

$$8\overline{)27.000}$$

$$\begin{array}{r} 3.375 \\ \hline 8)\overline{27.000} \\ \underline{24} \\ 30 \\ \underline{24} \\ 60 \\ \underline{56} \\ 40 \\ \underline{40} \\ 0 \end{array}$$

$$\frac{27}{8} = 3.375$$

Now Try:

11. Write each fraction as a decimal.

     a. $\dfrac{7}{20}$

                     ____

           9

b. $\dfrac{26}{9}$ **b.** $\dfrac{32}{9}$

$$\begin{array}{r} 2.888...\\ 9\overline{)26.000...}\\ \underline{18}\\ 80\\ \underline{72}\\ 80\\ \underline{72}\\ 80\\ \underline{72}\\ 8 \end{array}$$

$$\frac{26}{9} = 2.888...$$

The remainder is never 0. Because 8 is always left after the subtraction, this quotient is a repeating decimal. A convenient notation for a repeating decimal is a bar over the digit (or digits) that repeats.

$$\frac{26}{9} = 2.888...\ \text{ or } \ 2.\overline{8}$$

Objective 6 Practice Exercises

For extra help, see Example 11 on page 8 of your text.

Write each fraction as a decimal. For repeating decimals, write the answer two ways: using the bar notation and rounding to the nearest thousandth.

16. $\dfrac{3}{7}$ **16.** _______________

17. $\dfrac{4}{9}$ **17.** _______________

18. $\dfrac{151}{200}$ **18.** _______________

                Copyright © 2020 Pearson Education, Inc.

Objective 7   Write percents as decimals and decimals as percents.

Review these examples for Objective 7: | **Now Try:**

12. Convert each percent to a decimal and each decimal to a percent.

12. Convert each percent to a decimal and each decimal to a percent.

a. 54%

54% = 0.54

a. 91%

____

b. 390%

390% = 3.90, or 3.9

b. 430%

____

c. 3%

3% = 0.03

c. 6%

____

d. 0.29

0.29 = 29%

d. 0.43

____

e. 4.6

4.6 = 460%

e. 5.2

____

f. 0.004

0.004 = 0.4%

f. 0.008

____

Objective 7 Practice Exercises

For extra help, see Example 12 on page 9 of your text.

Convert each percent to a decimal and each decimal to a percent.

19.   362%

19. _________________

20.   0.4%

20. _________________

21.   0.084

21. _________________

Objective 8   Write percents as fractions and fractions as percents.

Review these examples for Objective 8:	Now Try:

Review these examples for Objective 8:

13. Write each percent as a fraction. Give answers in lowest terms.

   a. 12%

Recall writing 12% as a decimal.
$$12\% = 12 \div 100 = 0.12$$
Because 0.12 means 12 hundredths,
$$0.12 = \frac{12}{100} = \frac{12 \div 4}{100 \div 4} = \frac{3}{25}$$

   b. 350%

$$350\% = \frac{350}{100} = \frac{350 \div 50}{100 \div 50} = \frac{7}{2} = 3\frac{1}{2}$$

14. Write each fraction as a percent. Round to the nearest tenth as necessary.

   a. $\dfrac{3}{5}$

$$\frac{3}{5} = \left(\frac{3}{5}\right)(100\%) = \left(\frac{3}{5}\right)\left(\frac{100}{1}\%\right) = \left(\frac{3}{5}\right)\left(\frac{20\cdot 5}{1}\%\right)$$
$$= \frac{60}{1}\% = 60\%$$

   b. $\dfrac{1}{15}$

$$\frac{1}{15} = \left(\frac{1}{15}\right)(100\%) = \left(\frac{1}{15}\right)\left(\frac{100}{1}\%\right)$$
$$= \left(\frac{1}{3\cdot 5}\right)\left(\frac{5\cdot 20}{1}\%\right)$$
$$= \frac{20}{3}\% = 6\frac{2}{3}\%$$

$6\frac{2}{3}\%$ rounds to 6.7%.

Now Try:

13. Write each percent as a fraction. Give answers in lowest terms.

   a. 30%

____

   b. 125%

____

14. Write each fraction as a percent. Round to the nearest tenth as necessary.

   a. $\dfrac{3}{20}$

____

   b. $\dfrac{1}{18}$

____

Objective 8 Practice Exercises

For extra help, see Examples 13–14 on pages 9–10 of your text.

Convert each percent to a fraction and each fraction to a percent.

22.   140%   **22.** ________________

23.   55.6%   **23.** ________________

24. $\dfrac{11}{40}$ **24.** ________________

Chapter R REVIEW OF THE REAL NUMBER SYSTEM

R.2   Basic Concepts from Algebra

Learning Objectives

1   Write sets using set notation.
2   Use number lines.
3   Classify numbers.
4   Find additive inverses.
5   Use absolute value.
6   Use inequality symbols.

Key Terms

Use the vocabulary terms listed below to complete each statement in exercises 1−15.

> **set   elements   finite set   infinite set   empty set   variable**
>
> **set-builder notation   number line   coordinate   graph**
>
> **additive inverse   signed numbers   absolute value**
>
> **equation   inequality**

1. A ________________________ is a line with a scale to indicate the set of real numbers.

2. ________________________ is used to describe a set of numbers without listing them.

3. The number that corresponds to a point on the number line is the ________________________ of that point.

4. A collection of objects is a ________________________.

5. If the number of elements in a set can be listed or counted and the counting process comes to an end, then the set is a(n) ________________________.

6. The set with no elements is called the ________________________.

7. A mathematical statement that two quantities are not equal is a(n) ________________________.

8. Positive and negative numbers are ________________________.

9. If the number of elements in a set cannot be listed or counted, then the set is a(n) ________________________.

10. A letter used to represent a number or a set of numbers is a(n) ________________________.

11. The ________________________ of a number is its distance from 0 on the number line.

12. The ________________________ of a number a is $-a$.

13. A mathematical statement that two quantities are equal is a(n) ________________________.

14. The point of the number line that corresponds to a number is its ________________________.

15. The ________________________ of a set are the numbers or objects that make up the set.

Objective 1   Write sets using set notation.

Review these examples for Objective 1:

1. List the elements in the set.

    $\{x \mid x \text{ is a natural number less than 6}\}$

    The natural numbers less than 6 are 1, 2, 3, 4, 5. The set is $\{1, 2, 3, 4, 5\}$.

2. Use set-builder notation to describe each set.

    $\{2, 4, 6, 8, 10\}$

    One way to describe this set is $\{x \mid x \text{ is one of the first five even natural numbers}\}$.

Now Try:

1. List the elements in the set.

    $\{x \mid x \text{ is a whole number less than 6}\}$

    ____

2. Use set-builder notation to describe each set.
    $\{1, 2, 3, 4, 5\}$

    ____

Objective 1 Practice Exercises

For extra help, see Examples 1–2 on page 15 of your text.

Write the set by listing its elements.

1. $\{y \mid y \text{ is a natural number divisible by 5}\}$ 1. ________________

Write each set using set-builder notation. (More than one description is possible.)

2. $\{1, 3, 5, 7, 9\}$ 2. ________________

3. $\{7, 8, 9, 10\}$ 3. ________________

Objective 2 Use number lines.

Objective 2 Practice Exercises

For extra help, see pages 16–17 of your text.

Graph the elements of each set on a number line.

4. $\left\{-\dfrac{1}{3},\ 0,\ \dfrac{1}{3},\ \dfrac{2}{3}\right\}$

4.

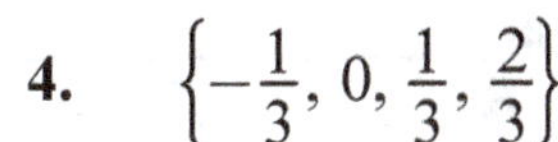

5. $\left\{-4,\ -2.5,\ 0,\ \sqrt{9}\right\}$

5.

6. $\left\{-4,\ -\dfrac{3}{2},\ \dfrac{1}{2},\ 5\right\}$

6.

Objective 3 Classify numbers.

Review these examples for Objective 3:

3. List the numbers in the following set that are elements of each set.

$$\left\{-6,-\sqrt{5},-\dfrac{7}{40},\ 0,\ 0.3,\ \dfrac{5}{9},\ 1.\overline{75},\ \sqrt{6},\ 4\right\}$$

 a. Integers

 $-6,\ 0,\ 4$

 b. Rational numbers

 $-6,-\dfrac{7}{40},\ 0,\ 0.3,\ \dfrac{5}{9},\ 1.\overline{75},\ 4$

 c. Irrational numbers

 $-\sqrt{5},\ \sqrt{6}$

 d. Real numbers

 All are real numbers.

Now Try:

3. List the numbers in the following set that are elements of each set.

$$\left\{-7,-\sqrt{3},-\dfrac{8}{9},0,0.4,\dfrac{3}{5},2.\overline{15},\sqrt{7},3\right\}$$

 a. Integers

 ____

 b. Rational numbers

 ____

 c. Irrational numbers

 ____

 d. Real numbers

 ____

Objective 3 Practice Exercises

For extra help, see Examples 3–4 on page 18 of your text.

Which elements of each set are (a) natural numbers, (b) whole numbers, (c) integers, (d) rational numbers, (e) irrational numbers, (f) real numbers?

7. $\left\{-4, -\sqrt{2}, -\frac{1}{3}, 0, \frac{4}{5}, \sqrt{7}, 5\right\}$

7. a._______________

   b._______________

   c._______________

   d._______________

   e._______________

   f. _______________

Decide if each statement is **true** *or* **false**.

8. All rational numbers are real numbers.

8. _______________

9. Some irrational numbers are rational numbers.

9. _______________

Objective 4   Find additive inverses.

Objective 4 Practice Exercises

For extra help, see pages 18–19 of your text.

Give the additive inverse of each number.

10. 0

10. _______________

11. $-\dfrac{5}{2}$

11. _______________

12. $\sqrt{2}$

12. _______________

                    Copyright © 2020 Pearson Education, Inc.

Objective 5   Use absolute value.

| **Review these examples for Objective 5:** | **Now Try:** |

Review these examples for Objective 5:

5. Find each value.

    a. $|-4|$

       $|-4| = -(-4) = 4$

    b. $-|7|$

       $-|7| = -(7) = -7$

    c. $-|-7|$

       $-|-7| = -(7) = -7$

    d. $|-4| + |3|$

       $|-4| + |3| = 4 + 3 = 7$

Now Try:

5. Find each value.

    a. $|-11|$

    b. $-|10|$

    c. $-|-10|$

    d. $|-6| + |3|$

6. The table below shows the changes in population for five cities.

City	1980–1990	1990–2000	2000–2009
New York	250,925	685,714	383,603
Los Angeles	518,548	209,422	137,048
Chicago	−221,346	112,290	−44,748
Houston	35,415	323,078	304,295
Philadelphia	−102,633	−68,027	29,747

Source: factfinder.census.gov

Which city had the greatest change in population in which year?

New York from 1990−2000.

6. Use the table shown at the left to determine which city has the least change in population in which year.

Objective 5 Practice Exercises

For extra help, see Examples 5–6 on page 20 of your text.

Find the value of each expression.

13. $\big|6\big| - \big|-1\big|$

13. _________________

14. $\big|-7\big| + \big|-8\big|$

14. _________________

Objective 6   Use inequality symbols.

Review these examples for Objective 6:	**Now Try:**

Review these examples for Objective 6:

7. Use a number line to compare -5 and -1, and to compare 0 and -1.

-5 is located to the left of -1. For this reason, $-5 < -1$. Also, $-1 > -5$. From the same number line, $-1 < 0$, or $0 > -1$.

8. Decide whether the statement is true or false.

$$7 \cdot 3 \le 6(4)$$

True. $21 < 24$

Now Try:

7. Insert $<$ or $>$ in the blank to make a true statement.
$$-5 \underline{\quad} -9$$

____

8. Decide whether the statement is true or false.
$$3 \cdot 8 > 4(6)$$

____

Objective 6 Practice Exercises

For extra help, see Examples 7–8 on pages 21–22 of your text.

Identify each inequality as **true** *or* **false**.

15. $-3 < -5$

15. _______________

16. $\dfrac{3}{4} > \dfrac{2}{3}$

16. _______________

17. $-|7-4| \le -4$

17. _______________

                Copyright © 2020 Pearson Education, Inc.

Chapter R REVIEW OF THE REAL NUMBER SYSTEM

R.3   Operations on Real Numbers

Learning Objectives
1   Add real numbers.
2   Subtract real numbers.
3   Find the distance between two points on a number line.
4   Multiply real numbers.
5   Find reciprocals and divide real numbers.

Key Terms

Use the vocabulary terms listed below to complete each statement in exercises 1−5.

     sum   **difference**   **product**   **reciprocals**   **quotient**

1.   The answer to a multiplication problem is called the ________________________.

2.   Pairs of numbers whose product is 1 are called ____________________.

3.   The answer to a subtraction problem is called the ____________________.

4.   The answer to a division problem is called the __________________.

5.   The answer to an addition problem is called the ______________________.

Objective 1   Add real numbers.

Review these examples for Objective 1:

1. Find the sum.

$$-13+(-9)$$

Because −13 and −9 have the same sign, add their absolute values.

$$-13+(-9) = -(|-13|+|-9|)$$
$$= -(22)$$
$$= -22$$

2. Find the sum.

$$5+(-3)$$

Subtract the absolute values, 5 and 3. Because 5 has the greater absolute value, the sum must be positive.

$$5+(-3) = 5-3 = 2$$

Now Try:

1. Find the sum.

$$-17+(-4)$$

____

2. Find the sum.

$$12+(-9)$$

____

Objective 1 Practice Exercises

For extra help, see Examples 1–2 on pages 26–27 of your text.

Find each sum.

1. $-5.1 + (-7.3)$ 1. ______________

2. $-16.32 + 2.27$ 2. ______________

3. $\dfrac{2}{11} + \left(-\dfrac{2}{3}\right)$ 3. ______________

Objective 2   Subtract real numbers.

Review these examples for Objective 2:

3.   Find the difference.

$$-13 - 5$$

$$-13 - 5 = -13 + (-5) = -18$$

4.   Perform the indicated operation.

$$12 - (-4) - 8 - 15$$

$$12 - (-4) - 8 - 15 = (12 + 4) - 8 - 15$$
$$= 16 - 8 - 15$$
$$= 8 - 15$$
$$= 8 + (-15)$$
$$= -7$$

Now Try:

3.   Find the difference.

$$-15 - 7$$

____

4.   Perform the indicated operation.

$$19 - (-9) - 8 - 7$$

____

Objective 2 Practice Exercises

For extra help, see Examples 3–4 on pages 28–29 of your text.

Find each difference.

4. $-3 - 7$ 4. ______________

5. $-6.25 - (-2.47)$ 5. ______________

6. $\dfrac{3}{5}-\left(-\dfrac{1}{3}\right)$

6. ______________________

Objective 3   Find the distance between two points on a number line.

Review this example for Objective 3:

5. Find the distance between the points –6 and 3.

Find the absolute value of the difference of the numbers, taken in either order.
$$|-6-3|=|-9|=9, \quad \text{or} \quad |3-(-6)|=|3+6|=9$$

Now Try:

5. Find the distance between the points –7 and 5.

____

Objective 3 Practice Exercises

For extra help, see Example 5 on page 29 of your text.

Find the distance between each pair of points.

7. –4 and –1

7. ______________________

8. 5 and –2

8. ______________________

9. –1 and 8

9. ______________________

Objective 4   Multiply real numbers.

Review these examples for Objective 4:

6. Find each product.

a. $-4(-6)$

The numbers have the same sign, so the product is positive.
$$-4(-6)=24$$

b. $\dfrac{3}{4}(-8)$

$$\dfrac{3}{4}(-8)=-6$$

Now Try:

6. Find each product.

a. $-7(-11)$

____

b. $\dfrac{5}{6}(-12)$

____

Objective 4 Practice Exercises

For extra help, see Example 6 on page 30 of your text.

Find each product.

10. $(-5)(7)$ 10. _______________

11. $-1.2(-3.27)$ 11. _______________

12. $\dfrac{13}{11}\left(-\dfrac{33}{26}\right)$ 12. _______________

Objective 5   Find reciprocals and divide real numbers.

Review these examples for Objective 5:

7.   Find each quotient.

   a. $\dfrac{-15}{5}$

The numbers have opposite signs, so the quotient is negative.

$$\dfrac{-15}{5} = -3$$

   b. $\dfrac{21}{-7}$

$$\dfrac{21}{-7} = -3$$

   d. $\dfrac{8}{0}$

$\dfrac{8}{0}$ is undefined.

   e. $\dfrac{0}{-1}$

$$\dfrac{0}{-1} = 0$$

Now Try:

7.   Find each quotient.

   a. $\dfrac{-16}{8}$

____

   b. $\dfrac{27}{-9}$

____

   d. $\dfrac{-6}{0}$

____

   e. $\dfrac{0}{-3}$

____

                Copyright © 2020 Pearson Education, Inc.

Name: Date:
Instructor: Section:

Objective 5 Practice Exercises

For extra help, see Example 7 on page 32 of your text.

Give the reciprocal of each number.

13. -5 13. _______________

14. $-\dfrac{11}{18}$ 14. _______________

Divide where possible.

15. $\dfrac{-7}{0}$ 15. _______________

Chapter R REVIEW OF THE REAL NUMBER SYSTEM

R.4   Exponents, Roots, and Order of Operations

Learning Objectives
1   Use exponents.
2   Find square roots.
3   Use the rules for order of operations.
4   Evaluate algebraic expressions for given values of variables.

Key Terms

Use the vocabulary terms listed below to complete each statement in exercises 1−6.

> **factors**   **exponent**   **base**   **exponential expression**
>
> **square root**   **algebraic expression**

1.   A number written with an exponent is an ________________________________.

2.   The ________________ is the number that is a repeated factor when written with an exponent.

3.   A(n) ________________ is a number that indicates how many times a factor is repeated.

4.   A ________________________________ of a number r is a number that can be multiplied by itself to obtain r.

5.   A collection of numbers, variables, operation symbols, and grouping symbols is a(n) ________________________________.

6.   Two or more numbers whose product is a third number are ________________ of that third number.

Objective 1   Use exponents.

Review these examples for Objective 1:
1.   Write using exponents.

    a. $5 \cdot 5 \cdot 5 \cdot 5$

    $5 \cdot 5 \cdot 5 \cdot 5 = 5^4$

    b. $\dfrac{5}{6} \cdot \dfrac{5}{6} \cdot \dfrac{5}{6}$

    $\dfrac{5}{6} \cdot \dfrac{5}{6} \cdot \dfrac{5}{6} = \left(\dfrac{5}{6}\right)^3$

Now Try:
1.   Write using exponents.

    a. $9 \cdot 9 \cdot 9$

    ____

    b. $\dfrac{3}{8} \cdot \dfrac{3}{8} \cdot \dfrac{3}{8} \cdot \dfrac{3}{8}$

    ____

    c. $(-7)(-7)(-7)$

        $(-7)(-7)(-7) = (-7)^3$

    e. $x \cdot x \cdot x \cdot x \cdot x$

        $x \cdot x \cdot x \cdot x \cdot x = x^5$

2. Evaluate.

    a. 6^2

        $6^2 = 6 \cdot 6 = 36$

    b. $\left(\dfrac{3}{4}\right)^3$

        $\left(\dfrac{3}{4}\right)^3 = \dfrac{3}{4} \cdot \dfrac{3}{4} \cdot \dfrac{3}{4} = \dfrac{27}{64}$

    d. $(-3)^4$

        $(-3)^4 = (-3)(-3)(-3)(-3) = 81$

    e. $(-5)^3$

        $(-5)^3 = (-5)(-5)(-5) = -125$

3. Evaluate.

    a. 3^4

    3^4 means $3 \cdot 3 \cdot 3 \cdot 3$ which equals 81.

    b. $(-3)^4$

    $(-3)^4$ means $(-3)(-3)(-3)(-3)$ which equals 81.

    c. -3^4

    -3^4 means $-3 \cdot 3 \cdot 3 \cdot 3$ which equals -81.

---

    c. $(-10)(-10)(-10)(-10)$

    ____

    e. $z \cdot z \cdot z \cdot z \cdot z \cdot z$

    ____

2. Evaluate.

    a. 7^2

    ____

    b. $\left(\dfrac{2}{5}\right)^3$

    ____

    c. $(-4)^4$

    ____

    d. $(-2)^5$

    ____

3. Evaluate.

    a. 5^4

    ____

    b. $(-5)^4$

    ____

    c. -5^4

    ____

Objective 1 Practice Exercises

For extra help, see Examples 1–3 on pages 37–38 of your text.

Write the expression using exponents.

    1. $b \cdot b \cdot b \cdot b \cdot b \cdot b \cdot b$ **1.** ______________

Evaluate each expression.

2. $-(-4)^2$ 2. _______________

3. $(-2)^3$ 3. _______________

Objective 2   Find square roots.

Review these examples for Objective 2:
4. Find each square root that is a real number.

    a. $\sqrt{64}$

       $\sqrt{64} = 8,$ since $8^2 = 64.$

    c. $\sqrt{\dfrac{4}{25}}$

       $\sqrt{\dfrac{4}{25}} = \dfrac{2}{5},$ since $\left(\dfrac{2}{5}\right)^2 = \dfrac{4}{25}.$

    f. $-\sqrt{400}$

    $-\sqrt{400} = -20$, since the negative sign is outside the radical symbol.

    g. $\sqrt{-400}$

    $\sqrt{-400}$ is not a real number since the negative sign is inside the radical symbol.

Now Try:
4. Find each square root that is a real number.

    a. $\sqrt{49}$

       ____

    c. $\sqrt{\dfrac{16}{81}}$

       ____

    f. $-\sqrt{900}$

       ____

    g. $\sqrt{-900}$

       ____

Objective 2 Practice Exercises

For extra help, see Example 4 on page 39 of your text.

Find each square root. If it is not a real number, say so.

4. $-\sqrt{\dfrac{4}{49}}$ 4. _______________

5. $\sqrt{10,000}$ 5. _______________

6. $\sqrt{-81}$ 6. _______________

Name: _______________________   Date: _______________________

Instructor: _________________   Section: _____________________

Objective 3   Use the rules for order of operations.

| **Review these examples for Objective 3:** | **Now Try:** |

Review these examples for Objective 3:

5. Simplify $7+3\cdot 4$.

$$7+3\cdot 4 = 7+12$$
$$= 19$$

6. Simplify $6\cdot 5^2 +3-|-7+4|$

$$6\cdot 5^2 +3-|-7+4| = 6\cdot 5^2 +3-|-3|$$
$$= 6\cdot 5^2 +3-3$$
$$= 6\cdot 25 +3-3$$
$$= 150+3-3$$
$$= 153-3$$
$$= 150$$

7. Simplify $\dfrac{9+5^3}{8\sqrt{16}-2^5}$.

$$\frac{9+5^3}{8\sqrt{16}-2^5} = \frac{9+125}{8\cdot 4-32}$$
$$= \frac{9+125}{32-32}$$
$$= \frac{134}{0}$$

Because division by 0 is undefined, the given expression is undefined.

Now Try:

5. Simplify $8+5\cdot 6$.

____

6. Simplify $8\cdot 2^2 +6-|-9+5|$

____

7. Simplify $\dfrac{-5(9)+4^2}{2\sqrt{81}-\dfrac{1}{3}(54)}$.

____

Objective 3 Practice Exercises

For extra help, see Examples 5–7 on pages 39–40 of your text.

Simplify each expression.

7. $4^3 \div 2^5 +3\sqrt{36}$

7. _______________________

8. $-4(-3)^3 -5(8-4)$

8. _______________________

9. $\dfrac{6(-3)+(-5)^2(-4)}{11-3^2}$

9. ______________________

Objective 4   Evaluate algebraic expressions for given values of variables.

Review these examples for Objective 4:	**Now Try:**

Review these examples for Objective 4:

8. Evaluate each expression for $m = -3$, $n = 6$, $p = -8$, and $q = 16$.

a. $12m - 8n$

Substitute –3 for m and 6 for n.
$$12m - 8n = 12(-3) - 8(6)$$
$$= -36 - 48$$
$$= -84$$

c. $-5n^2 - m^2\left(\sqrt{q}\right)$

$$-5n^2 - m^2\left(\sqrt{q}\right) = -5(6)^2 - (-3)^2\left(\sqrt{16}\right)$$
$$= -5(36) - (9)(4)$$
$$= -180 - 36$$
$$= -216$$

Now Try:

7. Evaluate each expression for $m = -5$, $n = 8$, $p = -10$, and $q = 36$.

a. $7m - 5n$

____

b. $-6m^2 + n^2\left(\sqrt{q}\right)$

____

Objective 4 Practice Exercises

For extra help, see Example 8 on page 41 of your text.

Evaluate each expression if a = –1, b = 3, and c = –5.

10. $2a^3 - b^2$

10. ______________________

11. $\dfrac{2a + 4b^2}{3c - 2a}$

11. ______________________

12. $\dfrac{6c - b}{7a^3 - 4c}$

12. ______________________

Chapter R REVIEW OF THE REAL NUMBER SYSTEM

R.5   Properties of Real Numbers

Learning Objectives	
1	Use the distributive property.
2	Use the identity properties.
3	Use the inverse properties.
4	Use the commutative and associative properties.

Key Terms

Use the vocabulary terms listed below to complete each statement in exercises 1−4.

term   coefficient   like terms   combining like terms

1.   A _________________________________ is the numerical factor of a term.

2.   A number, a variable, or a product or quotient of a number and one or more variables raised to powers is called a _________________________________.

3.   Terms with exactly the same variables, including the same exponents, are called _________________________________.

4.   Adding or subtracting like terms by using the properties of real numbers is called _________________________________.

Objective 1   Use the distributive property.

Review these examples for Objective 1:

1.   Use the distributive property to rewrite each expression.

   b. $-5(8+m)$

   $$-5(8+m) = -5(8) - 5(m)$$
   $$= -40 - 5m$$

   d. $6r - 11r$

   $$6r - 11r = 6r + (-11r)$$
   $$= [6 + (-11)]r$$
   $$= -5r$$

Now Try:

1.   Use the distributive property to rewrite each expression.

   b. $-4(3+q)$

   ____

   d. $9r - 15r$

   ____

Objective 1 Practice Exercises

For extra help, see Example 1 on page 46 of your text.

Use the distributive property to rewrite each expression.

1. $-3(z-7)$ 1. _______________

2. $6y-15y$ 2. _______________

Use the distributive property to calculate the expression mentally.

3. $24 \cdot 13 + 26 \cdot 13$ 3. _______________

Objective 2   Use the identity properties.

Review these examples for Objective 2:

2. Simplify each expression.

    a. $17m + m$

$$17m + m = 17m + 1m$$
$$= (17+1)m$$
$$= 18m$$

    c. $-(a-9z)$

$$-(a-9z) = -1(a-9z)$$
$$= -1(a) + (-1)(-9z)$$
$$= -a + 9z$$

Now Try:

2. Simplify each expression.

    a. $29q + q$

____

    c. $-(4-5c)$

____

Objective 2 Practice Exercises

For extra help, see Example 2 on page 47 of your text.

Complete each statement.

4. $-6 + 0 = $ _____   4. _______________

5.   _____ $+ 0 = -2.5$ 5. _______________

6. $-\dfrac{7}{3} \cdot 1 = $ _____   6. _______________

Objective 3   Use the inverse properties.

Objective 3 Practice Exercises

For extra help, see pages 47–48 of your text.

Complete each statement.

7. $5 \cdot \underline{\quad\quad} = 1$ 7. ______________________

8. $-\dfrac{3}{4}\left(-\dfrac{4}{3}\right) = \underline{\quad\quad}$ 8. ______________________

9. $\dfrac{1}{7} + \underline{\quad\quad} = 0$ 9. ______________________

Objective 4   Use the commutative and associative properties.

Review these examples for Objective 4:

3.   Simplify $-6x + 9x - 4 + 8x - 6$.

$$
\begin{aligned}
-6x + 9x - 4 + 8x - 6 &= (-6x + 9x) - 4 + 8x - 6 \\
&= (-6 + 9)x - 4 + 8x - 6 \\
&= 3x - 4 + 8x - 6 \\
&= [3x + (-4 + 8x)] - 6 \\
&= [3x + (8x - 4)] - 6 \\
&= [(3x + 8x) - 4] - 6 \\
&= (11x - 4) - 6 \\
&= 11x - 4 - 6 \\
&= 11x - 10
\end{aligned}
$$

4.   Simplify each expression.

   a. $7x + 5 - 6(x + 3) + 14$

$$
\begin{aligned}
7x + 5 - 6(x + 3) + 14 &= 7x + 5 - 6x - 18 + 14 \\
&= 7x - 6x + 5 - 18 + 14 \\
&= x + 1
\end{aligned}
$$

Now Try:

3.   Simplify $-7x + 3x - 9 + 2x - 7$.

   ____

4.   Simplify each expression.

   a. $4y + 5 - 3(y + 3) - 2$

   ____

d. $(4x)(7)(z)$ **d.** $(6x)(4)y$

$$(4x)(7)(z) = [(4x)(7)]z$$
$$= [4(x \cdot 7)]z$$
$$= [4(7x)]z$$
$$= [(4 \cdot 7)x]z$$
$$= [28x]z$$
$$= 28(xz)$$
$$= 28xz$$

____

Objective 4 Practice Exercises

For extra help, see Examples 3–4 on pages 49–50 of your text.

Simplify each expression.

10. $-(2w-7)+11+3(4w-6)-5w$ **10.** ________________

11. $2(x-3)-7-6(2x-5)+7x$ **11.** ________________

12. $8-2(4d-1)+3(d-6)+2d$ **12.** ________________

Chapter 1 LINEAR EQUATIONS, INEQUALITIES, AND APPLICATIONS

1.1   Linear Equations in One Variable

Learning Objectives

1.   Distinguish between expressions and equations.
2.   Identify linear equations.
3.   Solve linear equations by using the addition and multiplication properties of equality.
4.   Solve linear equations using the distributive property.
5.   Solve linear equations with fractions or decimals.
6.   Identify conditional equations, contradictions, and identities.

Key Terms

Use the vocabulary terms listed below to complete each statement in exercises 1−7.

linear (first-degree) equation in one variable   **solution**

solution set   **equivalent equations**

conditional equation   **identity**   **contradiction**

1. Equations that have exactly the same solution sets are called

   ___________________________.

2. An equation that can be written in the form $Ax + B = C$, where A, B, and C are real numbers and $A \neq 0$, is called a ___________________________________.

3. The set of all numbers that satisfy an equation is called its

   ___________________________.

4. An equation with no solution is called a(n) _______________________________.

5. A(n) _________________________________ is an equation that is true for some values of the variable and false for other values.

6. An equation that is true for all values of the variable is called a(n)

   ___________________________.

7. A _____________________ of an equation is a number that makes the equation true when substituted for the variable.

Name: Date:
Instructor: Section:

Objective 1 Distinguish between expressions and equations.

Review these examples for Objective 1:

1. Decide whether each of the following is an *expression* or an *equation*.

   a. $5x - 11 = 4$

   Because there is an equality symbol, this is an equation.

   b. $5x - 11$

   Because there is no equality symbol, this is an expression.

Now Try:

1. Decide whether each of the following is an *expression* or an *equation*.

   a. $19x + 11$

   ____

   b. $19x = 11$

   ____

Objective 1 Practice Exercises

For extra help, see Example 1 on page 56 of your text.

Decide whether each of the following is an **expression** *or an* **equation**.

1. $13 - k = 11$ 1. _______________

2. $\dfrac{1}{3}q - \dfrac{5}{3} = \dfrac{7}{3}$ 2. _______________

3. $6k - 4$ 3. _______________

Objective 2 Identify linear equations.

Objective 2 Practice Exercises

For extra help, see pages 56–57 of your text.

Decide whether each of the following is a **linear equation** *or a* **nonlinear equation**.

4. $13 - k = 11$ 4. _______________

5. $\dfrac{1}{q} - \dfrac{5}{3} = \dfrac{7}{3}$ 5. _______________

6. $x^2 + 2y = 9$ 6. _______________

Objective 3   Solve linear equations by using the addition and multiplication properties of equality.

Review this example for Objective 3:	**Now Try:**

2. Solve $5x - 3x - 6 = 9 + 4x + 3$.

$$5x - 3x - 6 = 9 + 4x + 3$$
$$2x - 6 = 12 + 4x$$
$$2x - 6 - 4x = 12 + 4x - 4x$$
$$-2x - 6 = 12$$
$$-2x - 6 + 6 = 12 + 6$$
$$-2x = 18$$
$$\frac{-2x}{-2} = \frac{18}{-2}$$
$$x = -9$$

Check   Substitute -9 for x in the original equation.

$$5x - 3x - 6 = 9 + 4x + 3$$
$$5(-9) - 3(-9) - 6 = 9 + 4(-9) + 3$$
$$-45 + 27 - 6 = 9 - 36 + 3$$
$$-24 = -24$$

2. Solve $6x - 3x - 7 = 8 + x - 5$.

____

Objective 3 Practice Exercises

For extra help, see Example 2 on pages 57–58 of your text.

Solve and check each equation.

7. $9r - 4r + 8r - 6 = 10r - 11 + 2r$

7. ______________

8. $9x - 4 + 6x - 1 = 12 - 2x$

8. ______________

9. $18 - 3y + 11 = 7y + 6y - 27 - 9y$

9. ______________

Objective 4   Solve linear equations using the distributive property.

Review these examples for Objective 4:

| | **Now Try:** |

3. Solve $5(k-4)-k=k-2$.

3. Solve $9(k-2)-k=k+10$.

Step 1 Since there are no fractions or decimals, Step 1 does not apply.

$$\text{Step 2} \quad 5(k-4)-k=k-2$$
$$5(k)+5(-4)-k=k-2$$
$$5k-20-k=k-2$$
$$4k-20=k-2$$
$$\text{Step 3} \quad 4k-20-k=k-2-k$$
$$3k-20=-2$$
$$3k-20+20=-2+20$$
$$3k=18$$
$$\text{Step 4} \qquad \frac{3k}{3}=\frac{18}{3}$$
$$k=6$$

Step 5 Check by substituting 6 for k in the original equation.
$$5(k-4)-k=k-2$$
$$5(6-4)-6\overset{?}{=}6-2$$
$$5(2)-6\overset{?}{=}4$$
$$10-6\overset{?}{=}4$$
$$4=4 \quad \text{True}$$

The solution, 6, checks, so the solution set is $\{6\}$.

4. Solve $3(4-7x)=39-9(x+3)$.

4. Solve
$$6(3-2x)=42-8(x+6).$$

$$3(4-7x)=39-9(x+3)$$
$$\text{Step 2} \qquad 12-21x=39-9x-27$$
$$12-21x=12-9x$$
$$\text{Step 3} \quad 12-21x+9x=12-9x+9x$$
$$12-12x=12$$
$$12-12x-12=12-12$$
$$-12x=0$$
$$\text{Step 4} \qquad \frac{-12x}{-12}=\frac{0}{-12}$$
$$x=0$$

$$\textit{Step 5} \quad 3(4-7x)=39-9(x+3)$$

$$3[4-7(0)] \overset{?}{=} 39-9(0+3)$$

$$3(4-0) \overset{?}{=} 39-9(3)$$

$$3(4) \overset{?}{=} 39-27$$

$$12 = 12 \quad \text{True}$$

The solution set is $\{0\}$.

Objective 4 Practice Exercises

For extra help, see Examples 3–4 on pages 58–59 of your text.

Solve and check each equation.

10. $5(2p+1)-(p+3)=7$ 10. _______________

11. $-[p-(4p+2)]=3+(4p+7)$ 11. _______________

12. $-9w-(4+3w)=-(2w-1)-5$ 12. _______________

Objective 5   Solve linear equations with fractions or decimals.

Review these examples for Objective 5: | **Now Try:**

5.   Solve $\dfrac{x+8}{3}+\dfrac{3x-5}{6}=6.$

5.   Solve $\dfrac{2x+7}{5}+\dfrac{3x-9}{10}=-3.$

$$\frac{x+8}{3}+\frac{3x-5}{6}=6$$

$$\textit{Step 1} \quad 6\left(\frac{x+8}{3}+\frac{3x-5}{6}\right)=6(6)$$

$$6\left(\frac{x+8}{3}\right)+6\left(\frac{3x-5}{6}\right)=6(6)$$

$$2(x+8)+3x-5=36$$

$$2x+16+3x-5=36$$

$$5x+11=36$$

____

$Step$ 2 $5x+11-11=36-11$

$Step$ 3 $\dfrac{5x}{5}=\dfrac{25}{5}$

$x=5$

$Step$ 4 $\dfrac{x+8}{3}+\dfrac{3x-5}{6}=6$

$$\dfrac{5+8}{3}+\dfrac{3(5)-5}{6}\overset{?}{=}6$$

$$\dfrac{13}{3}+\dfrac{15-5}{6}\overset{?}{=}6$$

$$\dfrac{13}{3}+\dfrac{10}{6}\overset{?}{=}6$$

$$\dfrac{13}{3}+\dfrac{5}{3}\overset{?}{=}6$$

$$6=6 \quad \text{True}$$

The solution set is $\{5\}$.

6. Solve $0.2x+0.04(10-x)=0.06(4)$.

To clear decimals, multiply by 100.

$$0.2x+0.04(10-x)=0.06(4)$$
$$100[0.2x+0.04(10-x)]=100[0.06(4)]$$
$$100(0.2x)+100[0.04(10-x)]=100[0.06(4)]$$
$$20x+4(10)+4(-x)=24$$
$$20x+40-4x=24$$
$$16x+40=24$$
$$16x+40-40=24-40$$
$$16x=-16$$
$$\dfrac{16x}{16}=\dfrac{-16}{16}$$
$$x=-1$$

A check confirms that $\{-1\}$ is the solution set.

Objective 5 Practice Exercises

For extra help, see Examples 5–6 on pages 60–61 of your text.

Solve and check each equation.

13. $\dfrac{x-5}{2}-\dfrac{x+6}{3}=-4$

6. Solve
$$0.5x+0.04(5-8x)=0.07(8).$$

____

13. _____________

14. $\dfrac{2x+5}{5} - \dfrac{3x+1}{2} = \dfrac{7-x}{2}$ 14. _______________

15. $0.35(140) + 0.15w = 0.05(w + 1100)$ 15. _______________

Objective 6   Identify conditional equations, contradictions, and identities.

Review these examples for Objective 6:

7. Solve each equation. Decide whether it is a *conditional equation*, an *identity*, or a *contradiction*.

   a. $6(2x+7)-2=4(x+10)$

$$6(2x+7)-2=4(x+10)$$
$$12x+42-2=4x+40$$
$$12x+40=4x+40$$
$$12x-4x+40-40=4x-4x+40-40$$
$$\frac{8x}{8}=\frac{0}{8}$$
$$x=0$$

   The solution 0 checks, so the solution set is $\{0\}$. Since the solution set has only one element, $6(2x+7)-2=4(x+10)$ is a conditional equation.

   b. $9x-11=9(x-2)$

$$9x-11=9(x-2)$$
$$9x-11=9x-18$$
$$9x-11-9x=9x-18-9x$$
$$-11=-18 \quad \text{False}$$

   Since the result is false, the equation has no solution. The solution set is $\varnothing$, so the equation $9x-11=9(x-2)$ is a contradiction.

Now Try:

7. Solve each equation. Decide whether it is a *conditional equation*, an *identity*, or a *contradiction*.

   a. $5x-19=5(7x-4)+1$

   ____

   b. $10x-4=5(2x+2)-3$

   ____

                39

c. $8x+5=8(x+1)-3$

$$8x+5=8(x+1)-3$$
$$8x+5=8x+8-3$$
$$8x+5=8x+5$$
$$8x+5-8x-5=8x+5-8x-5$$
$$0=0 \quad \text{True}$$

The final line gives a true statement, which indicates that the solution set is {all real numbers}. The equation $8x+5=8(x+1)-3$ is an identity.

c. $6x-13=6(x-3)+5$

____

Objective 6 Practice Exercises

For extra help, see Example 7 on page 62 of your text.

Decide whether each equation is a **conditional equation**, *an* **identity**, *or a* **contradiction**. *Give the solution set.*

16. $7(2-5b)-32=10b-3(6+15b)$

16. ________________

17. $7(3-4q)-10(q-2)=19(5-2q)$

17. ________________

18. $13p-9(3-2p)=3(10p-9)+1$

18. ________________

Chapter 1 LINEAR EQUATIONS, INEQUALITIES, AND APPLICATIONS

1.2   Formulas and Percent

Learning Objectives
1   Solve a formula for a specified variable.
2   Solve applied problems using formulas.
3   Solve percent problems.
4   Solve problems involving percent increase or decrease.
5   Solve problems from the health care industry.

Key Terms

Use the vocabulary terms listed below to complete each statement in exercises 1−3.

mathematical model   **formula**   **percent**

1. An equation in which variables are used to describe a relationship is a

   ______________________.

2. A ______________________________ is an equation or inequality that describes a real situation.

3. ______________________ means "one per hundred."

Objective 1   Solve a formula for a specified variable.

Review these examples for Objective 1:

1. Solve the formula $V = LWH$ for L.

$$V = LWH$$
$$V = L(WH)$$
$$\frac{V}{WH} = \frac{L(WH)}{WH}$$
$$\frac{V}{WH} = L$$

2. Solve $F = \frac{9}{5}C + 32$, for C.

   Step 1 Multiply by 5 to clear fractions.

$$5(F) = 5\left(\frac{9}{5}C + 32\right)$$
$$5F = 5\left(\frac{9}{5}C\right) + 5(32)$$
$$5F = 9C + 160$$

   Step 2 $5F - 160 = 9C + 160 - 160$
$$5F - 160 = 9C$$

Now Try:

1. Solve the formula $A = LW$ for L.

   ____

2. Solve $V = \frac{1}{3}Bh$, for h.

   ____

$$\textit{Step 3} \quad \frac{5F-160}{9} = \frac{9C}{9}$$

$$\frac{5F-160}{9} = C$$

$$C = \frac{5F-160}{9}$$

$$C = \frac{5F}{9} - \frac{160}{9}$$

Objective 1 Practice Exercises

For extra help, see Examples 1–4 on pages 66–68 of your text.

Solve each formula or equation for the specified variable.

1. $V = \frac{1}{3}Bh$ for B

    1. _______________

2. $r = \frac{7}{x+5}$ for x

    2. _______________

3. $\frac{2+x}{3} = \frac{y}{6}$ for y

    3. _______________

Objective 2   Solve applied problems using formulas.

Review this example for Objective 2:   **Now Try:**

5. It takes Grace $\frac{2}{3}$ hr to drive 30 miles. What is her average rate?

5. It takes Tyler $\frac{1}{4}$ hr to drive 15 miles. What is his average rate?

    ____

We use the formula $d = rt$ solving for r.

$$\frac{d}{t} = \frac{rt}{t}$$

$$\frac{d}{t} = r, \text{ or } r = \frac{d}{t}$$

Substitute the values for d and t.

$$r = \frac{30}{\frac{2}{3}} = 30 \cdot \frac{3}{2} = 45$$

Her average rate is 45 mph.

Objective 2 Practice Exercises

For extra help, see Example 5 on page 68 of your text.

Solve each problem using the appropriate formula.

4. A cord of wood contains 128 cubic feet of wood. A stack of wood is 4 feet high and 8 feet long. How wide must it be to contain a cord?

4. ______________

5. If $700 earns $112 simple interest in 2 years, find the rate of interest.

5. ______________

6. The Fahrenheit temperature is 104°. Find the Celsius temperature.

6. ______________

Objective 3   Solve percent problems.

Review these examples for Objective 3:

6.

a. An alcohol and water mixture measures 45 liters. The mixture contains 9 liters of alcohol. What percent of the mixture is alcohol?

Let x represent the percent of alcohol in the mixture.

$$x = \frac{9}{45} \quad \begin{array}{l} \leftarrow \text{part} \\ \leftarrow \text{whole} \end{array}$$

$x = 0.20$ or 20%

The mixture is 20% alcohol.

Now Try:

6.

a. A certificate of deposit pays $117 simple interest in one year on a principal of $4500. What interest rate is being paid on this deposit?

____

b. The purchase price of a new car is \$12,500. In order to finance the car a purchaser is required to make a minimum down payment of 20% of the purchase price. What is the minimum down payment required?

Let x represent the amount of down payment.

$$\frac{x}{12,500} = 0.20$$

$$12,500 \cdot \frac{x}{12,500} = 12,500(0.20)$$

$$x = 2500$$

The down payment is \$2500.

7. The graph shows spending by department at a college. The college's total budget is \$280 million. How much was budgeted for the business department?

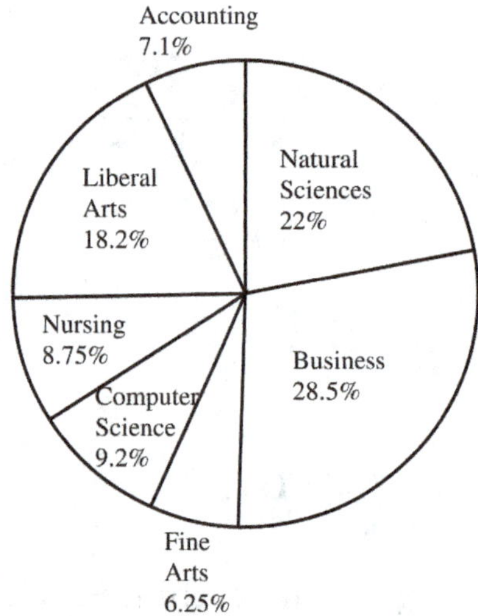

Since 28.5% was budgeted for the business department, let x = this amount in millions of dollars. $28.5\% = 0.285$.

$$\frac{x}{280} = 0.285$$

$$x = 0.285(280)$$

$$x = 79.8$$

\$79.8 million was budgeted for the business department.

b. A salesperson earned \$33,250 on annual sales of \$950,000. What is her rate of commission?

———————

7. Refer to the graph at the left. If the college's total budget is \$330 million, how much is budgeted for fine arts?

———————

Objective 3 Practice Exercises

For extra help, see Examples 6–7 on pages 69–70 of your text.

Solve each problem.

7. Twelve percent of a college student body has a grade   7. _______________
    point average of 3.0 or better. If there are 1250
    students enrolled in the college, how many have a
    grade point average of less than 3.0?

8. In a class of freshmen and sophomores only, there   8. _______________
    are 85 students. If 34 are freshmen, what percent of
    the class are sophomores?

9. At a large university, 40% of the student body is   9. _______________
    from out of state. If the total enrollment at the
    university is 25,000 students, how many are from
    in-state?

Objective 4   Solve problems involving percent increase or decrease.

Review these examples for Objective 4:

8. Use the percent change equation in each
    application.

    a. Over the last three years, Calvin's salary has
    increased from $2700 per month to $3200. What
    is the percent increase?

    Let x represent the percent increase.

    $$\text{percent increase} = \frac{\text{amount of increase}}{\text{original amount}}$$

    $$x = \frac{3200 - 2700}{2700}$$

    $$x = \frac{500}{2700}$$

    $$x \approx 0.185, \text{ or } 18.5\%$$

    Calvin's salary increased by approximately
    18.5%.

Now Try:

8. Use the percent change equation
    in each application.

    a. Students at Withrow's
    College were charged $1560 for
    tuition this semester. If the
    tuition was $1480 last semester,
    find the percent of increase.

    ____

b. The number of days employees of Prodex Manufacturing Company were absent from their jobs decreased from 96 days last month to 72 days this month. Find the percent of decrease.

Let x represent the percent decrease.

$$\text{percent decrease} = \frac{\text{amount of decrease}}{\text{original amount}}$$

$$x = \frac{96-72}{96}$$

$$x = \frac{24}{96}$$

$$x = 0.25, \text{ or } 25\%$$

The absentee rate decreased by 25%.

b. During a sale, the price of a futon was cut from $1250 to $999. Find the percent of decrease in price.

____

Objective 4 Practice Exercises

For extra help, see Example 8 on pages 70–71 of your text.

Solve each problem. Round money answer to the nearest cent and percent answer to the nearest tenth, if necessary.

10. The sale price of a piece of furniture is $1020. This represents 15% off the regular price. What is the regular price?

10. _______________

11. Mark Johnson invested $2500 in stock one year ago. During the year the stock increased in value by $162.50. What interest rate has Mark's investment earned?

11. _______________

12. Tracy works as a massage therapist. During July, she cut her price on massages from $54 to $45.50. By what percent did she decrease the price?

12. _______________

Objective 5   Solve problems from the health care industry.

Review these examples for Objective 5:

9. If a child weighs k kilograms, then the child's body surface area S in square meters (m^2) is determined by the formula

$$S = \frac{4k+7}{k+90}.$$

Use the formula to find the body surface area, to the nearest hundredth, of a child who weighs 30 lb. Use $k = 13.608$.

Use the value for k in the formula.

$$S = \frac{4(13.608)+7}{13.608+90}$$

$$S = 0.59$$

The child's body surface area is 0.59 m^2.

10. If D represents the usual adult dose of a medication in milligrams, the corresponding child's dose in milligrams is calculated by using the formula

$$C = \frac{\text{body surface area in square meters}}{1.7} \times D$$

Determine the appropriate dose, to the nearest unit, for a child whose body surface area is 0.67 m^2, if the usual adult dose is 100 mg.

Use the values for $S = 0.67$ m^2 and $D = 100$ mg in the formula.

$$C = \frac{0.67}{1.7} \times 100$$

$$C = 39$$

The child's dose area is 39 mg.

Now Try:

9. If a child weighs k kilograms, then the child's body surface area S in square meters (m^2) is determined by the formula

$$S = \frac{4k+7}{k+90}.$$

Use the formula to find the body surface area, to the nearest hundredth, of a child who weighs 35 lb. Use $k = 15.876$.

____

10. If D represents the usual adult dose of a medication in milligrams, the corresponding child's dose in milligrams is calculated by using the formula

$$C = \frac{\text{body surface area in m}^2}{1.7} \times D$$

Determine the appropriate dose, to the nearest unit, for a child whose body surface area is 0.59 m^2, if the usual adult dose is 50 mg.

____

     47

Objective 5 Practice Exercises

For extra help, see Examples 9–10 on pages 71–72 of your text.

Solve each problem.

13. If a child weighs k kilograms, then the child's body surface area S in square meters (m^2) is determined by the formula

$$S = \frac{4k+7}{k+90}.$$

Use the formula to find the body surface area, to the nearest hundredth, of a child who weighs 30 kg.

13. ______________________

14. If D represents the usual adult dose of a medication in milligrams, the corresponding child's dose in milligrams is calculated by using the formula

$$C = \frac{\text{body surface area in square meters}}{1.7} \times D$$

Determine the appropriate dose, to the nearest unit, for a child whose body surface area is $0.87\ m^2$, if the usual adult dose is 100 mg.

14. ______________________

Chapter 1 LINEAR EQUATIONS, INEQUALITIES, AND APPLICATIONS

1.3   Applications of Linear Equations

Learning Objectives
1   Translate from words to mathematical expressions.
2   Write equations from given information.
3   Distinguish between simplifying expressions and solving equations.
4   Use the six steps in solving an applied problem.
5   Solve percent problems.
6   Solve investment problems.
7   Solve mixture problems.

Key Terms

Use the vocabulary terms listed below to complete each statement in exercises 1−4.

sum	of	product	increased by
quotient	less than	per	double
decreased by	more than	difference	ratio

1.   ________________________, ________________________, and

      ________________________ are words that mean addition.

2.   ________________________, ________________________, and

      ________________________ are words that mean multiplication.

3.   ________________________, ________________________, and

      ________________________ are words that mean division.

4.   ________________________, ________________________, and

      ________________________ are words that mean subtraction.

Objective 1   Translate from words to mathematical expressions.

For extra help, see pages 78–79 of your text.

Translate each verbal phrase into a mathematical expression. Use x to represent the unknown number.

1.   The product of a number and 7 subtracted from 13   1. ________________

2.   −7 increased by 4 times a number   2. ________________

3.   The quotient of 4 more than a number and 9   3. ________________

Name: _______________________

Instructor: ___________________

Date: _______________________

Section: _____________________

Objective 2   Write equations from given information.

Review these examples for Objective 2:	**Now Try:**

1. Translate each verbal sentence into an equation.

1. Translate each verbal sentence into an equation.

a. If the product of a number and 24 is decreased by 9, the result is 63.

$$24x - 9 = 63$$

a. If the product of a number and 3 is decreased by 100, the result is 197.

____

b. The quotient of a number and the number minus 6 is 32.

$$\frac{x}{x-6} = 32$$

b. The quotient of a number and the number plus 5 is 23.

____

Objective 2 Practice Exercises

For extra help, see Example 1 on page 80 of your text.

Use the variable x for the unknown, and write an equation representing the verbal sentence. Do not solve.

4. 18 minus a number is equal to the number times 4.

4. _____________________

5. The ratio of a number and the difference between the number and 3 is 17.

5. _____________________

6. Twice a number is 3 times the sum of the number and 7.

6. _____________________

Objective 3   Distinguish between simplifying expressions and solving equations.

Review these examples for Objective 3:	**Now Try:**

2. Decide whether each is an *expression* or an *equation*. Simplify the expression, and solve the equation.

2. Decide whether each is an *expression* or an *equation*. Simplify the expression, and solve the equation.

a. $5(7 + x) - 3x + 11$

This is an expression.

$$5(7 + x) - 3x + 11$$
$$= 35 + 5x - 3x + 11$$
$$= 2x + 46$$

a. $9(x - 5) - 8x = 20$

____

Copyright © 2020 Pearson Education, Inc.

b. $6(4+x)-5x+8=-4$

This is an equation.
$$6(4+x)-5x+8=-4$$
$$24+6x-5x+8=-4$$
$$32+x=-4$$
$$x=-36$$
The solution set is $\{-36\}$.

b. $20x-4(x-7)+3$

____

Objective 3 Practice Exercises

For extra help, see Example 2 on page 80 of your text.

Decide whether each is an expression or an equation. Simplify the expression, and solve the equation.

7. $3(z-9)+8(2z-4)$

7. _______________

8. $3(z-9)=8(2z-4)$

8. _______________

9. $\dfrac{k}{3}-\dfrac{k+5}{4}=12$

9. _______________

Objective 4   Use the six steps in solving an applied problem.

Review this example for Objective 4:

3. The width of a rectangle is $\frac{1}{4}$ of the difference between the length and 1 meter. If the perimeter of the rectangle is 62 meters, find the length and width.

Step 1 Read the problem. Find the length and width.

Step 2 Assign a variable. Let $L=$ length.
Then $\frac{1}{4}(L-1)=$ width.

Step 3 Write an equation. Use the formula for the perimeter of a rectangle.
$$P=2L+2W$$
$$62=2L+2\left[\frac{1}{4}(L-1)\right]$$

Now Try:

3. In a triangle, the shortest side is 3 meters less than the longest side. The length of the third side is 11 meters less than twice the length of the longest side. If the perimeter is 22 meters, find the lengths of the three sides of the triangle.

____

Step 4 Solve.

$$62 = 2L + \frac{1}{2}(L-1)$$

$$2(62) = 2\left(2L + \frac{1}{2}(L-1)\right)$$

$$124 = 4L + L - 1$$

$$124 = 5L - 1$$

$$125 = 5L$$

$$\frac{125}{5} = \frac{5L}{5}$$

$$25 = L$$

Step 5 State the answer. The length is 25 m and the width is $\frac{1}{4}(25-1) = 6$ m.

Step 6 Check. The perimeter is $2(25) + 2(6) = 62$ m.

Objective 4 Practice Exercises

For extra help, see Examples 3–4 on pages 81–82 of your text.

Solve each problem.

10.   Mrs. Henry has 36 feet of fencing, which she plans to install to form a square-shaped yard for her dog. If the back of her house becomes one side of the square with the other sides requiring fencing, how long should each side be?

10. ________________

11.   In the controversial 2000 presidential election, George W. Bush and Al Gore received a total of 537 electoral votes. Bush received five more votes than Gore. How many votes did each candidate receive?

11. ________________

____

12.   The sum of two numbers is 41. The larger number is 1 more than 4 times the smaller number. What are these numbers?

12. ________________

____

Name: Date:
Instructor: Section:

Objective 5   Solve percent problems.

Review this example for Objective 5:

Now Try:

5. A merchant in a small store took in $508.80 in one day. This included merchandise sales and the 6% sales tax. Find the amount of merchandise sold.

5. Jim wishes to make $235,000 selling his home. If his realtor must earn a 6% commission, for what price should Jim sell his home?

____

Step 1 Read the problem. Find the amount sold.

Step 2 Assign a variable. Let $x =$ the amount sold.

Step 3 Write an equation.
$$x + 0.06x = 508.80$$

Step 4 Solve.
$$1.06x = 508.80$$
$$\frac{1.06x}{1.06} = \frac{508.80}{1.06}$$
$$x = 480$$

Step 5 State the answer. The merchant sold $480 of merchandise.

Step 6 Check. $480 + 0.06(480) = 508.80$
The answer checks.

Objective 5 Practice Exercises

For extra help, see Example 5 on page 83 of your text.

Solve each problem.

13. A savings account earns interest at an annual rate of 4%. At the end of one year Sheri would like to have a total of $5200 in the account. How much should she deposit now?

13. ___________________

14. At the end of a day, a storekeeper had $642 in the cash registers, counting both the sale of goods and the sales tax of 7%. Find the amount of the tax.

14. ___________________

15. A certificate of deposit earns $2\frac{3}{4}\%$ annually. If $6200 is invested, what is the certificate of deposit worth in one year?

15. ______________________

Objective 6   Solve investment problems.

Review this example for Objective 6:

6. Larry invested some money at 8% simple interest and $700 less than this amount at 7%. His total annual income from the interest was $584. How much was invested at each rate?

Step 1 Read the problem. Find the two amounts.

Step 2 Assign a variable.
Let x = the amount at 8%.
Let $x - 700$ = the amount at 7%.

Step 3 Write an equation.
$$0.08x + 0.07(x - 700) = 584$$

Step 4 Solve.
$$0.08x + 0.07x - 49 = 584$$
$$0.15x - 49 = 584$$
$$0.15x = 633$$
$$x = 4220$$

Step 5 State the answer. Larry must invest $4220 at 8%, and $4220 - 700 = $3520 at 7%.

Step 6 Check. The sum of the two amounts should be $584.
$$0.08(4220) + 0.07(3520) = 584$$

Now Try:

6. Tracy invested some money at 5% and $300 more than twice this amount at 7%. Her total annual income from the two investments is $325. How much is invested at each rate?

____

Objective 6 Practice Exercises

For extra help, see Example 6 on page 84 of your text.

Solve each problem.

16. Louisa invested $16,000 in bonds paying 7% simple interest. How much additional money should she invest at 4% simple interest so that the average return on the two investments is 6%?

16. _________________

17. Desiree invested some money at 9% and $100 less than three times that amount at 7%. Her total annual interest was $83. How much did she invest at each rate?

17. 7%: _____________

     9%: _____________

18. Jacob has $48,000 invested in stocks paying 6%. How much additional money should he invest in certificates of deposit paying 2.5% so that the average return on the two investments is 4%?

18. _________________

Name: Date:

Instructor: Section:

Objective 7   Solve mixture problems.

Review these examples for Objective 7:

7. How many gallons of a 20% alcohol solution must be mixed with 15 gallons of a 12% alcohol solution to obtain a 14% alcohol solution?

Step 1 Read the problem. Find the amount of the 20% alcohol solution.

Step 2 Assign a variable. Let $x =$ the amount of the 20% alcohol solution.

Step 3 Write an equation.
$$0.20x + 0.12(15) = 0.14(x + 15)$$

Step 4 Solve.
$$0.20x + 1.8 = 0.14x + 2.1$$
$$0.06x = 0.3$$
$$x = 5$$

Step 5 State the answer. 5 gallons of the 20% alcohol solution must be added.

Step 6 Check.
$$0.20(5) + 0.12(15) \overset{?}{=} 0.14(5 + 15)$$
$$2.8 = 2.8$$
The answer checks.

Now Try:

7. How many ounces of a 35% solution of acid must be mixed with a 60% solution to get 20 ounces of a 50% solution?

____

Objective 7 Practice Exercises

For extra help, see Examples 7–8 on pages 85–86 of your text.

Solve each problem.

19. How many pounds of peanuts worth $3 per pound must be mixed with mixed nuts worth $5.50 per pound to make 40 pounds of a mixture worth $5 per pound?

19. _______________

20. How many pounds of candy worth $7 per pound must be mixed with candy worth $4.50 per pound to make 100 pounds of candy worth $6 per pound?

20. _______________

Chapter 1 LINEAR EQUATIONS, INEQUALITIES, AND APPLICATIONS

1.4   Further Applications of Linear Equations

Learning Objectives
1   Solve problems about different denominations of money.
2   Solve problems about uniform motion.
3   Solve problems about angles.
4   Solve problems about consecutive integers.

Key Terms

Use the vocabulary terms listed below to complete each statement in exercises 1−4.

vertical angles   complementary angles   supplementary angles

consecutive integers

1.   Two angles whose measures sum to 180° are _______________________________.

2.   Two angles whose measures sum to 90° are _______________________________.

3.   Angles formed by two intersecting lines are called _______________________________.

4.   Two integers that differ by 1 are _______________________________.

Objective 1   Solve problems about different denominations of money.

Review this example for Objective 1:

1.   A collection of dimes and nickels has a total value of $2.70. The number of nickels is 2 more than twice the number of dimes. How many of each type of coin are in the collection?

Step 1 Read the problem. Find the number of dimes and the number of nickels.

Step 2 Assign a variable.
Let d = the number of dimes.
Then $2d + 2$ = the number of nickels.

Step 3 Write an equation.
$$0.10d + 0.05(2d + 2) = 2.70$$

Step 4 Solve.
$$10d + 5(2d + 2) = 270$$
$$10d + 10d + 10 = 270$$
$$20d + 10 = 270$$
$$20d = 260$$
$$d = 13$$

Now Try:

1.   Erica's piggy bank has quarters and nickels in it. The total number of coins in the piggy bank is 50. Their total value is $8.90. How many of each type are in the piggy bank?

____

Step 5 State the answer. There are 13 dimes and $2(13)+2=28$ nickels.

Step 6 Check.
$0.10(13)+0.05(28)=1.30+1.40=2.70$
The answer is correct.

Objective 1 Practice Exercises

For extra help, see Example 1 on pages 92–93 of your text.

Solve each problem.

1. Anthony sells two different size jars of peanut butter. The large size sells for \$2.60 and the small size sells for \$1.80. He has 80 jars worth \$164. How many of each size jar does he have?

1.
large _________________
small _________________

2. Stan has 14 bills in his wallet worth \$95 altogether. If the wallet contains only \$5 and \$10 bills, how many bills of each denomination does he have?

2.
\$5 bills _________________
\$10 bills _________________

3. Twice as many general admission tickets to a basketball game were sold as reserved seat tickets. General admission tickets cost \$10 and reserved seat tickets cost \$15. If the total value of both kinds of tickets was \$26,250, how many tickets of each kind were sold?

3.
general admission_______
reserved seats _________

Objective 2   Solve problems about uniform motion.

Review these examples for Objective 2:

2. Bill and Hillary start in Washington and fly in opposite directions. At the end of 4 hours, they are 4896 kilometers apart. If Bill flies 60 kilometers per hour faster than Hillary, what are their speeds?

Step 1 Read the problem. Find the rate of each plane.

Step 2 Assign a variable.
Let x = the rate of Hillary's plane.
Then $x + 60$ = the rate of Bill's plane.
Using $d = rt$, Hillary's plane goes a distance of $4x$ and Bill's plane goes a distance of $4(x + 60)$.

Step 3 Write an equation for the total distance.
$$4x + 4(x + 60) = 4896$$

Step 4 Solve.
$$4x + 4x + 240 = 4896$$
$$8x + 240 = 4896$$
$$8x = 4656$$
$$x = 582$$

Step 5 State the answer. Hillary's plane is flying at 582 kph, and Bill's plane is flying at $582 + 60 = 642$ kph.

Step 6 Check.
$$4(582) + 4(642) = 4896$$
The answer is correct.

3. John left Louisville at noon on the same day that Mike left Louisville at 1 P.M. Both were traveling in the same direction. At 5 P.M., Mike was 62 miles behind John. If John was traveling 2 miles per hour faster than Mike, what were their speeds?

Step 1 Read the problem. Find the speed for Mike and the speed for John.

Step 2 Assign a variable. Let x = Mike's speed.
Then $x + 2$ = John's speed.
Using $d = rt$, John goes a distance of $5(x + 2)$ and Mike goes a distance of $4x$.

Now Try:

2. Two planes left Philadelphia traveling in opposite directions. Plane A left 15 minutes before plane B. After plane B had been flying for 1 hour, the planes were 860 miles apart. What were the speeds of the two planes if plane A was flying 40 miles per hour faster than plane B?

____

3. In a run for charity Ted runs at a speed of 5 miles per hour. Bob leaves 10 minutes after Ted and runs at 6 miles per hour. How long will it take Bob to catch up with Ted? (Hint: Change minutes to hours.)

____

      59

Step 3 Write an equation.
$$5(x+2)-4x=62$$

Step 4 Solve.
$$5x+10-4x=62$$
$$x+10=62$$
$$x=52$$

Step 5 State the answer. Mike's speed was 52 mph, and John's speed was 52 + 2 = 54 mph.

Step 6 Check. $5(54)-4(52)=62$ The answer checks.

Objective 2 Practice Exercises

For extra help, see Examples 2–3 on pages 93–95 of your text.

Solve each problem.

4.   It takes a kayak $1\frac{1}{2}$ hours to go 24 miles downstream and 4 hours to return. Find the speed of the current and the speed of the kayak in still water.

4.
kayak speed ___________

current speed ___________

5.   A plane can travel 300 miles per hour with the wind and 230 miles per hour against the wind. Find the speed of the wind and the speed of the plane in still air.

5.
plane speed___________

wind speed ___________

6.   At the beginning of a fund-raising walk, Steve and Vic are 30 miles apart. If they leave at the same time and walk in the same direction, Steve would overtake Vic in 15 hours. If they walked toward each other, they would meet in 3 hours. What are their speeds?

6.
Steve___________

Vic ___________

Objective 3   Solve problems about angles.

Review this example for Objective 3:	**Now Try:**
4. Find the measure of each angle.	**4.** Find the measure of each angle.

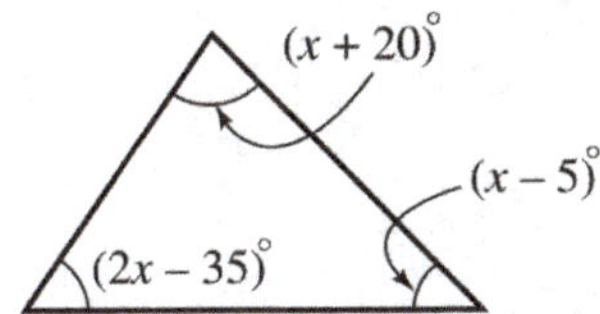

Step 1 Read the problem. Find the measure of each angle.

Step 2 Assign a variable. The variables have been assigned.

Step 3 Write an equation. The sum of the three angles must be 180°.
$$(2x - 35) + (x + 20) + (x - 5) = 180$$

Step 4 Solve.
$$2x - 35 + x + 20 + x - 5 = 180$$
$$4x - 20 = 180$$
$$4x = 200$$
$$x = 50$$

Step 5 State the answer. The measures of the angles are $2(50) - 35 = 65°$, $50 + 20 = 70°$, and $50 - 5 = 45°$.

Step 6 Check. $65° + 70° + 45° = 180°$.

Objective 3 Practice Exercises

For extra help, see Example 4 on pages 95–96 of your text.

Solve each problem.

7.   One angle of a triangle measures 30° larger than a second angle. The third angle measures 4 times the second angle. Find the measure of each angle.

7. _______________

8.   The measure of an angle is 15° larger than 10 times its supplement. Find the measure of each angle.

8. _______________

9. The measure of one angle of a triangle is 30° more than that of a second angle. The measure of the third angle is half the measure of the second angle. Find the measure of each angle.

9. _______________________

Objective 4   Solve problems about consecutive integers.

Review this example for Objective 4:

5. Find three consecutive integers such that the sum of the first and third, increased by 2, is 13 more than the second.

Step 1 Read the problem. Find the measure of each integer.

Step 2 Assign a variable. Let
x = the least of the three consecutive integers.
Then $x + 1$ = the middle integer,
and $x + 2$ = the greatest

Step 3 Write an equation. The sum of the first and third, increased by 2, is 13 more than the second.
$$x + (x+2) + 2 = x + 1 + 13$$

Step 4 Solve.
$$x + (x+2) + 2 = x + 1 + 13$$
$$2x + 4 = x + 14$$
$$x = 10$$

Step 5 State the answer. The solution is 10, so the first integer is $x = 10$, the second is $10 + 1 = 11$, and the third is $10 + 2 = 12$. The three integers are 10, 11, and 12.

Step 6 Check. The sum of the first and third is $10 + 12 = 22$. If this is increased by 2, the result is $22 + 2 = 24$, which is 13 more than the second (11). The answer is correct.

Now Try:

5. Find three consecutive integers such that the sum of the first and second, decreased by 7, is 45 more than the third.

____

Name: Date:
Instructor: Section:

Objective 4 Practice Exercises

For extra help, see Example 5 on page 96 of your text.

Solve each problem.

10. Find three consecutive odd integers whose sum is   10. _________________
     363.

11. The sum of four consecutive even integers is 4. Find   11. _________________
     the integers.

Chapter 1 LINEAR EQUATIONS, INEQUALITIES, AND APPLICATIONS

1.5   Linear Inequalities in One Variable

Learning Objectives
1   Graph intervals on a number line.
2   Solve linear inequalities using the addition property.
3   Solve linear inequalities using the multiplication property.
4   Solve linear inequalities with three parts.
5   Solve applied problems using linear inequalities.

Key Terms

Use the vocabulary terms listed below to complete each statement in exercises 1−5.

interval   **interval notation**   **inequality**

linear inequality in one variable   **equivalent inequalities**

1.   The ______________________________ for $a \le x < b$ is $[a, b)$.

2.   A(n) ______________________________ can be written in the form $Ax + B > C$, $Ax + B \ge C$, $Ax + B < C$, or $Ax + B \le C$, where A, B, and C are real numbers with $A \ne 0$.

3.   An algebraic expression related by $>, \ge, <,$ or $\le$ is called a(n) ______________________.

4.   An ______________________________ is a portion of a number line.

5.   Inequalities with the same solution set are ______________________________.

Objective 1   Graph intervals on a number line.

Review these examples for Objective 1:

1.   Write each inequality in interval notation and graph the interval.

   a. Graph $x > -3$.

   The statement $x > -3$ says that x can represent any value greater than -3, but cannot equal -3, written $(-3, \infty)$. We graph this interval by placing a parenthesis at -3 and drawing an arrow to the right. The parenthesis indicates that -3 is not part of the graph.

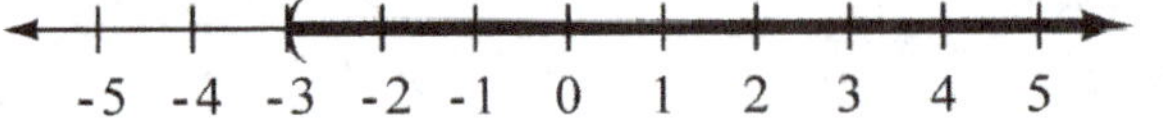

Now Try:

1.   Write each inequality in interval notation and graph the interval.

   a. Graph $x > -1$.

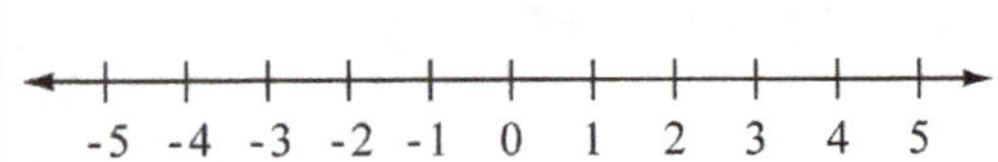

b. Write the inequality $-4 \leq x < 3$ in interval notation, and graph the interval.

$-4 \leq x < 3$
x is between –4 and 3 (excluding 3).
In interval notation, we write $\left[-4,\ 3\right)$.

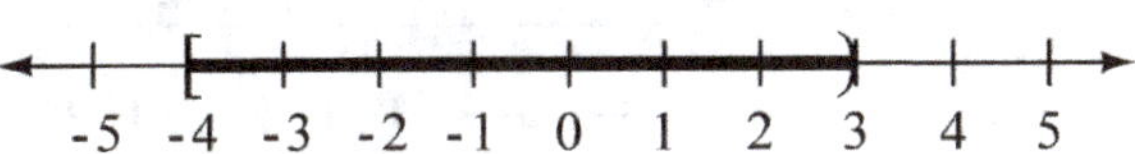

b. Write the inequality $-5 < x \leq -1$ in interval notation, and graph the interval.

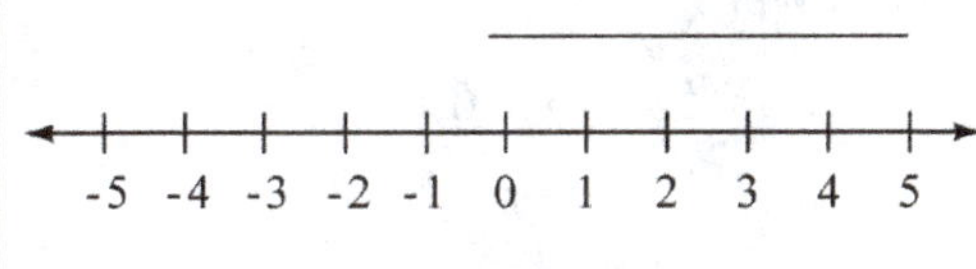

Objective 1 Practice Exercises

For extra help, see Example 1 on pages 103–104 of your text.

Write each inequality in interval notation and graph the interval.

1. $3 < a$

1.   _______________________

2. $y \geq -2$

2.   _______________________

3. $-3 < a \leq 2$

3.   _______________________

Objective 2   Solve linear inequalities using the addition property.

Review this example for Objective 2:

2.   Solve $x - 6 < -11$ and graph the solution set.

$x - 6 < -11$
$x - 6 + 6 < -11 + 6$
$x < -5$

A check confirms that $(-\infty, -5)$, graphed below, is the solution set.

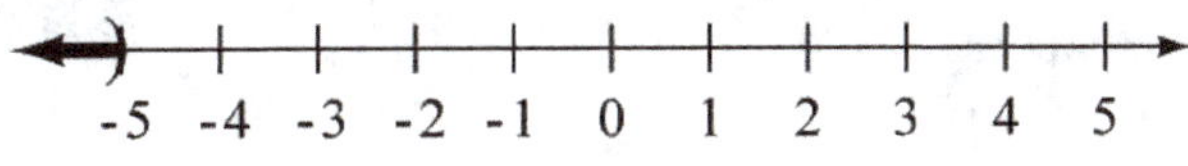

Now Try:

2.   Solve $x + 3 < -1$ and graph the solution set.

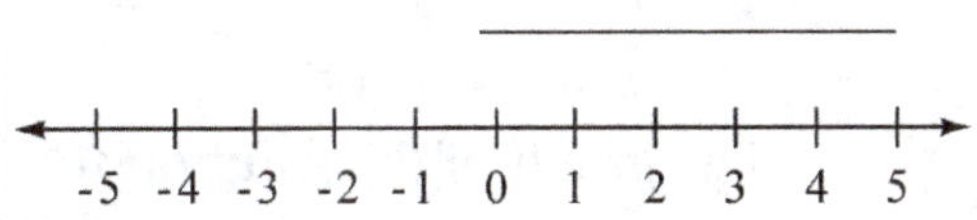

Objective 2 Practice Exercises

For extra help, see Examples 2–3 on pages 105–106 of your text.

Solve each inequality, giving its solution set in both interval and graph forms. Check your answers.

4. $5a + 3 \le 6a$

4. _______________________

5. $6 + 3x < 4x + 4$

5. _______________________

6. $3 + 5p \le 4p + 3$

6. _______________________

Objective 3 Solve linear inequalities using the multiplication property.

Review these examples for Objective 3:

4. Solve the inequality, and graph the solution set.

$$6x < -24$$

We divide each side by 6.

$$6x < -24$$

$$\frac{6x}{6} < \frac{-24}{6}$$

$$x < -4$$

The graph of the solution set $(-\infty, -4)$, is shown below.

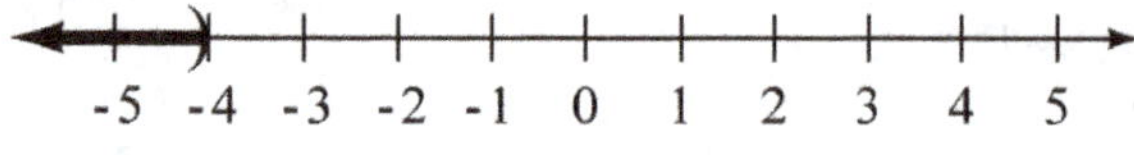

Now Try:

4. Solve the inequality, and graph the solution set.

$$8x \le -40$$

____

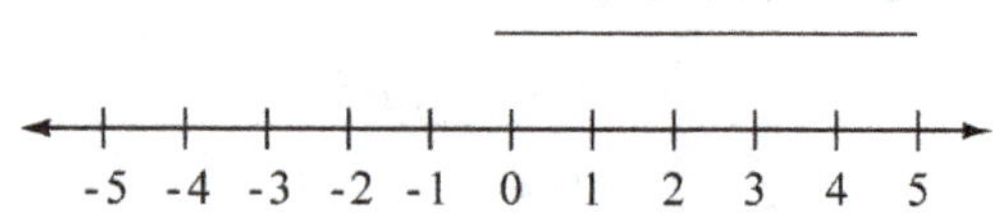

   Copyright © 2020 Pearson Education, Inc.

5. Solve $-5(x+3)+3 \geq 8-3x$, and graph the solution set.

> *Step 1* $-5(x+3)+3 \geq 8-3x$
>
> $\qquad -5x-15+3 \geq 8-3x$
>
> $\qquad -5x-12 \geq 8-3x$
>
> *Step 2* $-5x-12+3x \geq 8-3x+3x$
>
> $\qquad -2x-12 \geq 8$
>
> $\qquad -2x-12+12 \geq 8+12$
>
> $\qquad -2x \geq 20$
>
> *Step 3* $\dfrac{-2x}{-2} \leq \dfrac{20}{-2}$
>
> $\qquad x \leq -10$

The solution set is $(-\infty, -10]$. The graph is shown below.

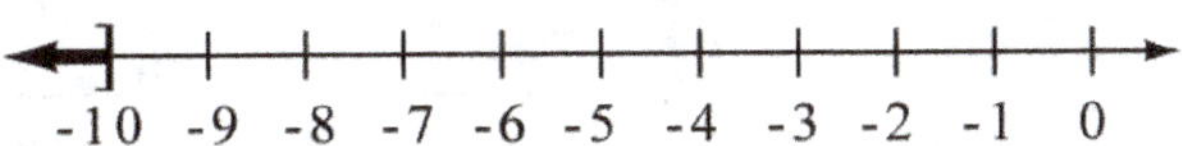

5. Solve $-6(x+4)+1 \geq 9-2x$, and graph the solution set.

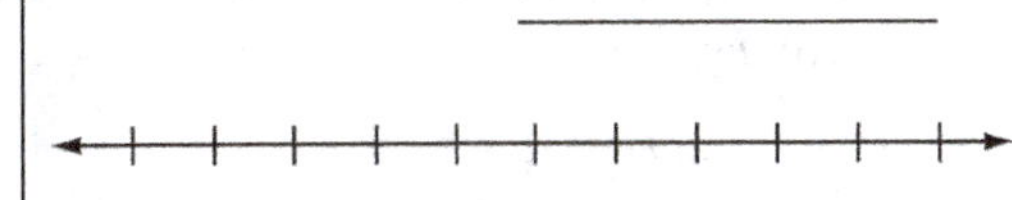

6. Solve $-\dfrac{2}{5}(r-4)-\dfrac{1}{3} < \dfrac{1}{3}(4-r)$, and graph the solution set.

> *Step 1* $15\left[-\dfrac{2}{5}(r-4)-\dfrac{1}{3}\right] < 15\left[\dfrac{1}{3}(4-r)\right]$
>
> $\qquad 15\left[-\dfrac{2}{5}(r-4)\right]-15\left(\dfrac{1}{3}\right) < 15\left[\dfrac{1}{3}(4-r)\right]$
>
> $\qquad -6(r-4)-5 < 5(4-r)$
>
> $\qquad -6r+24-5 < 20-5r$
>
> $\qquad -6r+19 < 20-5r$
>
> *Step 2* $-6r+19+5r < 20-5r+5r$
>
> $\qquad -r+19 < 20$
>
> $\qquad -r+19-19 < 20-19$
>
> $\qquad -r < 1$
>
> *Step 3* $-1(-r) > -1(1)$
>
> $\qquad r > -1$

The solution set is $(-1, \infty)$. The graph is shown below.

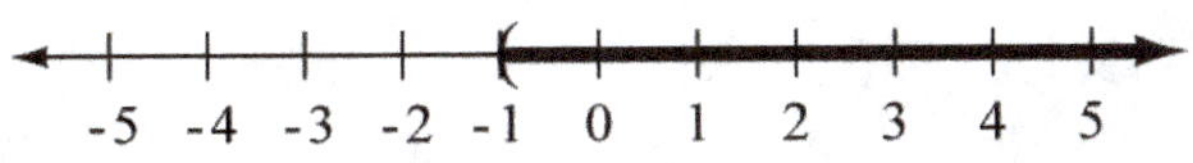

6. Solve $-\dfrac{1}{2}(r-5)+\dfrac{1}{3} < \dfrac{1}{3}(5-r)$, and graph the solution set.

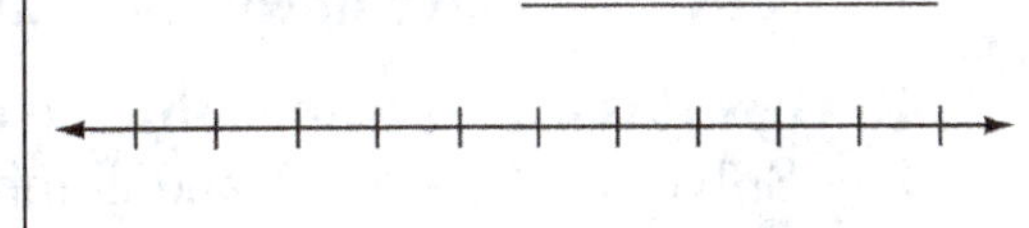

Objective 3 Practice Exercises

For extra help, see Examples 4–6 on pages 107–109 of your text.

Solve each inequality, giving its solution set in both interval and graph forms. Check your answers.

7. $-2s < 4$

7. ______________________

8. $4k \geq -16$

8. ______________________

9. $-9m \geq -36$

9. ______________________

Objective 4   Solve linear inequalities with three parts.

Review this example for Objective 4:

8. Solve $3 \leq 4x - 5 < 7$ and graph the solution set.

$$3 \leq -4x - 5 < 7$$
$$3 + 5 \leq -4x < 7 + 5$$
$$8 \leq -4x < 12$$
$$-\frac{8}{4} \geq -\frac{4x}{4} > -\frac{12}{4}$$
$$-2 \geq x > -3$$
$$-3 < x \leq -2$$

The solution set is $(-3, -2]$. The graph is shown below.

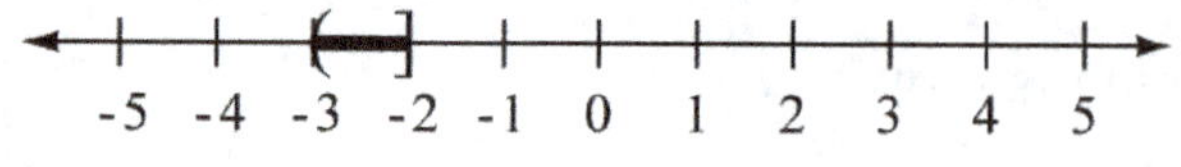

Now Try:

8. Solve $8 \leq -6x - 4 < 20$ and graph the solution set.

____

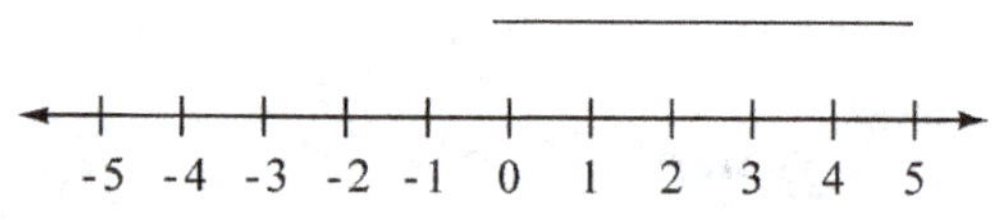

              Copyright © 2020 Pearson Education, Inc.

Objective 4 Practice Exercises

For extra help, see Examples 7–8 on page 110 of your text.

Solve each inequality, giving its solution set in both interval and graph forms. Check your answers.

10. $7 < 2x + 3 \leq 13$

10. _______________________

11. $-17 \leq 3x - 2 < -11$

11. _______________________

12. $1 < 3z + 4 < 19$

12. _______________________

Objective 5   **Solve applied problems using linear inequalities.**

Review this example for Objective 5:

10. Ruth tutors mathematics in the evenings in an office for which she pays \$600 per month rent. If rent is her only expense and she charges each student \$40 per month, how many students must she teach to make a profit of at least \$1600 per month?

Step 1   Read the problem again.

Step 2   Assign a variable.
    Let x = the number of students.

Step 3   Write an inequality.
    $40x - 600 \geq 1600$

Step 4   Solve.
$$40x - 600 + 600 \geq 1600 + 600$$
$$40x \geq 2200$$
$$\frac{40x}{40} \geq \frac{2200}{40}$$
$$x \geq 55$$

Now Try:

10. Two sides of a triangle are equal in length, with the third side 8 feet longer than one of the equal sides. The perimeter of the triangle cannot be more than 38 feet. Find the largest possible value for the length of the equal sides.

              ____

Step 5 State the answer. Ruth must have 55 or more students to have at least $1600 profit.

Step 6 Check. $40(55) - 600 = 1600$ Also, any number greater than 55 makes the profit greater than $1600.

Objective 5 Practice Exercises

For extra help, see Examples 9–10 on pages 111–112 of your text.

Solve each problem.

13.   Lauren has grades of 98 and 86 on her first two chemistry quizzes. What must she score on her third quiz to have an average of at least 91 on the three quizzes?

13. ______________

14.   Nina has a budget of $230 for gifts for this year. So far she has bought gifts costing $47.52, $38.98, and $26.98. If she has three more gifts to buy, find the average amount she can spend on each gift and still stay within her budget.

14. ______________

15.   Andrew must have an average of 80 of the points on four exams to receive a B in the class. He has earned 78, 83, and 75 on the first three exams. What is the lowest score he can earn on a 100-point test to guarantee a B in the class?

15. ______________

Chapter 1 LINEAR EQUATIONS, INEQUALITIES, AND APPLICATIONS

1.6   Set Operations and Compound Inequalities

Learning Objectives
1   Recognize set intersection and union.
2   Find the intersection of two sets.
3   Solve compound inequalities with the word *and*.
4   Find the union of two sets.
5   Solve compound inequalities with the word *or*.

Key Terms

Use the vocabulary terms listed below to complete each statement in exercises 1−3.

intersection   **compound inequality**   **union**

1.   The ___________________ of two sets, A and B, is the set of elements that belong to either A or B or both.

2.   A _____________________________________ is formed by joining two inequalities with a connective word such as *and* or *or*.

3.   The ___________________ of two sets, A and B, is the set of elements that belong to both A and B.

Objective 2   Find the intersection of two sets.

Review this example for Objective 2:

1.   Let $A = \{6,\ 7,\ 8,\ 9\}$ and $B = \{6,\ 8,\ 10\}$. Find $A \cap B$.

The set $A \cap B$ contains those elements that belong to both A and B.

$$A \cap B = \{6,\ 7,\ 8,\ 9\} \cap \{6,\ 8,\ 10\}$$
$$= \{6,\ 8\}$$

Now Try:

1.   Let $A = \{20,\ 30,\ 40,\ 50\}$ and $B = \{30,\ 50,\ 70\}$. Find $A \cap B$.

____

Objective 2 Practice Exercises

For extra help, see Example 1 on page 116 of your text.

Let $A = \{0, 1, 2, 3, 4, 5\}$, $B = \{2, 4, 6, 8, 10\}$, $C = \{1, 3, 5, 7, 9\}$, and $D = \{0, 2, 4\}$. Specify each set.

1. $A \cap D$ 1. _______________

2. $B \cap C$ **2.** _______________

3. $A \cap C$ **3.** _______________

Objective 3 Solve compound inequalities with the word *and*.

Review these examples for Objective 3:

2. Solve $x + 6 \le 15$ and $x + 3 \ge 7$, and graph the solution set.

Step 1 Solve each inequality individually.

$$x + 6 \le 15 \qquad \text{and} \qquad x + 3 \ge 7$$
$$x + 6 - 6 \le 15 - 6 \quad \text{and} \quad x + 3 - 3 \ge 7 - 3$$
$$x \le 9 \qquad \text{and} \qquad x \ge 4$$

Step 2 The solution set of the compound inequality includes all numbers that satisfy both inequalities from Step 1.

The graphs of $x \le 9$ and $x \ge 4$ are shown below.

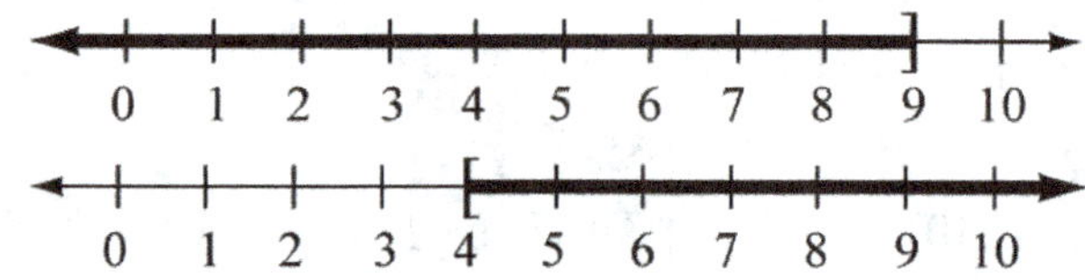

The intersection of these two graphs is the solution set [4, 9].

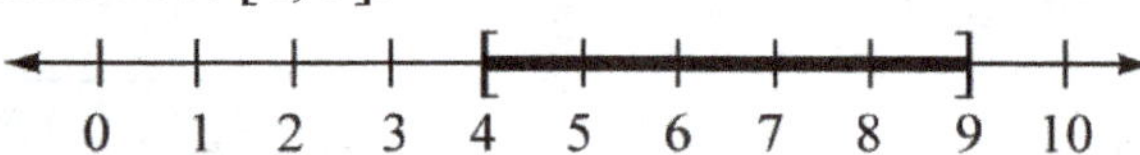

3. Solve $-5x - 6 > 8$ and $6x - 3 \le -21$, and graph the solution set.

Step 1 Solve each inequality individually.
$$-5x - 6 > 8 \qquad \text{and} \quad 6x - 3 \le -21$$
$$-5x > 14 \qquad \text{and} \qquad 6x \le -18$$
$$x < -\frac{14}{5} \quad \text{and} \qquad x \le -3$$

The graphs of $x < -\dfrac{14}{5}$ and $x \le -3$ are shown

Now Try:

2. Solve $x + 4 \le 20$ and $x - 3 \ge 9$ and graph the solution set.

____

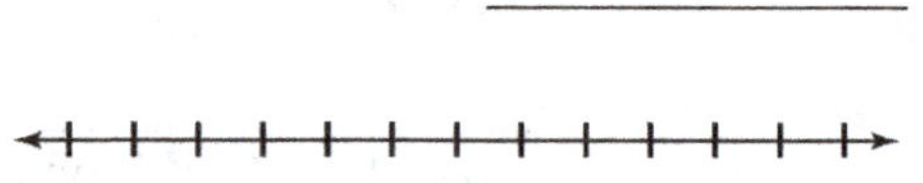

3. Solve $-8x + 6 > 30$ and $4x - 7 \le 11$, and graph the solution set.

____

below. $-\frac{14}{5}$

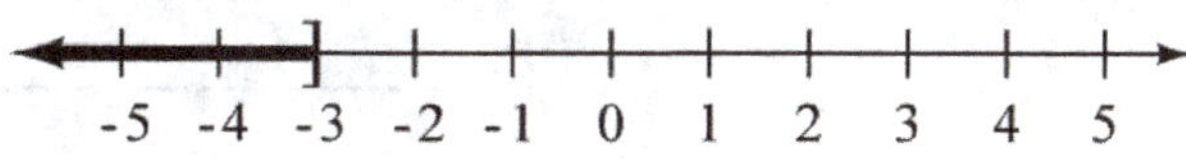

Step 2 Now find all the values of x that are less than $-\dfrac{14}{5}$ and also less than or equal to -3. The solution set is $\left(-\infty, -3\right]$.

Objective 3 Practice Exercises

For extra help, see Examples 2–4 on pages 117–118 of your text.

For each compound inequality, give the solution set in both interval and graph forms.

4. $1 - 2s \leq 7$ and $2s + 7 \geq 11$

4.   ______________________

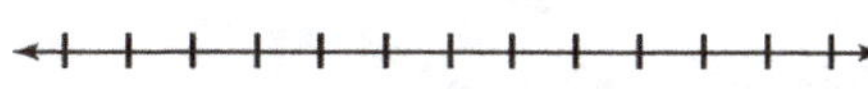

5. $3x + 2 < 11$ and $2 - 3x \leq 14$

5.   ______________________

6. $5t > 0$ and $5t + 4 \leq 9$

6.   ______________________

Objective 4   Find the union of two sets.

Review this example for Objective 4:

5.   Let $A = \{6, 7, 8, 9\}$ and $B = \{6, 8, 10\}$. Find $A \cup B$.

The set $A \cup B$ contains those elements that belong to either A or B (or both).

$$A \cup B = \{6, 7, 8, 9\} \cup \{6, 8, 10\}$$
$$= \{6, 7, 8, 9, 10\}$$

Now Try:

5.   Let $A = \{20, 30, 40, 50\}$ and $B = \{30, 50, 70\}$. Find $A \cup B$.

____

Objective 4 Practice Exercises

For extra help, see Example 5 on page 118 of your text.

Let A = {0, 1, 2, 3, 4, 5}, B = {2, 4, 6, 8, 10}, C = {1, 3, 5, 7, 9}, D = {0, 2, 4}, and E = {0}. Specify each set.

7. $A \cup D$ 7. ________________

8. $B \cup C$ 8. ________________

9. $A \cup E$ 9. ________________

Objective 5   Solve compound inequalities with the word *or*.

Review these examples for Objective 5:

6. Solve the compound inequality, and graph the solution set.

$$7x - 6 < 4x \quad \text{or} \quad -4x \leq -16$$

Step 1 Solve each inequality individually.
$$7x - 6 < 4x \quad \text{or} \quad -4x \leq -16$$
$$3x < 6$$
$$x < 2 \quad \text{or} \quad x \geq 4$$

The graphs of $x < 2$ and $x \geq 4$ are shown below.

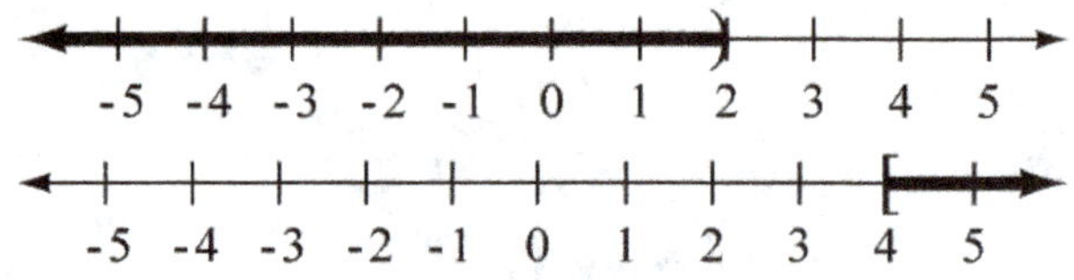

Step 2 Since the inequalities are joined with *or*, find the union of the two solution sets. The union is shown below. $(-\infty, 2) \cup [4, \infty)$

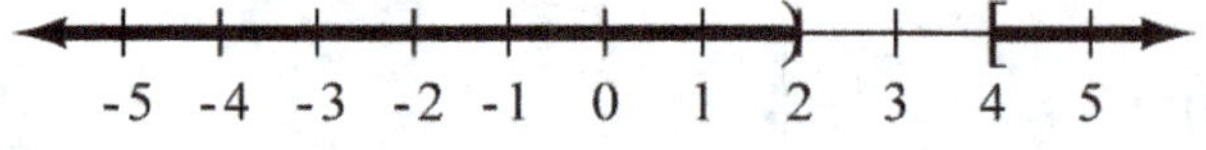

Now Try:

6. Solve the compound inequality, and graph the solution set.
$$5x - 6 < 2x \quad \text{or} \quad -7x \leq -35$$

____

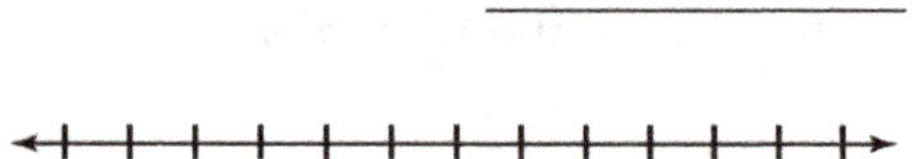

8. Solve the compound inequality, and graph the solution set.

$$-5x + 4 \geq -16 \quad \text{or} \quad 6x - 11 \geq -29$$

Solve each inequality.

$$-5x + 4 \geq -16 \quad \text{or} \quad 6x - 11 \geq -29$$
$$-5x \geq -20 \quad \text{or} \quad 6x \geq -18$$
$$x \leq 4 \quad \text{or} \quad x \geq -3$$

Graph each inequality.

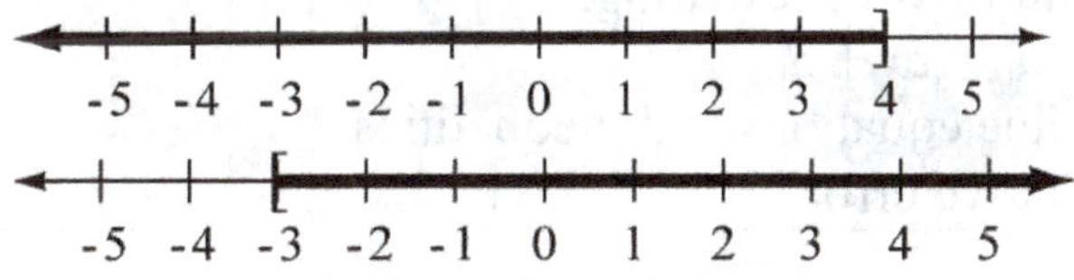

The solution is the union of the two sets,

$$(-\infty, \infty).$$

8. Solve the compound inequality, and graph the solution set.

$$-9x + 3 \geq -15 \text{ or } 8x - 7 \geq -15$$

____

Objective 5 Practice Exercises

For extra help, see Examples 6–9 on pages 119–121 of your text.

For each compound inequality, give the solution set in both interval and graph forms.

10. $q + 3 > 7$ or $q + 1 \leq -3$

10. __________

11. $3 > 4m + 2$ or $4m - 3 \geq -2$

11. __________

12. $2r + 4 \geq 8$ or $4r - 3 < 1$

12. __________

Chapter 1 LINEAR EQUATIONS, INEQUALITIES, AND APPLICATIONS

1.7   Absolute Value Equations and Inequalities

Learning Objectives	
1	Use the distance definition of absolute value.
2	Solve equations of the form $|ax + b| = k$, for $k > 0$.
3	Solve inequalities of the form $|ax + b| < k$ and of the form $|ax + b| > k$, for $k > 0$.
4	Solve absolute value equations that involve rewriting.
5	Solve equations of the form $|ax + b| = |cx + d|$.
6	Solve special cases of absolute value equations and inequalities.
7	Solve an application involving relative error.

Key Terms

Use the vocabulary terms listed below to complete each statement in exercises 1−2.

absolute value equation   **absolute value inequality**

1. An ____ is an equation that involves the absolute value of a variable expression.

2. An ____ is an inequality that involves the absolute value of a variable expression.

Objective 2   Solve equations of the form $|ax + b| = k$, for $k > 0$.

Review this example for Objective 2:

1. Solve $|3x + 2| = 14$. Graph the solution set.

    This is Case 1.
    $$3x + 2 = 14 \quad \text{or} \quad 3x + 2 = -14$$
    $$3x = 12 \quad \text{or} \quad 3x = -16$$
    $$x = 4 \quad \text{or} \quad x = -\frac{16}{3}$$

    Check

    $$\text{Let } x = 4 \qquad \text{Let } x = -\frac{16}{3}$$

    $$3(4) + 2 \overset{?}{=} 14 \quad \bigg| \quad 3\left(-\frac{16}{3}\right) + 2 \overset{?}{=} -14$$

    $$12 + 2 \overset{?}{=} 14 \quad \bigg| \quad -16 + 2 \overset{?}{=} -14$$

    $$14 = 14 \quad \bigg| \quad -14 = -14$$

    The check confirms that the solution set is
    $\left\{-\frac{16}{3}, 4\right\}$.

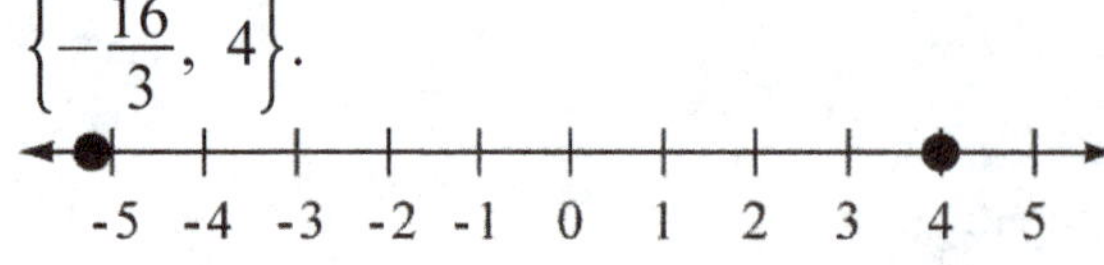

Now Try:

1. Solve $|5x + 4| = 11$. Graph the solution set.

    ____

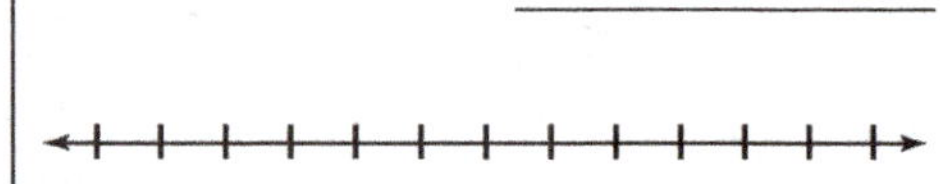

              Copyright © 2020 Pearson Education, Inc.

Objective 2 Practice Exercises

For extra help, see Example 1 on pages 126–127 of your text.

Solve each equation.

1. $|2x+3| = 10$

1. _________________

2. $|5r-15| = 0$

2. _________________

3. $\left|\dfrac{1}{2}x - 3\right| = 4$

3. _________________

Objective 3   Solve inequalities of the form $|ax + b| < k$ and of the form $|ax + b| > k$, for $k > 0$.

Review these examples for Objective 3:

2. Solve $|4x+2| > 10$. Graph the solution set.

This is Case 2.
$$4x+2 > 10 \quad \text{or} \quad 4x+2 < -10$$
$$4x > 8 \quad \text{or} \quad 4x < -12$$
$$x > 2 \quad \text{or} \quad x < -3$$
A check confirms that the solution set is
$(-\infty, -3) \cup (2, \infty)$.

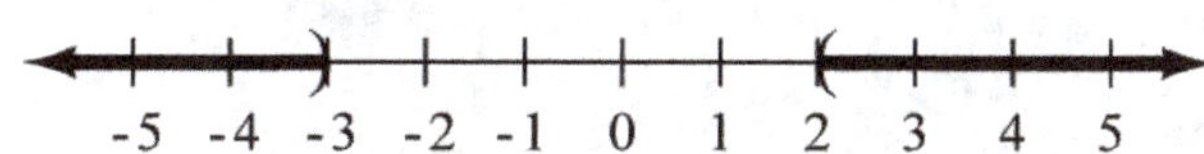

3. Solve $|10x+5| < 25$. Graph the solution set.

This is Case 3.
$$-25 < 10x+5 < 25$$
$$-30 < 10x < 20$$
$$-3 < x < 2$$
A check confirms that the solution set is $(-3, 2)$.

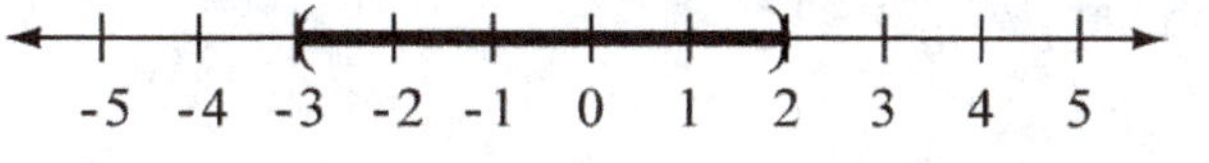

Now Try:

2. Solve $|8x+4| > 12$. Graph the solution set.

____

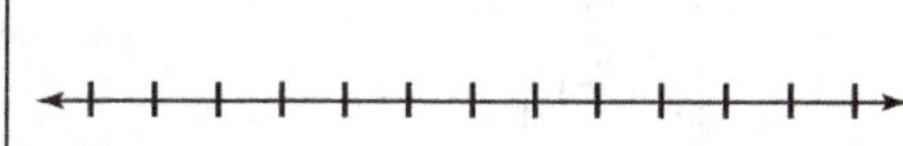

3. Solve $|6x+3| < 15$. Graph the solution set.

____

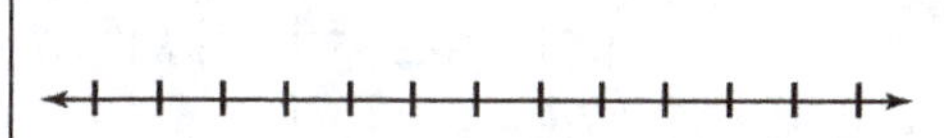

Name: Date:
Instructor: Section:

Objective 3 Practice Exercises

For extra help, see Examples 2–4 on pages 127–128 of your text.

Solve each inequality and graph the solution set.

4. $|x-2|>8$

4.   _________________________

                                                                                       ←+—+—+—+—+—+—+—+—+—+—+—+—+—+→

5. $|2r-9|\ge 23$

5.   _________________________

   ←+—+—+—+—+—+—+—+—+—+—+—+—+—+→

6. $|5r+2|<18$

6.   _________________________

   ←+—+—+—+—+—+—+—+—+—+—+—+—+—+→

Objective 4   Solve absolute value equations that involve rewriting.

Review this example for Objective 4:

Now Try:

5.   Solve $|x-7|+3=15$.

5.   Solve $|x-3|+4=11$.

Isolate the absolute value.

$$|x-7|+3=15$$
$$|x-7|+3-3=15-3$$
$$|x-7|=12$$

Now solve, using Case 1.

$$x-7=12 \quad \text{or} \quad x-7=-12$$
$$x=19 \quad \text{or} \quad x=-5$$

Check

    Let $x=19$ Let $x=-5$

$$|19-7|+3\overset{?}{=}15 \quad | \quad |-5-7|+3\overset{?}{=}15$$
$$|12|+3\overset{?}{=}15 \quad | \quad |-12|+3\overset{?}{=}15$$
$$15=15 \quad | \quad 15=15$$

The check confirms that the solution set
is {−5, 19}.

6. Solve each inequality.

   a. $|x-7|+2 \geq 15$

$$|x-7|+2 \geq 15$$
$$|x-7| \geq 13$$
$$x-7 \geq 13 \quad \text{or} \quad x-7 \leq -13$$
$$x \geq 20 \quad \text{or} \quad x \leq -6$$

The solution set is $(-\infty, -6] \cup [20, \infty)$.

   b. $|x-7|+2 \leq 15$

$$|x-7|+2 \leq 15$$
$$|x-7| \leq 13$$
$$-13 \leq x-7 \leq 13$$
$$-6 \leq x \leq 20$$

The solution set is [–6, 20].

6. Solve each inequality.

   a. $|x+9|-2 \geq 11$

____

   b. $|x+9|-2 \leq 11$

____

Objective 4 Practice Exercises

For extra help, see Examples 5–6 on page 129 of your text.

Solve each equation.

7. $\quad |2w-1|+7=12$

7. ______________

8. $\quad \left|2-\dfrac{1}{2}x\right|-5=18$

8. ______________

9. $\quad |4t+3|+8=10$

9. ______________

Objective 5   **Solve equations of the form $|ax + b| = |cx + d|$.**

Review this example for Objective 5:

7.   Solve $|z+5|=|3z-9|$.

$$z+5=3z-9 \quad \text{or} \quad z+5=-(3z-9)$$
$$z+14=3z \quad\quad \text{or} \quad z+5=-3z+9$$
$$14=2z \quad\quad \text{or} \quad\quad 4z=4$$
$$z=7 \quad\quad \text{or} \quad\quad z=1$$

Check

Let $z=7$ Let $z=1$

$$|7+5|\overset{?}{=}|3(7)-9| \quad |1+5|\overset{?}{=}|3(1)-9|$$
$$|12|\overset{?}{=}|21-9| \quad\quad |6|\overset{?}{=}|-6|$$
$$12=12 \quad\quad\quad\quad 6=6$$

The check confirms that the solution set
is $\{1, 7\}$.

Now Try:

7.   Solve $|z+7|=|2z+8|$.

____

Objective 5 Practice Exercises

For extra help, see Example 7 on page 130 of your text.

Solve each problem.

10. $|y+5|=|3y+1|$ **10.** ______________

11. $|2p-4|=|7-p|$ **11.** ______________

12. $|3x-2|=|5x+8|$ **12.** ______________

Objective 6   Solve special cases of absolute value equations and inequalities.

Review these examples for Objective 6:	**Now Try:**

Review these examples for Objective 6:

8. Solve each equation.

   a. $|9x-1|=-13$

The absolute value of an expression can never be negative, so there are no solutions for this equation. The solution set is $\varnothing$.

   b. $|8x-12|=0$

$$8x-12=0$$
$$8x=12$$
$$x=\frac{3}{2}$$

The solution set is $\left\{\frac{3}{2}\right\}$.

9. Solve each inequality.

   a. $|x|\geq-12$

The absolute value of a number is always greater than or equal to 0. Thus, $|x|\geq-12$ is always true, and the solution set is $(-\infty,\ \infty)$.

   b. $|x-11|+7<2$

$$|x-11|+7<2$$
$$|x-11|<-5$$

There is no number whose absolute value is less than -5, so this inequality has no solution. The solution set is $\varnothing$.

   c. $|2x-8|-5\leq-5$

$$|2x-8|-5\leq-5$$
$$|2x-8|\leq0$$

The value of $|2x-8|$ will never be less than zero. However, $|2x-8|$ will equal 0.

$$2x-8=0$$
$$2x=8$$
$$x=4$$

The solution set is $\{4\}$.

Now Try:

8. Solve each equation.

   a. $|6x+13|=-5$

   ____

   b. $|7x+35|=0$

   ____

9. Solve each inequality.

   a. $|x|\geq-2$

   ____

   b. $|x+12|+6<3$

   ____

   c. $|5x-20|+7\leq7$

   ____

Objective 6 Practice Exercises

For extra help, see Examples 8–9 on pages 130–131 of your text.

Solve each problem.

13. $\left|7+\dfrac{1}{2}x\right|=0$

13. ______________________

14. $\left|m-2\right|\geq-1$

14. ______________________

15. $\left|k+5\right|\leq-2$

15. ______________________

Objective 7   Solve an application involving relative error.

Review this example for Objective 7:

10. Suppose a machine filling 16.9 oz water bottles is set for a relative error that is no greater than 0.025 oz. How many ounces may a filled water bottle contain?

$$\left|\frac{16.9-x}{16.9}\right|\leq 0.025$$

$$-0.025\leq\frac{16.9-x}{16.9}\leq 0.025$$

$$-0.4225\leq 16.9-x\leq 0.4225$$

$$-17.3225\leq\ -x\ \leq-16.4775$$

$$17.3225\geq\ \ x\ \ \geq 16.4775$$

$$16.4775\leq\ \ x\ \ \leq 17.3225$$

The bottle may contain between 16.4775 and 17.3225 oz, inclusive.

Now Try:

10. Suppose a machine filling 16.9 oz water bottles is set for a relative error that is no greater than 0.05 oz. How many ounces may a filled water bottle contain?

____

          Copyright © 2020 Pearson Education, Inc.

Objective 7 Practice Exercises

For extra help, see Example 10 on page 131 of your text.

Determine the number of ounces a filled 16.9 oz water bottle may contain for the given relative error.

16.   no greater than 0.04 oz   **16.** _______________

17.   no greater than 0.015 oz   **17.** _______________

18.   no greater than 0.03 oz   **18.** _______________

Chapter 2 LINEAR EQUATIONS, GRAPHS, AND FUNCTIONS

2.1   Linear Equations in Two Variables

<table>
<tr><td colspan="2">Learning Objectives</td></tr>
<tr><td>1</td><td>Interpret a line graph.</td></tr>
<tr><td>2</td><td>Plot ordered pairs.</td></tr>
<tr><td>3</td><td>Find ordered pairs that satisfy a given equation.</td></tr>
<tr><td>4</td><td>Graph lines.</td></tr>
<tr><td>5</td><td>Find x- and y-intercepts.</td></tr>
<tr><td>6</td><td>Recognize equations of horizontal and vertical lines.</td></tr>
<tr><td>7</td><td>Find the midpoint of a line segment.</td></tr>
</table>

Key Terms

Use the vocabulary terms listed below to complete each statement in exercises 1−14.

> **ordered pair**   **origin**   ***x*-axis**   ***y*-axis**
>
> **rectangular (Cartesian) coordinate system**   **plot**
>
> **components**   **coordinate**   **quadrant**   **graph of an equation**
>
> **first-degree equation**   **linear equation in two variables**
>
> ***y*-intercept**   ***x*-intercept**

1. If a graph intersects the y-axis at k, then the ___________________ is $(0, k)$.

2. An equation that can be written in the form $Ax + By = C$, where A, B, and C are real numbers and $A, B \neq 0$, is called a ___________________________________.

3. Each number in an ordered pair represents a ___________________________ of the corresponding point.

4. The axis lines in a coordinate system intersect at the ___________________.

5. If a graph intersects the x-axis at k, then the ___________________ is $(k, 0)$.

6. In a rectangular coordinate system, the horizontal number line is called the

   ___________________________.

7. A ___________________________________ is one of the four regions in the plane determined by a rectangular coordinate system.

8. A pair of numbers written between parentheses in which order is important is called a(n) ___________________________.

9. In a rectangular coordinate system, the vertical number line is called the

   ___________________________.

10. The two numbers in an ordered pair are the _________________________ of the ordered pair.

11. To _________________ an ordered pair is to locate the corresponding point on a coordinate system.

12. Together, the x-axis and the y-axis form a _________________________________.

13. The _________________________________ is the set of points corresponding to all ordered pairs that satisfy the equation.

14. A _________________________________ has no term with a variable to a power greater than one.

Objective 3   Find ordered pairs that satisfy a given equation.

Review these examples for Objective 3:

1. In parts (a)–(d), complete each ordered pair for $5x + 4y = 20$. Then part (e), write the results as a table of ordered pairs.

a. $(0, \underline{\quad})$

$$5x + 4y = 20$$
$$5(0) + 4y = 20$$
$$4y = 20$$
$$y = 5$$

The ordered pair is $(0, 5)$.

b. $(\underline{\quad}, 0)$

$$5x + 4y = 20$$
$$5x + 4(0) = 20$$
$$5x = 20$$
$$x = 4$$

The ordered pair is $(4, 0)$.

c. $(-4, \underline{\quad})$

$$5x + 4y = 20$$
$$5(-4) + 4y = 20$$
$$-20 + 4y = 20$$
$$4y = 40$$
$$y = 10$$

The ordered pair is $(-4, 10)$.

Now Try:

1. In parts (a)–(d), complete each ordered pair for $2x - 5y = 10$. Then part (e), write the results as a table of ordered pairs.

a. $(0, \underline{\quad})$

____

b. $(\underline{\quad}, 0)$

____

c. $(-5, \underline{\quad})$

____

d. (____,–5)

$$5x + 4y = 20$$
$$5x + 4(-5) = 20$$
$$5x - 20 = 20$$
$$5x = 40$$
$$x = 8$$

The ordered pair is (8,–5).

e. Write the ordered pairs in a table.

x	y
0	5
4	0
–4	10
8	–5

d. (____, 2)

____

e. Write the ordered pairs in a table.

____

Objective 3 Practice Exercises

For extra help, see Example 1 on page 152 of your text.

For each of the given equations, complete the ordered pairs beneath it.

1. $5x + 4y = 10$

     (a)   (2,)

     (b)   (4,)

     (c)   (,3)

     (d)   (0,)

     (e)   (,2)

1.

(a) _______________

(b) _______________

(c) _______________

(d) _______________

(e) _______________

Complete each table of values. Write the results as ordered pairs.

2. $-7x + 2y = -14$

x	y
	0
0	
3	
	7

2. _______________

3. $y - 4 = 0$ **3.** ____________________

x	y
-4	
0	
6	
-12	

Objective 5   Find x- and y-intercepts.

Review these examples for Objective 5:

2.   Find the x- and y-intercepts of $3x + y = 6$ and graph the equation.

To find the y-intercept, let $x = 0$.
To find the x-intercept, let $y = 0$.

$$3(0) + y = 6 \quad \big| \quad 3x + 0 = 6$$
$$0 + y = 6 \quad \big| \quad 3x = 6$$
$$y = 6 \quad \big| \quad x = 2$$

The intercepts are $(0, 6)$ and $(2, 0)$. To find a third point, as a check, we let $x = 1$.

$$3(1) + y = 6$$
$$3 + y = 6$$
$$y = 3$$

This gives the ordered pair $(1, 3)$.

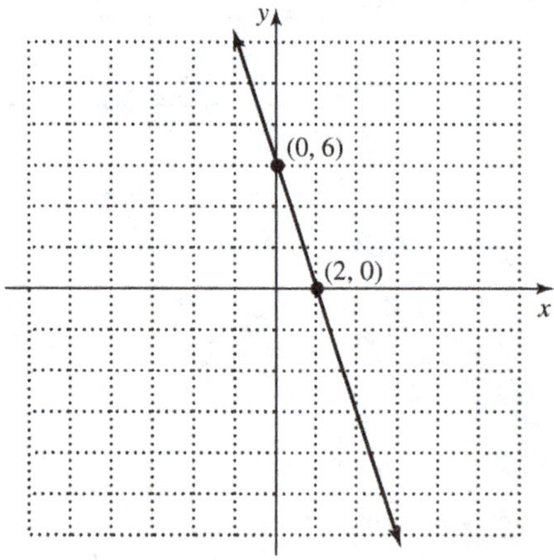

Now Try:

2.   Find the x- and y-intercepts of $5x - 2y = -10$ and graph the equation.

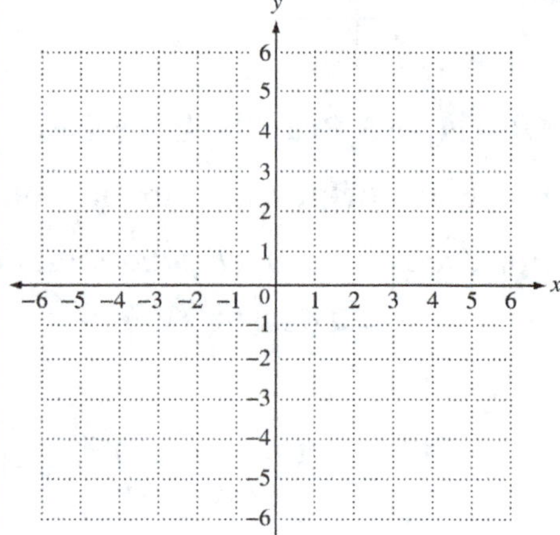

3. Graph $x + 5y = 0$.

To find the y-intercept, let $x = 0$.
To find the x-intercept, let $y = 0$.

$$0 + 5y = 0 \quad | \quad x + 5(0) = 0$$
$$5y = 0 \quad | \quad x + 0 = 0$$
$$y = 0 \quad | \quad x = 0$$

The x- and y-intercepts are the same point $(0, 0)$. We must select two other values for x or y to find two other points. We choose $y = 1$ and $y = -1$.

$$x + 5(1) = 0 \quad | \quad x + 5(-1) = 0$$
$$x + 5 = 0 \quad | \quad x - 5 = 0$$
$$x = -5 \quad | \quad x = 5$$

We use $(-5, 1)$, $(0, 0)$, and $(5, -1)$ to draw the graph.

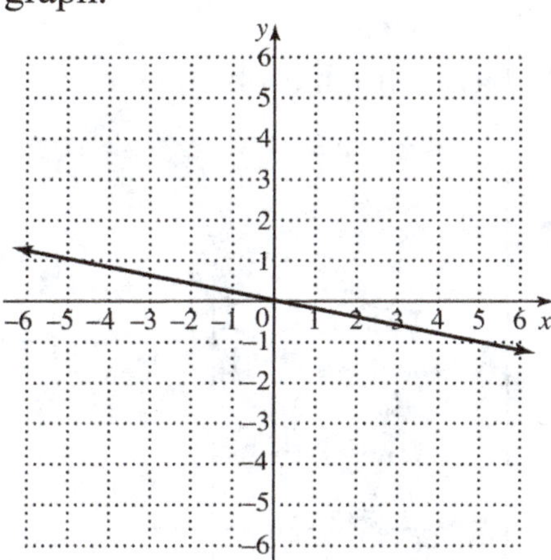

3. Graph $3x - y = 0$.

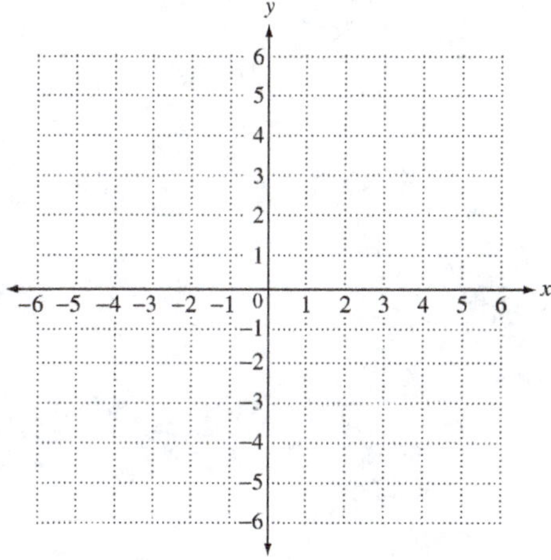

Objective 5 Practice Exercises

For extra help, see Examples 2–3 on pages 154–155 of your text.

Find the intercepts, then graph the equation.

4. $4x - y = 4$

4. _______________________

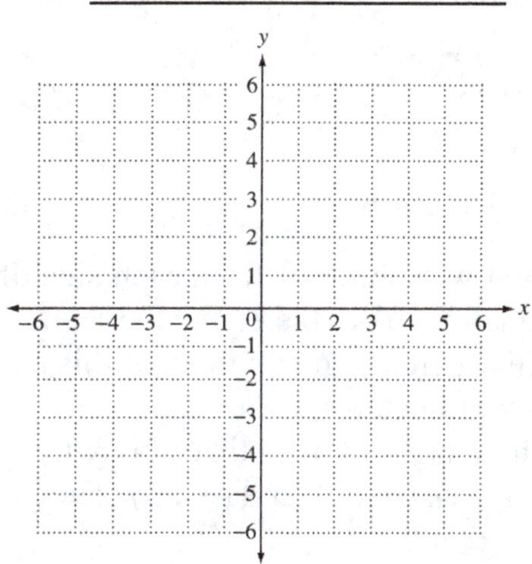

      89

5. $2x - 3y = 6$ **5.**   ______________________

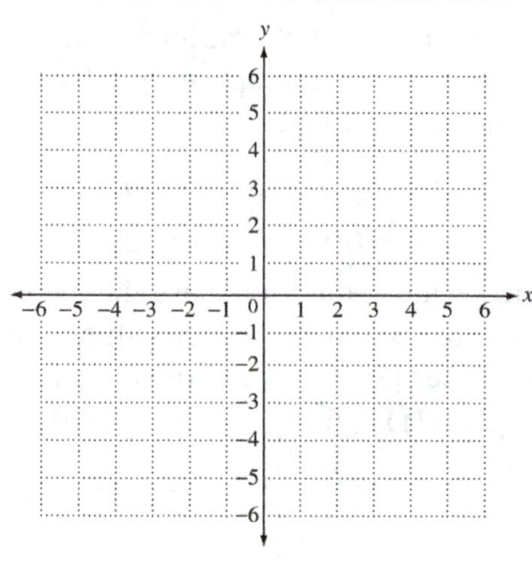

Objective 6   Recognize equations of horizontal and vertical lines.

Review these examples for Objective 6:

4.   Graph each equation.

   a. $y = -2$

For any value of x, y is always –2. Three
ordered pairs that satisfy the equation are
(–4, –2), (0, –2) and (2, –2). Drawing a line
through these points gives the horizontal line.
The y-intercept is (0, –2). There is no x-intercept.

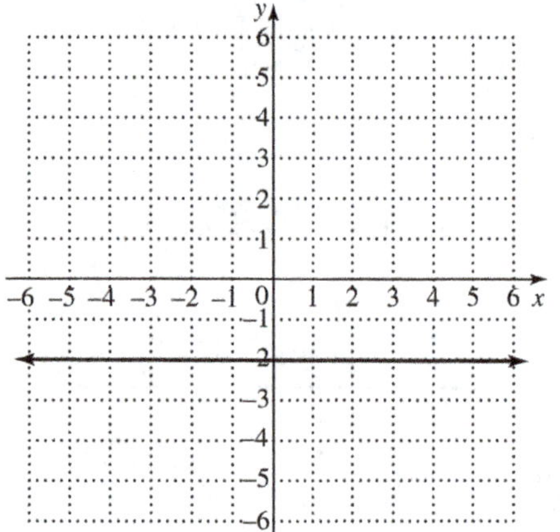

   b. $x + 4 = 0$

First we subtract 4 from each side of the
equation to get the equivalent equation $x = -4$.
All ordered-pair solutions of this equation have
x-coordinate –4.
Three ordered pairs that satisfy the equation are
(–4, –1), (–4, 0), and (–4, 3). The graph is a

Now Try:

4.   Graph each equation.

   a. $y = 4$

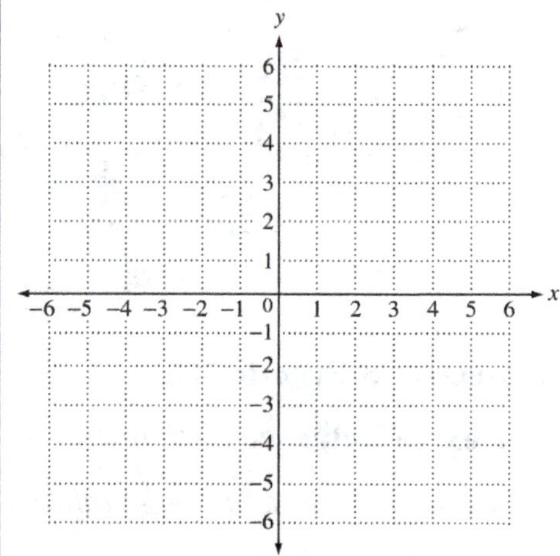

   b. $x = 0$

                    Copyright © 2020 Pearson Education, Inc.

vertical line. The *x*-intercept is (–4, 0). There is no *y*-intercept.

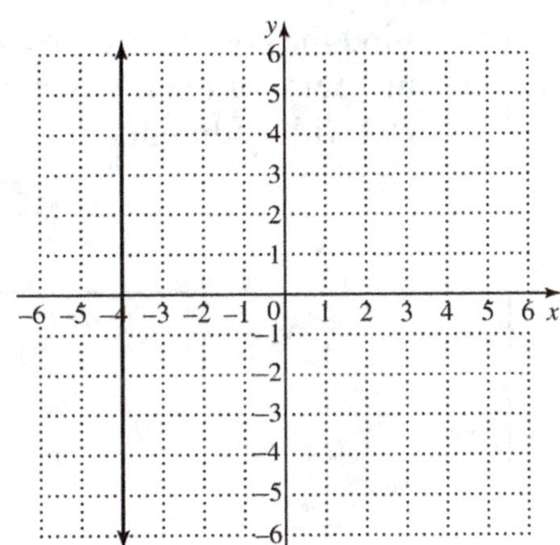

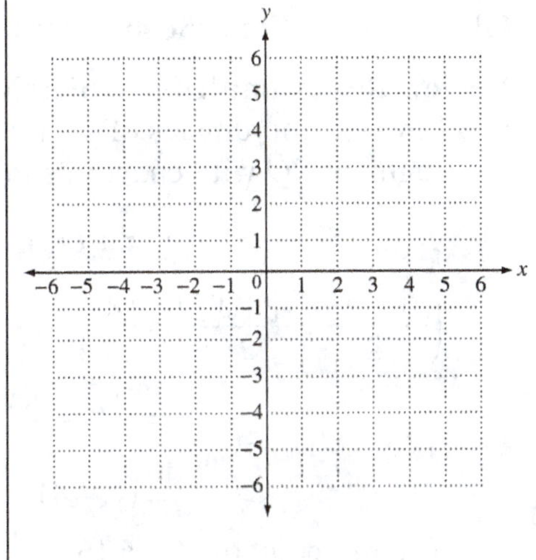

Objective 6 Practice Exercises

For extra help, see Example 4 on pages 155 of your text.

Find the intercepts, and graph the line.

6. $x - 1 = 0$

6.

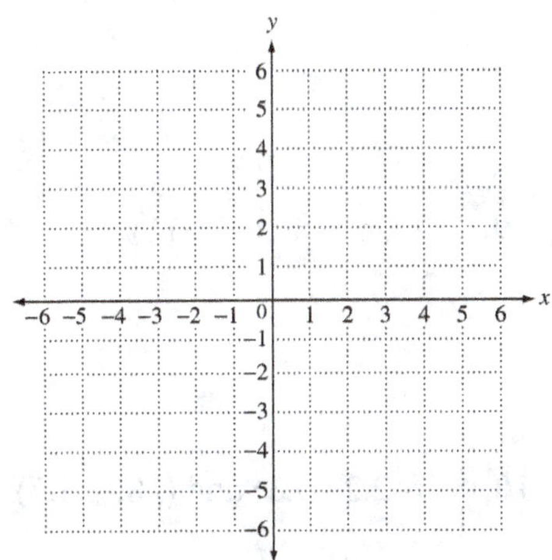

7. $y + 3 = 0$

7.

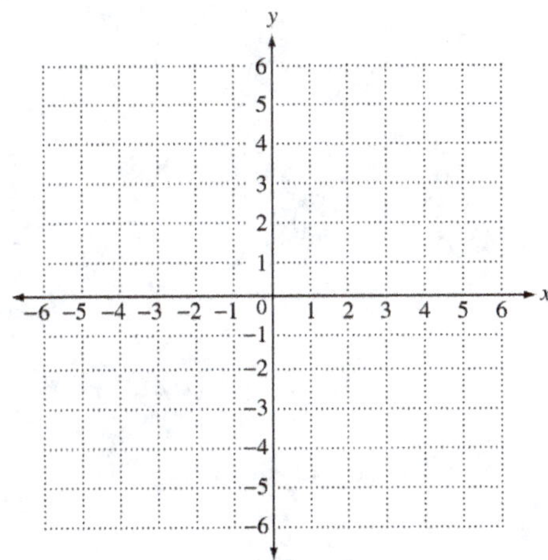

Objective 7   Find the midpoint of a line segment.

Review this example for Objective 7:

5. Find the coordinates of the midpoint of the line segment PQ with endpoints $P(8,-5)$ and $Q(4,-3)$.

$$P(8,-5)=(x_1,\ y_1) \text{ and } Q(4,-3)=(x_2,\ y_2)$$

$$\left(\frac{x_1+x_2}{2},\frac{y_1+y_2}{2}\right)=\left(\frac{8+4}{2},\frac{-5+(-3)}{2}\right)$$

$$=\left(\frac{12}{2},\frac{-8}{2}\right)$$

$$=(6,-4)$$

The midpoint of PQ is $(6,-4)$.

Now Try:

5. Find the coordinates of the midpoint of the line segment PQ with endpoints $P(7,-6)$ and $Q(3, 2)$.

____

Objective 7 Practice Exercises

For extra help, see Example 5 on page 157 of your text.

Find the midpoint of each segment with the given endpoints.

8. $(-4, 8)$ and $(8,-4)$

8. _______________

9. $(8, 5)$ and $(-3,-11)$

9. _______________

10. $(-2.2,-9.3)$ and $(-8.4, 5.7)$

10. _______________

Chapter 2 LINEAR EQUATIONS, GRAPHS, AND FUNCTIONS

2.2   The Slope of a Line

Learning Objectives
1   Find the slope of a line, given two points on the line.
2   Find the slope of a line, given an equation of the line.
3   Graph a line given its slope and a point on the line.
4   Determine whether two lines are parallel, perpendicular, or neither using slope.
5   Solve problems involving average rate of change.

Key Terms

Use the vocabulary terms listed below to complete each statement in exercises 1−3.

rise   **run**   **slope**

1.   The _________________ of a line is the ratio of the change in y compared to the change in x when moving along the line from one point to another.

2.   The vertical change between two different points on a line is called the _________________.

3.   The horizontal change between two different points on a line is called the _________________.

Objective 1   Find the slope of a line, given two points on the line.

Review this example for Objective 1:

1.   Find the slope of the line passing through $(-5, 4)$ and $(2, -6)$

Apply the slope formula.
$$(x_1, y_1) = (-5,\ 4) \text{ and } (x_2, y_2) = (2, -6)$$

$$\text{slope } m = \frac{y_2 - y_1}{x_2 - x_1} = \frac{-6 - 4}{2 - (-5)}$$

$$= \frac{-10}{7}, \text{ or } -\frac{10}{7}$$

Now Try:

1.   Find the slope of the line passing through $(-6, 7)$ and $(3, -9)$

____

Objective 1 Practice Exercises

For extra help, see Example 1 on page 162 of your text.

Find the slope of the line through the given points.

1. $(4, 3)$ and $(3, 5)$ 1. _________________

2. (5,–2) and (2, 7)

2. _______________

3. (7, 2) and (–7, 3)

3. _______________

Objective 2   Find the slope of a line, given an equation of the line.

Review these examples for Objective 2:

2. Find the slope of the line $5x + y = 10$.

The intercepts can be used as the two points needed to find the slope. Let $y = 0$ to find that the x-intercept is $(2, 0)$. Then let $x = 0$ to find that the y-intercept is $(0, 10)$.

$$m = \frac{y_2 - y_1}{x_2 - x_1} = \frac{10 - 0}{0 - 2} = \frac{10}{-2} = -5$$

3. Find the slope of each line.

a. $y = -2$

The graph of $y = -2$ is a horizontal line.

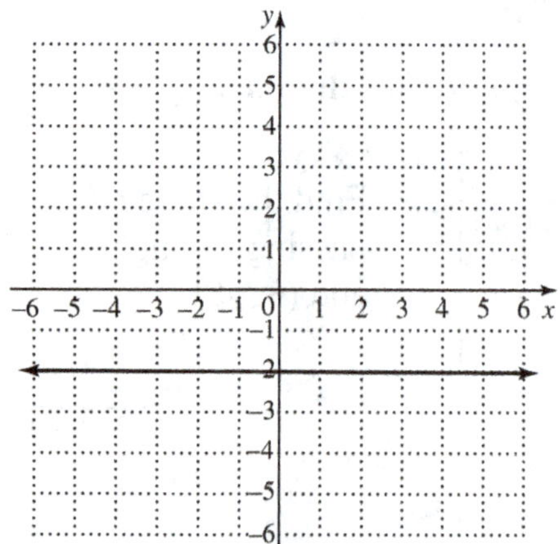

Select two different points on the line, such as $(-4,-2)$ and $(2,-2)$ and use the slope formula.

$$m = \frac{y_2 - y_1}{x_2 - x_1} = \frac{-2 - (-2)}{2 - (-4)} = \frac{0}{6} = 0$$

The slope is 0.

Now Try:

2. Find the slope of the line $7x - y = 6$.

____

3. Find the slope of each line.

a. $y = 4$

____

b. $x + 4 = 0$ **b.** $x = 0$

The graph of $x + 4 = 0$ is a vertical line.

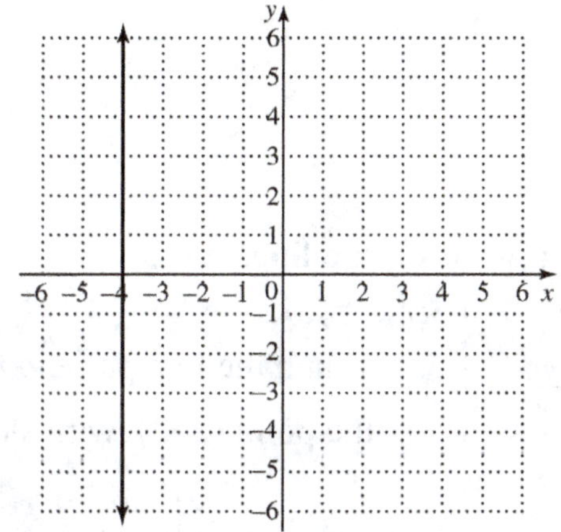

Select two different points on the line, such as $(-4, 3)$ and $(-4, -2)$ and use the slope formula.

$$m = \frac{y_2 - y_1}{x_2 - x_1} = \frac{-2 - 3}{-4 - (-4)} = \frac{-5}{0} \quad \text{undefined}$$

The slope is undefined.

4. Find the slope of the graph of $4x - 3y = 7$.

Solve the equation for y.
$$4x - 3y = 7$$
$$-3y = -4x + 7$$
$$y = \frac{4}{3}x - \frac{7}{3}$$

The slope is given by the coefficient of x, so the slope is $\frac{4}{3}$.

4. Find the slope of the graph of $7x - 4y = 8$.

____

Objective 2 Practice Exercises

For extra help, see Examples 2–4 on pages 163–164 of your text.

Find the slope of each line.

4. $x = 0$ 4. ___________________

5. $3y = 2x - 1$ 5. ___________________

6. $2x + 7y = 7$ **6.** _________________

Objective 3   Graph a line given its slope and a point on the line.

Review this example for Objective 3:

5.   Graph the line passing through the point $(1,-3)$, with slope $-\frac{5}{2}$.

First, locate the point $(1, -3)$. Then write the slope $-\frac{5}{2}$ as

$$\text{slope } m = \frac{\text{change in } y \text{ (rise)}}{\text{change in } x \text{ (run)}} = \frac{5}{-2}.$$

Locate another point on the line by counting up 5 units from $(1,-3)$, and then to the left 2 units. Finally, draw the line through this new point, $(-1, 2)$.

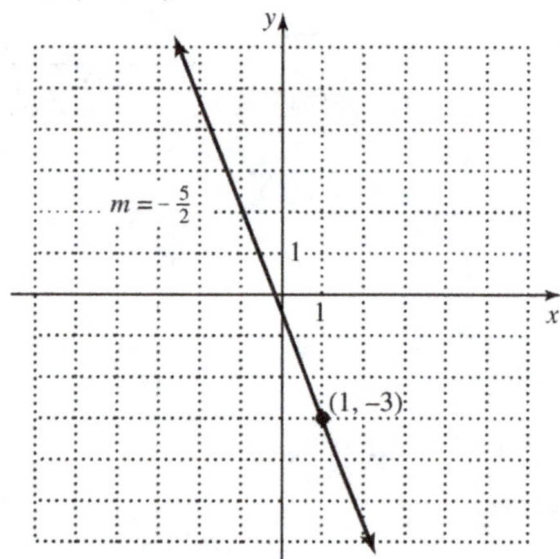

Now Try:

5.   Graph the line passing through the point $(2, 2)$, with slope $\frac{1}{3}$.

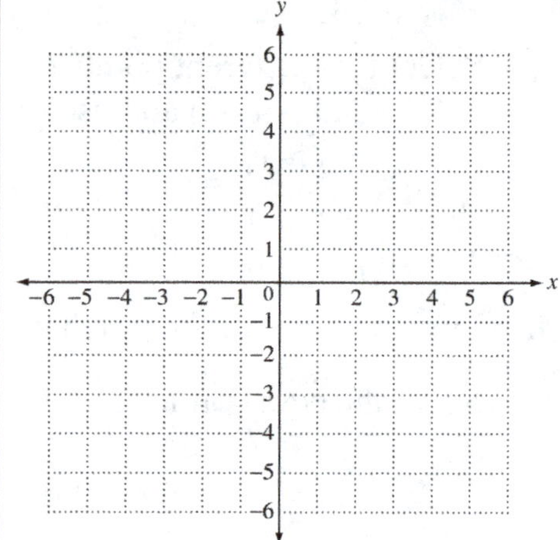

                   Copyright © 2020 Pearson Education, Inc.

Objective 3 Practice Exercises

For extra help, see Example 5 on page 165 of your text.

Graph the line passing through the given point and having the given slope.

7. $(4,-2);\ m=-1$

7.

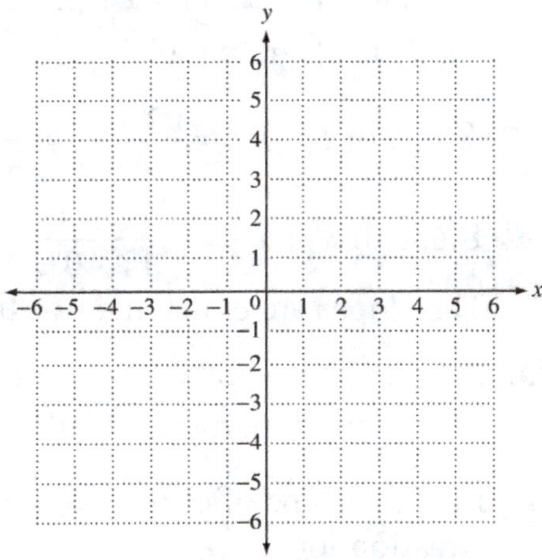

8. $(-3,-2);\ m=\frac{2}{3}$

8.

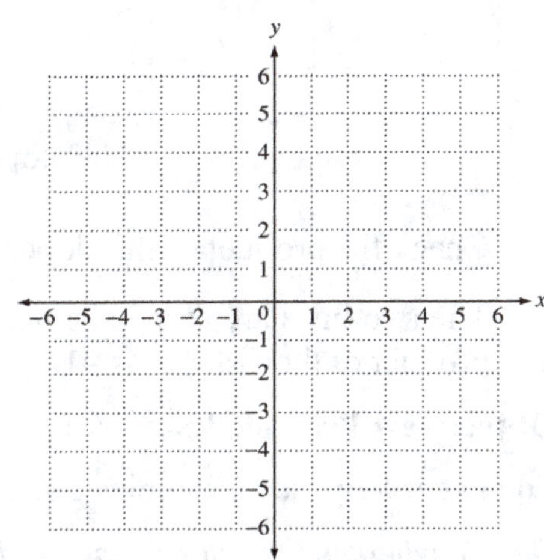

9. $(-3,-1);$ undefined slope

9.

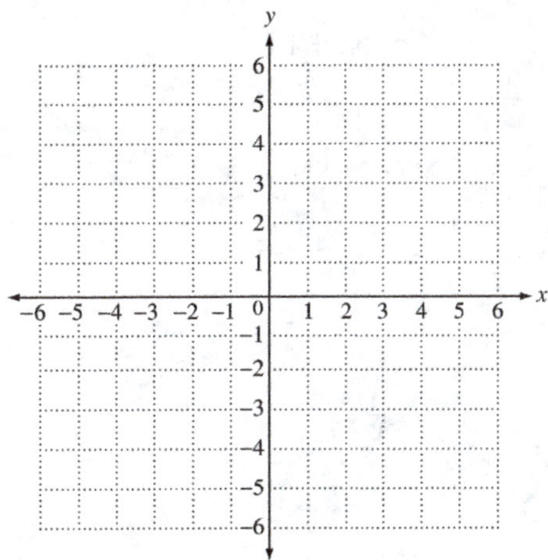

Objective 4   Determine whether two lines are parallel, perpendicular, or neither using slope.

Review these examples for Objective 4:

6. Determine whether the lines L_1 passing through (7, 6) and (–4, 2) and L_2 passing through (0,–4) and (11, 0), are parallel.

$$\text{Slope of } L_1 : \ m_1 = \frac{2-6}{-4-7} = \frac{-4}{-11} = \frac{4}{11}$$

$$\text{Slope of } L_2 : \ m_2 = \frac{0-(-4)}{11-0} = \frac{4}{11}$$

The slopes are equal, the two lines are parallel.

7a. Are the lines with equations $x + 3y = 8$ and $-3x + y = 5$ perpendicular?

Find the slope of each line by first solving each equation for y.

$$3y = -x + 8 \qquad\qquad y = 3x + 5$$

$$y = -\frac{1}{3}x + \frac{8}{3}$$

The slope is $-\dfrac{1}{3}$.   The slope is 3.

Check the product of the slopes: $-\dfrac{1}{3}(3) = -1$.

The two lines are perpendicular because the product of their slopes is –1.

Now Try:

6. Determine whether the lines L_1 passing through (6, 3) and (–4, 5) and L_2 passing through (0, 3) and (15, 0), are parallel.

____

7a. Are the lines with equations $9x - y = 7$ and $x + 9y = 11$ perpendicular?

____

Objective 4 Practice Exercises

For extra help, see Examples 6–7 on pages 166–168 of your text.

Decide whether the lines in each pair are **parallel***,* **perpendicular***, or* **neither***.*

10. $y = -5x - 2$

        $y = 5x + 11$

10. _______________

11. $-x + y = -7$

        $x - y = -3$

11. _______________

12. $2x + 2y = 7$ 12. ______________

$2x - 2y = 5$

Objective 5   Solve problems involving average rate of change.

Review these examples for Objective 5:

8. A small company had the following sales during their first three years of operation.

Year	Sales
2005	$82,250
2006	$89,790
2007	$96,100

a. What was the rate of change in 2005–2006?
b. What was the rate of change in 2006–2007?
c. What was the rate of change in 2005–2007?

a. We use the ordered pairs (2005, 82,250) and (2006, 89,790).

$$\text{average rate of change} = \frac{89,790 - 82,250}{2006 - 2005}$$

$$= \frac{7540}{1} = 7540$$

This means sales increased by an average of $7540 from 2005 to 2006.

b. We use the ordered pairs (2006, 89,790) and (2007, 96,100).

$$\text{average rate of change} = \frac{96,100 - 89,790}{2007 - 2006}$$

$$= \frac{6310}{1} = 6310$$

This means sales increased by an average of $6310 from 2006 to 2007.

c. We use the ordered pairs (2005, 82,250) and (2007, 96,100).

$$\text{average rate of change} = \frac{96,100 - 82,250}{2007 - 2005}$$

$$= \frac{13,850}{2} = 6925$$

This means sales increased by an average of $6925 from 2005 to 2007.

Now Try:

8. A plane had an altitude of 8500 feet at 4:02 P.M. and 12,700 feet at 4:39 P.M. What was the average rate of change in the altitude in feet per minute?

____

10. Enrollment in a college was 11,500 two years ago, 10,975 last year, and 10,800 this year. What is the average rate of change in enrollment per year for this 3-year period?

We use the ordered pairs (1, 11,500) and (3, 10,800).

$$\text{average rate of change} = \frac{10,800 - 11,500}{3 - 1}$$

$$= \frac{-700}{2}$$

$$= -350$$

The enrollment decreases at a rate of 350 students per year.

9. A company had 44 employees during the first year of operation. During their eighth year, the company had 79 employees. What was the average rate of change in the number of employees per year?

____

Objective 5 Practice Exercises

For extra help, see Examples 8–9 on pages 169 of your text.

Solve each problem.

13. Suppose in 2005, the sales of a company were $1,625,000. In 2010, the company had sales of $2,250,000. Find the average rate of change in the sales per year.

13. ______________

14. A state had a population of 755,000 in 2000 and a population of 809,000 in 2012. Find the average rate of change in population per year.

14. ______________

15. Suppose a man's salary was $45,750 in 1995 and $60,000 in 2010. Find the average rate of change in the salary per year.

15. ______________

Chapter 2 LINEAR EQUATIONS, GRAPHS, AND FUNCTIONS

2.3   Writing Equations of Lines

Learning Objectives
1. Write an equation of a line given its slope and y-intercept.
2. Graph a line using its slope and y-intercept.
3. Write an equation of a line given its slope and a point on the line.
4. Write an equation of a line given two points on the line.
5. Write equations of horizontal and vertical lines.
6. Write an equation of a line parallel or perpendicular to a given line.
7. Write an equation of a line that models real data.

Key Terms

Use the vocabulary terms listed below to complete each statement in exercises 1−3.

    slope-intercept form   **point-slope form**   **standard form**

1. A linear equation in the form $y - y_1 = m(x - x_1)$ is written in

   _______________________________.

2. A linear equation in the form $Ax + By = C$ is written in

   _______________________________.

3. A linear equation in the form $y = mx + b$ is written in

   _______________________________.

Objective 1   Write an equation of a line given its slope and y-intercept.

Review this example for Objective 1:

1. Write an equation of the line with slope $\frac{5}{7}$ and y-intercept $(0, -6)$.

   Here, $m = \frac{5}{7}$ and $b = -6$, so we can write the following equation.
   $$y = mx + b$$
   $$y = \frac{5}{7}x + (-6), \text{ or } y = \frac{5}{7}x - 6$$

Now Try:

1. Write an equation of the line with slope $\frac{7}{9}$ and y-intercept $(0, 8)$.

   ____

Objective 1 Practice Exercises

For extra help, see Example 1 on page 176 of your text.

Write the slope-intercept form equation of the line with the given slope and y-intercept.

1. $m = \dfrac{3}{2};\ b = -\dfrac{2}{3}$

1. _________________

2. $m = -7;\ b = -2$

2. _________________

3. Slope: $-\dfrac{6}{5}$; y-intercept $\left(0,\ \dfrac{2}{5}\right)$

3. _________________

Objective 2   Graph a line using its slope and y-intercept.

Review this example for Objective 2:

2. Graph the equation by using the slope and y-intercept.

$$2x - 3y = 6$$

Solve for y to write the equation in slope-intercept form.

$$2x - 3y = 6$$
$$-3y = -2x + 6$$
$$y = \frac{2}{3}x - 2$$

The y-intercept is $(0, -2)$. Graph this point.

The slope is $\dfrac{2}{3}$. By definition,

$$\text{slope } m = \frac{\text{change in } y \text{ (rise)}}{\text{change in } x \text{ (run)}} = \frac{2}{3}$$

From the y-intercept, count up 2 units and to the right 3 units to obtain the point $(3, 0)$.

Draw the line through the points $(0, -2)$ and $(3, 0)$ to obtain the graph.

Now Try:

2. Graph the equation by using the slope and y-intercept.

$$2x - 3y = 0$$

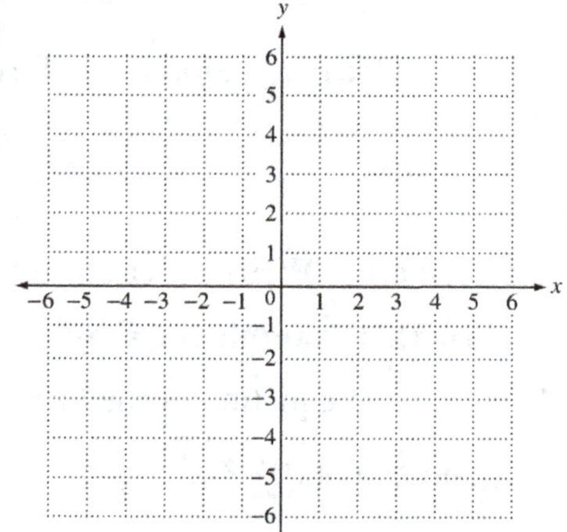

     Copyright © 2020 Pearson Education, Inc.

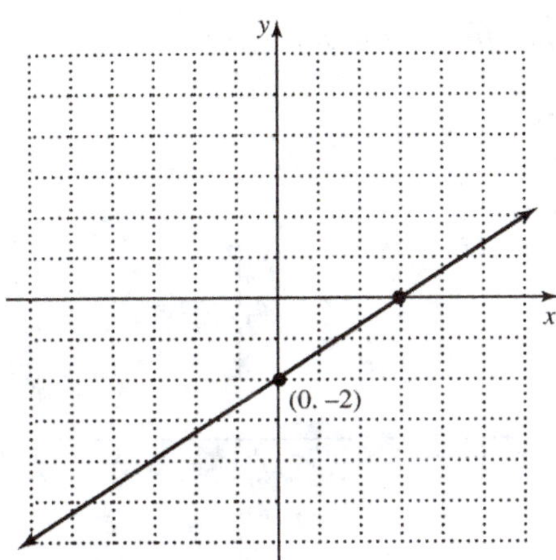

Objective 2 Practice Exercises

For extra help, see Example 2 on page 177 of your text.

Graph each equation by using the slope and y-intercept.

4. $4x - y = 4$

4.

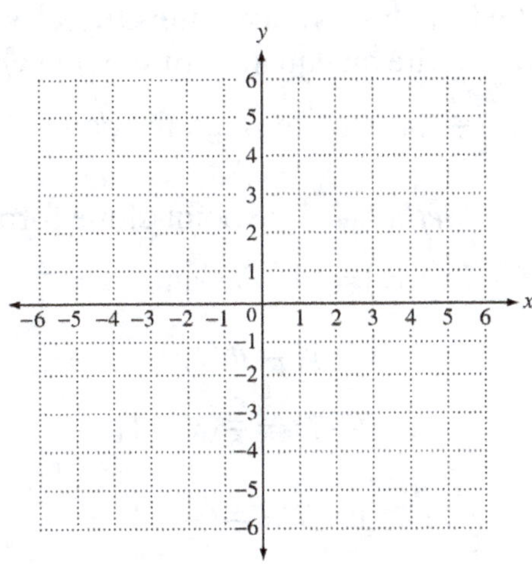

5. $y = -3x + 6$

5.

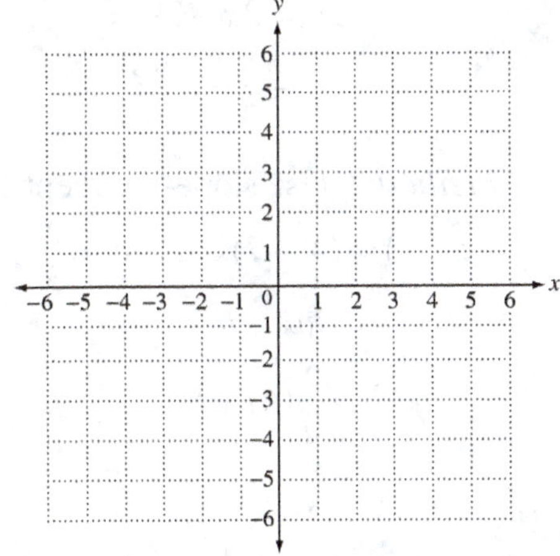

                    103

Graph the line passing through the given point and having the given slope.

6. $(-2,-2)$; $m = 0$

6.

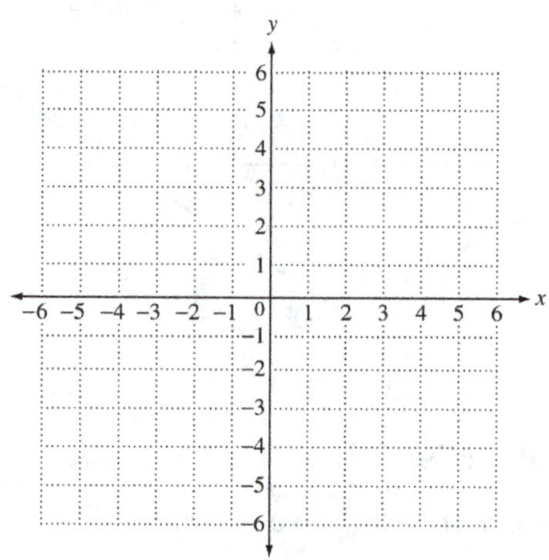

Objective 3   Write an equation of a line given its slope and a point on the line.

Review this example for Objective 3:

3.   Write an equation of the line with slope $\frac{2}{3}$ passing through the point $(4,-7)$.

Method 1 Use point-slope form, with $(x_1, y_1) = (4,-7)$ and $m = \frac{2}{3}$.

$$y - y_1 = m(x - x_1)$$

$$y - (-7) = \frac{2}{3}(x - 4)$$

$$y + 7 = \frac{2}{3}(x - 4)$$

$$3y + 21 = 2x - 8$$

$$3y = 2x - 29$$

$$y = \frac{2}{3}x - \frac{29}{3}$$

Method 2 Use slope-intercept form, with $(x_1, y_1) = (4,-7)$ and $m = \frac{2}{3}$.

$$y = mx + b$$

$$-7 = \frac{2}{3}(4) + b$$

$$-7 = \frac{8}{3} + b$$

$$-\frac{29}{3} = b, \text{ or } b = -\frac{29}{3}$$

Now Try:

3.   Write an equation of the line with slope $\frac{4}{5}$ passing through the point $(6,-4)$.

____

Knowing $m = \frac{2}{3}$ and $b = -\frac{29}{3}$ gives the

equation $y = \frac{2}{3}x - \frac{29}{3}$, same as Method 1.

Objective 3 Practice Exercises

For extra help, see Example 3 on page 178 of your text.

Write the equation in standard form of the line satisfying the given conditions.

7. $(-3, 4)$; $m = -\frac{3}{5}$ 7. _______________

8. $(-4, -7)$; $m = \frac{4}{3}$ 8. _______________

9. $(-1, 2)$; $m = \frac{2}{3}$ 9. _______________

Objective 4   Write an equation of a line given two points on the line.

Review this example for Objective 4:

4.   Write the equation of the line passing through the point (6, 8) and (–3, 5). Give the final answer in slope-intercept form and then in standard form.

First, find the slope of the line.

$(x_1, y_1) = (6,\ 8)$ and $(x_2, y_2) = (-3,\ 5)$

$$\text{slope } m = \frac{y_2 - y_1}{x_2 - x_1} = \frac{5 - 8}{-3 - 6} = \frac{-3}{-9} = \frac{1}{3}$$

Now use (x_1, y_1), here (6, 8) and point-slope

Now Try:

4.   Write the equation of the line passing through the point (7, 15) and (15, 9). Give the final answer in slope-intercept form and then in standard form.

____

form.

$$y - y_1 = m(x - x_1)$$

$$y - 8 = \frac{1}{3}(x - 6)$$

$$y - 8 = \frac{1}{3}x - 2$$

$$y = \frac{1}{3}x + 6 \quad \text{Slope-intercept form}$$

$$3y = x + 18$$

$$-x + 3y = 18$$

$$x - 3y = -18 \qquad \text{Standard form}$$

Objective 4 Practice Exercises

For extra help, see Example 4 on page 179 of your text.

Write the equation in standard form of the line through the given points.

10. (3, 7), (5, 4)   10. _______________

11. (2, –1), (5, –2)   11. _______________

12. (–1, –4), (–2, –3)   12. _______________

Objective 5   Write equations of horizontal and vertical lines.

Review these examples for Objective 5:

5. Write an equation of the line passing through the point (2, –2) that satisfies the given condition.

a. The line has slope 0.

Since the slope is 0, this is a horizontal line.
$$y = -2.$$

b. The line has undefined slope.

This is a vertical line, since the slope is undefined.
$$x = 2$$

Now Try:

5. Write an equation of the line passing through the point (–5, 5) that satisfies the given condition.

a. The line has slope 0.

____

b. The line has undefined slope.

____

Objective 5 Practice Exercises

For extra help, see Example 5 on page 180 of your text.

Write the equation in standard form of the line through the given points.

13. $(-1, -7)$, $(-1, 8)$ **13.** ______________________

14. $(0, 2)$, $(0, -6)$ **14.** ______________________

15. $(4, -5)$, $(8, -5)$ **15.** ______________________

Objective 6   **Write an equation of a line parallel or perpendicular to a given line.**

Review these examples for Objective 6:	**Now Try:**
6. Write an equation in slope-intercept form of the line passing through the point $(-4, 5)$ that satisfies the given condition.	**6.** Write an equation in slope-intercept form of the line passing through the point $(-6, 8)$ that satisfies the given condition.
a. The line is parallel to $5x + 2y = 10$.	**a.** The line is parallel to $3x + 4y = 12$.

First, find the slope of the given line.
$$5x + 2y = 10$$
$$2y = -5x + 10$$
$$y = -\frac{5}{2}x + 5$$

The slope is $-\frac{5}{2}$.

Use point-slope form with $(x_1, y_1) = (-4, 5)$ and $m = -\frac{5}{2}$.

____

         107

$$y - y_1 = m(x - x_1)$$
$$y - 5 = -\frac{5}{2}(x - (-4))$$
$$y - 5 = -\frac{5}{2}(x + 4)$$
$$y - 5 = -\frac{5}{2}x - 10$$
$$y = -\frac{5}{2}x - 5$$

b. The line is perpendicular to $5x + 2y = 10$.

From part (a), the line in slope-intercept form is
$y = -\frac{5}{2}x + 5.$

The line perpendicular to this line must have

slope $\frac{2}{5}$, the negative reciprocal of $-\frac{5}{2}$.

Use point-slope form with $(x_1, y_1) = (-4,\ 5)$ and

$m = \frac{2}{5}.$

$$y - y_1 = m(x - x_1)$$
$$y - 5 = \frac{2}{5}(x - (-4))$$
$$y - 5 = \frac{2}{5}(x + 4)$$
$$y - 5 = \frac{2}{5}x + \frac{8}{5}$$
$$y = \frac{2}{5}x + \frac{33}{5}$$

b. The line is perpendicular to $3x + 4y = 12$.

____

Objective 6 Practice Exercises

For extra help, see Example 6 on pages 180–181 of your text.

Write the equation in standard form of the line satisfying the given conditions.

16.   parallel to $2x + 3y = -12$, through $(9, -3)$

16. _______________

17.   parallel to $4x - 3y = 8$, through $(-2, 3)$.

17. _______________

18.   perpendicular to $x - 3y = 0$, through $(-10, 2)$

18. _______________

Objective 7 Write an equation of a line that models real data.

Review these examples for Objective 7:

8. The table and scatter graph shows the number of internet users in the world from 1998 to 2005, where year 0 represents 1998.

Year	Number of Internet Users (millions)
0	147
2	361
4	587
6	817
8	1093

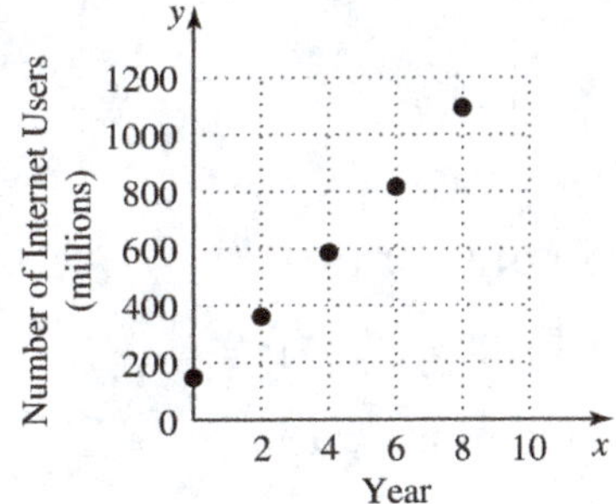

a. Find an equation that models the data.

The points appear to lie approximately in a straight line. y represents the number of internet users in year x. To find an equation of the line, we choose the ordered pairs (0, 147) and (8, 1093) from the table and find the slope of the line through these points.

$$(x_1, y_1) = (0,\ 147) \text{ and } (x_2, y_2) = (8,\ 1093)$$

$$\text{slope } m = \frac{y_2 - y_1}{x_2 - x_1} = \frac{1093 - 147}{8 - 0} = \frac{946}{8}$$

$$= 118.25$$

Now Try:

8. The table and scatter graph shows the average annual telephone expenditures for residential and pay telephones from 2001 to 2006, where year 0 represents 2001.

Year	Annual Telephone Expenditures
0	$686
2	$620
3	$592
4	$570
5	$542

a. Find an equation that models the data.

____

Use the slope, 118.25, and the point (0, 147) in slope-intercept form.

$$y = mx + b$$

$$147 = 118.25(0) + b$$

$$147 = b$$

Thus, $m = 118.25$ and $b = 147$, so the equation of the line is $y = 118.25x + 147$.

b. Find and interpret the ordered pair associated with the equation for $x = 5$.

If $x = 5$, then

$$y = 118.25(5) + 147$$

$$= 591.25 + 147$$

$$= 738.25$$

In 2003, there were 738.25 million internet users.

b. Find the ordered pair associated with the equation for $x = 1$.

____

Objective 7 Practice Exercises

For extra help, see Examples 7–9 on pages 182–185 of your text.

Solve each problem.

19. To run a newspaper ad, there is a $25 set up fee plus a charge of $1.25 per line of type in the ad. Let x represent the number of lines in the ad so that y represents the total cost of the ad (in dollars).
a. Write an equation in the form $y = mx + b$.
b. Give three ordered pairs associated with the equation for x-values 0, 5, and 10.

19.

a. _________________

b. _________________

20. The table and scatter graph shows the U.S. municipal solid waste recycling percent since 1985, where year 0 represents 1985.

a. Find an equation that models the data.

b. Use the equation from part (a) to predict the percent of municipal solid waste recycling in the year 2015.

20.

a. _______________

b. _______________

Year	Recycling Percent
0	10.1
5	16.2
10	26.0
15	29.1
20	32.5

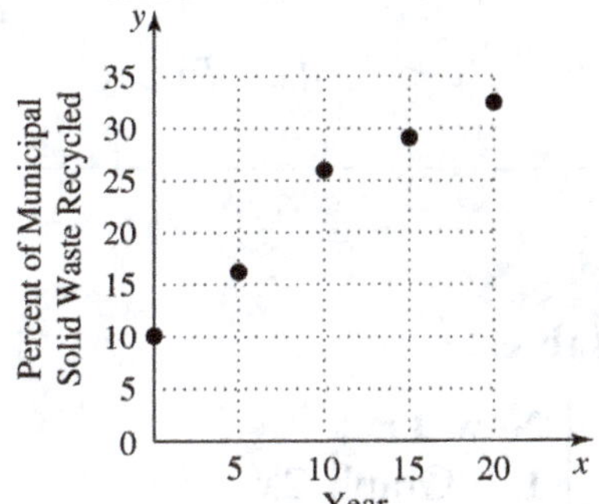

Chapter 2 LINEAR EQUATIONS, GRAPHS, AND FUNCTIONS

2.4   Linear Inequalities in Two Variables

Learning Objectives
1   Graph linear inequalities in two variables.
2   Graph the intersection of two linear inequalities.
3   Graph the union of two linear inequalities.

Key Terms

Use the vocabulary terms listed below to complete each statement in exercises 1–2.

linear inequality in two variables   **boundary line**

1.   In the graph of a linear inequality, the ________________________________
separates the region that satisfies the inequality from the region that does not
satisfy the inequality.

2.   An inequality that can be written in the form $Ax + By < C$, $Ax + By > C$,
$Ax + By \leq C$, or $Ax + By \geq C$ is called a________________________________.

Objective 1   Graph linear inequalities in two variables.

Review these examples for Objective 1:

1.   Graph $3x - 2y \leq 6$.

The inequality $3x - 2y \leq 6$ means that
     $3x - 2y < 6$ or $3x - 2y = 6$.
We begin by graphing the line $3x - 2y = 6$ with
intercepts $(0, -3)$ and $(2, 0)$. This boundary line
divides the plane into two regions, one of which
satisfies the inequality. We use the test point
$(0, 0)$ to see whether the resulting statement is
true or false, thereby determining whether the
point is in the shaded region or not.

$$3x - 2y \leq 6$$
$$3(0) - 2(0) \overset{?}{\leq} 6$$
$$0 - 0 \overset{?}{\leq} 6$$
$$0 \leq 6 \quad \text{True}$$

Since the last statement is true, we shade the
region that includes the test point $(0, 0)$. The
shaded region, along with the boundary line, is
the desired graph.

Now Try:

1.   Graph $2x + 5y \leq -8$.

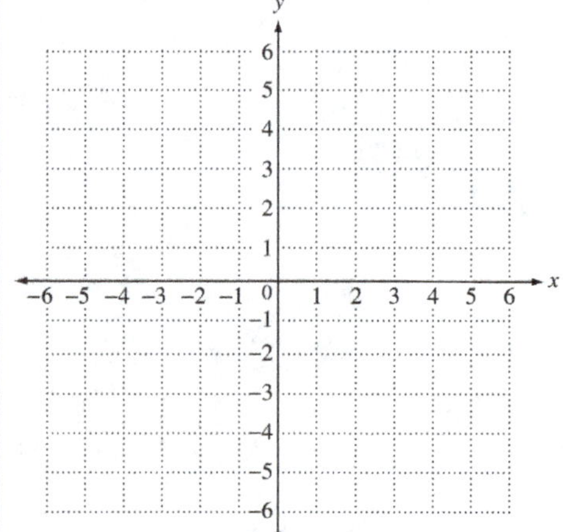

        Copyright © 2020 Pearson Education, Inc.

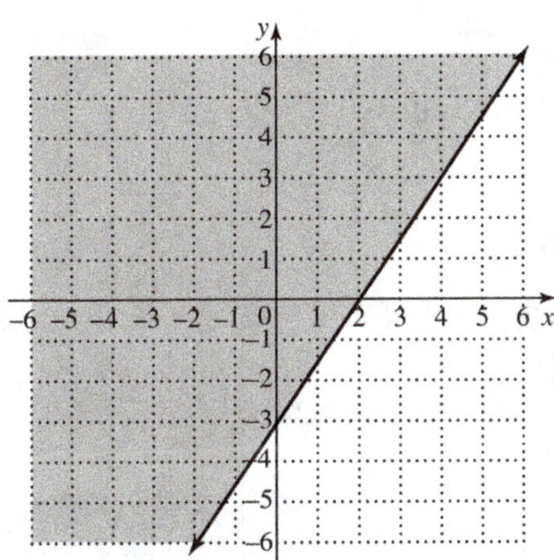

2. Graph $y \geq 3x$.

We graph $y = 3x$ using a solid line through $(0, 0)$, $(1, 3)$ and $(2, 6)$. Because $(0, 0)$ is on the line $y \geq 3x$, it cannot be used as a test point. Instead we choose a test point off the line, say $(3, 0)$.

$$0 \overset{?}{\geq} 3(3)$$

$$0 \geq 9 \quad \text{False}$$

Because $0 \geq 9$ is false, shade the other region.

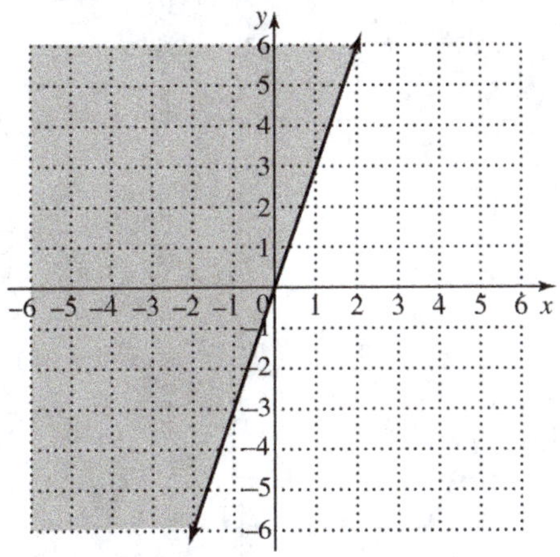

2. Graph $y \geq x$.

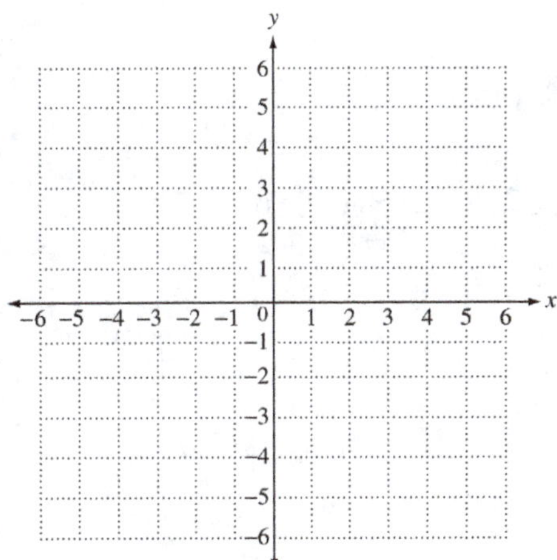

                    113

Objective 1 Practice Exercises

For extra help, see Examples 1–3 on pages 193–194 of your text.

Graph each linear inequality.

1. $x - y < 5$

1.

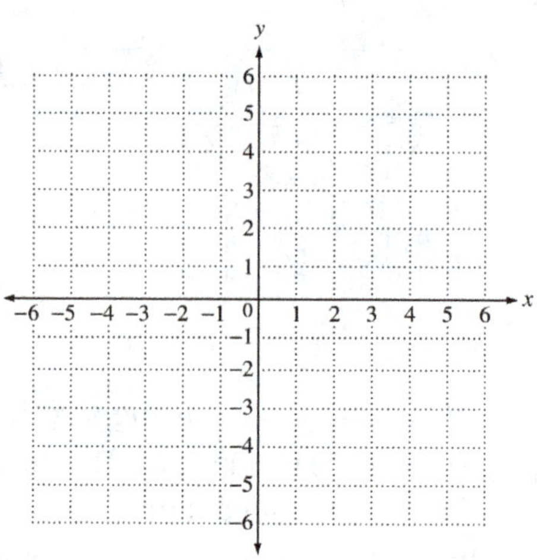

2. $2x + 3y \geq 6$

2.

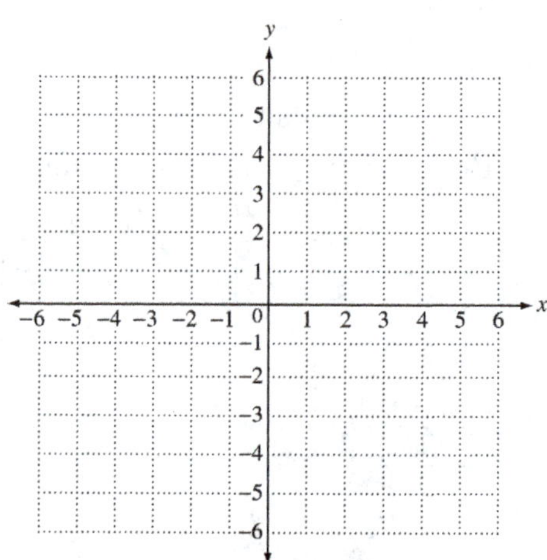

3. $x \leq 4y$

3.

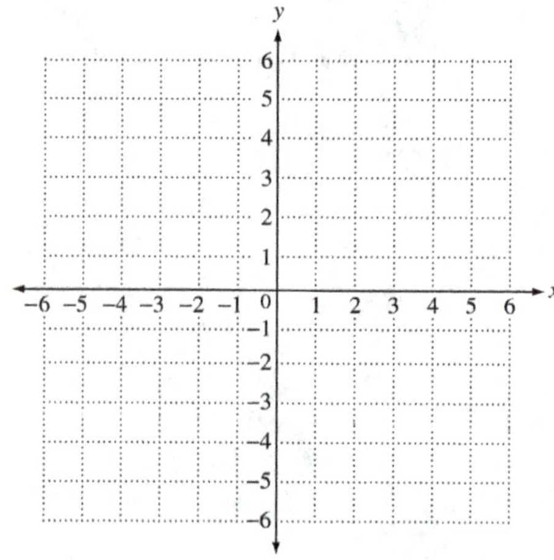

Objective 2   Graph the intersection of two linear inequalities.

Review this example for Objective 2:

4.   Graph $x + y \le 4$ and $x > 2$.

A pair of inequalities joined with the word *and* is interpreted as the intersection of the solution sets of the inequalities. The graph of the intersection of two or more inequalities is the region of the plane where all points satisfy all of the inequalities at the same time.

Begin by graphing each of the two inequalities separately.

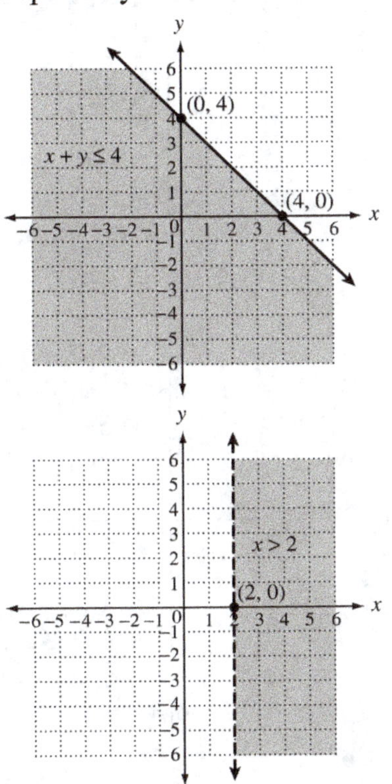

The intersection of the two graphs is the graph we are seeking.

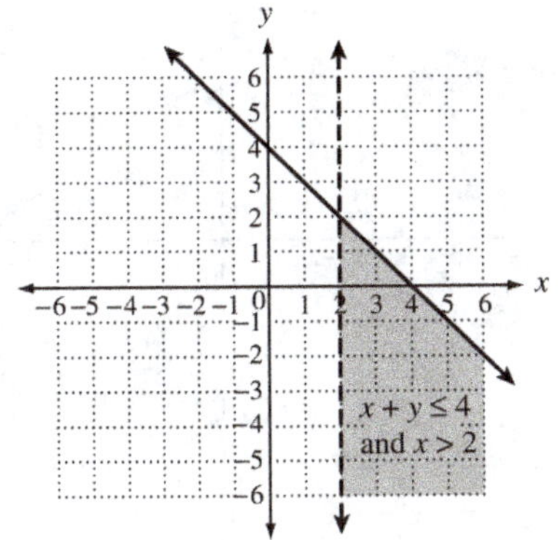

Now Try:

4.   Graph $3x - 4y < 12$ and $y > -4$.

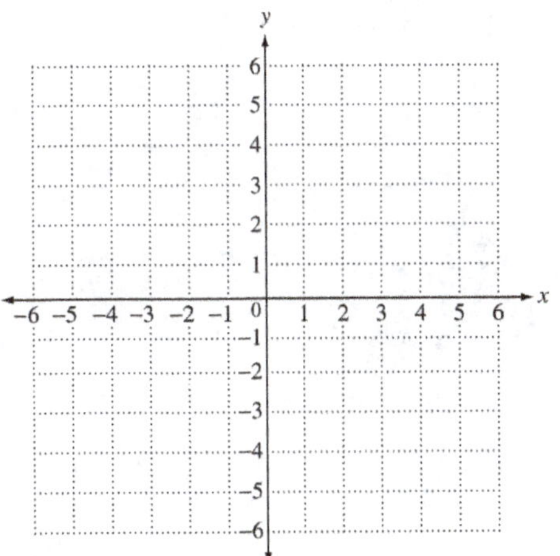

Objective 2 Practice Exercises

For extra help, see Example 4 on page 195 of your text.

Graph the intersection of each pair of linear inequalities.

4. $x - y < 3$ and $x + y > -2$

4.

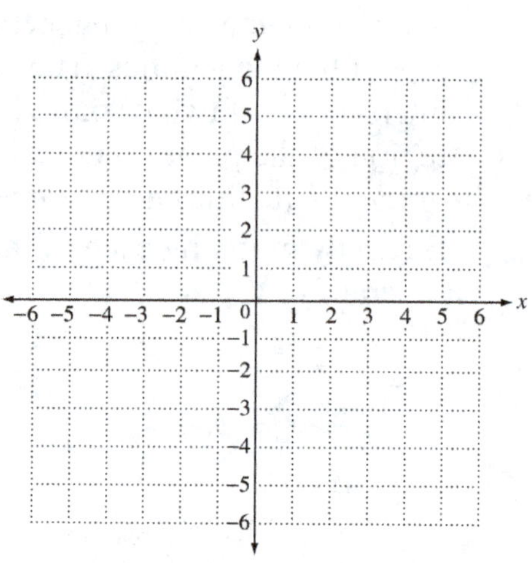

5. $4x + y \leq 4$ and $x - 2y \leq -2$

5.

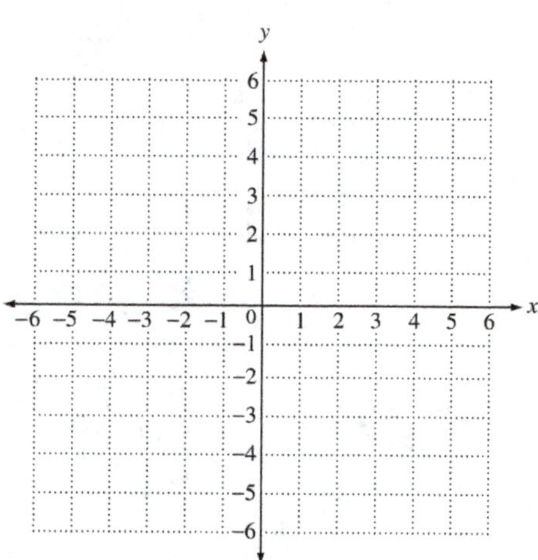

6. $x > 2$ and $2x - 3y < 6$

6.

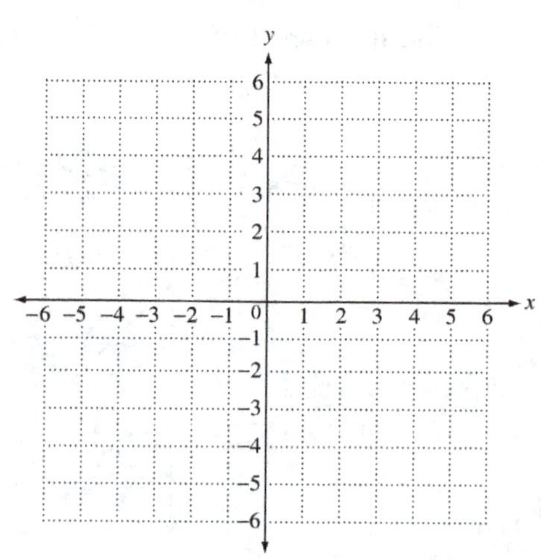

   Copyright © 2020 Pearson Education, Inc.

Objective 3 Graph the union of two linear inequalities.

Review this example for Objective 3:

5. Graph $x + y \le 4$ or $x > 2$.

When two inequalities are joined by the word *or*, we must find the union of the graphs of the inequalities. The graph of the union of two inequalities includes all of the points that satisfy either inequality.

Begin by graphing each of the two inequalities separately.

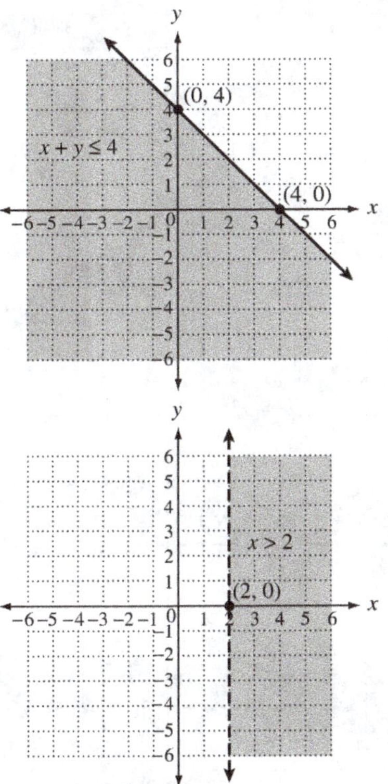

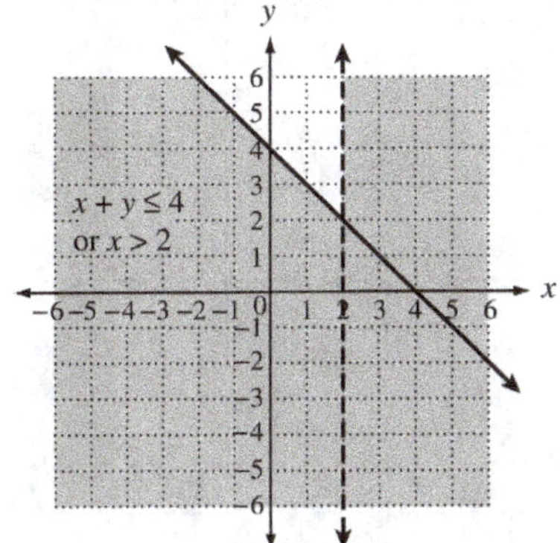

The union of the two graphs is the graph we are seeking.

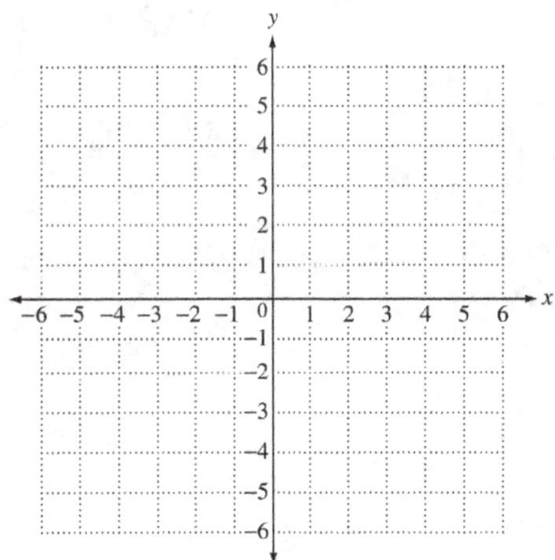

Now Try:

5. Graph $4x + 3y < 12$ or $x \ge 2$.

Objective 3 Practice Exercises

For extra help, see Example 5 on page 196 of your text.

Graph the union of each pair of linear inequalities.

7. $4x - 2y \geq -4$ or $x \geq 1$

7. 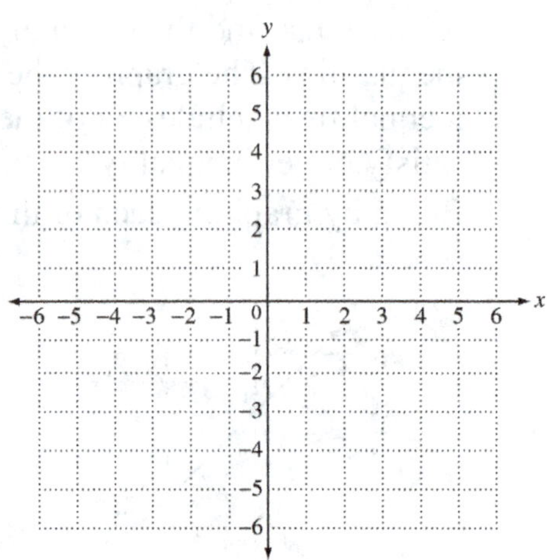

8. $x \geq 4$ or $y < -3$

8. 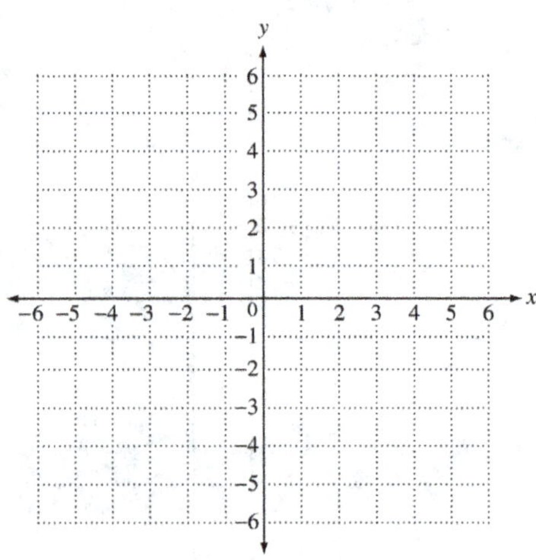

9. $x + y \geq 0$ or $x - y \geq 0$

9. 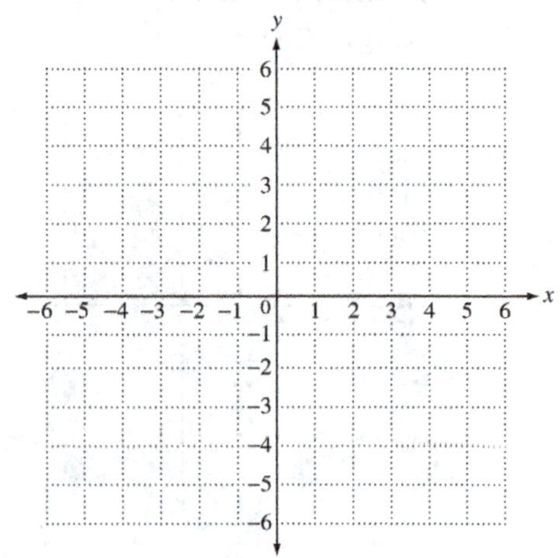

          Copyright © 2020 Pearson Education, Inc.

Chapter 2 LINEAR EQUATIONS, GRAPHS, AND FUNCTIONS

2.5   Introduction to Relations and Functions

Learning Objectives
1   Define and identify relations and functions.
2   Find domain and range.
3   Identify functions defined by graphs and equations.

Key Terms

Use the vocabulary terms listed below to complete each statement in exercises 1−6.

dependent variable	independent variable	relation
function	domain	range

1. The _______________________ of a relation is the set of second components (y-values) of the ordered pairs of the relation.

2. A _______________________ is a set of ordered pairs of real numbers.

3. If the quantity y depends on x, then y is called the _______________________ in a relation between x and y.

4. The _______________________ of a relation is the set of first components (x-values) of the ordered pairs of the relation.

5. A _______________________ is a set of ordered pairs in which each value of the first component, x, corresponds to exactly one value of the second component, y.

6. If the quantity y depends on x, then x is called the _______________________ in a relation between x and y.

Objective 1   Define and identify relations and functions.

Review these examples for Objective 1:

1. Write the relation as a set of ordered pairs.

Number of Hours Worked	Paycheck Amount (in dollars)
8	96
16	192
24	288
32	384

{(8, 96), (16, 192), (24, 288), (32, 384)}

Now Try:

1. Write the relation as a set of ordered pairs.

Number of Hours Fishing	Number of Fish Caught
1	3
2	4
3	6
5	7

____

2. Determine whether each relation defines a function.

    a. $G = \{(-5,-2),\ (-2,\ 6),\ (6,\ 8),\ (8,\ 11),$
        $(11,\ 11)\}$

Relation G is a function. Although the last two ordered pairs have the same y-value, this does not violate the definition of a function.

    b. $H = \{(-9,\ 2),\ (-6,\ 2),\ (-6,\ 9)\}$

In relation H, the last two ordered pairs have the same x-value pair with different y-values. H is a relation, but not a function.

2. Determine whether each relation defines a function.

    a. $\{(10,\ 1),\ (100,\ 2),\ (70,\ 2)\}$

    ____

    b. $\{(0,\ 2),\ (2,\ 6),\ (6,\ 3),\ (0,\ 7)\}$

    ____

Objective 1 Practice Exercises

For extra help, see Examples 1–2 on pages 199–200 of your text.

Write the relation as a set of ordered pairs.

1.

x	y
1	3
1	4
2	-1
3	7

1. ________________

Decide whether each relation is a function.

2. $\{(2,-2,),\ (3,-3),\ (4,-4)\}$

2. ________________

3. $\{(3,\ 4),\ (5,\ 2),\ (4,\ 3),\ (5,\ 3),\ (-2,\ 2)\}$

3. ________________

Objective 2   Find domain and range.

Review these examples for Objective 2:

3. Give the domain and range of each relation. Tell whether the relation defines a function.

$\{(15, 2),\ (20, 3),\ (6, 10),\ (-1, 2)\}$

The domain is the set of x-values $\{-1, 6, 15, 20\}$.
The range is the set of y-values $\{2, 3, 10\}$.
The relation is a function, because each x-value corresponds to exactly one y-value.

Now Try:

3. Give the domain and range of each relation. Tell whether the relation defines a function.

$\{(13, -1),\ (13, -2),\ (13, 4)\}$

domain: _____________

range: _____________

____

4.   Give the domain and range of each relation.

a.

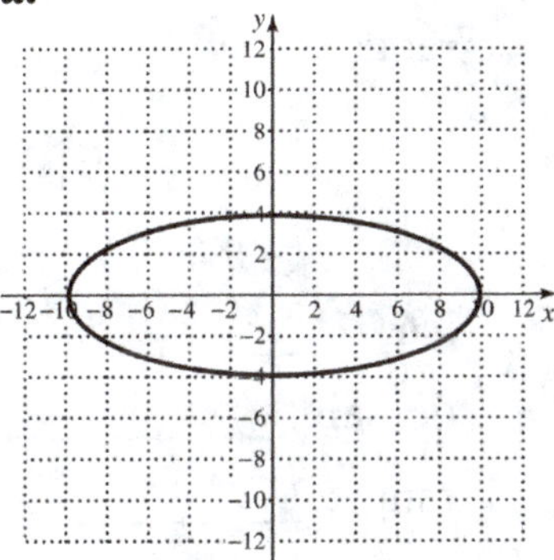

The *x*-values of the points on the graph include all numbers between –10 and 10, inclusive. The *y*-values include all numbers between –4 and 4, inclusive.
The domain is [–10, 10]. The range is [–4, 4].

b.

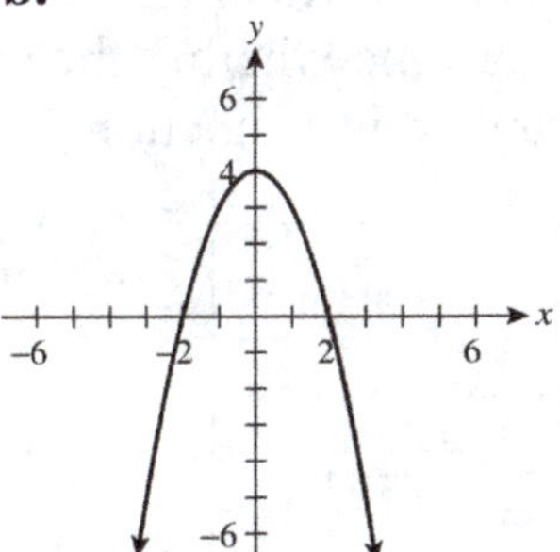

The graph extends indefinitely left and right, as well as downward. The domain is $(-\infty, \infty)$. Because there is a greatest *y*-value, 4, the range includes all numbers less than or equal to 4, written $(-\infty, 4]$.

4.   Give the domain and range of each relation.

a.

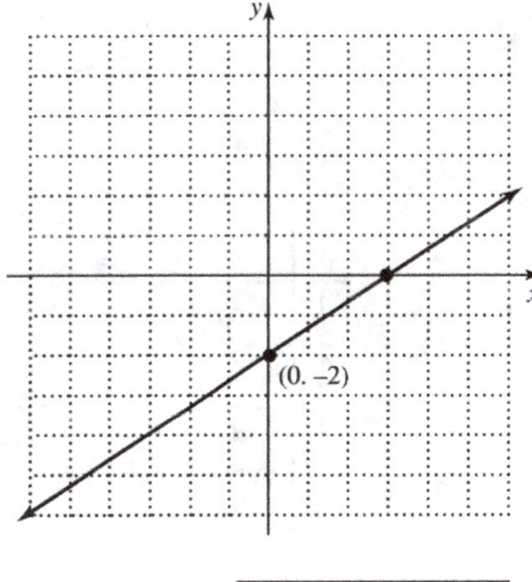

____

____

b.

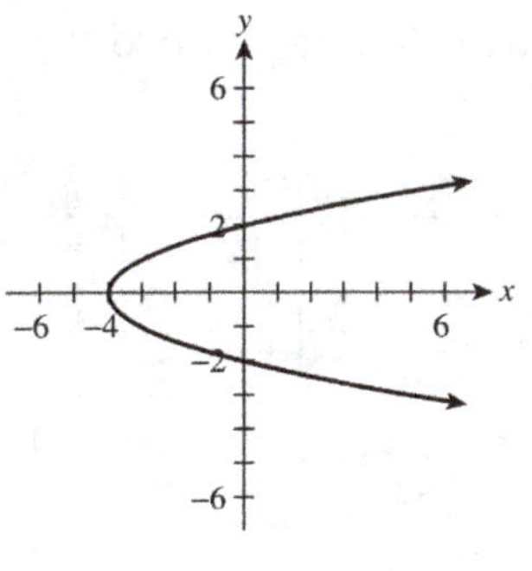

____

____

Objective 2 Practice Exercises

For extra help, see Examples 3–4 on pages 202–203 of your text.

Decide whether the relation is a function, and give the domain and range of the relation.

4. $\{(5,\ 2),\ (3, -1),\ (1, -3),\ (-1, -5)\}$

4. _______________

domain:_______________

range: _______________

5.

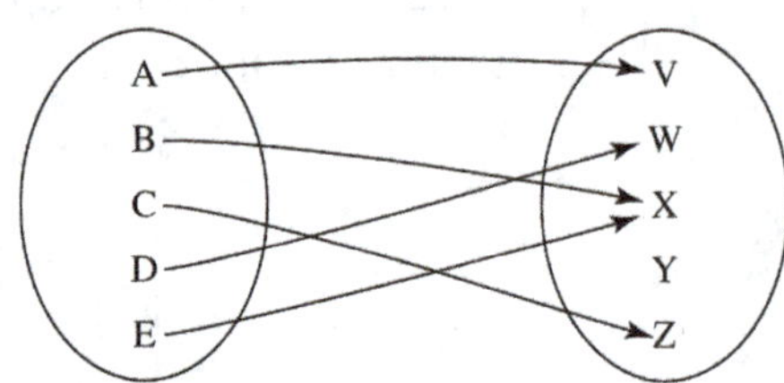

5. _______________

domain: _______________

range: _______________

6.

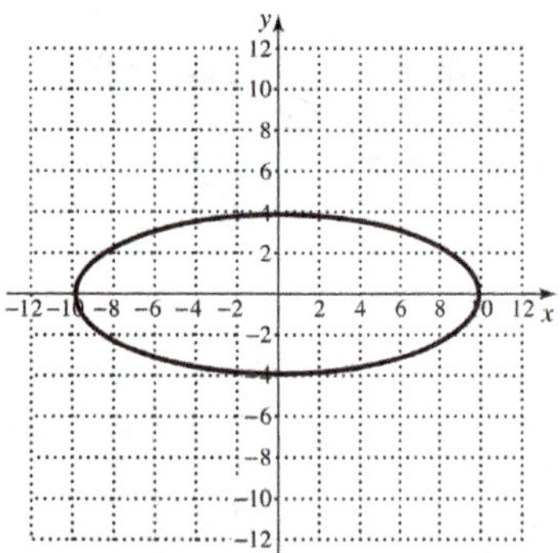

x	y
1	3
2	−1
−1	4
1	4

6. _______________

domain: _______________

range: _______________

Objective 3   Identify functions defined by graphs and equations.

Review these examples for Objective 3:

5. Use the vertical line test to determine whether the relation graphed is a function.

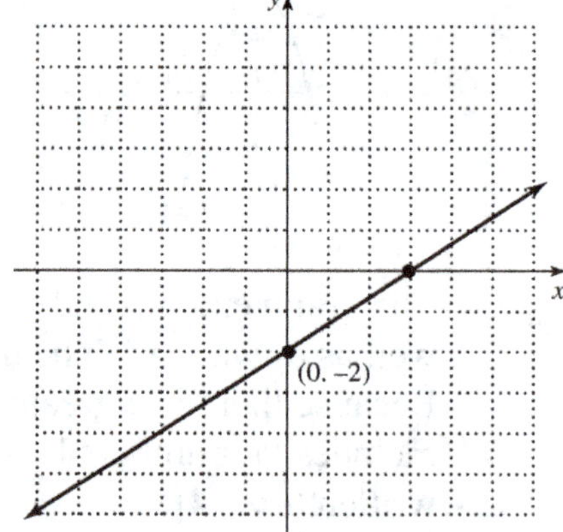

The graph is not a function.

6. Decide whether the relation defines y as a function of x. Give the domain.

$$y = 2x - 4$$

Each x value corresponds to just one y-value and the relation defines a function. Since x can be any real number, the domain is $(-\infty, \infty)$.

Now Try:

5. Use the vertical line test to determine whether the relation graphed is a function.

____

6. Decide whether the relation defines y as a function of x. Give the domain.

$$y = 3x - 1$$

____

____

                    Copyright © 2020 Pearson Education, Inc.

Objective 3 Practice Exercises

For extra help, see Examples 5–6 on pages 204–206 of your text.

Use the vertical line test to determine whether the relation graphed is a function.

7.

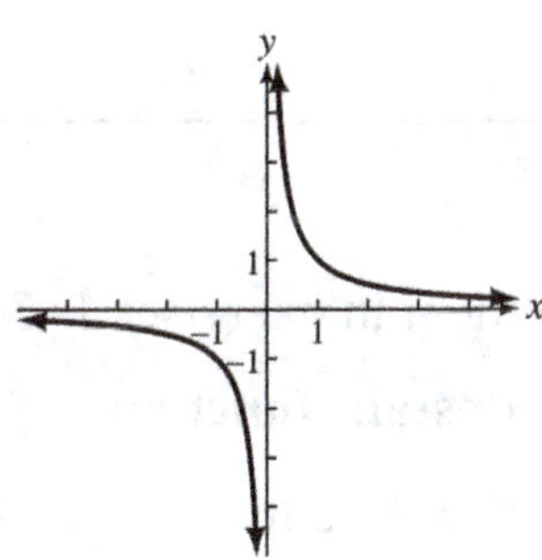

7. _______________

Decide whether each equation defines y as a function of x. Give the domain.

8. $y^2 = x + 1$

8. _______________

9. $y = \dfrac{3}{x+6}$

9. _______________

Chapter 2 LINEAR EQUATIONS, GRAPHS, AND FUNCTIONS

2.6   Function Notation and Linear Functions

Learning Objectives
1   Use function notation.
2   Graph linear and constant functions.

Key Terms

Use the vocabulary terms listed below to complete each statement in exercises 1−3.

      function notation   linear function   constant function

1. A function defined by an equation of the form $f(x) = ax + b$, for real numbers a and b, is a ______________________________.

2. ______________________________ $f(x)$ represents the value of the function at x, that is, the y-value that corresponds to x.

3. A ______________________________ is a linear function of the form $f(x) = b$, for a real number b.

Objective 1   Use function notation.

Review these examples for Objective 1:

1. Let $f(x) = 7x - 3$. Evaluate the function f for the following.

    $x = 4$

    Start with the given function. Replace x with 4.
$$f(x) = 7x - 3$$
$$f(4) = 7(4) - 3$$
$$f(4) = 28 - 3$$
$$f(4) = 25$$
    Thus, $f(4) = 25$.

2. Let $f(x) = -x^2 + 6x - 8$. Find the following.

    a. $f(5)$

    Replace x with 5.
$$f(x) = -x^2 + 6x - 8$$
$$f(5) = -5^2 + 6 \cdot 5 - 8$$
$$f(5) = -25 + 30 - 8$$
$$f(5) = -3$$

Now Try:

1. Let $f(x) = 8x - 7$. Evaluate the function f for the following.

    $x = 3$

    ____

2. Let $f(x) = -x^2 - 7x + 10$. Find the following.

    a. $f(-6)$

    ____

b. $f(p)$

Replace x with p.
$$f(x) = -x^2 + 6x - 8$$
$$f(p) = -p^2 + 6p - 8$$

3. Let $g(x) = 5x + 6$. Find and simplify $g(n+8)$.

Replace x with $n + 8$.
$$g(x) = 5x + 6$$
$$g(n+8) = 5(n+8) + 6$$
$$g(n+8) = 5n + 40 + 6$$
$$g(n+8) = 5n + 46$$

4. For the function, find $f(5)$.

$$f = \{(7, -27),\ (5, -25),\ (3, -23),\ (1, -21)\}$$

From the ordered pair $(5, -25)$, we have
$f(5) = -25$.

5. Refer to the function graphed below.

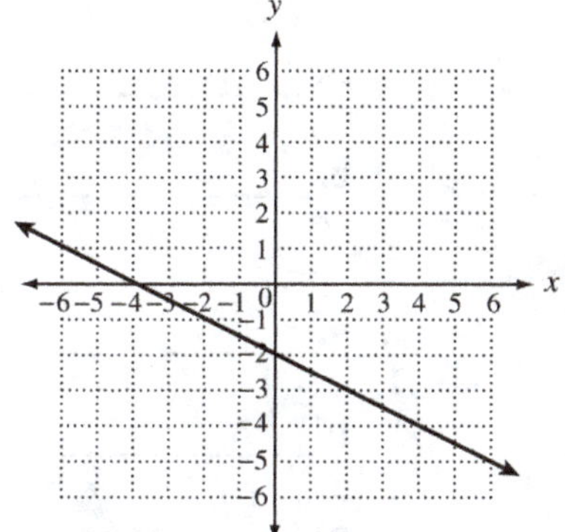

a. Find $f(-2)$.

Locate -2 on the x-axis. Moving down to the graph of f and over to the y-axis gives -1 for the corresponding y-value. Thus, $f(-2) = -1$.

b. For what value of x is $f(x) = -5$?

Locate -5 on the y-axis. Moving across the graph of f and up to the x-axis gives $x = 6$.
Thus, $f(6) = -5$.

---

b. $f(c)$

____

3. Let $g(x) = 4x - 7$. Find and simplify $g(a-1)$.

____

4. For the function, find $f(-6)$.

$$f = \{(-2,\ 11),\ (-4,\ 17),$$
$$(-6,\ 21),\ (-8,\ 24)\}$$

____

5. Refer to the function graphed below.

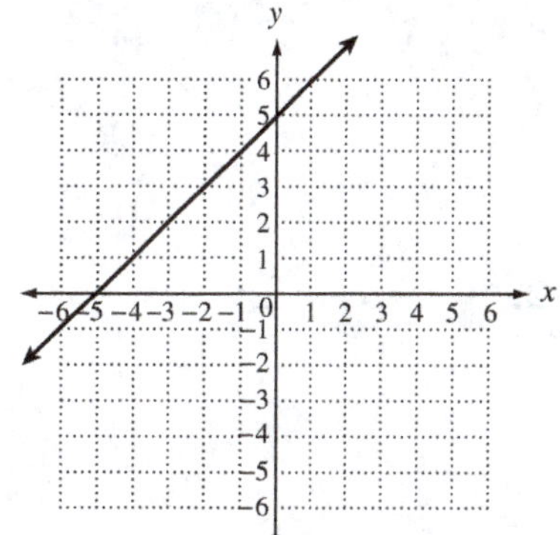

a. Find $f(-2)$.

____

b. For what value of x is $f(x) = 2$?

____

                125

6. Write the equation using function notation $f(x)$. Then find $f(-5)$.

$$x - 5y = 8$$

Step 1 $x - 5y = 8$

$$-5y = -x + 8$$

$$y = \frac{1}{5}x - \frac{8}{5}$$

Step 2 $f(x) = \frac{1}{5}x - \frac{8}{5}$

$$f(-5) = \frac{1}{5}(-5) - \frac{8}{5}$$

$$f(-5) = -\frac{13}{5}$$

6. Write the equation using function notation $f(x)$. Then find $f(-3)$.

$$2x + 3y = 7$$

____

Objective 1 Practice Exercises

For extra help, see Examples 1–6 on pages 210–213 of your text.

For each function f, find (a) $f(-2)$, (b) $f(0)$, and (c) $f(-x)$.

1. $f(x) = 3x - 7$

1. a.____________

b.____________

c.____________

2. $f(x) = 2x^2 + x - 5$

2. a.____________

b.____________

c.____________

3. $f(x) = 9$

3. a.____________

b.____________

c.____________

Objective 2　Graph linear and constant functions.

Review this example for Objective 2:

7. Graph the function $f(x) = -2x - 3$. Give the domain and range.

   The graph of the function has slope -2 and y-intercept -3. To graph this function, plot the y-intercept $(0, -3)$ and use the definition of slope as $\dfrac{\text{rise}}{\text{run}}$ to find a second point on the line. Since the slope is -2, move down two units and right one unit to the point $(1, -5)$. Draw the straight line through the points to obtain the graph. The domain and range are both $(-\infty, \infty)$.

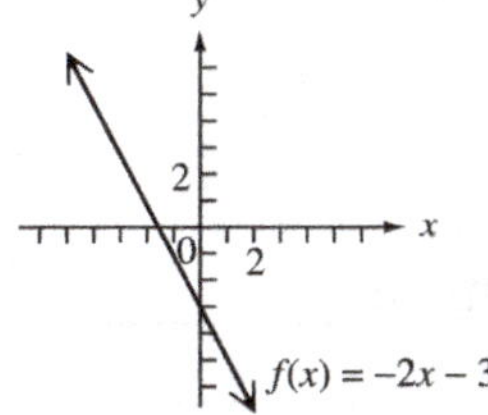

Now Try:

7. Graph the function $f(x) = \dfrac{1}{2}x + \dfrac{1}{2}$. Give the domain and range.

   ____

   ____

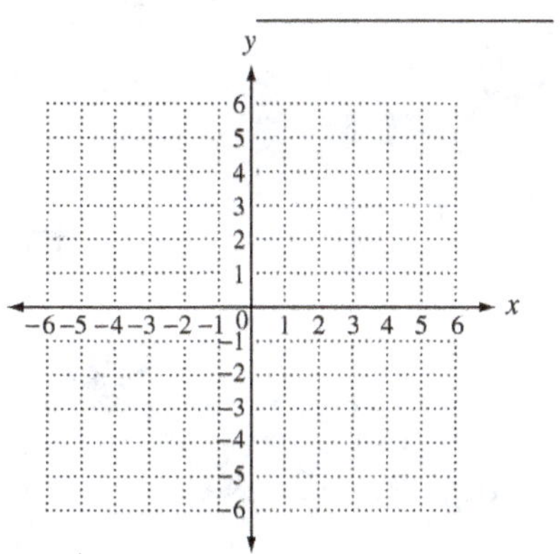

Objective 2 Practice Exercises

For extra help, see Example 7 on page 214 of your text.

Graph each function. Give the domain and range.

4. $2x - y = -2$

4. domain ____________

   range ____________

5. $y + \dfrac{1}{2}x = -2$

5. domain ________

range __________

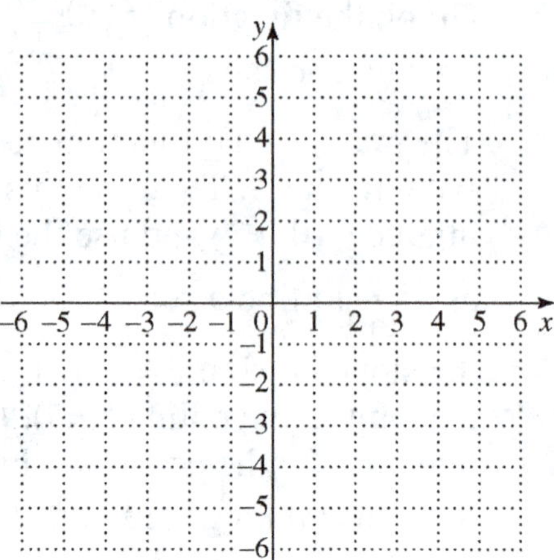

6. $y = 2$

6. domain ________

range __________

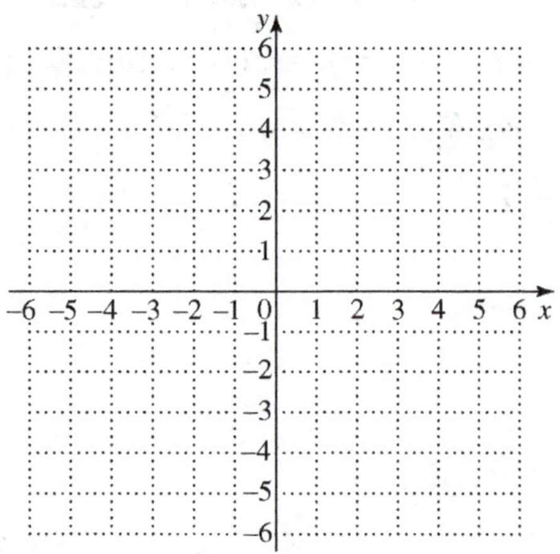

Chapter 3 SYSTEMS OF LINEAR EQUATIONS

3.1   Systems of Linear Equations in Two Variables

Learning Objectives
1   Determine whether an ordered pair is a solution of a linear system.
2   Solve linear systems by graphing.
3   Solve linear systems (with two equations and two variables) by substitution.
4   Solve linear systems (with two equations and two variables) by elimination.
5   Solve special systems.

Key Terms

Use the vocabulary terms listed below to complete each statement in exercises 1−7.

system of equations	**linear system**
solution set of a system	**consistent system** **inconsistent system**
independent equations	**dependent equations**

1.   Equations of a system that have different graphs are called

     ________________________________.

2.   A system of equations with at least one solution is a

     ________________________________.

3.   Two or more equations that are to be solved at the same time form a

     ____.

4.   The ____ of linear equations includes all
     the ordered pairs that make all the equations of the system true at the same time.

5.   Equations of a system that have the same graph (because they are different forms
     of the same equation) are called ________________________________.

6.   A system with no solution is called a(n) ________________________________.

7.   A(n) ________________________________ consists of two or more linear
     equations with the same variables.

                          129

Name: Date:
Instructor: Section:

Objective 1   Determine whether an ordered pair is a solution of a linear system.

Review this example for Objective 1:

1. Decide whether the given ordered pair is a solution of the given system.

$$x + y = 8; \quad (3, 5)$$
$$2x - y = 1$$

Replace x with 3 and y with 5 in each equation of the system.

$$
\begin{array}{c|c}
x + y = 8 & 2x - y = 1 \\
\overset{?}{3 + 5 = 8} & \overset{?}{2(3) - 5 = 1} \\
\text{True} \quad 8 = 8 & \text{True} \quad 1 = 1
\end{array}
$$

Since (3, 5) makes both equations true, (3, 5) is a solution of the system.

Now Try:

1. Decide whether the given ordered pair is a solution of the given system.

$$x - 5y = 11; \quad (6, -1)$$
$$3x + 8y = 10$$

____

Objective 1 Practice Exercises

For extra help, see Example 1 on pages 230–231 of your text.

Decide whether the given ordered pair is a solution of the given system. Write solution *or* not a solution.

1. (4, 1)
    $$2x + 3y = 11$$
    $$3x - 2y = 9$$

1. _____________

2. $(-3, -1)$
    $$5x - 3y = -12$$
    $$2x + 3y = -9$$

2. _____________

3. (4, 0)
    $$4x + 3y = 16$$
    $$x - 4y = -4$$

3. _____________

Objective 2   Solve linear systems by graphing.

Review this example for Objective 2:

2. Solve the system of equations by graphing.

$$x - 2y = 6 \quad (1)$$

$$2x + y = 2 \quad (2)$$

To graph these linear equations, we plot several points for each line.

$x - 2y = 6$

x	y
0	-3
6	0
4	-1

$2x + y = 2$

x	y
0	2
1	0
-1	4

From the graph we see that the solution is (2,–2).

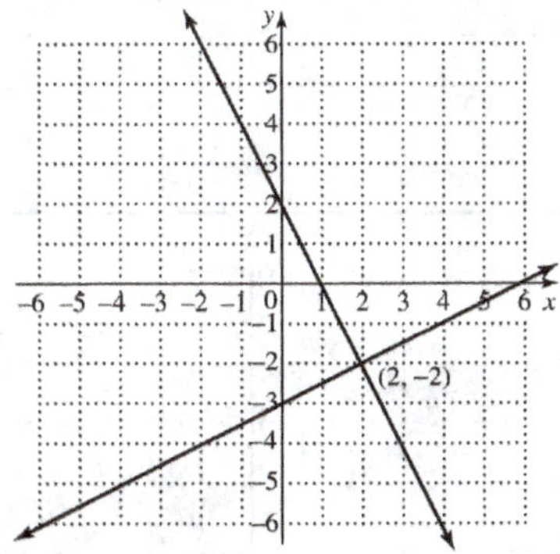

To check, substitute 2 for x and –2 for y in each equation.

$$x - 2y = 6$$
$$2 - 2(-2) \overset{?}{=} 6$$
$$2 + 4 \overset{?}{=} 6$$
$$\text{True } 6 = 6$$

$$2x + y = 2$$
$$2(2) + (-2) \overset{?}{=} 2$$
$$4 - 2 \overset{?}{=} 2$$
$$\text{True } 2 = 2$$

The solution set is {(2,–2)}.

Now Try:

2. Solve the system of equations by graphing.

$$2x - 5y = 8 \quad (1)$$

$$5x - 4y = 3 \quad (2)$$

____

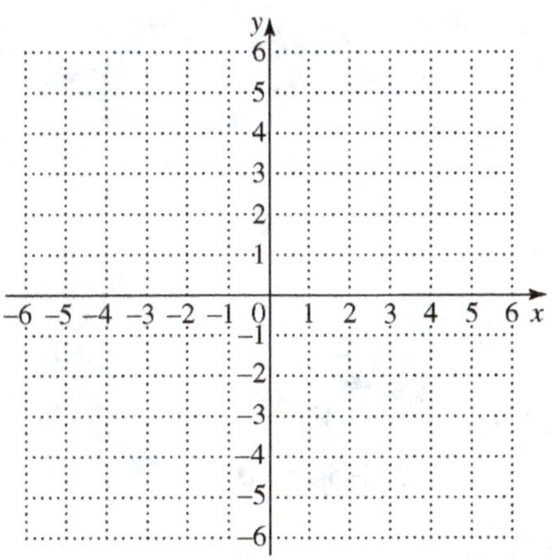

Objective 2 Practice Exercises

For extra help, see Example 2 on page 231 of your text.

Solve the system by graphing.

4. $2x + 3y = 5$

  $3x - y = 13$

4. ________________________________

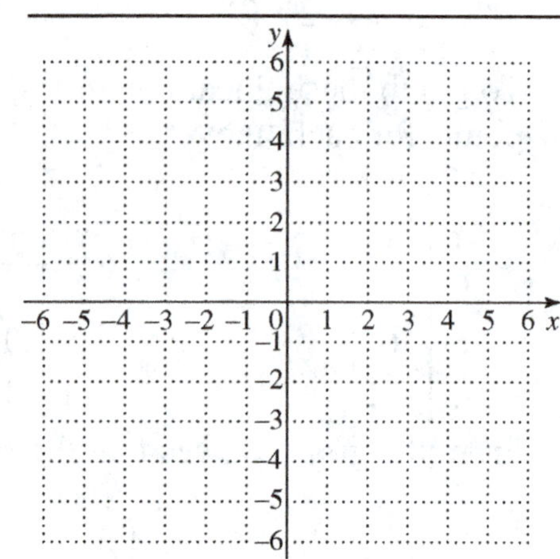

5. $2x = y$

  $5x + 3y = 0$

5. ________________________________

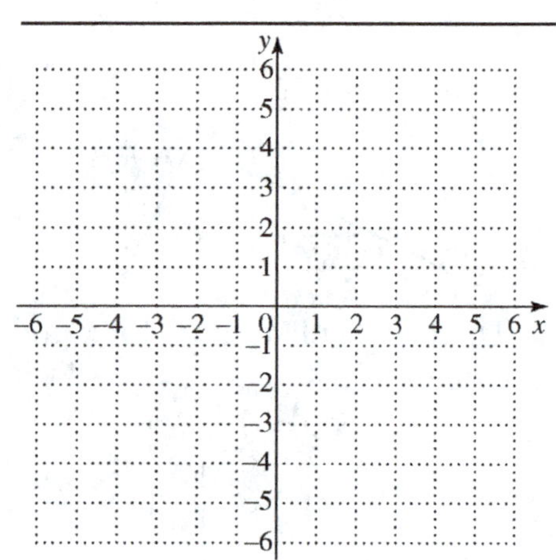

6. $y - 2 = 0$

  $3x - 4y = -17$

6. ________________________________

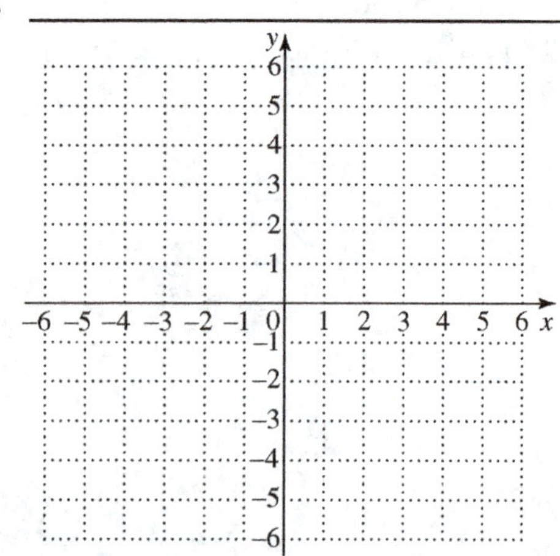

          Copyright © 2020 Pearson Education, Inc.

Objective 3 Solve linear systems (with two equations and two variables) by substitution.

Review this example for Objective 3:

4. Solve the system by substitution.
$$3x - 2y = 16 \quad (1)$$
$$x - 7y = -1 \quad (2)$$

Step 1 First, solve equation (2) for x.
$$x - 7y = -1$$
$$x = 7y - 1$$

Step 2 Substitute $7y - 1$ for x in equation (1).
$$3x - 2y = 16$$
$$3(7y - 1) - 2y = 16$$

Step 3 Solve for y.
$$21y - 3 - 2y = 16$$
$$19y - 3 = 16$$
$$19y = 19$$
$$y = 1$$

Step 4 Find x. From Step 1, $x = 7y - 1$.
Substitute 1 for y.
$$x = 7(1) - 1 = 6$$

Step 5 Check the solution (6, 1) in both equations (1) and (2).

$$
\begin{array}{c|c}
3x - 2y = 16 & x - 7y = -1 \\
\overset{?}{3(6) - 2(1) = 16} & \overset{?}{6 - 7(1) = -1} \\
\overset{?}{18 - 2 = 16} & \overset{?}{6 - 7 = -1} \\
16 = 16 \ \text{True} & -1 = -1 \ \text{True}
\end{array}
$$

The solution set is $\{(6, 1)\}$.

Now Try:

4. Solve the system by substitution.
$$4x + 9y = 11 \quad (1)$$
$$2x - y = 11 \quad (2)$$

____

Objective 3 Practice Exercises

For extra help, see Examples 3–5 on pages 232–235 of your text.

Solve each system by substitution.

7. $3x - 2y = -1$

          $x = \dfrac{3}{4}y$

7. _______________

8. $2x + y = 6$

          $y = 5 - 3x$

8. _______________

9. $y = 11 - 2x$

          $x = 18 - 3y$

9. _______________

Objective 4   Solve linear systems (with two equations and two variables) by elimination.

Review this example for Objective 4:

7. Solve the system by elimination.

$$4x + 3y = 41 \quad (1)$$
$$3x - 2y = 1 \quad (2)$$

Step 1 Both equations are in standard form.

Step 2 Multiply equation (1) by 2 and equation (2) by 3.

$$8x + 6y = 82 \quad \text{Multiply (1) by 2.}$$
$$9x - 6y = 3 \quad \text{Multiply (2) by 3.}$$

Now Try:

7. Solve the system by elimination.

$$4x + 5y = 11 \quad (1)$$
$$3x - 2y = -9 \quad (2)$$

____

Step 3 Now add.

$$8x + 6y = 82$$
$$9x - 6y = 3$$
$$\overline{17x \quad\;\; = 85}$$

Step 4 Solve for x. $x = 5$ *Step 5*

To find y, substitute 5 for x in either equation (1) or (2).

$$3x - 2y = 1 \quad (2)$$
$$3(5) - 2y = 1$$
$$15 - 2y = 1$$
$$-2y = -14$$
$$y = 7$$

Step 6 Check the solution (5, 7) in both equations (1) and (2).

$$4x + 3y = 41 \qquad\qquad 3x - 2y = 1$$
$$\overset{?}{4(5) + 3(7)} = 41 \qquad \overset{?}{3(5) - 2(7)} = 1$$
$$\overset{?}{20 + 21} = 41 \qquad\qquad \overset{?}{15 - 14} = 1$$
$$41 = 41 \;\; \text{True} \qquad\qquad\quad 1 = 1 \;\; \text{True}$$

The solution set is $\{(5,\, 7)\}$.

Objective 4 Practice Exercises

For extra help, see Examples 6–7 on pages 235–237 of your text.

Solve each system by elimination.

10. $3x - y = 11$

   $x + y = 5$

10. ________________

11. $\dfrac{1}{2}x + \dfrac{1}{4}y = 5$

   $\dfrac{1}{2}x - \dfrac{3}{4}y = -3$

11. ________________

12. $x + 2y = 7$ **12.** _______________

       $x - y = -2$

Objective 5   Solve special systems.

Review these examples for Objective 5:

8.   Solve the system.
$$5x + 5y = 9 \quad (1)$$
$$x + y = 3 \quad (2)$$

We multiply equation (2) by –5 and then add the result to equation (1).

$$5x + 5y = 9 \quad (1)$$
$$\underline{-5x - 5y = -15} \quad \text{Multiply (2) by } -5.$$
$$0 = -6 \quad \text{False}$$

The system is inconsistent.
Since the statement is false, there are no ordered pairs that satisfy both equations.
The solution set is $\varnothing$.

9.   Solve the system.
$$x + 5y = 9 \quad (1)$$
$$3x + 15y = 27 \quad (2)$$

We multiply equation (1) by –3 and then add the result to equation (2).

$$-3x - 15y = -27 \quad \text{Multiply (1) by } -3.$$
$$\underline{3x + 15y = 27 \quad (2)}$$
$$0 = 0 \quad \text{True}$$

The equations are dependent.
Since the statement is true, the solution set is the set of all points on the line with equation $x + 5y = 9$, written as $\{(x, y) \mid x + 5y = 9\}$.

Now Try:

8.   Solve the system.
$$4x + 3y = 12 \quad (1)$$
$$8x + 6y = -24 \quad (2)$$

____

9.   Solve the system.
$$12x + 9y = 36 \quad (1)$$
$$4x + 3y = 12 \quad (2)$$

____

10. Write the pair of equations in slope-intercept form, and use the results to tell how many solutions the system has.

$$5x - y = 4$$
$$15x - 3y = 12$$

Solve each equation for y.

$5x - y = 4$	$15x - 3y = 12$
$-y = -5x + 4$	$5x - y = 4$
$y = 5x - 4$	Same result

The lines have the same slope and same y-intercept, indicating that they coincide. There are infinitely many solutions.

10. Write the pair of equations in slope-intercept form, and use the results to tell how many solutions the system has.

$$2x - 5y = 7$$
$$6x - 15y = 21$$

____

____

____

Objective 5 Practice Exercises

For extra help, see Examples 8–10 on pages 237–239 of your text.

Solve each system of equations using any method.

13. $8x + 4y = -1$
        $4x + 2y = 3$

13. _______________

14. $x + 2y = 4$
        $8y = -4x + 16$

14. _______________

15. $-3x + 2y = 6$
        $-6x + 4y = 12$

15. _______________

Chapter 3 SYSTEMS OF LINEAR EQUATIONS

3.2   Systems of Linear Equations in Three Variables

Learning Objectives

1 Understand the geometry of systems of three equations in three variables.
2 Solve linear systems (with three equations and three variables) by elimination.
3 Solve linear systems (with three equations and three variables) in which some of the equations have missing terms.
4 Solve special systems.

Key Terms

Use the vocabulary terms listed below to complete each statement in exercises 1−3.

ordered triple   inconsistent system   dependent system

1. The solution of a linear system of equations in three variables is written as a(n) ______________________.

2. A system of equations in which all solutions of the first equation are also solutions of the second equation is a(n) ______________________.

3. A system of equations that has no common solution is called a(n) ______________________.

Objective 1   Understand the geometry of systems of three equations in three variables.

Objective 1 Practice Exercises

For extra help, see pages 245–246 of your text.

Answer each question.

1. If a system of linear equations in three variables has a single solution, how do the planes that are the graphs of the equations intersect?

1. ______________________

2. If a system of linear equations in three variables has no solution, how do the planes that are the graphs of the equations intersect?

2. ______________________

                        Copyright © 2020 Pearson Education, Inc.

Name: Date:

Instructor: Section:

**Objective 2   Solve linear systems (with three equations and three variables) by
elimination.**

Review this example for Objective 2:

1. Solve the system.

$$x - 2y + 5z = -7 \quad (1)$$
$$2x + 3y - 4z = 14 \quad (2)$$
$$3x - 5y + z = 7 \quad (3)$$

Step 1: Since x in equation (1) has coefficient 1, choose x as the focus variable and (1) as the working equation.

Step 2: Multiply working equation (1) by -2 and add the result to equation (2) to eliminate focus variable x.

$$-2x + 4y - 10z = 14 \quad \text{Multiply (1) by } -2$$
$$\underline{2x + 3y - 4z = 14 \quad (2)}$$
$$7y - 14z = 28 \quad (4)$$

Step 3: Multiply working equation (1) by -3 and add the result to equation (3) to eliminate focus variable x.

$$-3x + 6y - 15z = 21 \quad \text{Multiply (1) by } -3$$
$$\underline{3x - 5y + z = 7 \quad (3)}$$
$$y - 14z = 28 \quad (5)$$

Step 4: Write equations (4) and (5) as a system, then solve the system.

$$7y - 14z = 28 \quad (4)$$
$$y - 14z = 28 \quad (5)$$

We will eliminate z.

$$-7y + 14z = -28 \quad \text{Multiply (4) by } -1$$
$$\underline{y - 14z = 28 \quad (5)}$$
$$-6y = 0 \quad \text{Add.}$$
$$y = 0 \quad \text{Divide by } -6.$$

Substitute 0 for y in either equation to find z.

$$y - 14z = 28 \quad (5)$$
$$0 - 14z = 28 \quad \text{Let } y = 0.$$
$$-14z = 28$$
$$z = -2$$

Now Try:

1. Solve the system.

$$2x + y + 2z = -1$$
$$3x - y + 2z = -6$$
$$3x + y - z = -10$$

____

Step 5: Now substitute $y = 0$ and $z = -2$ in working equation (1) to find the value of the remaining variable, focus variable x.

$$x - 2y + 5z = -7 \quad (1)$$
$$x - 2(0) + 5(-2) = -7 \quad \text{Let } y = 0 \text{ and } z = -2.$$
$$x - 10 = -7$$
$$x = 3$$

Step 6: It appears that the ordered triple $(3, 0, -2)$ is the solution of the system. We must check that the solution satisfies all three original equations of the system.

$$x - 2y + 5z = -7 \quad (1)$$
$$3 - 2(0) + 5(-2) \stackrel{?}{=} -7$$
$$3 - 0 - 10 \stackrel{?}{=} -7$$
$$-7 = -7$$

Because $(3, 0, -2)$ also satisfies equations (2) and (3), the solution set is $\{(3, 0, -2)\}$.

Objective 2 Practice Exercises

For extra help, see Example 1 on pages 247–248 of your text.

Solve each system of equations.

3. $\quad x + y + z = 2$
    $\quad x - y + z = -2$
    $\quad x - y - z = -4$

3. _________________

4. $\quad 2x + y - z = 9$
    $\quad x + 2y + z = 3$
    $\quad 3x + 3y - z = 14$

4. _________________

5.
$$2x - 5y + 2z = 30$$
$$x + 4y + 5z = -7$$
$$\tfrac{1}{2}x - \tfrac{1}{4}y + z = 4$$

5. _______________

Objective 3   **Solve linear systems (with three equations and three variables) in which some of the equations have missing terms.**

Review this example for Objective 3:

2. Solve the system.
$$3x \quad\;\; - 4z = -23 \quad (1)$$
$$y + 5z = \;\; 24 \quad (2)$$
$$x - 3y \quad\;\;\;\; = \;\;\; 2 \quad (3)$$

Since equation (1) is missing the variable y, one way to begin is to eliminate y again, using equations (2) and (3).
$$3y + 15z = 72 \quad \text{Multiply (2) by 3}$$
$$\underline{x - 3y \qquad\;\; = \;\; 2 \quad (3)}$$
$$x \qquad + 15z = 74 \quad (4)$$

Now solve the system composed of equations (1) and (4).
$$3x \;\; - 4z = \;\; -23 \quad (1)$$
$$\underline{-3x - 45z = -222 \quad \text{Multiply (4) by } -3}$$
$$-49z = -245$$
$$z = 5$$

Substitute 5 for z in (1) and solve for x.
$$3x - 4z = -23 \quad (1)$$
$$3x - 4(5) = -23 \quad \text{Let } z = 5.$$
$$3x - 20 = -23$$
$$3x = -3$$
$$x = -1$$

Now Try:

2. Solve the system.
$$x + 5y \qquad\;\; = -23$$
$$4y - 3z = -29$$
$$2x \qquad + 5z = \;\; 19$$

____

Substitute 5 for z in (2) and solve for y.

$$y + 5z = 24 \quad (2)$$
$$y + 5(5) = 24 \quad \text{Let } z = 5.$$
$$y + 25 = 24$$
$$y = -1$$

A check verifies that the solution set is $\{(-1, -1, 5)\}$.

Objective 3 Practice Exercises

For extra help, see Example 2 on pages 248–249 of your text.

Solve each system of equations.

6.
$$7x \qquad + z = -1$$
$$3y - 2z = 8$$
$$5x + y \qquad = 2$$

6. _______________

7.
$$2x + 5y \qquad = 18$$
$$3y + 2z = 4$$
$$\frac{1}{4}x - y \qquad = -1$$

7. _______________

8.
$$5x \qquad - 2z = 8$$
$$4y + 3z = -9$$
$$\frac{1}{2}x + \frac{2}{3}y \qquad = -1$$

8. _______________

Objective 4 Solve special systems.

Review these examples for Objective 4:

3. Solve the system.
$$x - y + z = 7 \quad (1)$$
$$2x + 5y - 4z = 2 \quad (2)$$
$$-x + y - z = 4 \quad (3)$$

Since x in equation (1) has coefficient 1, choose x as the focus variable and (1) as the working equation. Using equations (1) and (3), we have

$$\begin{aligned} x - y + z &= 7 \quad (1) \\ -x + y - z &= 4 \quad (3) \\ \hline 0 &= 11 \quad \text{False} \end{aligned}$$

The resulting false statement indicates that equations (1) and (3) have no common solution. Thus, the system is inconsistent and the solution set is $\varnothing$. The graph of this system would show that three planes are parallel to each other as shown below.

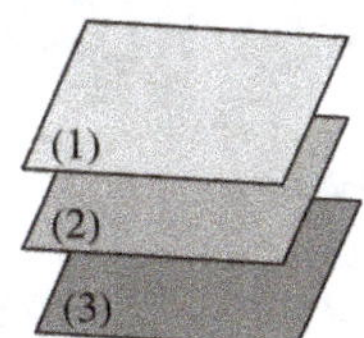

4. Solve the system.
$$3x - 2y + 5z = 4 \quad (1)$$
$$-6x + 4y - 10z = -8 \quad (2)$$
$$\frac{3}{2}x - y + \frac{5}{2}z = 2 \quad (3)$$

Multiplying each side of equation (1) by -2 gives equation (2). Multiplying each side of equation (1) by $\frac{1}{2}$ gives equation (3). Thus, the equations are dependent, and all three equations have the same graph as shown below. The solution set is written
$$\{(x, y, z) \mid 3x - 2y + 5z = 4\}.$$

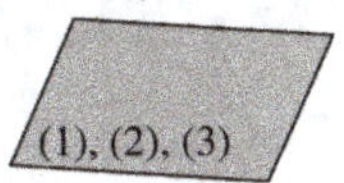

Now Try:

3. Solve the system.
$$-4x - 2y + z = -19$$
$$-6x + 2y - 6z = -8$$
$$-4x + 2y - 5z = -6$$

____

4. Solve the system.
$$x - 5y + 2z = 0$$
$$-x + 5y - 2z = 0$$
$$\frac{1}{2}x - \frac{5}{2}y + z = 0$$

____

5.   Solve the system.

$$3x + 2y + z = 3 \quad (1)$$
$$-6x - 4y - 2z = -6 \quad (2)$$
$$x + \frac{2}{3}y + \frac{1}{3}z = 4 \quad (3)$$

Multiplying each side of equation (1) by -2 gives equation (2), so these two equations are dependent. Multiplying each side of equation (1) by $\frac{1}{3}$ does not give equation (3). Instead, we obtain two equations with the same coefficients, but with different constant terms. The graphs of

5.   Solve the system.

$$3x + 2y + z = 3$$
$$-6x - 4y - 2z = -6$$
$$9x + 6y + 3z = 1$$

____

Objective 4 Practice Exercises

For extra help, see Examples 3–5 on pages 249–250 of your text.

Solve each system of equations.

9.
$$8x - 7y + 2z = 1$$
$$3x + 4y - z = 6$$
$$-8x + 7y - 2z = 5$$

9. ________________

10.
$$3x - 2y + 4z = 5$$
$$-3x + 2y - 4z = -5$$
$$\frac{3}{2}x - y + 2z = \frac{5}{2}$$

10. ________________

11.
$$-x + 5y - 2z = 3$$
$$2x - 10y + 4z = -6$$
$$-3x + 15y - 6z = 9$$

11. ________________

Chapter 3 SYSTEMS OF LINEAR EQUATIONS

3.3   Applications of Systems of Linear Equations

Learning Objectives
1   Solve geometry problems using two variables.
2   Solve money problems using two variables.
3   Solve mixture problems using two variables.
4   Solve distance-rate-time problems using two variables.
5   Solve problems with three variables using a system of three equations.

Key Terms

Use the vocabulary terms listed below to complete each statement in exercises 1–2.

elimination method   **substitution**

1.   Using the addition property to solve a system of equations is called the

_______________________________.

2.   _______________________________ is being used when one expression is replaced by
another.

Objective 1   Solve geometry problems using two variables.

Review this example for Objective 1:

1.   The length of a rectangular field is 5 m more
than the width. Find the length and width if the
perimeter is 70 m.

Step 1 Read the problem. We must find the
dimensions of the field.

Step 2 Assign variables. Let L = length and
W = width.

Step 3 Write a system of equations. We use the
equation for perimeter.
   $2L + 2W = 70$
A second equation uses the information given.
   $L = W + 5$
The system of equations is
   $2L + 2W = 70$ (1)
   $L = W + 5$ (2)

Step 4 Solve the system. Since equation (2) is
solved for L, we use substitution.

Now Try:

1.   The length of a rectangle is 7 ft
more than the width. The
perimeter is 54 ft. Find the
dimensions of the rectangle.

____

$$2L + 2W = 70$$
$$2(W + 5) + 2W = 70$$
$$2W + 10 + 2W = 70$$
$$4W + 10 = 70$$
$$4W = 60$$
$$W = 15$$

Let $W = 15$ in equation (2) to find L.
$$L = 15 + 5 = 20$$

Step 5 State the answer. The length is 20 m and the width is 15 m.

Step 6 Check.
$$2(20) + 2(15) = 70$$
$$20 = 15 + 5$$

The answer is correct.

Objective 1 Practice Exercises

For extra help, see Example 1 on pages 254–255 of your text.

Solve each problem.

1. The side of a square is 5 centimeters shorter than the side of an equilateral triangle. The perimeter of the square is 7 centimeters less than the perimeter of the triangle. Find the lengths of a side of the square and of a side of the triangle.

    1. square: ___________

       triangle: _________

2. The perimeter of a rectangle is 96 inches. If the width were tripled, the width would be 36 inches more than the length. Find the length and width of the rectangle.

    2. length: ___________

       width: ___________

                Copyright © 2020 Pearson Education, Inc.

3. The perimeter of a triangle is 70 centimeters. Two   3. ___________________
    sides of the triangle have the same length. The third
    side is 7 centimeters longer than either of the equal
    sides. Find the length of the equal sides of the
    triangle.

Objective 2   Solve money problems using two variables.

Review this example for Objective 2:

2. The total receipts for a basketball game were
    $4690.50. There were 723 tickets sold, some for
    children and some for adults. If the adult tickets
    cost $9.50 and the children's tickets cost $4, how
    many of each type were there?

    Step 1 Read the problem. There are two
    unknowns.

    Step 2 Assign variables.
        Let a = the number of adult tickets sold.
        Let c = the number of child tickets sold.

    Step 3 Write a system of equations. We write
    one equation using the total number of tickets.
        $a + c = 723$
    We write another equation using the cost.
        $9.50a + 4c = 4690.50$
    The system of equations is
        $a + c = 723$ (1)
        $9.50a + 4c = 4690.50$ (2)

    Step 4 Solve the system. To eliminate c,
    multiply equation (1) by -4, and add.
        $-4a - 4c = -2892$ Multiply (1) by -4.
        $\underline{9.5a + 4c = 4690.5}$ (2)
        $5.5a \quad\quad = 1798.5$ Add.
        $\quad\quad a = 327$ Divide by 5.5.

    To find the value of c, let $a=327$ in equation (1).
        $327 + c = 723$
        $\quad\quad c = 396$

Now Try:

2. The Garden Center ordered 6
    ounces of marigold seed and 8
    ounces of carnation seed, paying
    $214.54. They later ordered
    another 12 ounces of marigold
    seed and 18 ounces of carnation
    seed, paying $464.28. Find the
    price per ounce for each type of
    seed.

    marigold   ___________

    carnation   ___________

Step 5 State the answer. The number of adult tickets sold is 327 and the number of child tickets sold is 396.

Step 6 Check.

$$327 + 396 = 723$$

$$9.50(327) + 4(396) = 4690.50$$

The answer is correct.

Objective 2 Practice Exercises

For extra help, see Example 2 on page 256 of your text.

Solve each problem.

4. Pablo has some $10-bills and some $20-bills. The total value of the money is $650, with a total of 40 bills. How many of each are there?

4. $10-bills __________

   $20-bills __________

5. Big Giant Super Market will sell 5 large jars and 2 small jars of their peanut butter for $36. They will also sell 2 large jars and 5 small jars for $27. What is the price of each jar?

5. small __________

   large __________

6. A taxi charges a flat rate plus a certain charge per mile. A trip of 7 miles costs $5.30, while a trip of 3 miles costs $3.70. Find the flat rate and the charge per mile.

6. flat rate __________

   per mile __________

Objective 3   Solve mixture problems using two variables.

Review this example for Objective 3:

3. A 75% solution will be mixed with a 55% solution to get 70 liters of 63% solution? How many liters of the 55% and 75% solutions should be used?

Step 1 Read the problem. There are two solution strengths. We are looking for an "in between" strength.

Step 2 Assign variables.
Let x = the number of liters of 75% solution.
Let y = the number of liters of 55% solution.

Step 3 Write a system of equations.
Write one equation using the total amount.
$$x + y = 70$$
Write each percent as a decimal and multiply each solution by its concentration.
$$0.75x + 0.55y = 0.63(70)$$
The system of equations is
$$x + y = 70 \qquad (1)$$
$$0.75x + 0.55y = 44.1 \quad (2)$$

Step 4 Solve the system. Multiply equation (2) by 100. Multiply equation (1) by –55 to eliminate y.

$$
\begin{array}{ll}
-55x - 55y = -3850 & \text{Multiply (1) by } -55. \\
\underline{75x + 55y = 4410} & \text{Multiply (2) by 100.} \\
20x = 560 & \text{Add.} \\
x = 28 & \text{Divide by 20.}
\end{array}
$$

Substitute the 28 for x in equation (1) to find the value of y.
$$28 + y = 70$$
$$y = 42$$

Step 5 State the answer. The desired mixture will contain 28 liters of 75% solution and 42 liters of 55% solution.

Step 6 Check.
Total amount: $28 + 42 = 70$
Total concentration: $0.75(28) + 0.55(42) = 44.1$
The answer is correct.

Now Try:

3. How many liters of water should be added to 25% antifreeze solution to get 30 liters of a 20% solution? How many liters of 25% solution are needed?

water   _______________

25% solution _______________

Objective 3 Practice Exercises

For extra help, see Example 3 on pages 257–258 of your text.

Solve each problem.

7. Jorge wishes to make 150 pounds of coffee blend that can be sold for $8 per pound. The blend will be a mixture of coffee worth $6 per pound and coffee worth $12 per pound. How many pounds of each kind of coffee should be used in the mixture?

7.
$6 coffee______________

$12 coffee______________

8. Bags of coffee worth $90 a bag must be mixed with coffee worth $75 a bag to get 50 bags worth $87 a bag. How many bags of each are needed?

8.
$90 coffee______________

$75 coffee______________

9. A pharmacist wants to add water to a solution that contains 80% medicine. She wants to obtain 12 oz. of a solution that is 20% medicine. How much water and how much of the 80% solution should she use?

9.
water________________

80% solution __________

Objective 4   Solve distance-rate-time problems using two variables.

Review this example for Objective 4:	**Now Try:**

Review this example for Objective 4:

4. A train travels 600 kilometers in the same time that a truck travels 520 kilometers. Find the speed of the train and the truck if the train's average speed is 8 kilometers per hour faster than the truck's.

Step 1 Read the problem. We need to find the rate of each vehicle.

Step 2 Assign variables.
Let $x =$ the rate of the train.
Let $y =$ the rate of the truck.

	d	r	t
Train	600	x	$\dfrac{600}{x}$
Truck	520	y	$\dfrac{520}{y}$

Step 3 Write a system of equations. From comparing the two speeds we have an equation.
$$x = y + 8$$
Since both vehicles travel for the same time, we have a second equation.
$$\frac{600}{x} = \frac{520}{y}$$
Multiplying both sides by xy, we have
$$600y = 520x.$$
The system of equations is
$$x = y + 8 \qquad (1)$$
$$600y = 520x \quad (2)$$

Step 4 Solve the system. We solve the system by substitution. Replace x with $y + 8$ in equation (2).
$$600y = 520(y + 8)$$
$$600y = 520y + 4160$$
$$80y = 4160$$
$$y = 52$$
Because $x = y + 8$,
$$x = 52 + 8 = 60.$$

Step 5 State the answer. The train's rate is 60 km per hr. The truck's rate is 52 km per hr.

Now Try:

4. Ashley walks 10 miles in the same time that Taylor walks 6 miles. If Ashley walks 1 mile per hour less than twice Taylor's rate, what is the rate at which each walks?

Ashley   _____________

Taylor   _____________

Step 6 Check.

Train: $\dfrac{600}{60} = 10$ hr   Truck: $\dfrac{520}{52} = 10$ hr

The rate of the train is 8 km more than the rate of the truck.

Objective 4 Practice Exercises

For extra help, see Examples 4–5 on pages 258–260 of your text.

Solve each problem.

10. Two cars start together and travel in the same direction, one going twice as fast as the other. At the end of 3 hours, they are 96 miles apart. How fast is each traveling?

10.
slower_________________

faster_________________

11. Travis and his sister Kate jog to school daily. Travis jogs at 9 miles per hour, and Kate jogs at 5 miles per hour. When Travis reaches school, Kate is $\frac{1}{2}$ mile from the school. How far do Travis and Kate live from their school? How long does it take Travis to jog to school?

11. distance_________

time_________

Objective 5 Solve problems with three variables using a system of three equations.

Review this example for Objective 5:

7. Lee has some $5, $10, and $20-bills. He has a total of 51 bills, worth $795. The number of $5-bills is 25 less than the number of $20-bills. Find the number of each type of bill he has.

Step 1: Read the problem again. There are three unknowns.

Step 2: Assign variables.
Let x = the number of $5 bills,
let y = the number of $10 bills,
let z = the number of $20 bills.

Step 3: Write a system of three equations.
There are a total of 51 bills, so
$$x + y + z = 51 \quad (1)$$
The bills amounted to $795, so
$$5x + 10y + 20z = 795 \quad (2)$$
The number of $5-bills is 25 less than the number of $20-bills, so $x = z - 25$ or
$$x - z = -25 \quad (3)$$
The system is
$$\begin{aligned} x + y + z &= 51 \quad (1) \\ 5x + 10y + 20z &= 795 \quad (2) \\ x \quad\quad - z &= -25 \quad (3) \end{aligned}$$

Step 4: Solve the system.
Eliminate y.
$$\begin{aligned} -10x - 10y - 10z &= -510 \quad \text{Multiply (1) by } -10 \\ 5x + 10y + 20z &= 795 \quad (2) \\ \hline -5x \quad\quad + 10z &= 285 \quad (4) \end{aligned}$$

Solve the system consisting of equations (3) and (4).
$$\begin{aligned} 5x - 5z &= -125 \quad \text{Multiply (3) by 5} \\ -5x + 10z &= 285 \quad (4) \\ \hline 5z &= 160 \\ z &= 32 \end{aligned}$$

Substitute 32 for z in equation (3) and solve for x.
$$\begin{aligned} x - z &= -25 \quad (3) \\ x - 32 &= -25 \quad \text{Let } z = 32. \\ x &= 7 \end{aligned}$$

Now Try:

7. The manager of the Sweet Candy Shop wishes to mix candy worth $4 per pound, $6 per pound, and $10 per pound to get 100 pounds of a mixture worth $7.60 per pound. The amount of $10 candy must equal the total amounts of the $4 and the $6 candy. How many pounds of each must be used?

$4 candy ___________

$6 candy ___________

$10 candy ___________

Substitute 7 for x and 32 for z in equation (1) and solve for y.

$$x + y + z = 51 \quad (1)$$
$$7 + y + 32 = 51 \quad \text{Let } x = 7, z = 32.$$
$$y + 39 = 51$$
$$y = 12$$

Step 5: State the answer. Lee has 7 $5-bills, 12 $10-bills, and 32 $20-bills.

Step 6: Check that the total value of the bills is $795 and that the number of $5-bills is 25 less than the number of $20-bills.

Objective 5 Practice Exercises

For extra help, see Examples 6–7 on pages 261–263 of your text.

Solve each problem involving three unknowns.

12. Julie has $80,000 to invest. She invests part at 5%, one fourth this amount at 6%, and the balance 7%. Her total annual income from interest is $4700. Find the amount invested at each rate.

12. 5% _______________

6% _______________

7% _______________

13. A merchant wishes to mix gourmet coffee selling for $8 per pound, $10 per pound, and $15 per pound to get 50 pounds of a mixture that can be sold for $11.70 per pound. The amount of the $8 coffee must be 3 pounds more than the amount of the $10 coffee. Find the number of pounds of each that must be used.

13. $8/lb _______________

$10/lb _______________

$15/lb _______________

14. A boy scout troop is selling popcorn. There are three different kinds of popcorn in three different arrangements. Arrangement I contains 1 bag of cheddar cheese popcorn, 2 bags of caramel popcorn, and 3 bags of microwave popcorn. Arrangement II contains 3 bags of cheddar cheese popcorn, 1 bag of caramel popcorn, and 2 bags of microwave popcorn. Arrangement III contains 2 bags of cheddar cheese popcorn, 3 bags of caramel popcorn, and 1 bag of microwave popcorn. Jim needs 28 bags of cheddar cheese popcorn, 22 bags of caramel popcorn, and 22 bags of microwave popcorn to give as stocking stuffers for Christmas. How many of each arrangement should he buy?

14. I _______________

II _______________

III _______________

Chapter 4 EXPONENTS, POLYNOMIALS, AND POLYNOMIAL FUNCTIONS

4.1   Integer Exponents

Learning Objectives
1   Use the product rule for exponents.
2   Define 0 and negative exponents.
3   Use the quotient rule for exponents.
4   Use the power rules for exponents.
5   Simplify exponential expressions.

Key Terms

Use the vocabulary terms listed below to complete each statement in exercises 1−5.

> **exponent**   **base**   **product rule for exponents**
>
> **power rule for exponents**   **quotient rule for exponents**

1.   The statement "If m and n are any integers, then $a^m \cdot a^n = a^{m+n}$" is an example of the _______________________________.

2.   The statement "If m and n are any integers and $a \neq 0$, then $\dfrac{a^m}{a^n} = a^{m-n}$" is an example of the _______________________________.

3.   The statement "If m and n are any integers, then $\left(a^m\right)^n = a^{mn}$" is an example of the _______________________________.

4.   In the expression a^m, a is the _______________ and m is the _______________.

Objective 1   Use the product rule for exponents.

Review these examples for Objective 1:	**Now Try:**
1. Apply the product rule for exponents, if possible, in each case.	1. Apply the product rule for exponents, if possible, in each case.
a. $5^6 \cdot 5^4$	**a.** $7^5 \cdot 7^3$
$5^6 \cdot 5^4 = 5^{6+4} = 5^{10}$	__________
b. $7^4 \cdot 7$	**b.** $6^{15} \cdot 6$
$7^4 \cdot 7 = 7^4 \cdot 7^1$	
$\qquad = 7^{4+1}$	__________
$\qquad = 7^5$	

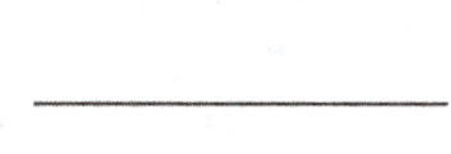

   c. $x^5 \cdot x^3 \cdot x^7$

$$x^5 \cdot x^3 \cdot x^7 = x^{5+3+7} = 5^{15}$$

   d. $(6y^3)(-5y^5)$

$$(6y^3)(-5y^5) = 6(-5)y^3 y^5$$
$$= -30 y^{3+5}$$
$$= -30 y^8$$

   e. $(9p^4 q)(4p^6 q^3)$

$$(9p^4 q)(4p^6 q^3) = 9(4)p^4 p^6 q^1 q^3$$
$$= 36 p^{10} q^4$$

   f. $a^3 \cdot b^2$

The bases are not the same, so the product rule
does not apply.

   c. $x^9 \cdot x^3$

   d. $(3x^4)(-8x^3)$

   e. $(4p^5 q^2)(6pq^4)$

   f. $k^5 \cdot m^4$

Objective 1 Practice Exercises

For extra help, see Example 1 on page 280 of your text.

Use the product rule to simplify each expression, if possible. Write each answer in exponential form.

1. $7^4 \cdot 7^3$

1. ______________

2. $(-2c^7)(-4c^8)$

2. ______________

3. $(3k^7)(-8k^2)(-2k^9)$

3. ______________

Objective 2   Define 0 and negative exponents.

Review these examples for Objective 2:

2. Evaluate.

   a. 9^0

$$9^0 = 1$$

   b. $(-9)^0$

$$(-9)^0 = 1$$

Now Try:

2. Evaluate.

   a. 25^0

   b. $(-25)^0$

c. -9^0

$$-9^0 = -(9^0) = -1$$

d. $-(-9)^0$

$$-(-9)^0 = -1$$

e. $(100k)^0$

$$(100k)^0 = 1,\ k \neq 0$$

f. $7^0 + 13^0$

$$7^0 + 13^0 = 1 + 1 = 2$$

3. Write using only positive exponents.

a. 3^{-2}

$$3^{-2} = \frac{1}{3^2}$$

b. 7^{-1}

$$7^{-1} = \frac{1}{7^1} = \frac{1}{7}$$

c. $(8z)^{-4}$

$$(8z)^{-4} = \frac{1}{(8z)^4},\ z \neq 0$$

d. $8z^{-4}$

$$8z^{-4} = 8\left(\frac{1}{z^4}\right) = \frac{8}{z^4},\ z \neq 0$$

e. $-n^{-8}$

$$-n^{-8} = -\frac{1}{n^8},\ n \neq 0$$

f. $(-n)^{-8}$

$$(-n)^{-8} = \frac{1}{(-n)^8},\ n \neq 0$$

c. -25^0

____

d. $-(-25)^0$

____

e. $(99k)^0$

____

f. $100^0 + 6^0$

____

3. Write using only positive exponents.

a. 5^{-4}

____

b. 9^{-1}

____

c. $(6z)^{-2}$

____

d. $12z^{-5}$

____

e. $-p^{-4}$

____

f. $(-p)^{-4}$

____

4. Evaluate.

    a. $6^{-1}+2^{-1}$

$$6^{-1}+2^{-1}=\frac{1}{6}+\frac{1}{2}=\frac{1}{6}+\frac{3}{6}=\frac{4}{6}=\frac{2}{3}$$

    b. $6^{-1}-9^{-1}$

$$6^{-1}-9^{-1}=\frac{1}{6}-\frac{1}{9}=\frac{3}{18}-\frac{2}{18}=\frac{1}{18}$$

    c. $\dfrac{1}{5^{-3}}$

$$\frac{1}{5^{-3}}=\frac{1}{\frac{1}{5^{3}}}=1\div\frac{1}{5^{3}}=1\cdot\frac{5^{3}}{1}=5^{3}=125$$

    d. $\dfrac{5^{-3}}{2^{-2}}$

$$\frac{5^{-3}}{2^{-2}}=\frac{\frac{1}{5^{3}}}{\frac{1}{2^{2}}}=\frac{1}{5^{3}}\div\frac{1}{2^{2}}=\frac{1}{5^{3}}\cdot\frac{2^{2}}{1}=\frac{2^{2}}{5^{3}}=\frac{4}{125}$$

4. Evaluate.

    a. $8^{-1}+6^{-1}$

    b. $5^{-1}-7^{-1}$

    c. $\dfrac{1}{7^{-2}}$

    d. $\dfrac{4^{-2}}{3^{-3}}$

Objective 2 Practice Exercises

For extra help, see Examples 2–4 on pages 281–282 of your text.

Evaluate the expression. Assume that all variables represent nonzero real numbers.

4. $2^{0}+6^{0}$

                                                       4. _______________

Evaluate or simplify each expression, and write it using only positive exponents. Assume that all variables represent nonzero real numbers.

5. $\dfrac{2}{r^{-7}}$ **5.** _______________

6. $-2k^{-4}$ **6.** _______________

Objective 3   Use the quotient rule for exponents.

Review these examples for Objective 3:

5. Apply the quotient rule for exponents, if possible, and write each result using only positive exponents.

a. $\dfrac{5^6}{5^4}$

$$\dfrac{5^6}{5^4} = 5^{6-4} = 5^2$$

b. $\dfrac{p^7}{p^3}$

$$\dfrac{p^7}{p^3} = p^{7-3} = p^4, \; p \neq 0$$

c. $\dfrac{k^5}{k^{13}}$

$$\dfrac{k^5}{k^{13}} = k^{5-13} = k^{-8} = \dfrac{1}{k^8}, \; k \neq 0$$

d. $\dfrac{4^9}{4^{-2}}$

$$\dfrac{4^9}{4^{-2}} = 4^{9-(-2)} = 4^{9+2} = 4^{11}$$

e. $\dfrac{9^{-3}}{9^7}$

$$\dfrac{9^{-3}}{9^7} = 9^{-3-7} = 9^{-10} = \dfrac{1}{9^{10}}$$

f. $\dfrac{7}{7^{-1}}$

$$\dfrac{7}{7^{-1}} = \dfrac{7^1}{7^{-1}} = 7^{1-(-1)} = 7^2$$

g. $\dfrac{z^{-3}}{z^{-9}}$

$$\dfrac{z^{-3}}{z^{-9}} = z^{-3-(-9)} = z^6, \; z \neq 0$$

Now Try:

5. Apply the quotient rule for exponents, if possible, and write each result using only positive exponents.

a. $\dfrac{10^7}{10^5}$

____

b. $\dfrac{q^8}{q^5}$

____

c. $\dfrac{m^6}{m^9}$

____

d. $\dfrac{8^7}{8^{-3}}$

____

e. $\dfrac{11^{-4}}{11^8}$

____

f. $\dfrac{9^2}{9^{-2}}$

____

g. $\dfrac{x^{-3}}{x^{-5}}$

____

h. $\dfrac{x^5}{y^4}$, $y \neq 0$ **h.** $\dfrac{y^4}{z^2}$

The quotient rule does not apply because the bases are different.   ______________

Objective 3 Practice Exercises

For extra help, see Example 5 on page 283 of your text.

Use the quotient rule to simplify each expression, if possible, and write it using only positive exponents. Assume that all variables represent nonzero real numbers.

7. $\dfrac{4k^7}{m^5}$ **7.** ______________

8. $\dfrac{3^{-1}}{3^4}$ **8.** ______________

9. $\dfrac{p^6}{p^{-2}}$ **9.** ______________

Objective 4   Use the power rules for exponents.

Review these examples for Objective 4:   **Now Try:**

6.   Simplify, using the power rules.   **6.**   Simplify, using the power rules.

a. $\left(p^7\right)^4$ **a.** $\left(q^5\right)^4$

$\left(p^7\right)^4 = p^{7 \cdot 4} = p^{28}$ __________

b. $(5y)^3$ **b.** $(7x)^3$

$(5y)^3 = 5^3 y^3 = 125y^3$ __________

c. $\left(\dfrac{3}{5}\right)^3$ **c.** $\left(\dfrac{2}{5}\right)^3$

$\left(\dfrac{3}{5}\right)^3 = \dfrac{3^3}{5^3} = \dfrac{27}{125}$ __________

d. $\left(8p^5\right)^2$ **d.** $\left(5r^5\right)^3$

$\left(8p^5\right)^2 = 8^2\left(p^5\right)^2 = 8^2 p^{5 \cdot 2} = 64p^{10}$ __________

e. $\left(-\dfrac{5n^4}{y}\right)^3$

$$\left(-\frac{5n^4}{y}\right)^3 = \frac{\left(-5n^4\right)^3}{y^3} = \frac{(-5)^3 n^{4\cdot3}}{y^3}$$

$$= \frac{-125n^{12}}{y^3}, \; y \neq 0$$

7. Write each expression with only positive exponents and then evaluate.

a. $\left(\dfrac{1}{3}\right)^{-4}$

$$\left(\frac{1}{3}\right)^{-4} = 3^4 = 81$$

b. $\left(\dfrac{4}{5}\right)^{-4}$

$$\left(\frac{4}{5}\right)^{-4} = \left(\frac{5}{4}\right)^4 = \frac{625}{256}$$

c. $\left(\dfrac{3}{8}\right)^{-2}$

$$\left(\frac{3}{8}\right)^{-2} = \left(\frac{8}{3}\right)^2 = \frac{64}{9}$$

e. $\left(-\dfrac{3p^4}{q}\right)^3$

____

7. Write each expression with only positive exponents and then evaluate.

a. $\left(\dfrac{1}{5}\right)^{-4}$

____

b. $\left(\dfrac{7}{9}\right)^{-2}$

____

c. $\left(\dfrac{3}{4}\right)^{-3}$

____

Objective 4 Practice Exercises

For extra help, see Examples 6–7 on pages 284–285 of your text.

Simplify each expression, and write it using only positive exponents. Assume that all variables represent nonzero real numbers.

10. $\left(-2w^3 z^7\right)^4$

10. __________

11. $\left(\dfrac{-2a}{b^2}\right)^7$

11. __________

12. $\left(\dfrac{2}{7}\right)^{-3}$

12. __________

Name: Date:
Instructor: Section:

Objective 5   Simplify exponential expressions.

Review these examples for Objective 5: | **Now Try:**

8. Simplify each expression so that no negative exponents appear in the final result. Assume that all variables represent nonzero real numbers.

8. Simplify each expression so that no negative exponents appear in the final result. Assume that all variables represent nonzero real numbers.

a. $6^3 \cdot 6^{-5}$

a. $8^4 \cdot 8^{-6}$

$$6^3 \cdot 6^{-5} = 6^{3+(-5)} = 6^{-2} = \frac{1}{6^2} = \frac{1}{36}$$

____

b. $x^{-5} \cdot x^{-3} \cdot x^4$

b. $x^{-7} \cdot x^6 \cdot x^{-2}$

$$x^{-5} \cdot x^{-3} \cdot x^4 = x^{-5+(-3)+4} = x^{-4} = \frac{1}{x^4}$$

____

c. $\left(7^{-3}\right)^{-4}$

c. $\left(5^{-2}\right)^{-3}$

$$\left(7^{-3}\right)^{-4} = 7^{(-3)(-4)} = 7^{12}$$

____

d. $\left(x^{-3}\right)^7$

d. $\left(x^{-8}\right)^3$

$$\left(x^{-3}\right)^7 = x^{(-3)7} = x^{-21} = \frac{1}{x^{21}}$$

____

e. $\dfrac{x^{-6}y^3}{x^3 y^{-4}}$

e. $\dfrac{x^{-7}y^3}{x^3 y^{-6}}$

$$\frac{x^{-6}y^3}{x^3 y^{-4}} = \frac{x^{-6}}{x^3} \cdot \frac{y^3}{y^{-4}}$$

$$= x^{-6-3}y^{3-(-4)}$$

$$= x^{-9}y^7$$

$$= \frac{y^7}{x^9}$$

____

f. $\left(5^2 x^{-4}\right)^{-3}$

f. $\left(3^2 x^{-5}\right)^{-2}$

$$\left(5^2 x^{-4}\right)^{-3} = \left(5^2\right)^{-3} \cdot \left(x^{-4}\right)^{-3}$$

$$= 5^{-6}x^{12}$$

$$= \frac{x^{12}}{5^6}, \text{ or } \frac{x^{12}}{15,625}$$

____

g. $\left(\dfrac{5x^3}{y^2}\right)^2\left(\dfrac{3x^4}{y^{-3}}\right)^{-1}$

$$\left(\frac{5x^3}{y^2}\right)^2\left(\frac{3x^4}{y^{-3}}\right)^{-1}=\frac{5^2\left(x^3\right)^2}{\left(y^2\right)^2}\cdot\frac{y^{-3}}{3x^4}$$

$$=\frac{25x^6}{y^4}\cdot\frac{y^{-3}}{3x^4}$$

$$=\frac{25}{3}x^{6-4}y^{-4-3}$$

$$=\frac{25}{3}x^2y^{-7}$$

$$=\frac{25x^2}{3y^7}$$

h. $\left(-\dfrac{7p^4q^6}{35pq^{-3}}\right)^{-2}$

$$\left(-\frac{7p^4q^6}{35pq^{-3}}\right)^{-2}=\left(\frac{p^{4-1}q^{6-(-3)}}{-5}\right)^{-2}$$

$$=\left(\frac{p^3q^9}{-5}\right)^{-2}$$

$$=\frac{\left(p^3\right)^{-2}\left(q^9\right)^{-2}}{(-5)^{-2}}$$

$$=\frac{p^{-6}q^{-18}}{(-5)^{-2}}$$

$$=\frac{(-5)^2}{p^6q^{18}}$$

$$=\frac{25}{p^6q^{18}}$$

g. $\left(\dfrac{2x^3}{y}\right)^2\left(\dfrac{3x^5}{y^{-4}}\right)^{-1}$

____

h. $\left(-\dfrac{6x^6y^7}{18x^2y^{-4}}\right)^{-2}$

____

     163

Objective 5 Practice Exercises

For extra help, see Example 8 on page 286 of your text.

Simplify each expression, and write it using only positive exponents. Assume that all variables represent nonzero real numbers.

13. $\dfrac{c^{10}\left(c^2\right)^3}{\left(c^3\right)^3\left(c^2\right)^{-9}}$

13. _________________

14. $\left(a^{-1}b^{-2}\right)^{-4}$

14. _________________

15. $\left(\dfrac{k^3 t^4}{k^2 t^{-1}}\right)^{-4}$

15. _________________

                Copyright © 2020 Pearson Education, Inc.

Chapter 4 EXPONENTS, POLYNOMIALS, AND POLYNOMIAL FUNCTIONS

4.2 Scientific Notation

Learning Objectives
1 Write numbers in scientific notation.
2 Convert numbers in scientific notation to standard notation.
3 Use scientific notation in calculations.

Key Terms

Use the vocabulary terms listed below to complete each statement in exercises 1−2.

exponent   **base**   **scientific notation**

1. A number written as $a \times 10^n$, where $1 \le |a| < 10$ and n is an integer, is written in

    _______________________________.

2. In the expression a^m, a is the _______________ and m is the _______________.

Objective 1 Write numbers in scientific notation.

Review these examples for Objective 1:

1. Write each number in scientific notation.

    a. 970,000

    Step 1 Place a caret to the right of the 9 to mark the new location of the decimal point.

      $9_\wedge 70,000$

    Step 2 Count from the decimal point, which is understood to be after the last 0, to the caret.

      $9_\wedge 70,000$

      count 5 places

    Step 3 Since 9.7 is to be made greater, the exponent on 10 is positive.

      $970,000 = 9.7 \times 10^5$

    b. 0.000064

      $0.00006_\wedge 4$ count 5 places

    Since 6.4 is to be made less, the exponent on 10 is negative.

      $0.000064 = 6.4 \times 10^{-5}$

    c. −0.000984

      $-0.000984 = -9.84 \times 10^{-4}$

Now Try:

1. Write each number in scientific notation.

    a. 3,946,000

    ____

    b. 0.00048

    ____

    c. −849,000

    ____

Objective 1 Practice Exercises

For extra help, see Example 1 on page 291 of your text.

Write each number in scientific notation.

1.   23,651

1. ________________

2. $-429,600,000,000$

2. ________________

3. -0.0002208

3. ________________

Objective 2   Convert numbers in scientific notation to standard notation.

Review these examples for Objective 2:

2. Write each number in standard notation.

a. 2.06×10^4

$$2.0600_\wedge$$

Move the decimal point 4 places to the right. Attach 0s as necessary.

$$2.06 \times 10^4 = 20,600$$

b. 3.41×10^{-5}

$$0_\wedge 00003.41$$

Move the decimal point 5 places to the left. Attach 0s as necessary.

$$3.41 \times 10^{-5} = 0.0000341$$

c. -9.24×10^0

$$-9.24 \times 10^0 = -9.24 \times 1 = -9.24$$

Now Try:

2. Write each number in standard notation.

a. 9.45×10^6

____

b. 8.04×10^{-5}

____

c. -7.89×10^0

____

Objective 2 Practice Exercises

For extra help, see Example 2 on page 291 of your text.

Write in standard notation.

4. 9×10^7

4. ________________

5. 4.2×10^{-5}

5. ________________

6. -4.02×10^4

6. ________________

Objective 3   Use scientific notation in calculations.

Review these examples for Objective 3:

3. Evaluate.

$$\frac{3,420,000 \times 0.00054}{0.000057 \times 27,000}$$

$$\frac{3,420,000 \times 0.00054}{0.000057 \times 27,000} = \frac{3.42 \times 10^6 \times 5.4 \times 10^{-4}}{5.7 \times 10^{-5} \times 2.7 \times 10^4}$$

$$= \frac{3.42 \times 5.4 \times 10^6 \times 10^{-4}}{5.7 \times 2.7 \times 10^{-5} \times 10^4}$$

$$= \frac{3.42 \times 5.4 \times 10^2}{5.7 \times 2.7 \times 10^{-1}}$$

$$= \frac{3.42 \times 5.4}{5.7 \times 2.7} \times 10^3$$

$$= 1.2 \times 10^3 = 1200$$

4. There are about 6×10^{23} atoms in a mole of atoms. About how many atoms are there in 81,000 moles?

First write 81,000 in scientific notation.

$$81,000 = 8.1 \times 10^4$$

Now multiply.

$$(6 \times 10^{23}) \times (8.1 \times 10^4) = 6 \times 8.1 \times 10^{23} \times 10^4$$

$$= 48.6 \times 10^{27}$$

$$= (4.86 \times 10^1) \times 10^{27}$$

$$= 4.86 \times 10^{28}$$

There are 4.86×10^{28} atoms in 81,000 moles.

Now Try:

3. Evaluate.

$$\frac{4,830,000 \times 0.00056}{0.00069 \times 4000}$$

____

4. The Milky Way galaxy is about 100,000 light-years in diameter. If one light-year is about 9.5×10^{15} meters, then about how many meters is the diameter of the Milky Way?

____

Objective 3 Practice Exercises

For extra help, see Examples 3–4 on page 292 of your text.

Evaluate. Write answer in scientific notation and in standard form.

7. $$\frac{0.0021 \times 4800}{1,600,000 \times 0.000007}$$

7. _________________

8. $(2.3 \times 10^4) \times (1.1 \times 10^{-2})$

8. _________________

Work the problem. Give answer in scientific notation.

9. There are about 6×10^{23} atoms in a mole of atoms. How many atoms are there in 8.1×10^{-5} mole?

9. _________________

Chapter 4 EXPONENTS, POLYNOMIALS, AND POLYNOMIAL FUNCTIONS

4.3   Adding and Subtracting Polynomials

Learning Objectives	
1	Define and classify polynomials.
2	Add and subtract polynomials.

Key Terms

Use the vocabulary terms listed below to complete each statement in exercises 1−12.

term	coefficient (numerical coefficient)	
algebraic expression	polynomial	polynomial in x
descending powers	trinomial	binomial
monomial	degree of a term	degree of a polynomial
negative of a polynomial		

1.   The ______________________________ is the sum of the exponents on the variables in that term.

2.   A polynomial in x is written in ______________________________ if the exponents on x decrease from left to right.

3.   A(n) ________________ is a number, a variable, or a product or quotient of a number and one or more variables raised to powers.

4.   A polynomial with exactly three terms is called a ______________________________.

5.   A(n) ______________________ is a term, or a finite sum of terms, in which all variables have whole number exponents and no variables appear in denominators.

6.   The numerical factor in a term is its ______________________________.

7.   A polynomial with exactly one term is called a ______________________.

8.   The ______________________________ is the greatest degree of any term of the polynomial.

9.   A(n) ________________________ is a polynomial with exactly two terms.

10.   A(n) ____________ ______________ is any combination of variables or constants joined by the basic operations of addition, subtraction, multiplication, and division (except by 0), or raising to powers or taking roots.

11. The _______________________ is obtained by changing the sign of every coefficient in the polynomial.

12. A polynomial containing only the variable x is a _______________________.

Objective 1   Define and classify polynomials.

Review these examples for Objective 1:

1. Write the polynomial in descending powers of the variable. Then give the leading term and the leading coefficient.

$$y^2 - 19y^4 + 7y^6 - 15y^3 + 17$$

Rewriting the expression, we have
$$7y^6 - 19y^4 - 15y^3 + y^2 + 17$$

The leading term is $7y^6$ and the leading coefficient is 7.

2. Identify each polynomial as a *monomial*, a *binomial*, a *trinomial*, or *none of these*. Also, give the degree.

a. $-5x^3 + 7x + 4$

This is a trinomial of degree 3.

b. $\frac{2}{5}x^2 y$

This is a monomial of degree 3 (because $2 + 1 = 3$).

c. $10m^4 + 27m^8$

This is a binomial of degree 8.

d. $p^7 - p^4 + 12p^2 - 1$

Polynomials of four terms or more do not have special names, so *none of these* is the answer that applies here. This polynomial has degree 7.

Now Try:

1. Write the polynomial in descending powers of the variable. Then give the leading term and the leading coefficient.
$$x^3 - 5x^5 + 11x^7 - 4$$

____

____

____

2. Identify each polynomial as a *monomial*, a *binomial*, a *trinomial*, or *none of these*. Also, give the degree.

a. $-y^3 + 7y^2 + 4$

____

b. $4x^9 y$

____

c. $15m^8 - m^{20}$

____

d. $p^6 - p^4 + 12p^2 - 15$

____

Objective 1 Practice Exercises

For extra help, see Examples 1–2 on pages 297–298 of your text.

Write the polynomial in descending powers. Give the leading term and coefficient.

1. $8 + 5y - 7y^2 + y^3$ 1. __________________

*Identify each polynomial as a **monomial**, a **binomial**, a **trinomial**, or **none of these**. Give the degree.*

2. $\frac{1}{2}x^2 - \frac{3}{4}x + \frac{1}{4}x$ 2. __________________

3. $p + 3p^4$ 3. __________________

Objective 2   Add and subtract polynomials.

Review these examples for Objective 2:	**Now Try:**
3. Combine like terms.	**3.** Combine like terms.
b. $8p + 7q - 10p + 3q$	**b.** $12m - 5n + 2m + 7n$
$8p + 7q - 10p + 3q = (8 - 10)p + (7 + 3)q$ $= -2p + 10q$	___________
c. $12x^3y - 9xy^3 + 4x^3y - 5xy^3$	**c.** $9x^2yz + 11xyz^2$ $-10x^2yz + 2xyz^2$
$12x^3y - 9xy^3 + 4x^3y - 5xy^3$ $= 12x^3y + 4x^3y - 9xy^3 - 5xy^3$ $= 16x^3y - 14xy^3$	___________
4. Add $(7a^4 - 8a^2 + 17a) + (-18a^4 + 3a^2 + 6)$.	**4.** Add $-7x^5 + 3x^3 - 4x$ $\underline{9x^5 - 17x^3 + 3x}$
$(7a^4 - 8a^2 + 17a) + (-18a^4 + 3a^2 + 6)$ $= 7a^4 - 18a^4 - 8a^2 + 3a^2 + 17a + 6$ $= -11a^4 - 5a^2 + 17a + 6$	___________

5. Subtract $(-7x^2-5x+10)-(-8x^2+3x-6)$.

$(-7x^2-5x+10)-(-8x^2+3x-6)$

$\quad = -7x^2-5x+10+8x^2-3x+6$

$\quad = -7x^2+8x^2-5x-3x+10+6$

$\quad = x^2-8x+16$

5. Subtract $(7y^3-4y^2+3y)$
$\qquad\qquad -(-3y^3+8y^2-10y)$

____

Objective 2 Practice Exercises

For extra help, see Examples 3–5 on pages 298–300 of your text.

Add.

4. $(x^2+6x-8)+(3x^2-10)$

4. _______________

Subtract.

5. $(5a^4-6a^2+9a)-(a^3-19a-1)$

5. _______________

Add or subtract as indicated.

6. $4ab+2bc-9ac+3ca-2cb-9ba$

6. _______________

                    171

Chapter 4 EXPONENTS, POLYNOMIALS, AND POLYNOMIAL FUNCTIONS

4.4   Polynomial Functions, Graphs, and Composition

Learning Objectives
1   Recognize and evaluate polynomial functions.
2   Use a polynomial function to model data.
3   Add and subtract polynomial functions.
4   Graph basic polynomial functions.
5   Find the composition of functions.

Key Terms

Use the vocabulary terms listed below to complete each statement in exercises 1−6.

polynomial function of degree n	**identity function**
squaring function	**cubing function**
composite function	**composition**

1. The polynomial function defined by $f(x) = x^3$ is called the

    _______________________________.

2. The function $g(f(x))$ is a _______________________________.

3. A function defined by $f(x) = a_n x^n + a_{n-1} x^{n-1} + \cdots + a_1 x + a_0$, where $a_n \neq 0$ and n is a whole number is a ____.

4. The polynomial function defined by $f(x) = x^2$ is called the

    _______________________________.

5. The simplest polynomial function is the _______________________________ defined by $f(x) = x$.

6. If f and g are functions, then the _______________________________ of g and f is defined by $(g \circ f)(x) = g(f(x))$ for all x in the domain of f such that $f(x)$ is in the domain of g.

Name: Date:

Instructor: Section:

Objective 1   Recognize and evaluate polynomial functions.

Review this example for Objective 1:

1. Let $f(x) = 6x^3 - 6x + 1$. Find $f(-3)$.

    Substitute -3 for x.

    $$f(-3) = 6(-3)^3 - 6(-3) + 1$$
    $$= 6(-27) - 6(-3) + 1$$
    $$= -162 + 18 + 1$$
    $$= -143$$

Now Try:

1. Let $p(x) = -x^4 + 3x^2 - x + 7$.
    Find $p(2)$.

    ____

Objective 1 Practice Exercises

For extra help, see Example 1 on page 303 of your text.

For each polynomial function, find (a) f(−2) and (b) f(3).

1. $f(x) = -x^2 - x - 5$

2. $f(x) = 2x^2 + 3x - 5$

3. $f(x) = 3x^4 - 5x^2$

1. (a) _______________

   (b) _______________

2. (a) _______________

   (b) _______________

3. (a) _______________

   (b) _______________

Objective 2   Use a polynomial function to model data.

Review this example for Objective 2:

2. The average undergraduate tuition, room, and board at public four-year colleges for the years 1989–2009 can be modeled by the function

$$f(x) = 112.38x^2 - 2949.32x + 29{,}488.08,$$

where $x = 1$ corresponds to 1989, $x = 2$ corresponds to 1990, etc. Use this model to estimate the tuition, room, and board in 2004. (Source: National Center for Education Statistics)

Since $x = 16$ corresponds to 2004, we must find $f(16)$.

$$f(16) = 112.38(16)^2 - 2949.32(16) + 29{,}488.08$$
$$= 11{,}068.24$$

According to the model, the average cost of undergraduate tuition, room, and board at four-year public colleges was $11,068.24 in 2004.

Now Try:

2. Use the function at the left to estimate the tuition, room, and board in 2000.

____

Objective 2 Practice Exercises

For extra help, see Example 2 on page 303 of your text.

Solve each problem.

4. A widget manufacturer estimates that her monthly revenue can be modeled by the function

$$f(x) = -0.006x^2 + 32x - 10{,}000.$$

  (a) Find the revenue if 2000 widgets are sold.

  (b) Find the revenue if 3500 widgets are sold.

4. a.________________

   b.________________

5. If a ball is batted at an angle of 35°, the distance that the ball travels is given approximately by

$$D = 0.029v^2 + 0.021v - 1,$$ where v is the bat speed in miles per hour and D is the distance traveled in feet. Find the distance a batted ball will travel if the ball is batted with a velocity of 90 miles per hour. Round your answer to the nearest whole number.

5. ________________

6. The unemployment rate in a certain community can be modeled by the equation

$y = 0.0248x^2 - 0.4810x + 7.8543$, where y is the unemployment rate (percent) and x is the month ($x = 1$ represents January, $x = 2$ represents February, etc.) Use the model to find the unemployment rate in August. Round your answer to the nearest tenth.

6. _______________

Objective 3   Add and subtract polynomial functions.

Review these examples for Objective 3:

3. For $f(x) = 2x^2 + 4x - 5$ and

$g(x) = -x^2 + 3x - 8$, find each of the following.

a. $(f + g)(x)$

$$(f + g)(x) = f(x) + g(x)$$
$$= (2x^2 + 4x - 5) + (-x^2 + 3x - 8)$$
$$= x^2 + 7x - 13$$

b. $(f - g)(x)$

$$(f - g)(x) = f(x) - g(x)$$
$$= (2x^2 + 4x - 5) - (-x^2 + 3x - 8)$$
$$= (2x^2 + 4x - 5) + (x^2 - 3x + 8)$$
$$= 3x^2 + x + 3$$

4. Find the following for polynomial functions f and g as defined.

$$f(x) = 9x^2 - 3x \quad \text{and} \quad g(x) = 5x.$$

$(f + g)(4)$

$$(f + g)(4) = f(4) + g(4)$$
$$= \left[9(4)^2 - 3(4)\right] + 5(4)$$
$$= [144 - 12] + 20$$
$$= 152$$

Alternatively, we could first find $(f + g)(x)$.

Now Try:

3. For $f(x) = 6x^2 - 7x + 12$ and

$g(x) = -3x^2 + x + 9$, find each of the following.

a. $(f + g)(x)$

____

b. $(f - g)(x)$

____

4. Find the following for polynomial functions f and g as defined.

$$f(x) = 7x^2 - 5x \quad \text{and}$$
$$g(x) = 4x.$$

$(f + g)(x)$ and $(f + g)(-2)$

____

$$(f+g)(x)=f(x)+g(x)$$
$$=(9x^2-3x)+5x$$
$$=9x^2+2x$$
$$(f+g)(4)=9(4)^2+2(4)$$
$$=152$$

Objective 3 Practice Exercises

For extra help, see Examples 3–4 on pages 304–305 of your text.

For the pair of functions, find (a) $(f+g)(x)$ and $(f-g)(x)$.

7. $f(x)=2x^2+4x-5,\ g(x)=-x^2+3x-8$

7. a.______________

b.______________

Let $f(x)=3x^2+2,\ g(x)=-5x,$ and $h(x)=x+2$. Find each of the following.

8. $(f-g)(-1)$

8. ______________

9. $(f+h)(2)$

9. ______________

Objective 4   Graph basic polynomial functions.

Review this example for Objective 4:

5.   Graph $f(x)=x^2+2$. Give the domain and range.

For each input, square it and then add 2.

x	$f(x)=x^2+2$
-2	6
-1	3
0	2
1	3
2	6

Now Try:

5.   Graph $f(x)=-x^3+1$. Give the domain and range.

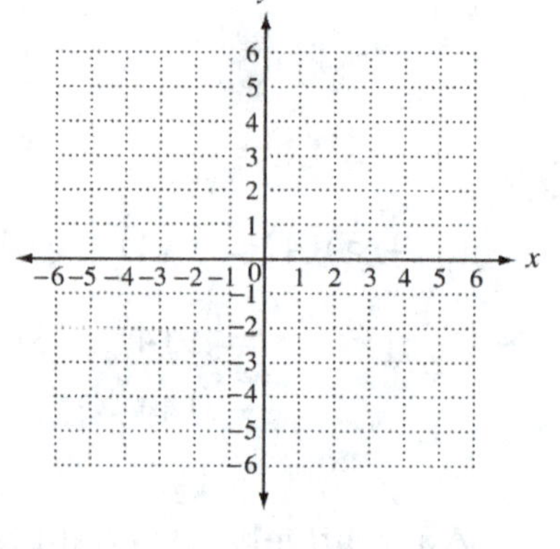

____

Name: ___________________________   Date: ___________________________

Instructor: _______________________   Section: ________________________

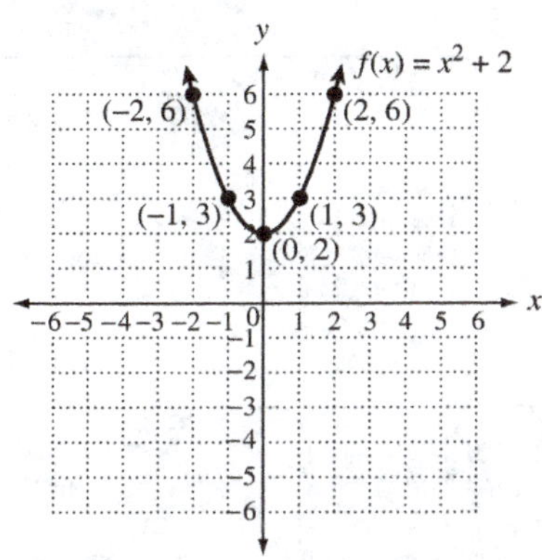

The domain is $(-\infty, \infty)$ and the range is $[2, \infty)$.

Objective 4 Practice Exercises

For extra help, see Example 5 on pages 306–307 of your text.

Graph each function by creating a table of ordered pairs. Give the domain and the range.

10. $f(x) = -3x + 2$

10. ______________________________

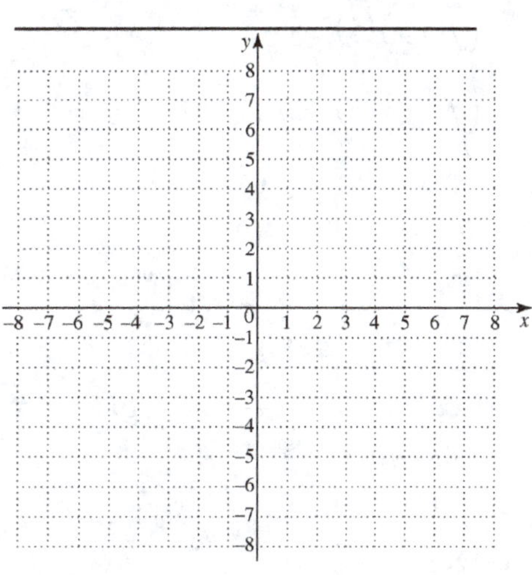

11. $f(x) = -2x^2$

11. ______________________________

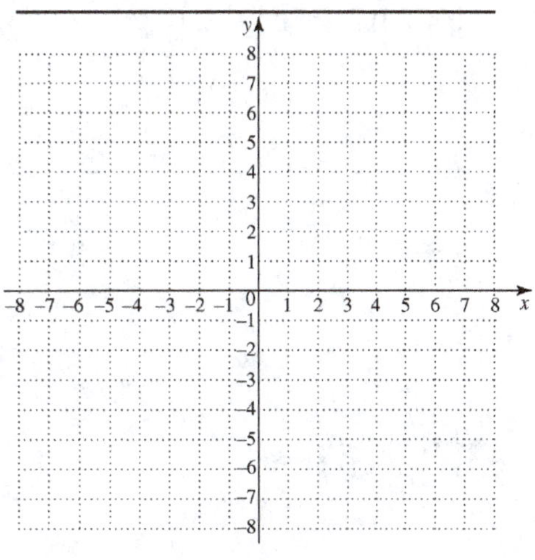

12. $f(x) = x^3 + 2$

12. _______________________

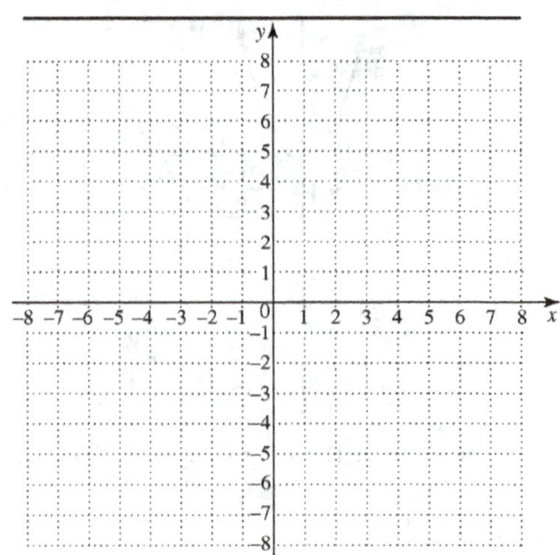

Objective 5 Find the composition of functions.

Review these examples for Objective 5:

6. Let $f(x) = 3x - 1$ and $g(x) = x^2 + 2$. Find $(f \circ g)(5)$.

$$(f \circ g)(5) = f(g(5))$$
$$= f(5^2 + 2)$$
$$= f(27)$$
$$= 3(27) - 1$$
$$= 80$$

8. Let $f(x) = 3x - 1$ and $g(x) = x^2 + 2$. Find the following.

$$(g \circ f)(2)$$
$$(g \circ f)(2) = g(f(2))$$
$$= (f(2))^2 + 2$$
$$= (3(2) - 1)^2 + 2$$
$$= 5^2 + 2$$
$$= 27$$

Now Try:

6. Let $f(x) = -3x - 3$ and $g(x) = x^2 - 5$. Find $(f \circ g)(-3)$.

____

8. Let $f(x) = -3x - 3$ and $g(x) = x^2 - 5$. Find the following.
$(g \circ f)(2)$

____

Objective 5 Practice Exercises

For extra help, see Examples 6–8 on pages 308–309 of your text.

Find the following.

13. Let $f(x) = 4x - 3$ and $g(x) = 2x^2 - 1$. Find the following.

    a. $(f \circ g)(2)$

    b. $(g \circ f)(-1)$

    c. $(f \circ g)(x)$

13. a.______________

    b.______________

    c.______________

14. Let $f(x) = \dfrac{1}{x}$ and $g(x) = 3x^2 - 4x + 1$. Find the following.

    a. $(f \circ g)(2)$

    b. $(g \circ f)\left(\dfrac{1}{3}\right)$

    c. $(g \circ f)(x)$

14. a.______________

    b.______________

    c.______________

Chapter 4 EXPONENTS, POLYNOMIALS, AND POLYNOMIAL FUNCTIONS

4.5   Multiplying Polynomials

Learning Objectives
1   Multiply terms.
2   Multiply any two polynomials.
3   Multiply binomials.
4   Find the product of a sum and difference of two terms.
5   Find the square of a binomial.
6   Multiply polynomial functions.

Key Terms

Use the vocabulary terms listed below to complete each statement in exercises 1–3.

FOIL   outer product   inner product

1.   The _________________________________ of $(2y-5)(y+8)$ is $-5y$.

2.   _________________________________ is a shortcut method for finding the product of two binomials.

3.   The _________________________________ of $(2y-5)(y+8)$ is $16y$.

Objective 1   Multiply terms.

Review this example for Objective 1:

1.   Find the product.

$$3p^3q^4(4p^4q^5)$$

$$3p^3q^4(4p^4q^5) = 3(4)p^3 \cdot p^4 \cdot q^4 \cdot q^5$$
$$= 12p^7q^9$$

Now Try:

1.   Find the product.

$$5x^4y^2(2x^5y)$$

____

Objective 1 Practice Exercises

For extra help, see Example 1 on page 315 your text.

Find each product.

1. $-2y^3(-8y^4)$ 1. _________________________

2. $3y^2z^3(6yz^4)$ 2. _________________________

                Copyright © 2020 Pearson Education, Inc.

3. $-12r^4s^7(-9r)$ **3.** ________________

Objective 2   Multiply any two polynomials.

Review these examples for Objective 2:   **Now Try:**

2. Find each product.   **2.** Find each product.

b. $6x^3(-5x^3+7x-8)$ **b.** $7x^2(-4x^2-8x-6)$

$$6x^3(-5x^3+7x-8)$$
$$= 6x^3(-5x^3)+6x^3(7x)+6x^3(-8)$$
$$= -30x^6+42x^4-48x^3$$

d. $5x^3(x+4)(x-7)$ **d.** $3x^2(x-1)(x+6)$

$$5x^3(x+4)(x-7)$$
$$= 5x^3\left[(x+4)(x)+(x+4)(-7)\right]$$
$$= 5x^3\left[x^2+4x-7x-28\right]$$
$$= 5x^3(x^2-3x-28)$$
$$= 5x^5-15x^4-140x^3$$

3. Find each product.   **3.** Find each product.

a. $(6a-5b)(2a+b)$ **a.** $(8x-7)(4x+5)$

$$\begin{array}{r} 6a-5b \\ \underline{2a+b} \\ 6ab-5b^2 \\ \underline{12a^2-10ab} \\ 12a^2-4ab-5b^2 \end{array}$$

b. $(6m^3-7m^2+3)(4m-8)$ **b.** $(4x^3-5x^2+9)(3x-7)$

$$\begin{array}{r} 6m^3-7m^2+3 \\ \underline{4m-8} \\ -48m^3+56m^2-24 \\ \underline{24m^4-28m^3+12m} \\ 24m^4-76m^3+56m^2+12m-24 \end{array}$$

Objective 2 Practice Exercises

For extra help, see Examples 2–3 on pages 315–316 of your text.

Find each product.

4. $7b^2(-5b^2 + 1 - 4b)$ 4. ___________________

5. $(3m - 5)(2m + 4)$ 5. ___________________

6. $3m^3 + 2m^2 - 4m$ 6. ___________________
    $$\underline{2m^2 + 1}$$

Objective 3   Multiply binomials.

Review this example for Objective 3: | **Now Try:**

4. Use the FOIL method to find the product.

$(9x - 4y)(7x + 6y)$

$(9x - 4y)(7x + 6y)$

   First   Outer   Inner   Last

$= 63x^2 + 54xy - 28xy - 24y^2$

$= 63x^2 + 26xy - 24y^2$

Now Try:

4. Use the FOIL method to find the product.

$(6z - 4)(9z - 5)$

____

Objective 3 Practice Exercises

For extra help, see Example 4 on pages 317–318 of your text.

Find each product.

7. $(3x + 2y)(2x - 3y)$ 7. ___________________

8. $(5a - b)(4a + 3b)$ 8. ___________________

9. $(x-5)(x+3)$ 9. ___________

Objective 4   Find the product of a sum and difference of two terms.

Review these examples for Objective 4: | **Now Try:**

5. Find each product.

5. Find each product.

c. $(8m+5n)(8m-5n)$

c. $(3m+8y)(3m-8y)$

$$(8m+5n)(8m-5n) = (8m)^2 - (5n)^2$$
$$= 64m^2 - 25n^2$$

d. $5x(x+4)(x-4)$

d. $10x^3(x+5)(x-5)$

$$5x(x+4)(x-4) = 5x(x^2-16)$$
$$= 5x^3 - 80x$$

Objective 4 Practice Exercises

For extra help, see Example 5 on page 318 of your text.

Find each product.

10. $(8k+5p)(8k-5p)$ 10. ___________

11. $(7x-3y)(7x+3y)$ 11. ___________

12. $(9-4y)(9+4y)$ 12. ___________

Objective 5   Find the square of a binomial.

Review these examples for Objective 5:   **Now Try:**

6. Find each product.

6. Find each product.

b. $(p-8)^2$

b. $(p-3)^2$

$$(p-8)^2 = p^2 - 2 \cdot p \cdot 8 + 8^2$$
$$= p^2 - 16p + 64$$

c. $(5p+4q)^2$

$$(5p+4q)^2 = (5p)^2 + 2(5p)(4q) + (4q)^2$$
$$= 25p^2 + 40pq + 16q^2$$

c. $(4p-7q)^2$

____

7. Use special products to find each product.

7. Use special products to find each product.

a. $[(4p-5)+6q][(4p-5)-6q]$

a. $[(7p-3)+2q][(7p-3)-2q]$

$$[(4p-5)+6q][(4p-5)-6q]$$
$$= (4p-5)^2 - (6q)^2$$
$$= 16p^2 - 40p + 25 - 36q^2$$

____

d. $(a+3b)^4$

d. $(5a+b)^4$

$$(a+3b)^4$$
$$= (a+3b)^2(a+3b)^2$$
$$= (a^2+6ab+9b^2)(a^2+6ab+9b^2)$$
$$= a^2(a^2+6ab+9b^2) + 6ab(a^2+6ab+9b^2)$$
$$\quad + 9b^2(a^2+6ab+9b^2)$$
$$= a^4 + 6a^3b + 9a^2b^2 + 6a^3b + 36a^2b^2 + 54ab^3$$
$$\quad + 9a^2b^2 + 54ab^3 + 81b^4$$
$$= a^4 + 12a^3b + 54a^2b^2 + 108ab^3 + 81b^4$$

____

Objective 5 Practice Exercises

For extra help, see Examples 6–7 on pages 319–320 of your text.

Find each square.

13. $(5y-3)^2$

13. ________________

14. $(2p+3q)^2$

14. ________________

15. $[(2x+3)-y]^2$

15. ________________

Objective 6   Multiply polynomial functions.

Review this example for Objective 6:

8. For $f(x) = 5x + 2$ and $g(x) = 3x^2 + 4x$, find $(fg)(x)$ and $(fg)(-2)$.

$$(fg)(x) = f(x) \cdot g(x)$$
$$= (5x + 2)(3x^2 + 4x)$$
$$= 15x^3 + 20x^2 + 6x^2 + 8x$$
$$= 15x^3 + 26x^2 + 8x$$
$$(fg)(-2) = 15(-2)^3 + 26(-2)^2 + 8(-2)$$
$$= -120 + 104 - 16$$
$$= -32$$

Now Try:

8. For $f(x) = 6x + 5$ and $g(x) = 7x^2 + 2x$, find $(fg)(x)$ and $(fg)(-3)$.

____

Objective 6 Practice Exercises

For extra help, see Example 8 on page 320 of your text.

For the pair of functions, find the product $(fg)(x)$.

16. $f(x) = x + 2$, $g(x) = 3x - 2$

16. ________________

Let $f(x) = 3x^2 + 2$, $g(x) = -5x$, and $h(x) = x + 2$. Find each of the following.

17. $(fg)(-1)$

17. ________________

18. $(fh)(2)$

18. ________________

                    185

Chapter 4   EXPONENTS, POLYNOMIALS, AND POLYNOMIAL FUNCTIONS

4.6   Dividing Polynomials

Learning Objectives
1   Divide a polynomial by a monomial.
2   Divide a polynomial by a polynomial of two or more terms.
3   Divide polynomial functions.

Key Terms

Use the vocabulary terms listed below to complete each statement in exercises 1−3.

quotient   dividend   divisor

1. In the division $\dfrac{5x^5-10x^3}{5x^2}=x^3-2x$, the expression $5x^5-10x^3$ is the

   ___________________.

2. In the division $\dfrac{5x^5-10x^3}{5x^2}=x^3-2x$, the expression x^3-2x is the

   ___________________.

3. In the division $\dfrac{5x^5-10x^3}{5x^2}=x^3-2x$, the expression $5x^2$ is the

   ___________________.

Objective 1   Divide a polynomial by a monomial.

Review these examples for Objective 1:

1. Divide.

   a. $\dfrac{16x^2-8x+24}{4}$

   $$\frac{16x^2-8x+24}{4}=\frac{16x^2}{4}-\frac{8x}{4}+\frac{24}{4}$$
   $$=4x^2-2x+6$$

   Check: $4(4x^2-2x+6)=16x^2-8x+24$

   b. $\dfrac{8x^3-24x^2+32x}{8x^2}$

   $$\frac{8x^3-24x^2+32x}{8x^2}=\frac{8x^3}{8x^2}-\frac{24x^2}{8x^2}+\frac{32x}{8x^2}$$
   $$=x-3+\frac{4}{x}$$

Now Try:

1. Divide.

   a. $\dfrac{18x^2+9x-27}{9}$

   ____

   b. $\dfrac{7a^3-14a^2+28a}{7a^2}$

   ____

c. $\dfrac{10x^3y^5 + 8x^4y^4 + 6x^4y^5}{x^4y^5}$
               c. $\dfrac{3x^2y^3 - 7x^3y^2 + 5x^3y^3}{x^3y^3}$

$$\dfrac{10x^3y^5 + 8x^4y^4 + 6x^4y^5}{x^4y^5}$$

$$= \dfrac{10x^3y^5}{x^4y^5} + \dfrac{8x^4y^4}{x^4y^5} + \dfrac{6x^4y^5}{x^4y^5}$$

$$= \dfrac{10}{x} - \dfrac{8}{y} + 6$$

Objective 1 Practice Exercises

For extra help, see Example 1 on page 324 of your text.

Perform each division.

1. $\dfrac{12x^6 + 28x^5 + 20x^3}{4x^2}$
               1. _______________

2. $\dfrac{24w^8 + 12w^6 - 18w^4}{-6w^5}$
              2. _______________

3. $\dfrac{9r^2s + 18rs^2 - 27s^3}{-27rs^2}$
              3. _______________

         187

Objective 2 Divide a polynomial by a polynomial of two or more terms.

Review these examples for Objective 2: | **Now Try:**

2. Divide $\dfrac{3m^2 - 11m - 20}{m - 5}$.

$$
\begin{array}{r}
3m + 4 \\
m - 5 \,\overline{\big)\, 3m^2 - 11m - 20} \\
\underline{3m^2 - 15m} \\
4m - 20 \\
\underline{4m - 20} \\
0
\end{array}
$$

The quotient is $3m + 4$.

2. Divide $\dfrac{5r^2 - 13r - 28}{r - 4}$.

____

3. Divide $4x^3 - 6x - 9$ by $x - 2$.

Add a term with 0 coefficient as a placeholder for the missing x^2-term.

$$
\begin{array}{r}
4x^2 + 8x + 10 \\
x - 2 \,\overline{\big)\, 4x^3 + 0x^2 - 6x - 9} \\
\underline{4x^3 - 8x^2} \\
8x^2 - 6x \\
\underline{8x^2 - 16x} \\
10x - 9 \\
\underline{10x - 20} \\
11
\end{array}
$$

The quotient is $4x^2 + 8x + 10 + \dfrac{11}{x - 2}$.

3. Divide $2x^3 - 15x - 7$ by $x - 3$.

____

4. Divide $8r^4 + 10r^3 - 6r^2 - 13r - 3$ by $2r^2 - 3$.

$$
\begin{array}{r}
4r^2 + 5r + 3 \\
2r^2 + 0r - 3 \,\overline{\big)\, 8r^4 + 10r^3 - 6r^2 - 13r - 3} \\
\underline{8r^4 + 0r^3 - 12r^2} \\
10r^3 + 6r^2 - 13r \\
\underline{10r^3 + 0r^2 - 15r} \\
6r^2 + 2r - 3 \\
\underline{6r^2 + 0r - 9} \\
2r + 6
\end{array}
$$

The quotient is $4r^2 + 5r + 3 + \dfrac{2r + 6}{2r^2 - 3}$.

4. Divide
$12r^4 - 6r^3 - 16r^2 - 20r - 32$ by $2r^2 - 5$.

____

5. Divide $4p^3 - 11p^2 + 14p - 16$ by $4p - 8$.

$$
\require{enclose}
\begin{array}{r}
p^2 - \dfrac{3}{4}p + 2 \\[4pt]
\end{array}
$$

$$
4p - 8 \,\overline{\smash{)}\, 4p^3 - 11p^2 + 14p - 16}
$$

$$
\begin{array}{r}
\underline{4p^3 - 8p^2} \\
-3p^2 + 14p \\
\underline{-3p^2 + 6p} \\
8p - 16 \\
\underline{8p - 16} \\
0
\end{array}
$$

The quotient is $p^2 - \dfrac{3}{4}p + 2.$

5. Divide $5p^3 - 13p^2 + 26p - 40$ by $5p - 10$.

____

Objective 2 Practice Exercises

For extra help, see Examples 2–5 on pages 325–327 of your text.

Perform each division.

4. $\dfrac{81a^2 - 1}{9a + 1}$

4. _______________

5. $\left(27p^4 - 36p^3 - 6p^2 + 23p - 20\right) \div \left(3p - 4\right)$

5. _______________

6. $\dfrac{6x^4 - 12x^3 + 13x^2 - 5x - 1}{2x^2 + 3}$

6. _______________

Objective 3   Divide polynomial functions.

Review this example for Objective 3:

6. For $f(x) = 4x^2 - 17x - 15$ and $g(x) = x - 5$, find $\left(\dfrac{f}{g}\right)(x)$ and $\left(\dfrac{f}{g}\right)(-2)$.

$$\left(\frac{f}{g}\right)(x) = \frac{f(x)}{g(x)} = \frac{4x^2 - 17x - 15}{x - 5}$$

$$
\begin{array}{r}
4x + 3 \\
x - 5 \overline{\smash{)}\, 4x^2 - 17x - 15} \\
\underline{4x^2 - 20x} \\
3x - 15 \\
\underline{3x - 15} \\
0
\end{array}
$$

$$\left(\frac{f}{g}\right)(x) = 4x + 3, \quad x \neq 5$$

$$\left(\frac{f}{g}\right)(-2) = 4(-2) + 3 = -5$$

Now Try:

6. For $f(x) = 2x^2 - 3x - 20$ and $g(x) = x - 4$, find $\left(\dfrac{f}{g}\right)(x)$ and $\left(\dfrac{f}{g}\right)(-2)$.

____

Objective 3 Practice Exercises

For extra help, see Example 6 on page 328 of your text.

For the pair of functions, find the quotient $\left(\frac{f}{g}\right)(x)$ and give any x-values that are not in the domain of the quotient function.

7. $f(x) = 4x^2 - 11x - 45,\ g(x) = x - 5$

7. _______________

Let $f(x) = 4x^2 - 81, g(x) = 3x,$ and $h(x) = 2x + 9$. Find each of the following.

8. $\left(\dfrac{f}{h}\right)(3)$

8. _______________

9. $\left(\dfrac{g}{h}\right)(-2)$

9. _______________

                    Copyright © 2020 Pearson Education, Inc.

Chapter 5 FACTORING

5.1   Greatest Common Factors and Factoring by Grouping

Learning Objectives
1   Factor out the greatest common factor.
2   Factor by grouping.

Key Terms

Use the vocabulary terms listed below to complete each statement in exercises 1−3.

greatest common factor   factor   factored form

1. An expression is in _______________________________ when it is written as a product.

2. The _______________________________ is the largest quantity that is a factor of each of a group of quantities.

3. An expression A is a _______________________ of an expression B if B can be divided by A with 0 remainder.

Objective 1   Factor out the greatest common factor.

Review these examples for Objective 1:

1. Factor out the greatest common factor.

    a. $7z - 35$

    $\text{GCF} = 7$
$$7z - 35 = 7 \cdot z - 7 \cdot 5$$
$$= 7(z - 5)$$

    b. $42m + 48p$

    $42m + 48p = 6(7m + 8p)$

    c. $9y + 7$

There is no common factor other than 1.

    d. $13 + 39z$

13 is the GCF.
$$13 + 39z = 13 \cdot 1 + 13 \cdot 3z$$
$$= 13(1 + 3z)$$

Now Try:

1. Factor out the greatest common factor.

    a. $6z - 54$

    ____

    b. $36m + 45p$

    ____

    c. $8y - 11$

    ____

    d. $14 + 28z$

    ____

2. Factor out the greatest common factor.

  a. $24x^2 + 36x^3$

The GCF is $12x^2$.
$$24x^2 + 36x^3 = 12x^2(2) + 12x^2(3x)$$
$$= 12x^2(2 + 3x)$$

  b. $28p^5 - 35p^4 + 63p^6$

$GCF = 7p^4$
$28p^5 - 35p^4 + 63p^6$
$$= 7p^4(4p) + 7p^4(-5) + 7p^4(9p^2)$$
$$= 7p^4(4p - 5 + 9p^2)$$

  c. $27p^2q^3 - 36pq^2 + 9p^3q^4$

$GCF = 9pq^2$
$27p^2q^3 - 36pq^2 + 9p^3q^4$
$$= 9pq^2(3pq) + 9pq^2(-4) + 9pq^2(p^2q^2)$$
$$= 9pq^2(3pq - 4 + p^2q^2)$$

  d. $4k^5 - 20k^7 + 44k^8$

$GCF = 4k^5$
$4k^5 - 20k^7 + 44k^8 = 4k^5(1 - 5k^2 + 11k^3)$

  e. $40m^3n^2 + 64m^5 - 56m^7n^4$

$GCF = 8m^3$
$40m^3n^2 + 64m^5 - 56m^7n^4$
$$= 8m^3(5n^2 + 8m^2 - 7m^4n^4)$$

3. Factor out the greatest common factor.

  a. $(x+4)(x+3) + (x+4)(3x+8)$

The greatest common factor is $x + 4$.
$(x+4)(x+3) + (x+4)(3x+8)$
$$= (x+4)[(x+3) + (3x+8)]$$
$$= (x+4)(4x+11)$$

---

2. Factor out the greatest common factor.

  a. $11x^2 + 44x$

____

  b. $36x^5 - 24x^4 + 48x^3$

____

  c. $6k^3 - 30k^5 + 42k^7$

____

  d. $45m^4n^2 - 15mn^4 + 10m^3n$

____

  e. $13x^2y + 26x^3y^2 - 39x^2y^4$

____

3. Factor out the greatest common factor.

  a. $(y+5)(y-9) + (y+5)(4y+5)$

____

b. $a^2(p-q)+b^2(p-q)$

GCF $= p-q$

$a^2(p-q)+b^2(p-q)$

$= (p-q)(a^2+b^2)$

c. $x(y+3z)^2 - w(y+3z)^3$

$x(y+3z)^2 - w(y+3z)^3$

$= (y+3z)^2[x-w(y+3z)]$

$= (y+3z)^2(x-wy-3wz)$

d. $(p-6)(p+8)-(p-6)(4p+5)$

$(p-6)(p+8)-(p-6)(4p+5)$

$= (p-6)[(p+8)-(4p+5)]$

$= (p-6)[p+8-4p-5]$

$= (p-6)[-3p+3]$

$= (p-6)[-3(p-1)]$

$= -3(p-6)(p-1)$

4. Factor $-c^4+5c^3-9c^2$ two ways.

First, c^2 could be used as the common factor.

$-c^4+5c^3-9c^2$

$= c^2(-c^2)+c^2(5c)+c^2(-9)$

$= c^2(-c^2+5c-9)$

Because of the leading negative sign, $-c^2$ could be used as the common factor.

$-c^4+5c^3-9c^2$

$= -c^2(c^2)-c^2(-5c)-c^2(9)$

$= -c^2(c^2-5c+9)$

b. $c^2(x-y)+d^2(x-y)$

____

c. $b(p+8q)^2 - d(p+8q)^3$

____

d. $(x-9)(x+7)-(x-9)(5x+3)$

____

4. Factor $-7b^4-8b^3+5b^2$ two ways.

____

Objective 1 Practice Exercises

For extra help, see Examples 1–4 on pages 342–344 of your text.

Factor out the greatest common factor, if possible.

1. $2x^2y^8+5p^3q$

1. _________________

2. $45a^2b^3 - 90ab + 15ab^2$ **2.** ________________

3. $(x+2)(2x+3) - (x+2)(x+1)$ **3.** ________________

Objective 2   Factor by grouping.

Review these examples for Objective 2:

5.   Factor $7x - 7y + ax - ay$.

Group the terms in pairs so that each pair has a common factor.

$7x - 7y + ax - ay$

$= (7x - 7y) + (ax - ay)$

$= 7(x - y) + a(x - y)$

$= (x - y)(7 + a)$

Check

$(x - y)(7 + a) = 7x + ax - 7y - ay$

$\qquad\qquad\qquad = 7x - 7y + ax - ay$

6.   Factor $8m - 8n - pm + pn$.

$8m - 8n - pm + pn$

$= (8m - 8n) + (-pm + pn)$

$= 8(m - n) - p(m - n)$

$= (m - n)(8 - p)$

Check by multiplying.

7.   Factor $5ax + 15ay + x + 3y$.

$5ax + 15ay + x + 3y$

$= (5ax + 15ay) + (x + 3y)$

$= 5a(x + 3y) + 1(x + 3y)$

$= (x + 3y)(5a + 1)$

Now Try:

5.   Factor $9a - 9b + ra - rb$.

____

6.   Factor $ay - 5y - 3a + 15$.

____

7.   Factor $9px + 36qx + p + 4q$.

____

8. Factor $x^2y^2 - 12 + 4x^2 - 3y^2$.

$$x^2y^2 - 12 + 4x^2 - 3y^2$$
$$= (x^2y^2 - 3y^2) + (4x^2 - 12)$$
$$= y^2(x^2 - 3) + 4(x^2 - 3)$$
$$= (x^2 - 3)(y^2 + 4)$$

8. Factor $a^2b^2 - 48 - 6b^2 + 8a^2$.

____

Objective 2 Practice Exercises

For extra help, see Examples 5–9 on pages 344–346 of your text.

Factor by grouping.

4. $3x^3 + 3xy^2 + 4x^2y + 4y^3$

4. _______________

5. $x - 8y^2 + 2xy^2 - 4$

5. _______________

6. $-3x - 6 + 2y + xy$

6. _______________

Chapter 5 FACTORING

5.2   Factoring Trinomials

Learning Objectives
1   Factor trinomials when the coefficient of the second-degree term is 1.
2   Factor trinomials by grouping when the coefficient of the second-degree term is not 1.
3   Factor trinomials using the FOIL method when the coefficient of the second-degree term is not 1.
4   Factor using substitution.

Key Terms

Use the vocabulary terms listed below to complete each statement in exercises 1–2.

prime polynomial   factoring

1.   The process of writing a polynomial as a product is called __________________.

2.   A polynomial that cannot be factored with integer coefficients is a

_______________________________.

Objective 1   Factor trinomials when the coefficient of the second-degree term is 1.

Review these examples for Objective 1:

1.   Factor each trinomial.

    a. $y^2 - 3y - 10$

Step 1 Find pairs of integers whose product is –10.

Step 1 Find pairs of integers whose product is -10.	*Step 2* Write sums of those pairs of integers.
$10(-1)$	$10 + (-1) = 9$
$-10(1)$	$-10 + 1 = -9$
$5(-2)$	$5 + (-2) = 3$
$-5(2)$	$-5 + 2 = -3$

The integers 2 and –5 have the necessary product and sum.

    $y^2 - 3y - 10$ factors as $(y - 5)(y + 2)$.

    b. $r^2 + 9r + 18$

Look for two integers with a product of 18 and a sum of 9. Only the pair 6 and 3 have a sum of 9.

    $r^2 + 9r + 18$ factors as $(r + 6)(r + 3)$.

Now Try:

1.   Factor each trinomial.

    a. $x^2 - 3x - 28$

    ____

    b. $p^2 + 10p + 21$

    ____

2. Factor $m^2 + 4m + 5$.

Look for factors of 5, 5 and 1, or –5 and –1.

Neither pair has a sum of 4, so $m^2 + 4m + 5$ cannot be factored with integer coefficients and is prime.

3. Factor $x^2 + 7ax - 18a^2$.

Look at this trinomial as a trinomial in the form $x^2 + bx + c$, where $b = 7a$ and $c = -18a^2$.

Step 1 Find pairs of integers whose product is $-18a^2$.	*Step* 2 Write sums of those pairs of integers.
$18a(-a)$	$18a + (-a) = 17a$
$-18a(a)$	$-18a + a = -17a$
$9a(-2a)$	$9a + (-2a) = 7a$
$-9a(2a)$	$-9a + 2a = -7a$
$6a(-3a)$	$6a + (-3a) = 3a$
$-6a(3a)$	$-6a + 3a = -3a$

The expressions $9a$ and $–2a$ have the necessary product and sum.

$$x^2 + 7ax - 18a^2 \text{ factors as } (x + 9a)(x - 2a).$$

4. Factor $18y^3 + 18y^2 - 360y$.

First factor out the GCF.

$$18y^3 + 18y^2 - 360y$$
$$= 18y(y^2 + y - 20)$$

To factor $y^2 + y - 20$, look for two integers whose product is –20 and whose sum is 1. The necessary integers are 5 and –4.

$$= 18y(y + 5)(y - 4)$$

2. Factor $x^2 + 10x + 13$.

2. _______________

3. Factor $x^2 + 11ax - 26a^2$.

3. _______________

4. Factor $15y^3 - 30y^2 - 120y$.

4. _______________

Objective 1 Practice Exercises

For extra help, see Examples 1–4 on pages 349–350 of your text.

Factor completely. If a polynomial cannot be factored, write prime.

1. $x^2 + 11x + 18$

1. _______________

2. $x^2 + 14x - 49$ **2.** _________________

3. $r^2s^2 + 4rs - 21$ **3.** _________________

Objective 2 **Factor trinomials by grouping when the coefficient of the second-degree term is not 1.**

Review this example for Objective 2: | **Now Try:**

5. Factor $15x^2 - x - 2$.   **5.** Factor $18x^2 + 9x - 5$.

Since $a = 15$, $b = -1$, and $c = -2$, the product ac is -30. The two integers whose product is -30 and whose sum is $b = -1$, are -6 and 5.

$$15x^2 - x - 2$$
$$= 15x^2 + 5x - 6x - 2$$
$$= 5x(3x + 1) - 2(3x + 1)$$
$$= (3x + 1)(5x - 2)$$

Objective 2 Practice Exercises

For extra help, see Example 5 on page 351 of your text.

Factor completely. If a polynomial cannot be factored, write prime.

4. $3y^2 + 13y + 4$ **4.** _________________

5. $20r^2 - 28r - 3$ **5.** _________________

6. $20x^2 + 39x - 11$ **6.** _________________

Objective 3   **Factor trinomials using the FOIL method when the coefficient of the second-degree term is not 1.**

Review these examples for Objective 3:

6. Factor the trinomial.

$$5x^2 + 8x + 3$$

Addition signs are used, since all the signs in the trinomial are addition. The first two expressions have a product of $5x^2$, so they must be $5x$ and x.

$$(5x + \underline{\quad})(x + \underline{\quad})$$

The product of the last two terms must be 3, which means the numbers must be 3 and 1.

$(5x + 1)(x + 3)$ gives the wrong middle term

$$15x + x = 16x$$

$(5x + 3)(x + 1)$ gives the correct middle term

$$3x + 5x = 8x$$

Therefore, $5x^2 + 8x + 3$ factors as $(5x + 3)(x + 1)$.

7. Factor $24x^2 + 7xy - 6y^2$.

There is no common factor (except 1). Try 6 and 4 for 24 and 3 and –2, or –3 and 2 for 6.

$(6x - 2y)(4x + 3y)$ Wrong: common factor

$(6x - 3y)(4x + 2y)$ Wrong: common factor

Try 8 and 3 for 24 with 3 and –2, or –3 and 2.

$(8x - 2y)(3x + 3y)$ Wrong: common factor

$(8x + 3y)(3x - 2y)$ Wrong middle term

$$-16xy + 9xy = -7xy$$

The last result differs from the correct middle term only in sign, so interchange the signs of the second terms in the factors.

$24x^2 + 7xy - 6y^2$ factors as $(8x - 3y)(3x + 2y)$.

8. Factor $-5x^2 + 22x + 15$.

Factor out –1 first.

$$-5x^2 + 22x + 15 = -1(5x^2 - 22x - 15)$$
$$= -1(5x + 3)(x - 5)$$
$$= -(5x + 3)(x - 5)$$

Now Try:

6. Factor the trinomial.

$$8x^2 - 6x - 5$$

____

7. Factor $15x^2 + xy - 6y^2$.

____

8. Factor $-7x^2 - 9x + 10$.

____

9. Factor $30y^3 + 42y^2 - 36y$.

$$30y^3 + 42y^2 - 36y$$
$$= 6y(5y^2 + 7y - 6)$$
$$= 6y(5y - 3)(y + 2)$$

9. Factor $36x^3 + 63x^2 - 135x$.

____

Objective 3 Practice Exercises

For extra help, see Examples 6–9 on pages 351–353 of your text.

Factor completely. If a polynomial cannot be factored, write prime.

7. $\quad 6p^2 - p - 15$

7. _______________

8. $\quad 6x^2 - 5xy - y^2$

8. _______________

9. $\quad 2a^3b - 10a^2b^2 + 12ab^3$

9. _______________

Objective 4   Factor by substitution.

Review these examples for Objective 4:

Now Try:

10. Factor $3(x+6)^2 - 4(x+6) - 15$.

10. Factor $12(z+4)^2 - 11(z+4) - 15$.

We let a substitution variable t represent $x + 6$.
$$3(x+6)^2 - 4(x+6) - 15$$
$$= 3t^2 - 4t - 15$$
$$= (3t + 5)(t - 3)$$
$$= [3(x+6) + 5][(x+6) - 3]$$
$$= (3x + 18 + 5)(x + 6 - 3)$$
$$= (3x + 23)(x + 3)$$

____

11. Factor $10y^4 + 29y^2 - 21$.

$$10y^4 + 29y^2 - 21.$$
$$= 10(y^2)^2 + 29y^2 - 21$$
$$= 10t^2 + 29t - 21 \qquad \text{Let } t = y^2.$$
$$= (5t - 3)(2t + 7)$$
$$= (5y^2 - 3)(2y^2 + 7)$$

11. Factor $8y^4 - 6y^2 - 5$.

____

Objective 4 Practice Exercises

For extra help, see Examples 10–11 on page 354 of your text.

Factor.

10. $8(p+5)^2 + 2(p+5) - 15$

10. _________________

11. $8(5-z)^2 - 14(5-z) + 3$

11. _________________

12. $4t^4 + 69t^2 + 17$

12. _________________

Chapter 5 FACTORING

5.3   Special Factoring

Learning Objectives	
1	Factor a difference of squares.
2	Factor a perfect square trinomial.
3	Factor a difference of cubes.
4	Factor a sum of cubes.

Key Terms

Use the vocabulary terms listed below to complete each statement in exercises 1–2.

> **perfect square trinomial   difference of squares**

1.   A ________________________________ is a binomial that can be factored as the product of the sum and difference of two terms.

2.   A ________________________________ is a trinomial that can be factored as the square of a binomial.

Objective 1   Factor a difference of squares.

Review these examples for Objective 1:

1.   Factor each polynomial.

   a. $t^2 - 25$

   $$t^2 - 25 = t^2 - 5^2$$
   $$= (t+5)(t-5)$$

   b. $5a^2 - 180$

   First factor out the common factor.
   $$5a^2 - 180 = 5(a^2 - 36)$$
   $$= 5(a+6)(a-6)$$

   c. $9m^2 - 121p^2$

   $$9m^2 - 121p^2 = (3m)^2 - (11p)^2$$
   $$= (3m+11p)(3m-11p)$$

Now Try:

1.   Factor each polynomial.

   a. $t^2 - 169$

   ____

   b. $4a^2 - 100$

   ____

   c. $25p^2 - 64q^2$

   ____

d. $64k^2 - (a+7)^2$

$$64k^2 - (a+7)^2 = (8k)^2 - (a+7)^2$$
$$= (8k + a + 7)(8k - [a+7])$$
$$= (8k + a + 7)(8k - a - 7)$$

e. $x^2 - 256$

$$x^2 - 256 = (x^2 + 16)(x^2 - 16)$$
$$= (x^2 + 16)(x + 4)(x - 4)$$

d. $49r^2 - (a-6)^2$

____

e. $x^4 - 1$

____

Objective 1 Practice Exercises

For extra help, see Example 1 on page 357 of your text.

Factor each binomial completely. If a binomial cannot be factored, write prime.

1. $25a^2 - 36$

1. __________________

2. $16y^4 - 81$

2. __________________

3. $q^2 - (2r + 3)^2$

3. __________________

Objective 2   Factor a perfect square trinomial.

Review these examples for Objective 2:
2.   Factor each polynomial.

a. $121p^2 - 66p + 9$

Here, $121p^2 = (11p)^2$, and $9 = 3^2$. Determine
twice the product of the two terms to see if this is
a perfect square trinomial.
$$2(11p)(-3) = -66p$$
This is the middle term of the trinomial.
$$121p^2 - 66p + 9 \text{ factors as } (11p - 3)^2.$$

Now Try:
2.   Factor each polynomial.

a. $169x^2 - 130x + 25$

____

                    203

b. $16m^2 + 70m + 81$

If this is a perfect square trinomial, it will equal $(8m+9)^2$. The middle term is

$\quad 2(8m)(9) = 144m$, which does not equal $70m$.

The polynomial is prime.

c. $(r+4)^2 + 8(r+4) + 16$

$(r+4)^2 + 8(r+4) + 16$

$\quad = [(r+4)+4]^2$

$\quad = (r+8)^2$

d. $x^2 - 10x + 25 - y^2$

Since there are four terms, we use factoring by grouping.

$x^2 - 10x + 25 - y^2$

$\quad = (x^2 - 10x + 25) - y^2$

$\quad = (x-5)^2 - y^2$

$\quad = (x-5+y)(x-5-y)$

b. $49a^2 + 56ab + 64b^2$

____

c. $(r+7)^2 + 12(r+7) + 36$

____

d. $m^2 - 18m + 81 - b^2$

____

Objective 2 Practice Exercises

For extra help, see Example 2 on page 358 of your text.

Factor each polynomial completely.

4. $16q^2 - 40q + 25$

4. _____________

5. $64p^4 + 48p^2q^2 + 9q^4$

5. _____________

6. $(m-n)^2 - 12(m-n) + 36$

6. _____________

Objective 3   Factor a difference of cubes.

Review these examples for Objective 3:

3. Factor each polynomial.

   a. $a^3 - 64$

$$a^3 - 64 = a^3 - 4^3$$
$$= (a - 4)(a^2 + 4a + 16)$$

   b. $125a^3 - 8b^3$

$$125a^3 - 8b^3$$
$$= (5a)^3 - (2b)^3$$
$$= (5a - 2b)[(5a)^2 + (5a)(2b) + (2b)^2]$$
$$= (5a - 2b)(25a^2 + 10ab + 4b^2)$$

   c. $216k^3 - 343m^3$

$$216k^3 - 343m^3$$
$$= (6k)^3 - (7m)^3$$
$$= (6k - 7m)[(6k)^2 + (6k)(7m) + (7m)^2]$$
$$= (6k - 7m)(36k^2 + 42km + 49m^2)$$

Now Try:

3. Factor each polynomial.

   a. $1000x^3 - y^3$

   ____

   b. $27a^3 - 125b^3$

   ____

   c. $64k^3 - 27n^3$

   ____

Objective 3 Practice Exercises

For extra help, see Example 3 on page 359 of your text.

Factor.

7. $8r^3 - 27s^3$

7. _______________

8. $216m^3 - 125p^6$

8. _______________

9. $8a^3 - 125b^3$

9. _______________

Objective 4   Factor a sum of cubes.

Review these examples for Objective 4:

4.   Factor each polynomial.

   a. $r^3 + 64$

$$r^3 + 64 = r^3 + 4^3$$
$$= (r+4)(r^2 - 4r + 16)$$

   b. $1000z^3 + 27$

$$1000z^3 + 27 = (10z)^3 + 3^3$$
$$= (10z+3)(100z^2 - 30z + 9)$$

   c. $8t^3 + 343r^6$

$$8t^3 + 343r^6$$
$$= (2t)^3 + (7r^2)^3$$
$$= (2t + 7r^2)[(2t)^2 - (2t)(7r^2) + (7r^2)^2]$$
$$= (2t + 7r^2)(4t^2 - 14tr^2 + 49r^4)$$

   d. $4x^3 + 500$

$$4x^3 + 500 = 4(x^3 + 125)$$
$$= 4(x^3 + 5^3)$$
$$= 4(x+5)(x^2 - 5x + 25)$$

   e. $(x+5)^3 + k^3$

$$(x+5)^3 + k^3$$
$$= [(x+5) + k][(x+5)^2 - (x+5)k + k^2]$$
$$= (x+5+k)(x^2 + 10x + 25 - xk - 5k + k^2)$$

Now Try:

4.   Factor each polynomial.

   a. $r^3 + 125$

   ____

   b. $216z^3 + 1$

   ____

   c. $8t^3 + 125r^6$

   ____

   d. $7x^3 + 7000$

   ____

   e. $(a-5)^3 + b^3$

   ____

Objective 4 Practice Exercises

For extra help, see Example 4 on page 360 of your text.

Factor.

10. $8a^3 + 64b^3$

10. _______________

11. $125p^3 + q^3$

11. _______________

12. $64x^3 + 343y^3$

12. _______________

Chapter 5 FACTORING

5.4   A General Approach to Factoring

Learning Objectives
1   Factor any polynomial.

Key Terms

Use the vocabulary terms listed below to complete each statement in exercises 1–2.

FOIL   **factoring by grouping**

1.   When there are more than three terms in a polynomial, use a process called
_________________________________ to factor the polynomial.

2.   ____________________ is a shortcut method for finding the product of two
binomials.

Objective 1   Factor any polynomial.

Review these examples for Objective 1:

1.   Factor each polynomial.

a. $6x + 48$

$$6x + 48 = 6(x + 8)$$
$$GCF = 6$$

b. $15n^3 p^3 + 5n^2 p$

$$15n^3 p^3 + 5n^2 p = 5n^2 p(3np^2 + 1)$$
$$GCF = 5n^2 p$$

c. $9x(a + c) - z(a + c)$

Factor out $(a + c)$
$$9x(a + c) - z(a + c) = (a + c)(9x - z)$$

d. $(x - 3)(x + 4) + (x - 3)(3x - 1)$

Factor out $(x - 3)$
$$(x - 3)(x + 4) + (x - 3)(3x - 1)$$
$$= (x - 3)[(x + 4) + (3x - 1)]$$
$$= (x - 3)(4x + 3)$$

Now Try:

1.   Factor each polynomial.

a. $18x + 54$

____

b. $18r^4 s^2 + 6r^3 s$

____

c. $7x(y + z) - 5(y + z)$

____

d. $(x + 1)(x + 2) + (x + 1)(x - 5)$

____

2. Factor each binomial if possible.

   a. $81m^2 - 49n^2$

Difference of squares
$$81m^2 - 49n^2 = (9m)^2 - (7n)^2$$
$$= (9m + 7n)(9m - 7n)$$

   b. $216p^3 - 125z^3$

Difference of cubes
$$216p^3 - 125z^3$$
$$= (6p)^3 - (5z)^3$$
$$= (6p - 5z)\left[(6p)^2 + (6p)(5z) + (5z)^2\right]$$
$$= (6p - 5z)(36p^2 + 30pz + 25z^2)$$

   c. $64x^3 + 343$

Sum of cubes
$$64x^3 + 343$$
$$= (4x)^3 + 7^3$$
$$= (4x + 7)\left[(4x)^2 - (4x)(7) + 7^2\right]$$
$$= (4x + 7)(16x^2 - 28x + 49)$$

   d. $16y^2 + 49$

$16y^2 + 49$ is prime. It is the sum of squares.
There is no common factor.

3. Factor each trinomial.

   a. $p^2 + 12p + 36$

Perfect square trinomial
$$p^2 + 12p + 36 = (p + 6)^2$$

   b. $64z^2 - 80z + 25$

Perfect square trinomial
$$64z^2 - 80z + 25 = (8z - 5)^2$$

   c. $3k^2 - 7k - 6$

$$3k^2 - 7k - 6 = (3k + 2)(k - 3)$$

2. Factor each binomial if possible.

   a. $121x^2 - 25y^2$

____

   b. $27t^3 - 64w^3$

____

   c. $8m^3 + 1$

____

   d. $36p^2 + 169$

____

3. Factor each trinomial.

   a. $p^2 - 18p + 81$

____

   b. $36z^2 - 84z + 49$

____

   c. $y^2 - 7y - 8$

____

d. $y^2 - 3y - 4$

$y^2 - 3y - 4 = (y-4)(y+1)$

e. $5x^2 - xy - 4y^2$

$$5x^2 - xy - 4y^2 = 5x^2 - 5xy + 4xy - 4y^2$$
$$= 5x(x-y) + 4y(x-y)$$
$$= (x-y)(5x+4y)$$

f. $36z^2 + 33z - 15$

$$36z^2 + 33z - 15 = 3(12z^2 + 11z - 5)$$
$$= 3(4z+5)(3z-1)$$

4. Factor each polynomial.

a. $4m^3 - 4m^2n + mn^2 - n^3$

Group the terms and factor each group.
$4m^3 - 4m^2n + mn^2 - n^3$
$$= (4m^3 - 4m^2n) + (mn^2 - n^3)$$
$$= 4m^2(m-n) + n(m-n)$$
$$= (m-n)(4m^2 + n^2)$$

c. $25b^2 + 10b + 1 - c^2$

Group the first three terms.
$25b^2 + 10b + 1 - c^2$
$$= (25b^2 + 10b + 1) - c^2$$
$$= (5b+1)^2 - c^2$$
$$= (5b+1+c)(5b+1-c)$$

d. $27x^3 - y^3 + 9x^2 - y^2$

Rearrange and group the terms.
$27x^3 - y^3 + 9x^2 - y^2$
$$= (27x^3 - y^3) + (9x^2 - y^2)$$
$$= (3x-y)(9x^2 + 3xy + y^2) + (3x-y)(3x+y)$$
$$= (3x-y)(9x^2 + 3xy + y^2 + 3x + y)$$

d. $4k^2 - 7k - 2$

____

e. $4x^2 + xy - 3y^2$

____

f. $30z^2 - 5z - 10$

____

4. Factor each polynomial.

a. $3c^3 - cd^2 + 3c^2d - d^3$

____

c. $36b^2 - 12b + 1 - c^2$

____

d. $125a^3 - b^3 + 25a^2 - b^2$

____

Name: Date:
Instructor: Section:

Objective 1 Practice Exercises

For extra help, see Examples 1–4 on pages 363–365 of your text.

Factor completely.

1. $12a^2b^2 + 3a^2b - 9ab^2$ 1. _______________

2. $2x^3y^4 - 72xy^2$ 2. _______________

3. $128x^3 - 2y^3$ 3. _______________

4. $2a^2 - 17a + 30$ 4. _______________

5. $x^3 - 3x^2 + 7x - 21$ 5. _______________

6. $a^2 - 6ab + 9b^2 - 25$ 6. _______________

Chapter 5   FACTORING

5.5   Solving Equations Using the Zero-Factor Property

Learning Objectives
1   Use the zero-factor property.
2   Solve applied problems that require the zero-factor property.
3   Solve a formula for a specified variable, where factoring is necessary.

Key Terms

Use the vocabulary terms listed below to complete each statement in exercises 1–4.

> **quadratic equation**   **standard form**   **double solution**
>
> **zero-factor property**

1.   An equation written in the form $ax^2 + bx + c = 0$ is written in the
______________________ of a quadratic equation.

2.   An equation that can written in the form $ax^2 + bx + c = 0$, with $a \neq 0$, is a

______________________.

3.   The ______________________ states that if two number have a product
of 0, then at least one of the numbers must be 0.

4.   When a quadratic equation has only one distinct solution, that number is a

______________________ .

Objective 1   Use the zero-factor property.

Review these examples for Objective 1:

1.   Solve $(x+7)(3x-5)=0$.

Use the zero-factor property.
$$(x+7)(3x-5)=0$$
$$x+7=0 \quad \text{or} \quad 3x-5=0$$
$$x=-7 \quad \text{or} \quad 3x=5$$
$$x=\frac{5}{3}$$

Check $(x+7)(3x-5)=0$

$$(-7+7)[3(-7)-5]\overset{?}{=}0 \quad \Big| \quad \left(\frac{5}{3}+7\right)\left[3\left(\frac{5}{3}\right)-5\right]\overset{?}{=}0$$

$$(0)(-26)\overset{?}{=}0 \quad \Big| \quad \frac{26}{3}(0)\overset{?}{=}0$$

$$\text{True} \quad 0=0 \quad \Big| \quad \text{True} \quad 0=0$$

The solution set is $\left\{-7, \frac{5}{3}\right\}$.

Now Try:

1.   Solve $(4x+5)(2x-3)=0$.

____

2. Solve $3x^2 + 7x = 6$.

Step 1 $\qquad\qquad\qquad 3x^2 + 7x = 6$

$\qquad\qquad\qquad\qquad 3x^2 + 7x - 6 = 0$

Step 2 $\qquad\qquad (x+3)(3x-2) = 0$

Step 3 $\quad x+3 = 0 \quad$ or $\quad 3x-2 = 0$

Step 4 $\qquad x = -3 \qquad$ or $\qquad x = \dfrac{2}{3}$

Step 5 Check each solution in the original equation.

Check $\qquad\qquad 3x^2 + 7x = 6$

$$3(-3)^2 + 7(-3) \overset{?}{=} 6 \quad\bigg|\quad 3\left(\frac{2}{3}\right)^2 + 7\left(\frac{2}{3}\right) \overset{?}{=} 6$$

$$27 - 21 \overset{?}{=} 6 \quad\bigg|\quad \frac{4}{3} + \frac{14}{3} \overset{?}{=} 6$$

$$\text{True} \quad 6 = 6 \quad\bigg|\quad \text{True} \quad 6 = 6$$

The solution set is $\left\{-3,\ \dfrac{2}{3}\right\}$.

3. Solve $36x^2 = 60x - 25$.

$$36x^2 - 60x + 25 = 0$$

$$(6x - 5)^2 = 0$$

$$6x - 5 = 0$$

$$6x = 5$$

$$x = \frac{5}{6}$$

There is only one distinct solution, which we call a double solution. The solution set is $\left\{\dfrac{5}{6}\right\}$.

4. Solve $7z^2 - 28z = 0$.

$$7z^2 - 28z = 0$$

$$7z(z - 4) = 0$$

$$7z = 0 \quad \text{or} \quad z - 4 = 0$$

$$z = 0 \quad \text{or} \quad z = 4$$

A check shows that the solution set is $\{0, 4\}$.

---

2. Solve $4x^2 + 15x = 4$.

____

3. Solve $49x^2 = 42x - 9$.

____

4. Solve $9z^2 - 45z = 0$.

____

5. Solve $4m^2 - 196 = 0$.

$$4(m^2 - 49) = 0$$
$$4(m + 7)(m - 7) = 0$$
$$m + 7 = 0 \quad \text{or} \quad m - 7 = 0$$
$$m = -7 \quad \text{or} \quad m = 7$$

A check shows the solution set is $\{-7, 7\}$.

6. Solve $(3q + 7)(q - 1) = 5(q + 1) - 2$.

$$(3q + 7)(q - 1) = 5(q + 1) - 2$$
$$3q^2 + 4q - 7 = 5q + 5 - 2$$
$$3q^2 + 4q - 7 = 5q + 3$$
$$3q^2 - q - 10 = 0$$
$$(3q + 5)(q - 2) = 0$$
$$3q + 5 = 0 \quad \text{or} \quad q - 2 = 0$$
$$q = -\frac{5}{3} \quad \text{or} \quad q = 2$$

A check shows the solution set is $\left\{-\dfrac{5}{3}, 2\right\}$.

7. Solve $-x^3 + x^2 = -20x$.

$$-x^3 + x^2 + 20x = 0$$
$$x^3 - x^2 - 20x = 0$$
$$x(x^2 - x - 20) = 0$$
$$x(x + 4)(x - 5) = 0$$
$$x = 0 \quad \text{or} \quad x + 4 = 0 \quad \text{or} \quad x - 5 = 0$$
$$x = -4 \qquad x = 5$$

A check shows that the solution set is $\{-4, 0, 5\}$.

5. Solve $8m^2 - 8 = 0$.

____

6. Solve
$$(3x + 3)(2x - 5) = 5(1 - x) - 6x.$$

____

7. Solve $-x^3 + 2x^2 = -35x$.

____

Objective 1 Practice Exercises

For extra help, see Examples 1–7 on pages 368–371 of your text.

Solve each equation.

1. $2x^2 - 3x - 20 = 0$

1. _______________

2. $15x^2 = x^3 + 56x$ **2.** ________________

3. $z^2 = 6z - 9$ **3.** ________________

Objective 2   Solve applied problems that require the zero-factor property.

Review this example for Objective 2:

8. A house has a floor area of 608 square meters. The floor has the shape of a rectangle whose length is 13 meters more than the width. Find the width and length of the floor.

Step 1 Read the problem again. There will be two answers.

Step 2 Assign a variable. Let w = the width. Then $w + 13$ = the length.

Step 3 Write an equation. The area is $A = lw$. Here, $608 = (w + 13)w$

Step 4 Solve.

$$608 = w^2 + 13w$$
$$0 = w^2 + 13w - 608$$
$$0 = (w + 32)(w - 19)$$
$$w + 32 = 0 \quad \text{or} \quad w - 19 = 0$$
$$w = -32 \quad \text{or} \quad w = 19$$

Step 5 State the answer. A distance cannot be negative, so reject –32 as a solution. The width is 19 m. The length is 19 + 13 = 32 m.

Step 6 Check. $32(19) = 608 \text{ m}^2$

Now Try:

8. Paul and Joan wish to buy floor covering that covers 150 square feet for their large recreation room. They wish to cover a rectangle that is 5 feet longer than it is wide. How wide should the rectangle be?

____

9. Jeff threw a stone straight upward at 46 feet per second from a dock 6 feet above a lake. The height of the stone above the lake t seconds after it is thrown is given by $h = -16t^2 + 46t + 6$. How long will it take for the stone to reach a height of 39 feet?

We let $h = 39$, and solve for t.

$$39 = -16t^2 + 46t + 6$$

$$16t^2 - 46t + 33 = 0$$

$$(8t - 11)(2t - 3) = 0$$

$$8t - 11 = 0 \quad \text{or} \quad 2t - 3 = 0$$

$$t = \frac{11}{8} \quad \text{or} \quad t = \frac{3}{2}$$

The stone reaches a height of 39 ft twice, on its way up at $1\frac{3}{8}$ seconds and again on its way down at $1\frac{1}{2}$ seconds.

9. If an object is propelled upward from a height of 16 feet with an initial velocity of 48 feet per second, its height h (in feet) t seconds later is given by the equation $h = -16t^2 + 48t + 16$. After how many seconds is the height 48 feet?

____

Objective 2 Practice Exercises

For extra help, see Examples 8–9 on pages 371–372 of your text.

Solve each problem.

4. The area of a triangle is 42 square centimeters. The base is 2 centimeters less than twice the height. Find the base and height of the triangle.

4. _____________

5. The Browns installed 96 feet of fencing around a rectangular play yard. If the yard covers 540 square feet, what are its dimensions?

5. _____________

____

6. A company determines that its daily revenue R (in dollars) for selling x items is modeled by the equation $R = x(150 - x)$. How many items must be sold for its revenue to be \$4400?

6. _______________

Objective 3 Solve a formula for a specified variable, where factoring is necessary.

Review this example for Objective 3:

10. Solve the formula for H.
$$S = 2HW + 2LW + 2LH$$
$$S - 2LW = 2HW + 2LH$$
$$S - 2LW = H(2W + 2L)$$
$$\frac{S - 2LW}{2W + 2L} = H$$

Now Try:

10. The formula for the surface area of an open box is $S = 2HW + LW + 2LH$, solve for W.

____

Objective 3 Practice Exercises

For extra help, see Example 10 on page 373 of your text.

Solve each equation for the specified variable.

7. $bd = c + ba$, for b

7. _______________

8. $n - mk = mt^2$, for m

8. _______________

9. $v - w = uvw$, for w

9. _______________

Chapter 6 RATIONAL EXPRESSIONS AND FUNCTIONS

6.1   Rational Expressions and Functions; Multiplying and Dividing

Learning Objectives
1   Define rational expressions.
2   Define rational functions and give their domains.
3   Write rational expressions in lowest terms.
4   Multiply rational expressions.
5   Find reciprocals of rational expressions.
6   Divide rational expressions.

Key Terms

Use the vocabulary terms listed below to complete each statement in exercises 1−2.

rational expression   **rational function**

1.   A ____ is a function that is defined by

a rational expression in the form $f(x) = \dfrac{P(x)}{Q(x)}$, where $Q(x) \neq 0$.

2.   The quotient of two polynomials with denominator not 0 is called a

________________________________.

Objective 1   Define rational expressions.

For extra help, see page 386 of your text.

Objective 2   Define rational functions and give their domains.

Review these examples for Objective 2:

1.   Give the domain of each rational function.

a. $f(x) = \dfrac{5}{3x - 9}$

Set the denominator equal to 0 and solve.
$$3x - 9 = 0$$
$$3x = 9$$
$$x = 3$$
The number 3 cannot be used for a replacement for x. The domain of f includes all real numbers except 3.

Set-builder notation: $\{x \mid x \neq 3\}$.

Interval notation: $(-\infty, 3) \cup (3, \infty)$

Now Try:

1.   Give the domain of each rational function.

a. $f(x) = \dfrac{9}{5x - 10}$

____

____

Name: Date:
Instructor: Section:

b. $g(x) = \dfrac{x}{x^2 - 3x + 2}$

Set the denominator equal to 0.

$$x^2 - 3x + 2 = 0$$

$$(x - 1)(x - 2) = 0$$

$$x - 1 = 0 \quad \text{or} \quad x - 2 = 0$$

$$x = 1 \quad \text{or} \quad x = 2$$

The domain of g includes all real numbers except 1 and 2, written $\{x \mid x \neq 1,\ 2\}$ in set-builder notation, and $(-\infty,\ 1) \cup (1,\ 2) \cup (2,\ \infty)$ in interval notation.

c. $h(x) = \dfrac{2x - 6}{5}$

The denominator, 5, can never be 0, so the domain of h includes all real numbers, written in set-builder notation as $\{x \mid x \text{ is a real number}\}$ and $(-\infty,\ \infty)$ in interval notation.

d. $f(x) = \dfrac{x - 6}{x^2 + 1}$

Setting $x^2 + 1$ equal to 0 leads to $x^2 = -1$. There is no real number whose square is -1. Therefore, any real number can be used as a replacement for x. The domain of f is $\{x \mid x \text{ is a real number}\}$ in set-builder notation and $(-\infty,\ \infty)$ in interval notation.

b. $g(x) = \dfrac{x + 2}{x^2 - 5x + 6}$

____

____

c. $h(x) = \dfrac{5x + 3}{7}$

____

____

d. $f(x) = \dfrac{x - 3}{x^2 + 9}$

____

____

Objective 2 Practice Exercises

For extra help, see Example 1 on pages 386–387 of your text.

Give the domain of each rational function using set-builder notation and interval notation.

1. $f(s) = \dfrac{8s + 7}{3s - 2}$

1. _______________

____

2. $f(x) = \dfrac{x - 6}{x^2 + 1}$

2. _______________

____

3. $f(q) = \dfrac{q+7}{q^2 - 3q + 2}$ **3.** _______________

____

Objective 3   Write rational expressions in lowest terms.

Review these examples for Objective 3:

2. Write the rational expression in lowest terms.

$$\frac{(x+7)(x-1)}{(x-1)(x+6)}$$

$$\frac{(x+7)(x-1)}{(x-1)(x+6)} = \frac{(x+7)(x-1)}{(x+6)(x-1)} = \frac{x+7}{x+6}$$

3. Write the rational expression in lowest terms.

$$\frac{p-25}{25-p}$$

$$\frac{p-25}{25-p} = \frac{p-25}{-1(p-25)} = \frac{1}{-1} = -1$$

Now Try:

2. Write the rational expression in lowest terms.

$$\frac{(x-4)(x+3)}{(x+3)(x-7)}$$

____

3. Write the rational expression in lowest terms.

$$\frac{w-6}{6-w}$$

____

Objective 3 Practice Exercises

For extra help, see Examples 2–3 on pages 388–389 of your text.

Write each rational expression in lowest terms.

4. $\dfrac{12k^3 + 12k^2}{3k^2 + 3k}$ **4.** _______________

5. $\dfrac{2y^2 - 3y - 5}{2y^2 - 11y + 15}$ **5.** _______________

6. $\dfrac{a^2-3a}{3a-a^2}$

6. _______________

Objective 4 Multiply rational expressions.

Review this example for Objective 4:

4. Multiply.

$$\frac{7r-21}{r}\cdot\frac{3r^2}{8r-24}$$

$$\frac{7r-21}{r}\cdot\frac{3r^2}{8r-24}=\frac{7(r-3)}{r}\cdot\frac{3r\cdot r}{8(r-3)}$$

$$=\frac{r(r-3)}{r(r-3)}\cdot\frac{7\cdot 3r}{8}$$

$$=\frac{21r}{8}$$

Now Try:

4. Multiply.

$$\frac{9x-36}{x}\cdot\frac{4x^2}{7x-28}$$

____

Objective 4 Practice Exercises

For extra help, see Example 4 on pages 390–391 of your text.

Multiply. Write each answer in lowest terms.

7. $\dfrac{x^2+x-12}{x^2+7x+10}\cdot\dfrac{x^2+3x-10}{x^2+2x-8}$

7. _______________

8. $\dfrac{x^2+10x+21}{x^2+14x+49}\cdot\dfrac{x^2+12x+35}{x^2-6x-27}$

8. _______________

9. $\dfrac{3m^2-m-10}{2m^2-7m-4}\cdot\dfrac{4m^2-1}{6m^2+7m-5}$

9. _______________

Objective 5   Find reciprocals of rational expressions.

Objective 5 Practice Exercises

For extra help, see page 391 of your text.

Find the reciprocal.

10. $\dfrac{r^2 + 2r}{5 + r}$ 10. ___________________

11. $\dfrac{7z + 7}{z^2 - 9}$ 11. ___________________

12. 0 12. ___________________

Objective 6   Divide rational expressions.

Review this example for Objective 6:

5. Divide.

a. $\dfrac{8x^2}{21} \div \dfrac{2x}{7}$

$$\dfrac{8x^2}{21} \div \dfrac{2x}{7} = \dfrac{8x^2}{21} \cdot \dfrac{7}{2x} = \dfrac{4x \cdot 2x}{3 \cdot 7} \cdot \dfrac{7}{2x} = \dfrac{4x}{3}$$

b. $\dfrac{3x + 7}{2x^3} \div \dfrac{6x + 14}{10x^6}$

$$\dfrac{3x + 7}{2x^3} \div \dfrac{6x + 14}{10x^6} = \dfrac{3x + 7}{2x^3} \cdot \dfrac{10x^6}{6x + 14}$$

$$= \dfrac{3x + 7}{2x^3} \cdot \dfrac{5x^3 \cdot 2x^3}{2(3x + 7)}$$

$$= \dfrac{5x^3}{2}$$

c. $\dfrac{2k^2 + 5k - 12}{2k^2 + k - 3} \div \dfrac{k^2 + 8k + 16}{2k^2 + k - 3}$

$$\dfrac{2k^2 + 5k - 12}{2k^2 + k - 3} \div \dfrac{k^2 + 8k + 16}{2k^2 + k - 3}$$

$$= \dfrac{2k^2 + 5k - 12}{2k^2 + k - 3} \cdot \dfrac{2k^2 + k - 3}{k^2 + 8k + 16}$$

$$= \dfrac{(2k - 3)(k + 4)}{(2k + 3)(k - 1)} \cdot \dfrac{(2k + 3)(k - 1)}{(k + 4)(k + 4)}$$

$$= \dfrac{2k - 3}{k + 4}$$

Now Try:

5. Divide.

a. $\dfrac{10p}{21} \div \dfrac{15p^2}{7}$

____

b. $\dfrac{2x - 5}{3x^2} \div \dfrac{6x - 15}{15x^3}$

____

c. $\dfrac{2a^2 - 5a - 12}{a^2 - 10a + 24} \div \dfrac{a^2 - 9}{a^2 - 9a + 18}$

____

Objective 6 Practice Exercises

For extra help, see Example 5 on page 392 of your text.

Divide. Write each answer in lowest terms.

13. $\dfrac{4m-12}{2m+10} \div \dfrac{9-m^2}{m^2-25}$ F

13. ______________________

14. $\dfrac{27-3k^2}{3k^2+8k-3} \div \dfrac{k^2-6k+9}{6k^2-19k+3}$

14. ______________________

15. $\dfrac{y^2+7y+10}{3y+6} \div \dfrac{y^2+2y-15}{4y-4}$

15. ______________________

Chapter 6 RATIONAL EXPRESSIONS AND FUNCTIONS

6.2 Adding and Subtracting Rational Expressions

Learning Objectives
1 Add and subtract rational expressions with the same denominator.
2 Find a least common denominator.
3 Add and subtract rational expressions with different denominators.

Key Terms

Use the vocabulary terms listed below to complete each statement in exercises 1−2.

least common denominator (LCD)   **equivalent expressions**

1. $\dfrac{24x-8}{9x^2-1}$ and $\dfrac{8}{3x+1}$ are __________________________________.

2.   The simplest expression that is divisible by all denominators is called the

     __________________________________.

Objective 1   Add and subtract rational expressions with the same denominator.

Review these examples for Objective 1:

1. Add or subtract as indicated.

b. $\dfrac{13}{3k^2} - \dfrac{19}{3k^2}$

$$\dfrac{13}{3k^2} - \dfrac{19}{3k^2} = \dfrac{13-19}{3k^2} = \dfrac{-6}{3k^2},\ \text{or}\ -\dfrac{2}{k^2}$$

c. $\dfrac{a}{a^2-b^2} + \dfrac{b}{a^2-b^2}$

$$\dfrac{a}{a^2-b^2} + \dfrac{b}{a^2-b^2} = \dfrac{a+b}{a^2-b^2}$$

$$= \dfrac{a+b}{(a+b)(a-b)}$$

$$= \dfrac{1}{a-b}$$

Now Try:

1. Add or subtract as indicated.

b. $\dfrac{16}{5z^2} - \dfrac{26}{5z^2}$

____

c. $\dfrac{c}{c^2-d^2} - \dfrac{d}{c^2-d^2}$

____

d. $\dfrac{5}{x^2+4x-5}+\dfrac{x}{x^2+4x-5}$

$$=\dfrac{5+x}{x^2+4x-5}$$

$$=\dfrac{5+x}{(x+5)(x-1)}$$

$$=\dfrac{1}{x-1}$$

d. $\dfrac{7}{x^2+6x-7}+\dfrac{x}{x^2+6x-7}$

____

Objective 1 Practice Exercises

For extra help, see Example 1 on pages 396–397 of your text.

Add or subtract as indicated. Write each answer in lowest terms.

1. $\dfrac{n}{m+3}-\dfrac{-3n+7}{m+3}$

1. _______________

2. $\dfrac{2x+3}{x^2+3x-10}+\dfrac{2-x}{x^2+3x-10}$

2. _______________

3. $\dfrac{k}{k^2-6k+8}-\dfrac{2}{k^2-6k+8}$

3. _______________

Objective 2   Find a least common denominator.

Review these examples for Objective 2:

2. Suppose that the given expressions are denominators of fractions. Find the LCD for each group of denominators.

a. $3x^2y^3,\ 7x^3y^2$

$$3x^2y^3 = 3\cdot x^2\cdot y^3$$
$$7x^3y^2 = 7\cdot x^3\cdot y^2$$
$$\text{LCD} = 3\cdot 7\cdot x^3\cdot y^3 = 21x^3y^3$$

Now Try:

2. Suppose that the given expressions are denominators of fractions. Find the LCD for each group of denominators.

a. $11x^2y,\ 3xy^4$

____

b. $m-4,\ m$

Each denominator is already factored. The LCD must be divisible by both $m-4$ and m.

$$m(m-4)$$

c. $m^2+7m+12,\ m^2+6m+9,\ 5m+20$

$$m^2+7m+12=(m+4)(m+3)$$
$$m^2+6m+9=(m+3)^2$$
$$5m+20=5(m+4)$$
$$\text{LCD}=5(m+4)(m+3)^2$$

b. $y+6,\ y$

____

c. $x^2-3x-4,\ x^2-8x+16,$ $3x+3$

____

Objective 2 Practice Exercises

For extra help, see Example 2 on pages 397–398 of your text.

Assume that the expressions given are denominators of fractions. Find the least common denominator (LCD) for each group.

4. $\quad q^2-36,\ (q+6)^2$

4. _________________

5. $\quad x^2+5x+6,\ 3x+6$

5. _________________

6. $\quad p-4,\ p^2-16,\ (p+4)^2$

6. _________________

Name:
Instructor:

Date:
Section:

Objective 3   Add and subtract rational expressions with different denominators.

Review these examples for Objective 3:

3. Add or subtract as indicated.

a. $\dfrac{2}{5r}+\dfrac{3}{10r}$

$$\dfrac{2}{5r}+\dfrac{3}{10r}=\dfrac{2\cdot 2}{5r\cdot 2}+\dfrac{3}{10r}$$

$$=\dfrac{4}{10r}+\dfrac{3}{10r}$$

$$=\dfrac{7}{10r}$$

b. $\dfrac{4}{x}-\dfrac{3}{x+4}$

$$\dfrac{4}{x}-\dfrac{3}{x+4}=\dfrac{4(x+4)}{x(x+4)}-\dfrac{3x}{x(x+4)}$$

$$=\dfrac{4x+16}{x(x+4)}-\dfrac{3x}{x(x+4)}$$

$$=\dfrac{4x+16-3x}{x(x+4)}$$

$$=\dfrac{x+16}{x(x+4)}$$

4. Subtract.

a. $\dfrac{19x}{5x+2}-\dfrac{4x-6}{5x+2}$

$$\dfrac{19x}{5x+2}-\dfrac{4x-6}{5x+2}=\dfrac{19x-(4x-6)}{5x+2}$$

$$=\dfrac{19x-4x+6}{5x+2}$$

$$=\dfrac{15x+6}{5x+2}$$

$$=\dfrac{3(5x+2)}{5x+2}$$

$$=3$$

Now Try:

3. Add or subtract as indicated.

a. $\dfrac{7}{9z}+\dfrac{5}{18z}$

____

b. $\dfrac{5}{s}-\dfrac{4}{s-6}$

____

4. Subtract.

a. $\dfrac{23x}{6x+1}-\dfrac{5x-3}{6x+1}$

____

b. $\dfrac{1}{x-5} - \dfrac{1}{x+5}$

$$\dfrac{1}{x-5} - \dfrac{1}{x+5} = \dfrac{1(x+5)}{(x-5)(x+5)} - \dfrac{1(x-5)}{(x-5)(x+5)}$$

$$= \dfrac{x+5-(x-5)}{(x-5)(x+5)}$$

$$= \dfrac{x+5-x+5}{(x-5)(x+5)}$$

$$= \dfrac{10}{(x-5)(x+5)}$$

b. $\dfrac{5}{x+3} - \dfrac{5}{x-3}$

———————

5. Add.

$\dfrac{6p}{p-4} + \dfrac{2}{4-p}$ The LCD is $p-4$.

Since the denominators are opposites, we multiply the second expression by -1.

$$= \dfrac{6p}{p-4} + \dfrac{2(-1)}{(4-p)(-1)}$$

$$= \dfrac{6p}{p-4} + \dfrac{-2}{p-4}$$

$$= \dfrac{6p-2}{p-4}$$

5. Add.

$\dfrac{x}{x-8} + \dfrac{12}{8-x}$

———————

6. Add and subtract as indicated.

$\dfrac{2}{x} - \dfrac{5}{x-2} + \dfrac{10}{x^2-2x}$ The LCD is $x(x-2)$.

$$= \dfrac{2}{x} - \dfrac{5}{x-2} + \dfrac{10}{x(x-2)}$$

$$= \dfrac{2(x-2)}{x(x-2)} - \dfrac{5x}{x(x-2)} + \dfrac{10}{x(x-2)}$$

$$= \dfrac{2x-4-5x+10}{x(x-2)}$$

$$= \dfrac{-3x+6}{x(x-2)}$$

$$= \dfrac{-3(x-2)}{x(x-2)}$$

$$= -\dfrac{3}{x}$$

6. Add and subtract as indicated.

$\dfrac{12}{x^2+3x} - \dfrac{3}{x} + \dfrac{4}{x+3}$

———————

7. Subtract.

$$\frac{7x}{x^2-4x+4}-\frac{2}{x^2-4}$$

$$=\frac{7x}{(x-2)^2}-\frac{2}{(x+2)(x-2)}$$

The LCD is $(x+2)(x-2)^2$.

$$=\frac{7x}{(x-2)^2}-\frac{2}{(x+2)(x-2)}$$

$$=\frac{7x(x+2)}{(x+2)(x-2)^2}-\frac{2(x-2)}{(x+2)(x-2)^2}$$

$$=\frac{7x(x+2)-2(x-2)}{(x+2)(x-2)^2}$$

$$=\frac{7x^2+14x-2x+4}{(x+2)(x-2)^2}$$

$$=\frac{7x^2+12x+4}{(x+2)(x-2)^2}$$

7. Subtract.

$$\frac{8x}{x^2-10x+25}-\frac{3}{x^2-25}$$

____

Objective 3 Practice Exercises

For extra help, see Examples 3–8 on pages 398–401 of your text.

Add or subtract as indicated. Write each answer in lowest terms.

7. $\dfrac{3}{n^2-16}-\dfrac{6n}{n^2+8n+16}$

7. _______________

8. $\dfrac{4z}{z^2+6z+8}+\dfrac{2z-1}{z^2+5z+6}$

8. _______________

9. $\dfrac{4y}{y^2+4y+3}-\dfrac{3y+1}{y^2-y-2}$

9. _______________

Chapter 6 RATIONAL EXPRESSIONS AND FUNCTIONS

6.3   Complex Fractions

Learning Objectives
1   Simplify complex fractions by simplifying the numerator and denominator (Method 1).
2   Simplify complex fractions by multiplying by a common denominator (Method 2).
3   Compare the two methods of simplifying complex fractions.
4   Simplify rational expressions with negative exponents.

Key Terms

Use the vocabulary terms listed below to complete each statement in exercises 1–2.

complex fraction   LCD

1. A ________________________ is a rational expression with one or more fractions in the numerator, denominator, or both.

2. To simplify a complex fraction, multiply the numerator and denominator by the ________________________ of all the fractions within the complex fraction.

Objective 1   Simplify complex fractions by simplifying the numerator and denominator (Method 1).

Review these examples for Objective 1:

1. Use Method 1 to simplify each complex fraction.

a. $\dfrac{\dfrac{x-2}{4x}}{\dfrac{x+3}{x}}$

$$\dfrac{\dfrac{x-2}{4x}}{\dfrac{x+3}{x}} = \dfrac{x-2}{4x} \div \dfrac{x+3}{x}$$

$$= \dfrac{x-2}{4x} \cdot \dfrac{x}{x+3}$$

$$= \dfrac{x(x-2)}{4x(x+3)}$$

$$= \dfrac{x-2}{4(x+3)}$$

Now Try:

1. Use Method 1 to simplify each complex fraction.

a. $\dfrac{\dfrac{x+1}{3x}}{\dfrac{x-1}{5x}}$

____

b. $\dfrac{\dfrac{4}{x}+3}{5-\dfrac{3}{x}}$

$$\dfrac{\dfrac{4}{x}+3}{5-\dfrac{3}{x}}=\dfrac{\dfrac{4}{x}+\dfrac{3x}{x}}{\dfrac{5x}{x}-\dfrac{3}{x}}$$

$$=\dfrac{\dfrac{4+3x}{x}}{\dfrac{5x-3}{x}}$$

$$=\dfrac{4+3x}{x}\cdot\dfrac{x}{5x-3}$$

$$=\dfrac{3x+4}{5x-3}$$

b. $\dfrac{2-\dfrac{7}{y}}{7-\dfrac{2}{y}}$

____

Objective 1 Practice Exercises

For extra help, see Example 1 on pages 405–406 of your text.

Use Method 1 to simplify each complex fraction.

1. $\dfrac{\dfrac{3a+4}{a}}{\dfrac{1}{a}+\dfrac{2}{5}}$

1. _______________

2. $\dfrac{\dfrac{2}{a+2}-4}{\dfrac{1}{a+2}-3}$

2. _______________

3. $\dfrac{\dfrac{a+2}{a-2}}{\dfrac{1}{a^2-4}}$

3. _______________

Objective 2   Simplify complex fractions by multiplying by a common denominator (Method 2).

Review these examples for Objective 2:

2. Use Method 2 to simplify each complex fraction.

a.
$$\dfrac{\dfrac{4}{x}+3}{5-\dfrac{3}{x}}$$

$$\dfrac{\dfrac{4}{x}+3}{5-\dfrac{3}{x}}=\dfrac{\left(\dfrac{4}{x}+3\right)\cdot x}{\left(5-\dfrac{3}{x}\right)\cdot x}$$

$$=\dfrac{\dfrac{4}{x}\cdot x+3\cdot x}{5\cdot x-\dfrac{3}{x}\cdot x}$$

$$=\dfrac{4+3x}{5x-3},\text{ or }\dfrac{3x+4}{5x-3}$$

b.
$$\dfrac{\dfrac{4}{x-4}+3x}{5x-\dfrac{3}{x-4}}$$

$$\dfrac{\dfrac{4}{x-4}+3x}{5x-\dfrac{3}{x-4}}=\dfrac{\left(\dfrac{4}{x-4}+3x\right)\cdot(x-4)}{\left(5x-\dfrac{3}{x-4}\right)\cdot(x-4)}$$

$$=\dfrac{\dfrac{4}{x-4}\cdot(x-4)+3x\cdot(x-4)}{5x\cdot(x-4)-\dfrac{3}{x-4}\cdot(x-4)}$$

$$=\dfrac{4-3x(x-4)}{5x(x-4)-3}$$

$$=\dfrac{4-3x^2+12x}{5x^2-20x-3},\text{ or }\dfrac{-3x^2+12x+4}{5x^2-20x-3}$$

Now Try:

2. Use Method 2 to simplify each complex fraction.

a.
$$\dfrac{2-\dfrac{7}{y}}{7-\dfrac{2}{y}}$$

____

b.
$$\dfrac{\dfrac{1}{a-2}+4a}{\dfrac{1}{a-2}+3a}$$

____

Objective 2 Practice Exercises

For extra help, see Example 2 on pages 406–407 of your text.

Use Method 2 to simplify each complex fraction.

4.
$$\dfrac{2x-y^2}{x+\dfrac{y^2}{x}}$$

4. _______________

5. $$\dfrac{r+\dfrac{3}{r}}{\dfrac{5}{r}+rt}$$

5. ______________

6. $$\dfrac{\dfrac{x-2}{x+2}}{\dfrac{x}{x-2}}$$

6. ______________

Objective 3   Compare the two methods of simplifying complex fractions.

Review these examples for Objective 3:

3. Use both Method 1 and Method 2 to simplify each complex fraction.

a. $$\dfrac{\dfrac{15}{10k+10}}{\dfrac{5}{3k+3}}$$

Using Method 1

$$\dfrac{\dfrac{15}{10k+10}}{\dfrac{5}{3k+3}}=\dfrac{\dfrac{15}{10(k+1)}}{\dfrac{5}{3(k+1)}}=\dfrac{15}{10(k+1)}\div\dfrac{5}{3(k+1)}$$

$$=\dfrac{15}{10(k+1)}\cdot\dfrac{3(k+1)}{5}$$

$$=\dfrac{9}{10}$$

Using Method 2

$$\dfrac{\dfrac{15}{10k+10}}{\dfrac{5}{3k+3}}=\dfrac{\dfrac{15}{10(k+1)}}{\dfrac{5}{3(k+1)}}=\dfrac{\dfrac{15}{10(k+1)}\cdot 30(k+1)}{\dfrac{5}{3(k+1)}\cdot 30(k+1)}$$

$$=\dfrac{45}{50}=\dfrac{9}{10}$$

Now Try:

3. Use both Method 1 and Method 2 to simplify each complex fraction.

a. $$\dfrac{\dfrac{1}{r}}{\dfrac{1+r}{1-r}}$$

____

                Copyright © 2020 Pearson Education, Inc.

b. $\dfrac{\dfrac{1}{m}-\dfrac{1}{n^2}}{\dfrac{1}{m^2}+\dfrac{1}{n}}$

Using Method 1

$$\frac{\dfrac{1}{m}-\dfrac{1}{n^2}}{\dfrac{1}{m^2}+\dfrac{1}{n}}=\frac{\dfrac{n^2}{mn^2}-\dfrac{m}{mn^2}}{\dfrac{n}{m^2n}+\dfrac{m^2}{m^2n}}=\frac{\dfrac{n^2-m}{mn^2}}{\dfrac{n+m^2}{m^2n}}$$

$$=\frac{n^2-m}{mn^2}\div\frac{n+m^2}{m^2n}$$

$$=\frac{n^2-m}{mn^2}\cdot\frac{m^2n}{n+m^2}$$

$$=\frac{m(n^2-m)}{n(n+m^2)}$$

Using Method 2

$$\frac{\dfrac{1}{m}-\dfrac{1}{n^2}}{\dfrac{1}{m^2}+\dfrac{1}{n}}=\frac{\left(\dfrac{1}{m}-\dfrac{1}{n^2}\right)\cdot m^2n^2}{\left(\dfrac{1}{m^2}+\dfrac{1}{n}\right)\cdot m^2n^2}$$

$$=\frac{\left(\dfrac{1}{m}\right)m^2n^2-\left(\dfrac{1}{n^2}\right)m^2n^2}{\left(\dfrac{1}{m^2}\right)m^2n^2+\left(\dfrac{1}{n}\right)m^2n^2}$$

$$=\frac{mn^2-m^2}{n^2+m^2n},\text{ or }\frac{m(n^2-m)}{n(n+m^2)}$$

Objective 3 Practice Exercises

For extra help, see Example 3 on pages 407–408 of your text.

Use either method to simplify each complex fraction.

7. $\dfrac{\dfrac{25k^2-m^2}{4k}}{\dfrac{5k+m}{7k}}$

---

b. $\dfrac{\dfrac{1}{a^2}+\dfrac{1}{b}}{\dfrac{1}{b^2}-\dfrac{1}{a}}$

____

7. _______________

8. $\dfrac{\dfrac{4}{p}-2p}{\dfrac{3-p^2}{6}}$

8. _______________

9. $\dfrac{\dfrac{4}{x}-\dfrac{1}{2}}{\dfrac{5}{x}+\dfrac{1}{3}}$

9. _______________

Objective 4   Simplify rational expressions with negative exponents.

Review this example for Objective 4:

4. Simplify, using only positive exponents in the answer.

$$\frac{2x^{-1}+y^2}{z^{-3}}$$

$$=\frac{\dfrac{2}{x}+y^2}{\dfrac{1}{z^3}}=\frac{xz^3\left(\dfrac{2}{x}+y^2\right)}{xz^3\left(\dfrac{1}{z^3}\right)}=\frac{2z^3+xy^2z^3}{x}$$

Now Try:

4. Simplify, using only positive exponents in the answer.

$$\frac{x^{-1}}{y-x^{-1}}$$

____

Objective 4 Practice Exercises

For extra help, see Example 4 on pages 408–409 of your text.

Simplify each expression, using only positive exponents in the answer.

10. $\dfrac{4x^{-2}}{2+6y^{-3}}$

10. _______________

Name:

Instructor:

Date:

Section:

11. $\dfrac{s^{-1}+r}{r^{-1}+s}$

11. _______________

12. $\dfrac{(m+n)^{-2}}{m^{-2}-n^{-2}}$

12. _______________

Chapter 6 RATIONAL EXPRESSIONS AND FUNCTIONS

6.4   Equations with Rational Expressions and Graphs

Learning Objectives
1   Determine the domain of the variable in a rational equation.
2   Solve rational equations.
3   Recognize the graph of a rational function.

Key Terms

Use the vocabulary terms listed below to complete each statement in exercises 1−4.

>   **domain of the variable in a rational equation**   **discontinuous**
>
>   **vertical asymptote**   **horizontal asymptote**

1.   A rational function in simplest form $f(x) = \dfrac{P(x)}{x-a}$ has the line $x = a$ as a

   _______________________________.

2.   The ____ is the intersection of the domains of the rational expressions in the equation.

3.   A graph of a function is _____________________ if there are one or more breaks in the graph.

4.   A horizontal line that a graph approaches as $|x|$ gets larger without bound is called a _________________________________.

Objective 1   Determine the domain of the variable in a rational equation.

Review these examples for Objective 1:

1.   Find the domain of the variable in each equation.

   a. $\dfrac{4}{5x} + \dfrac{3}{2x} = \dfrac{23}{50}$

   The domains of the three expressions are $\{x \mid x \neq 0\}$, $\{x \mid x \neq 0\}$, and $\{x \mid x \text{ is a real number}\}$. The intersection of these three domains is all real numbers except 0, written in set-builder notation as $\{x \mid x \neq 0\}$.

Now Try:

1.   Find the domain of the variable in each equation.

   a. $\dfrac{1}{x} + \dfrac{9}{4} = \dfrac{11}{2x}$

   ____

          Copyright © 2020 Pearson Education, Inc.

b. $\dfrac{r+5}{r^2-16} = \dfrac{3}{r-4} + \dfrac{1}{r+4}.$

The domains of the three expressions are respectively, $\{x \mid x \neq \pm 4\}$, $\{x \mid x \neq 4\}$, and $\{x \mid x \neq -4\}$. The domain of the variable is the intersection of the three domains, all real numbers except -4 and 4, written $\{x \mid x \neq \pm 4\}$.

b. $\dfrac{10}{x-7} + \dfrac{7}{x+8} = \dfrac{-3}{x^2+x-56}$

____

Objective 1 Practice Exercises

For extra help, see Example 1 on page 411 of your text.

(a) Without actually solving the equations below, list all possible numbers that would have to be rejected if they appeared as potential solutions. (b) Then give the domain using set notation.

1. $\dfrac{1}{x} + \dfrac{2}{x+1} = 0$

1. a._______________

   b._______________

2. $\dfrac{x}{6} - \dfrac{1}{2} = \dfrac{3}{x+1}$

2. a._______________

   b._______________

3. $\dfrac{3}{3+x} + \dfrac{5}{x^2-x} = \dfrac{9}{x}$

3. a._______________

   b._______________

Name: _______________________   Date: _______________________

Instructor: _________________   Section: ____________________

Objective 2   Solve rational equations.

Review these examples for Objective 2:

2. Solve $\dfrac{4}{5x}+\dfrac{3}{2x}=\dfrac{23}{50}$.

Step 1 The domain, which excludes 0, was found in Example 1(a).

Step 2 Multiply by the LCD, $50x$.

$$50x\left(\frac{4}{5x}+\frac{3}{2x}\right)=50x\left(\frac{23}{50}\right)$$

Step 3 $50x\left(\dfrac{4}{5x}\right)+50x\left(\dfrac{3}{2x}\right)=50x\left(\dfrac{23}{50}\right)$

$$40+75=23x$$
$$115=23x$$
$$5=x$$

Step 4 Check $\quad \dfrac{4}{5x}+\dfrac{3}{2x}=\dfrac{23}{50}$

$$\frac{4}{5(5)}+\frac{3}{2(5)}\overset{?}{=}\frac{23}{50}$$

$$\frac{4}{25}+\frac{3}{10}\overset{?}{=}\frac{23}{50}$$

$$\frac{23}{50}=\frac{23}{50}$$

The solution set is $\{5\}$.

3. Solve $\dfrac{10}{x^2-2x}+\dfrac{4}{x}=\dfrac{5}{x-2}$.

Step 1 The domain excludes 0 and 2.

Step 2 Multiply by the LCD, $x(x-2)$.

$$x(x-2)\left[\frac{10}{x^2-2x}+\frac{4}{x}\right]=x(x-2)\left(\frac{5}{x-2}\right)$$

Step 3

$$x(x-2)\left(\frac{10}{x^2-2x}\right)+x(x-2)\left(\frac{4}{x}\right)$$
$$=x(x-2)\left(\frac{5}{x-2}\right)$$

$$10+4(x-2)=5x$$
$$10+4x-8=5x$$
$$2+4x=5x$$
$$2=x$$

Now Try:

2. Solve $\dfrac{1}{x}+\dfrac{9}{4}=\dfrac{11}{2x}$.

____

3. Solve $\dfrac{4}{x-2}-\dfrac{3}{x+2}=\dfrac{16}{x^2-4}$.

____

   Copyright © 2020 Pearson Education, Inc.

Step 4 Since the proposed solution, 2, is not in the domain, it cannot be a solution of the equation.

$$\frac{10}{x^2 - 2x} + \frac{4}{x} = \frac{5}{x - 2}$$

$$\frac{10}{2^2 - 2(2)} + \frac{4}{2} \overset{?}{=} \frac{5}{2 - 2}$$

$$\frac{10}{0} + 2 \overset{?}{=} \frac{5}{0}$$

The equation has no solution The solution set is $\varnothing$.

4. Solve $\dfrac{r+5}{r^2 - 16} = \dfrac{3}{r - 4} + \dfrac{1}{r + 4}$.

The domain, which excludes ± 4, was found in Example 1(b). Multiply each side of the equation by the LCD, $(r + 4)(r - 4)$.

$$(r+4)(r-4)\left[\frac{r+5}{(r+4)(r-4)}\right]$$

$$= (r+4)(r-4)\left(\frac{3}{r-4} + \frac{1}{r+4}\right)$$

$$(r+4)(r-4)\left[\frac{r+5}{(r+4)(r-4)}\right]$$

$$= (r+4)(r-4)\left(\frac{3}{r-4}\right) + (r+4)(r-4)\left(\frac{1}{r+4}\right)$$

$$r+5 = 3(r+4) + (r-4)$$
$$r+5 = 3r + 12 + r - 4$$
$$r+5 = 4r + 8$$
$$-3 = 3r$$
$$-1 = r$$

Check $\quad \dfrac{r+5}{r^2 - 16} = \dfrac{3}{r - 4} + \dfrac{1}{r + 4}$

$$\frac{-1+5}{(-1)^2 - 16} \overset{?}{=} \frac{3}{-1 - 4} + \frac{1}{-1 + 4}$$

$$\frac{4}{1 - 16} \overset{?}{=} \frac{3}{-5} + \frac{1}{3}$$

$$\frac{4}{-15} = \frac{-4}{15}$$

The solution set is $\{-1\}$.

4. Solve
$$\frac{10}{x - 7} + \frac{7}{x + 8} = \frac{-3}{x^2 + x - 56}.$$

____

5. Solve $\dfrac{3}{x^2-1}-\dfrac{3x}{2x+2}=-\dfrac{2}{3}$.

Since the denominators cannot equal -1 and 1, they are excluded from the domain. Multiply by the LCD $6(x+1)(x-1)$.

$$6(x+1)(x-1)\left(\frac{3}{(x+1)(x-1)}-\frac{3x}{2(x+1)}\right)$$
$$=6(x+1)(x-1)\left(-\frac{2}{3}\right)$$

$$6(x+1)(x-1)\left(\frac{3}{(x+1)(x-1)}\right)$$
$$+6(x+1)(x-1)\left(-\frac{3x}{2(x+1)}\right)$$
$$=6(x+1)(x-1)\left(-\frac{2}{3}\right)$$

$$6(3)-3(x-1)(3x)=-4(x+1)(x-1)$$
$$18-9x^2+9x=-4x^2+4$$
$$-5x^2+9x+14=0$$
$$5x^2-9x-14=0$$
$$(5x-14)(x+1)=0$$
$$5x-14=0 \quad \text{or} \quad x+1=0$$
$$x=\frac{14}{5} \quad \text{or} \quad x=-1$$

Because -1 is not in the domain, it is not a solution. The solution set is $\left\{\dfrac{14}{5}\right\}$.

6. Solve $\dfrac{4}{x}-\dfrac{6}{x+1}=1$.

Exclude 0 and -1 from the domain.
Multiply by the LCD $x(x+1)$.

$$x(x+1)\left(\frac{4}{x}-\frac{6}{x+1}\right)=x(x+1)(1)$$

$$x(x+1)\left(\frac{4}{x}\right)-x(x+1)\left(\frac{6}{x+1}\right)=x(x+1)$$

$$(x+1)(4)-6x=x^2+x$$
$$4x+4-6x=x^2+x$$
$$x^2+3x-4=0$$
$$(x+4)(x-1)=0$$

5. Solve $\dfrac{q+12}{q^2-16}-\dfrac{3}{q-4}=\dfrac{1}{q+4}$.

____

6. Solve $\dfrac{3}{x}+\dfrac{4}{x+2}=1$.

____

$$x+4=0 \quad \text{or} \quad x-1=0$$
$$x=-4 \quad \text{or} \quad x=1$$

Both –4 and 1 are in the domain.
Check that the solution set is $\{-4, 1\}$.

Objective 2 Practice Exercises

For extra help, see Examples 2–6 on pages 412–414 of your text.

Solve each equation.

4. $\dfrac{8}{2m+4}+\dfrac{2}{3m+6}=\dfrac{7}{9}$

4. _______________________

5. $\dfrac{1}{b+2}-\dfrac{5}{b^2+9b+14}=\dfrac{-3}{b+7}$

5. _______________________

6. $\dfrac{-16}{n^2-8n+12}=\dfrac{3}{n-2}+\dfrac{n}{n-6}$

6. _______________________

                    241

Objective 3   Recognize the graph of a rational function.

Review this example for Objective 3:

7. Graph, and give the equations of the vertical and horizontal asymptotes.

$$f(x) = \frac{-2}{x+3}$$

Some ordered pairs that belong to the function are listed below.

x	$f(x) = \dfrac{-2}{x+3}$	x	$f(x) = \dfrac{-2}{x+3}$
-6	$\dfrac{2}{3}$	1	$-\dfrac{1}{2}$
-5	1	2	$-\dfrac{2}{5}$
-4	2	3	$-\dfrac{1}{3}$
-2	-2	4	$-\dfrac{2}{7}$
-1	-1	5	$-\dfrac{1}{4}$
0	$-\dfrac{2}{3}$	6	$-\dfrac{2}{9}$

There is no point on the graph for $x = -3$ because -3 is excluded from the domain of the rational function.

The dashed line $x = -3$ represents the vertical asymptote and is not part of the graph. The graph gets closer to the vertical asymptote as the x-values get closer to -3. Since the x-values approach 0 as $|x|$ increases, the horizontal asymptote is $y = 0$.

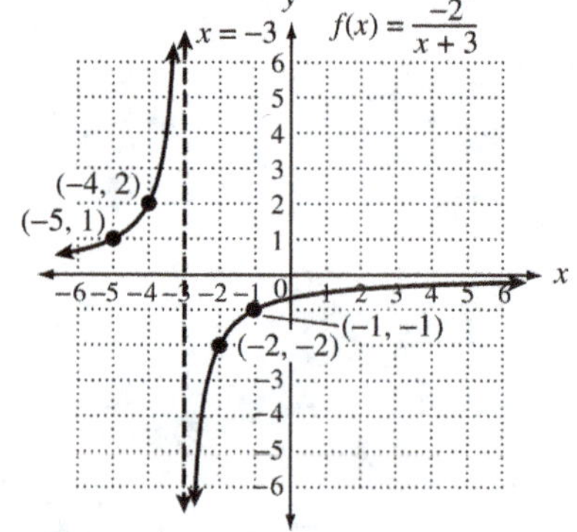

Now Try:

7. Graph, and give the equations of the vertical and horizontal asymptotes.

$$f(x) = -\frac{1}{x}$$

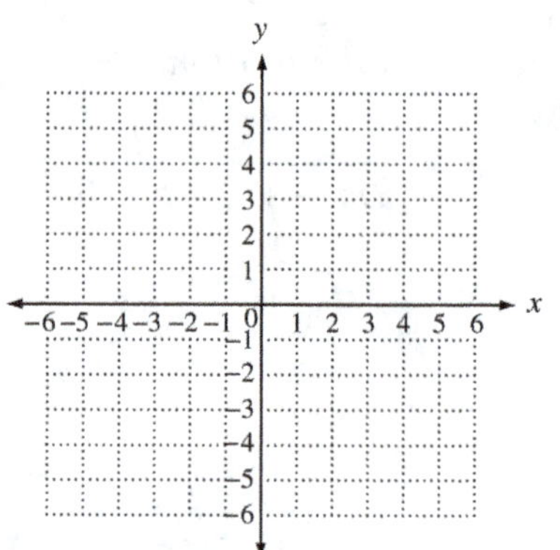

Vertical asymptote: ___________

Horizontal asymptote: _________

Objective 3 Practice Exercises

For extra help, see Example 7 on page 415 of your text.

Graph each rational function. Give the equations of the vertical and horizontal asymptotes.

7. $f(x) = \dfrac{5}{x}$

7.

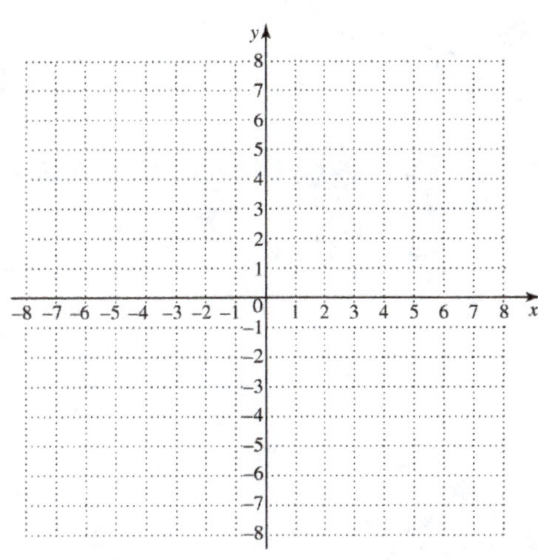

Vertical asymptote: ___________

Horizontal asymptote: __________

8. $f(x) = \dfrac{1}{x-4}$

8.

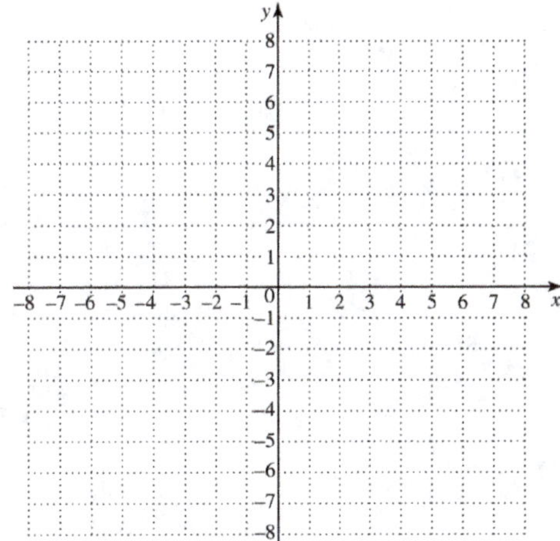

Vertical asymptote: ___________

Horizontal asymptote: __________

                    243

9. $f(x) = \dfrac{3}{x+2}$

9.

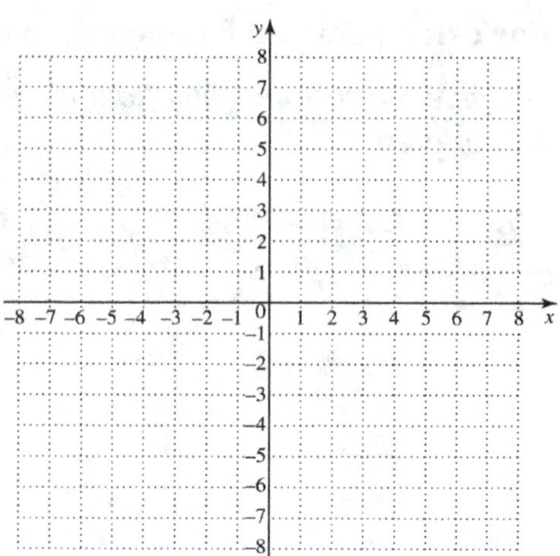

Vertical asymptote: _____________

Horizontal asymptote: _________

Chapter 6   RATIONAL EXPRESSIONS AND FUNCTIONS

6.5   Applications of Rational Expressions

Learning Objectives
1   Find the value of an unknown variable in a formula.
2   Solve a formula for a specified variable.
3   Solve applications using proportions.
4   Solve applications about distance, rate, and time.
5   Solve applications about work rates.

Key Terms

Use the vocabulary terms listed below to complete each statement in exercises 1–2.

ratio   proportion

1.   A comparison of two quantities using a quotient is a ____________________.

2.   A statement that two ratios are equal is a ____________________.

Objective 1   Find the value of an unknown variable in a formula.

Review this example for Objective 1:

1.   The height of a trapezoid is given by the formula $h = \dfrac{2A}{B+b}$. Find B, if $A = 40$, $h = 8$ and $b = 3$.

Substitute the given values into the equation and solve for B.

$$h = \frac{2A}{B+b}$$

$$8 = \frac{2(40)}{B+3}$$

$$8(B+3) = \frac{80}{B+3}(B+3)$$

$$8B+24 = 80$$

$$8B = 56$$

$$B = 7$$

Now Try:

1.   The formula for the slope of a line is $m = \dfrac{y_1 - y_2}{x_1 - x_2}$. Find x_2 if $m = 3$, $y_1 = 12$, $y_2 = 8$, and $x_1 = 5$.

____

Objective 1 Practice Exercises

For extra help, see Example 1 on page 421 of your text.

Find the value of the variable indicated.

1. If $F = \dfrac{GmM}{d^2}$, $F = 150$, $G = 32$, $M = 50$,
   and $d = 10$, find m.

   1. _______________

2. If $\dfrac{1}{f} = \dfrac{1}{d_0} + \dfrac{1}{d_i}$, $f = 10$, and $d_0 = 25$, find d_i.

   2. _______________

3. If $c = \dfrac{100b}{L}$, $c = 80$, and $b = 16$, find L.

   3. _______________

Objective 2   Solve a formula for a specified variable.

Review these examples for Objective 2:

2. Solve the following formula for R_1.
   $$\frac{1}{R} = \frac{1}{R_1} + \frac{1}{R_2}$$

   Start by multiplying by the LCD, RR_1R_2.

   $$RR_1R_2\left(\frac{1}{R}\right) = RR_1R_2\left(\frac{1}{R_1} + \frac{1}{R_2}\right)$$

   $$RR_1R_2\left(\frac{1}{R}\right) = RR_1R_2\left(\frac{1}{R_1}\right) + RR_1R_2\left(\frac{1}{R_2}\right)$$

   $$R_1R_2 = RR_2 + RR_1$$

   $$R_1R_2 - RR_1 = RR_2$$

   $$R_1(R_2 - R) = RR_2$$

   $$R_1 = \frac{RR_2}{R_2 - R}$$

Now Try:

2. Solve $\dfrac{1}{R} = \dfrac{1}{R_1} + \dfrac{1}{R_2}$ for R.

   ____

                Copyright © 2020 Pearson Education, Inc.

3. Solve the following formula for P.

$$r = \frac{A - P}{Pt}$$

$$r = \frac{A - P}{Pt}$$

$$Prt = A - P$$

$$Prt + P = A$$

$$P(rt + 1) = A$$

$$P = \frac{A}{1 + rt}$$

3. Solve $A = \dfrac{R_1 R_2}{R_1 + R_r}$ for R_1

____

Objective 2 Practice Exercises

For extra help, see Examples 2–3 on pages 421–422 of your text.

Solve each formula for the specified variable.

4. $\dfrac{V_1 P_1}{T_1} = \dfrac{V_2 P_2}{T_2}$ for T_2

4. _______________

5. $h = \dfrac{2A}{B + b}$ for B

5. _______________

6. $E = \dfrac{e(R + r)}{r}$ for r

6. _______________

Objective 3   Solve applications using proportions.

Review this example for Objective 3:

5. Ryan's car uses 5 gallons of gasoline to travel 120 miles. If he has 4 gallons of gas in the car, how much more gasoline will he need to travel 288 miles?

Step 1 Read the problem.

Step 2 Assign a variable. Let x = the additional number of gallons of gas needed.

Step 3 Write an equation. Use a proportion.

Now Try:

5. Beth's car uses 12 gallons of gasoline to travel 336 miles. If she has 5 gallons of gas in the car, how much more gasoline will she need to travel 252 miles?

____

$$\frac{\text{gallons} \rightarrow}{\text{miles} \rightarrow} \frac{5}{120} = \frac{4+x}{288} \frac{\leftarrow \text{gallons}}{\leftarrow \text{miles}}$$

Step 4 Solve.

$$\frac{5}{120} = \frac{4+x}{288}$$

$$(288)(5) = (120)(4+x)$$

$$1440 = 480 + 120x$$

$$960 = 120x$$

$$8 = x$$

Step 5 State the answer. Ryan will need 8 more gallons of gas.

Step 6 Check. The original 4 gallons plus 8 gallons equals 12 gallons.

$$\frac{5}{120} \overset{?}{=} \frac{4+8}{288}$$

$$\frac{1}{24} = \frac{1}{24}$$

Objective 3 Practice Exercises

For extra help, see Examples 4–5 on pages 422–423 of your text.

Use proportions to solve each problem.

7. Ryan's car travels 24 miles using 1 gallon of gas. If Ryan has 4 gallons of gas in his car, how much more gas will he need to travel 288 miles?

7. _________________

8. Connie paid $0.90 in sales tax on a purchase of $12.00. Later that day she purchased an item for which she paid $55.90 including sales tax, at the same tax rate. How much was the sales tax on that item?

8. _________________

                     Copyright © 2020 Pearson Education, Inc.

9. A certain city with a population of 400,000 had 1600 burglaries during the last year. If the number of burglaries increased by 200 this year and its burglary rate remained the same, by what number did its population increase?

9. _______________________

Objective 4   Solve applications about distance, rate, and time.

Review these examples for Objective 4:

6. Mark can row 5 miles per hour in still water. It takes him as long to row 4 miles upstream as 16 miles downstream. How fast is the current?

Step 1 Read the problem carefully. We must find the speed of the current.

Step 2 Assign a variable. Let x = the speed of the current. Traveling upstream (against the current) slows Mark down, so his rate is the difference between his rate in still water and the rate of the current, that is $5 - x$ mph. Traveling downstream (with the current) speeds him up, so his rate is the sum of his rate in still water and the rate of the current, $5 + x$ mph. We can summarize the given information in a table. Use the formula $d = rt$ or $t = \frac{d}{r}$.

	r	d	t
upstream	$5-x$	4	$\dfrac{4}{5-x}$
downstream	$5+x$	16	$\dfrac{16}{5+x}$

Step 3 Write an equation. The times are equal so we have $\dfrac{4}{5-x} = \dfrac{16}{5+x}$.

Step 4 Solve.

Now Try:

6. A boat goes 6 miles per hour in still water. It takes as long to go 40 miles upstream as 80 miles downstream. Find the speed of the current.

____

$$\frac{4}{5-x} = \frac{16}{5+x}$$

$$(5-x)(5+x)\left(\frac{4}{5-x}\right) = (5-x)(5+x)\left(\frac{16}{5+x}\right)$$

$$4(5+x) = (5-x)16$$

$$20+4x = 80-16x$$

$$20x = 60$$

$$x = 3$$

Step 5 State the answer.
The speed of the current is 3 miles per hour.

Step 6 Check the solution in the words of
the original problem. Mike can row upstream

$5 - 3 = 2$ miles per hour. It will take him $\frac{4}{2} = 2$

hours to row four miles upstream. Mike can row
downstream $5 + 3 = 8$ miles per hour. It will take

him $\frac{16}{8} = 2$ hours to row 16 miles downstream.

The time upstream equals the time downstream,
as required.

7. Bev and Mike drove 140 miles before stopping
at a rest area. After leaving the rest area, they
traveled an additional 90 miles at an average
speed that was 4 miles per hour faster than their
speed before stopping. If the entire trip took 4
hours of driving time, what was the average
speed during each part of the trip?

Step 1 Read the problem carefully. We must
find the average speed during each part of the
trip.

Step 2 Assign a variable.
Let $x =$ the rate during the first part of the trip.
Then $x + 4 =$ the rate during the second part of
the trip.

Express the times in terms of the known
distances and the variable rates. time for the first

part of the trip: $t = \frac{d}{r} = \frac{140}{x}$

time for the second part of the trip: $t = \frac{d}{r} = \frac{90}{x+4}$

7. Marlene ran 5 miles on her
treadmill followed by a 1 mile
cool-down walk. If she runs
twice as fast as she walks and
she spent 70 minutes on the
treadmill, what was her walking
rate in miles per hour?

____

     Copyright © 2020 Pearson Education, Inc.

	d	r	t
First part	140	x	$\dfrac{140}{x}$
Second part	90	$x+4$	$\dfrac{90}{x+4}$

Step 3 Write an equation.

$$\frac{140}{x} + \frac{90}{x+4} = 4$$

Step 4 Solve.

$$\frac{140}{x} + \frac{90}{x+4} = 4$$

$$x(x+4)\left(\frac{140}{x} + \frac{90}{x+4}\right) = x(x+4)(4)$$

$$x(x+4)\left(\frac{140}{x}\right) + x(x+4)\left(\frac{90}{x+4}\right) = x(x+4)(4)$$

$$140x + 560 + 90x = 4x^2 + 16x$$

$$4x^2 - 214x - 560 = 0$$

$$2x^2 - 107x - 280 = 0$$

$$(x-56)(2x+5) = 0$$

$$x - 56 = 0 \quad \text{or} \quad 2x + 5 = 0$$

$$x = 56 \quad \text{or} \qquad x = -\frac{5}{2}$$

We discard the negative answer since rate

Objective 4 Practice Exercises

For extra help, see Examples 6–7 on pages 424–426 of your text.

Solve each problem.

10. Lauren's boat can go 9 miles per hour in still water.   **10.** _______________
How far downstream can Lauren go if the river has a
current of 3 miles per hour and she must be back in 4
hours.

11. Pauline and Pete agree to meet in Columbia. Pauline   **11.** _______________
travels 120 miles, while Pete travels 80 miles. If
Pauline's speed is 20 miles per hour greater than
Pete's and they both spend the same amount of time
traveling, at what speed does each travel?

12. Olivia can ride her bike 4 miles per hour faster than Ted can ride his bike. If Olivia can go 30 miles in the same time that Ted can go 15 miles, what are their speeds?

12. _______________

Objective 5   Solve applications about work rates.

Review this example for Objective 5:

8. One pipe can fill a swimming pool in 8 hours and another pipe can fill the pool in 12 hours. How long will it take to fill the pool if both pipes are open?

Let x = the number of hours working together. The rate for the first pipe is $\frac{1}{8}$.

The rate for the second pipe is $\frac{1}{12}$.

The sum of the fractional part for each pipe multiplied by the time working together is the whole job.

$$\frac{1}{8}x + \frac{1}{12}x = 1$$

$$48\left(\frac{1}{8}x + \frac{1}{12}x\right) = 48(1)$$

$$48\left(\frac{1}{8}x\right) + 48\left(\frac{1}{12}x\right) = 48$$

$$6x + 4x = 48$$

$$10x = 48$$

$$x = \frac{48}{10} = \frac{24}{5} = 4\frac{4}{5}$$

Working together, it takes $4\frac{4}{5}$ hours to fill the pool. A check confirms this answer.

Now Try:

8. Chuck can weed the garden in $\frac{1}{2}$ hour, but David takes 2 hours. How long does it take them to weed the garden if they work together?

____

Name: Date:
Instructor: Section:

Objective 5 Practice Exercises

For extra help, see Example 8 on pages 426–427 of your text.

Solve each problem.

13.   Kelly can clean the house in 6 hours, but it takes
      Linda 4 hours. How long would it take them to clean
      the house if they worked together?

13. ______________

14.   A swimming pool can be filled by an inlet pipe in 18
      hours and emptied by an outlet pipe in 24 hours.
      How long will it take to fill the empty pool if the
      outlet pipe is accidentally left open at the same time
      as the inlet pipe is opened?

14. ______________

15.   Fred can seal an asphalt driveway in $\frac{1}{3}$ the time it
      takes John. Working together, it takes them $1\frac{1}{2}$
      hours. How long would it have taken Fred working
      alone?

15. ______________

Chapter 6   RATIONAL EXPRESSIONS AND FUNCTIONS

6.6   Variation

Learning Objectives
1   Write an equation expressing direct variation.
2   Find the constant of variation, and solve direct variation problems.
3   Solve inverse variation problems.
4   Solve joint variation problems.
5   Solve combined variation problems.

Key Terms

Use the vocabulary terms listed below to complete each statement in exercises 1–3.

     varies directly   **varies inversely**   **constant of variation**

1.   In the equations for direct and inverse variation, k is the ________________.

2.   If there exists a real number k such that $y = \dfrac{k}{x}$, then y ________________

      as x.

3.   If there exists a real number k such that $y = kx$, then y ________________

      as x.

Objective 1   Write an equation expressing direct variation.

For extra help, see page 433 of your text.

Objective 2   Find the constant of variation, and solve direct variation problems.

Review these examples for Objective 2:	Now Try:
1. If 12 gallons of gasoline cost \$34.68, how much does 1 gallon of gasoline cost? Write the variation equation. Let g represent the number of gallons of gasoline and let C represent the total cost of the gasoline. Then the variation equation is $C = kg$. $$C = kg$$ $$34.68 = 12k$$ $$2.89 = k$$ The cost per gallon is \$2.89. The variation equation is $C = 2.89g$.	1. One week a manufacturer sold 1200 items for a total profit of \$30,000. What was the profit for one item. Write the variation equation. ________________ ________________

2. A person's weight on the moon varies directly with the person's weight on Earth. A 120-pound person would weigh about 20 pounds on the moon. How much would a 150-pound person weigh on the moon?

If m represents the person's weight on the moon and w represents the person's weight on Earth.

$$m = kw$$

$20 = k \cdot 120$ Let $m = 20$ and $w = 120$.

$\dfrac{20}{120} = \dfrac{1}{6} = k$ Solve for k; lowest terms

Now, substitute $\dfrac{1}{6}$ for k and 150 for w in the variation equation.

$$m = kw$$

$$m = \frac{1}{6} \cdot 150 = 25$$

A 150-pound person will weigh 25 pounds on the moon.

3. The surface area of a sphere varies directly as the square of its radius. If the surface area of a sphere with a radius of 12 inches is 576π square inches, find the surface area of a sphere with a radius of 3 inches.

Step 1 A represents the surface area and r represents the radius.

$$A = kr^2$$

Step 2 Find the value of k when A is 576π and r is 12.

$$A = kr^2$$
$$576\pi = k \cdot 12^2$$
$$576\pi = 144k$$
$$\frac{576\pi}{144} = 4\pi = k$$

Step 3 Rewrite the variation equation.

$$A = 4\pi r^2$$

Step 4 Let $r = 3$ to find the surface area.

$$A = 4\pi \cdot 3^2 = 4\pi \cdot 9 = 36\pi$$

The surface area of a sphere with a radius of 3 inches is 36π square inches.

2. The pressure exerted by a certain liquid at a given point varies directly as the depth of the point beneath the surface of the liquid. The pressure at 10 feet is 50 pounds per square inch (psi). What is the pressure at 25 feet?

____

3. The area of a circle varies directly as the square of the radius. A circle with a radius of 5 centimeters has an area of 78.5 square centimeters. Find the area if the radius changes to 7 centimeters.

____

Name: Date:
Instructor: Section:

Objective 2 Practice Exercises

For extra help, see Examples 1–3 on pages 434–435 of your text.

Find the constant of variation, and write a direct variation equation.

1. $y = 13.75$ when $x = 55$ 1. _________________

Solve each problem.

2.   The circumference of a circle varies directly as the   2. _________________
     radius. A circle with a radius of 7 centimeters has a
     circumference of 43.96 centimeters. Find the
     circumference of the circle if the radius changes to
     11 centimeters.

3.   The force required to compress a spring varies   3. _________________
     directly as the change in length of the spring. If a
     force of 20 newtons is required to compress a spring
     2 centimeters in length, how much force is required
     to compress a spring of length 10 centimeters?

Objective 3   Solve inverse variation problems.

Review these examples for Objective 3:   **Now Try:**

4.   For a specified distance, time varies inversely   4. The length of a violin string
     with speed. If Ramona walks a certain distance   varies inversely with the
     on a treadmill in 40 minutes at 4.2 miles per   frequency of its vibrations. A
     hour, how long will it take her to walk the same   10-inch violin string vibrates at
     distance at 3.5 miles per hour?   a frequency of 512 cycles per
                                                               second. Find the frequency of an
     Let t = time and s = speed.   8-inch string.
     Since t varies inversely as s, there is a constant k

     such that $t = \dfrac{k}{s}$. Recall that $40 \text{ min} = \dfrac{40}{60} \text{ hr}$.

     $$t = \frac{k}{s}$$

     $$\frac{40}{60} = \frac{k}{4.2}$$

     $$2.8 = k$$ _____________

                     Copyright © 2020 Pearson Education, Inc.

Now use $t = \dfrac{k}{s}$ to find the value of t when $s = 3.5$.

$$t = \frac{2.8}{3.5} = \frac{4}{5}$$

It takes $\dfrac{4}{5}$ hr, or 48 min to walk the same distance.

5. With constant power, the resistance used in a simple electric circuit varies inversely as the square of the current. If the resistance is 120 ohms when the current is 12 amps, find the resistance if the current is reduced to 9 amps.

Let R represent resistance (in ohms) and $I =$ current (in amps). Then $R = \dfrac{k}{I^2}$.

First, we solve for the constant of variation by substituting 120 for R and 12 for I.

$$120 = \frac{k}{12^2}$$

$$k = 120 \cdot 12^2$$

Now use the value for k and 9 for I to find R.

$$R = \frac{120 \cdot 12^2}{9^2} \approx 213.3$$

The resistance is about 213.3 ohms when the current is 9 amps.

5. If y varies inversely as x^3, and $y = 9$ when $x = 2$, find y when $x = 4$.

____

Objective 3 Practice Exercises

For extra help, see Examples 4–5 on pages 436–437 of your text.

Solve each problem.

4. The illumination produced by a light source varies inversely as the square of the distance from the source. If the illumination produced 4 feet from a light source is 75 footcandles, find the illumination produced 9 feet from the same source.

4. ______________

5. The weight of an object varies inversely as the square of its distance from the center of Earth. If an object 8000 miles from the center of Earth weighs 90 pounds, find its weight when it is 12,000 miles from the center of Earth.

5. _______________________

6. The speed of a pulley varies inversely as its diameter. One kind of pulley, with a diameter of 3 inches, turns at 150 revolutions per minute. Find the speed of a similar pulley with diameter of 5 inches.

6. _______________________

Objective 4   Solve joint variation problems.

Review this example for Objective 4:

6. For a fixed interest rate, interest varies jointly as the principal and the time in years. If $5000 invested for 4 years earns $900, how much interest will $6000 invested for 3 years earn at the same interest rate?

Let I = the interest, p = the principal, and t = the time in years. Then, $I = kpt$.

$$I = kpt$$
$$900 = k \cdot 5000 \cdot 4 \quad \text{Substitute given values.}$$
$$\frac{900}{20{,}000} = \frac{9}{200} = k$$

Now use $k = \dfrac{9}{200}$.

$$I = \frac{9}{200} \cdot 6000 \cdot 3 = 810$$

$6000 invested for three years will earn $810 in interest.

Now Try:

6. The strength of a rectangular beam varies jointly as its width and the square of its depth. If the strength of a beam 2 inches wide by 10 inches deep is 1000 pounds per square inch, what is the strength of a beam 4 inches wide and 8 inches deep?

____

Name: Date:
Instructor: Section:

Objective 4 Practice Exercises

For extra help, see Example 6 on page 438 of your text.

Solve each problem.

7. Suppose d varies jointly as f^2 and g^2, and $d = 384$ when $f = 3$ and $g = 8$. Find d when $f = 6$ and $g = 2$.

7. ______________

8. The work w (in joules) done when lifting an object is jointly proportional to the product of the mass m (in kg) of the object and the height h (in meter) the object is lifted. If the work done when a 120 kg object is lifted 1.8 meters above the ground is 2116.8 joules, how much work is done when lifting a 100kg object 1.5 meters above the ground?

8. ______________

9. The absolute temperature of an ideal gas varies jointly as its pressure and its volume. If the absolute temperature is 250° when the pressure is 25 pounds per square centimeter and the volume is 50 cubic centimeters, find the absolute temperature when the pressure is 50 pounds per square centimeter and the volume is 75 cubic centimeters.

9. ______________

Objective 5   Solve combined variation problems.

Review this example for Objective 5:

7. The number of hours h that it takes w workers to assemble x machines varies directly as the number of machines and inversely as the number of workers. If four workers can assemble 12 machines in four hours, how many workers are needed to assemble 36 machines in eight hours?

The variation equation is $h = \dfrac{kx}{w}$.

To find k, let $h = 4$, $x = 12$, and $w = 4$.

$$4 = \frac{k \cdot 12}{4}$$

$$k = \frac{4 \cdot 4}{12}$$

$$k = \frac{4}{3}$$

Now find w when $x = 36$ and $h = 8$.

$$8 = \frac{\frac{4}{3} \cdot 36}{w}$$

$$8w = 48$$

$$w = 6$$

Eight workers are needed to assemble 36 machines in eight hours.

Now Try:

7. The volume of a gas varies directly as its temperature and inversely as its pressure. The volume of a gas at 85° C at a pressure of 12 kg/cm^2 is 300 cm^3. What is the volume when the pressure is 20 kg/cm^2 and the temperature is 30° C?

____

Objective 5 Practice Exercises

For extra help, see Example 7 on page 438 of your text.

Solve each problem.

10. The volume of a gas varies inversely as the pressure and directly as the temperature. If a certain gas occupies a volume of 1.3 liters at 300 K and a pressure of 18 kilograms per square centimeter, find the volume at 340 K and a pressure of 24 kilograms per square centimeter.

10. _______________

11. The time required to lay a sidewalk varies directly as its length and inversely as the number of people who are working on the job. If three people can lay a sidewalk 100 feet long in 15 hours, how long would it take two people to lay a sidewalk 40 feet long?

11. ___________________

12. When an object is moving in a circular path, the centripetal force varies directly as the square of the velocity and inversely as the radius of the circle. A stone that is whirled at the end of a string 50 centimeters long at 900 centimeters per second has a centripetal force of 3,240,000 dynes. Find the centripetal force if the stone is whirled at the end of a string 75 centimeters long at 1500 centimeters per second.

12. ___________________

Chapter 7   ROOTS, RADICALS, AND ROOT FUNCTIONS

7.1   Radical Expressions and Graphs

Learning Objectives
1   Find roots of numbers.
2   Find principal roots.
3   Graph functions defined by radical expressions.
4   Find nth roots of nth powers.
5   Use a calculator to find roots.

Key Terms

Use the vocabulary terms listed below to complete each statement in exercises 1−8.

square root	**principal square root**	**radicand**
radical	**radical expression**	**perfect square**
cube root	**index (order)**	

1. The number or expression inside a radical sign is called the _________________.

2. A number with a rational square root is called a _____________________.

3. In a radical of the form $\sqrt[n]{a}$, the number n is the _____________________.

4. The number b is a _____________________ of a if $b^2 = a$.

5. The expression $\sqrt[n]{a}$ is called a _________________________________.

6. The positive square root of a number is its _____________________________.

7. A _____________________ is a radical sign and the number or expression in it.

8. The number b is a _____________________ of a if $b^3 = a$.

Objective 1   Find roots of numbers.

Review these examples for Objective 1:	Now Try:
1. Simplify.	1. Simplify.
c. $\sqrt[4]{81}$	c. $\sqrt[4]{256}$
$\sqrt[4]{81} = 3$, because $3^4 = 81$.	_____________

e. $\sqrt[3]{\dfrac{125}{343}}$ **e.** $\sqrt[3]{\dfrac{64}{27}}$

$$\sqrt[3]{\dfrac{125}{343}} = \dfrac{5}{7}, \text{ because } \left(\dfrac{5}{7}\right)^3 = \dfrac{125}{343}.$$

____

Objective 1 Practice Exercises

For extra help, see Example 1 on page 456 of your text.

Find each root that is a real number. Use a calculator as necessary.

1. $\sqrt[4]{625}$ 1. ________________

2. $\sqrt[6]{64}$ 2. ________________

3. $\sqrt[4]{6561}$ 3. ________________

Objective 2   Find principal roots.

Review these examples for Objective 2:   **Now Try:**
2.   Find each root.   **2.**   Find each root.

b. $-\sqrt{144}$ **b.** $-\sqrt{121}$

$-\sqrt{144} = -12$

____

e. $\sqrt[4]{-256}$ **e.** $\sqrt[4]{-625}$

The index is even and the radicand is
negative, so $\sqrt[4]{-256}$ is not a real number.   _______________

g. $\sqrt[3]{-64}$ **g.** $\sqrt[3]{-125}$

$\sqrt[3]{-64} = -4$, because $(-4)^3 = -64$.

____

Objective 2 Practice Exercises

For extra help, see Example 2 on page 457 of your text.

Find each root.

4. $\sqrt[4]{-81}$ 4. ________________

5. $\sqrt[4]{256}$ 5. ________________

6. $\sqrt[7]{-1}$ 6. ________________

Objective 3 Graph functions defined by radical expressions.

Review these examples for Objective 3:

3. Graph each function by creating a table of values. Give the domain and the range.

a. $f(x) = \sqrt{x-1}$

Create a table of values.

x	$f(x) = \sqrt{x-1}$
1	$\sqrt{1-1} = 0$
5	$\sqrt{5-1} = 2$
10	$\sqrt{10-1} = 3$

For the radicand to be nonnegative, we must have $x-1 \geq 0$ or $x \geq 1$. Therefore, the domain is $[1, \infty)$. Function values are nonnegative, so the range is $[0, \infty)$.

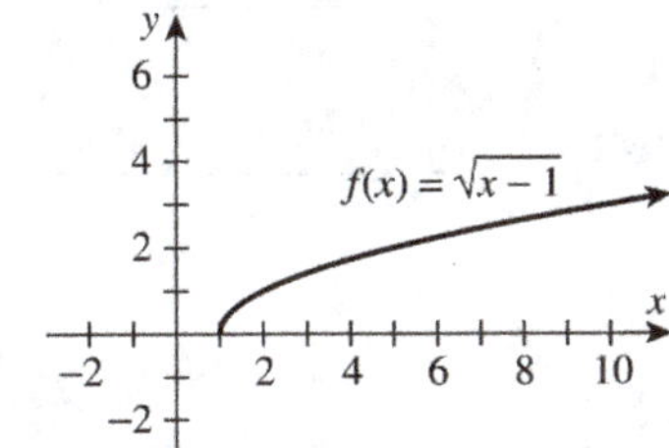

b. $f(x) = \sqrt[3]{x} + 1$

Create a table of values.

x	$f(x) = \sqrt[3]{x} + 1$
-8	$\sqrt[3]{-8} + 1 = -1$
-1	$\sqrt[3]{-1} + 1 = 0$
0	$\sqrt[3]{0} + 1 = 1$
1	$\sqrt[3]{1} + 1 = 2$
8	$\sqrt[3]{8} + 1 = 3$

Both the domain and range are $(-\infty, \infty)$.

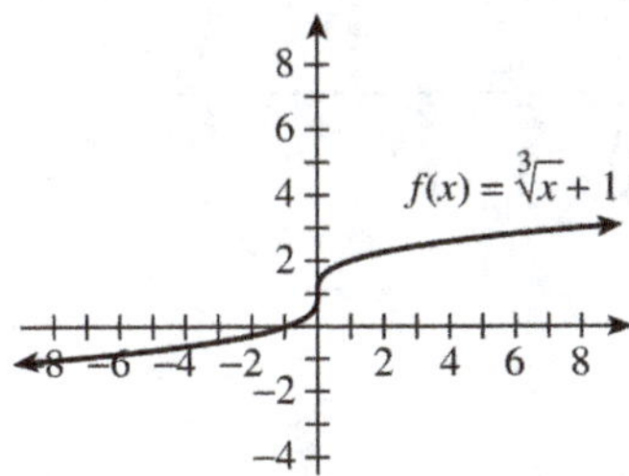

Now Try:

3. Graph each function by creating a table of values. Give the domain and the range.

a. $f(x) = \sqrt{x} - 1$

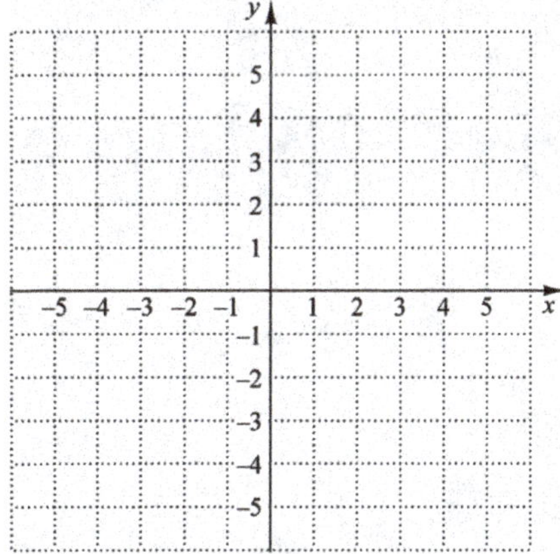

domain: _______________

range: _______________

b. $f(x) = \sqrt[3]{x} + 1$

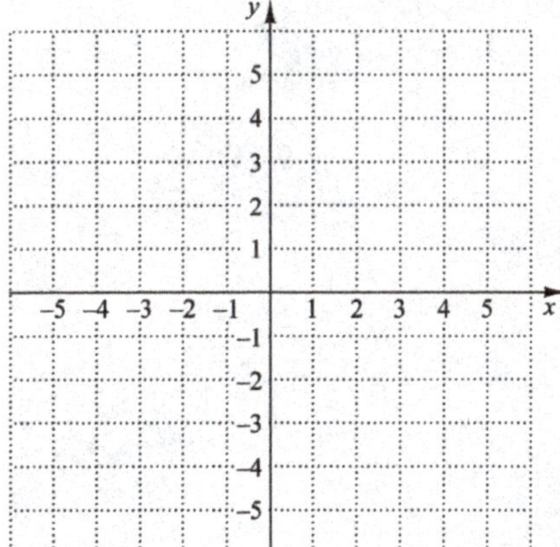

domain: _______________

range: _______________

Objective 3 Practice Exercises

For extra help, see Example 3 on page 458 of your text.

Graph each function and give its domain and its range.

7. $f(x) = \sqrt{x} + 2$

7.

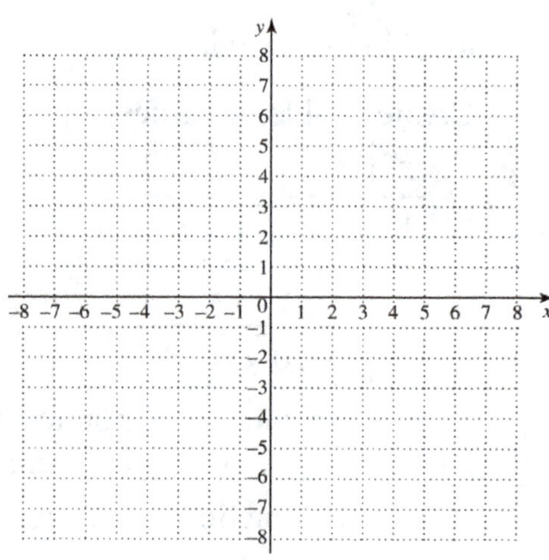

domain: _______________________

range: _______________________

8. $f(x) = \sqrt[3]{x} - 2$

8.

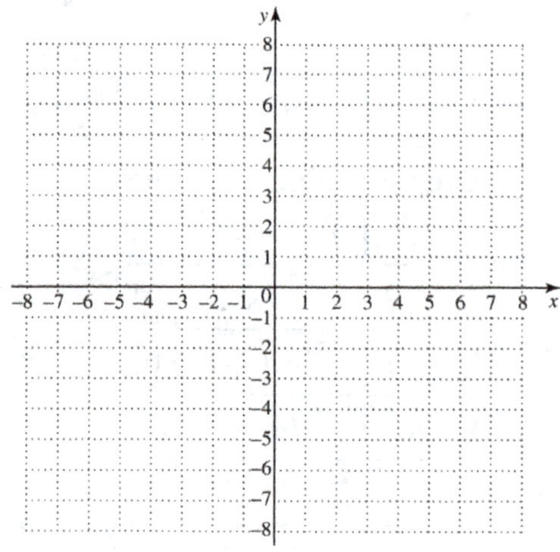

domain: _______________________

range: _______________________

9. $f(x) = \sqrt[3]{x} + 2$ **9.**

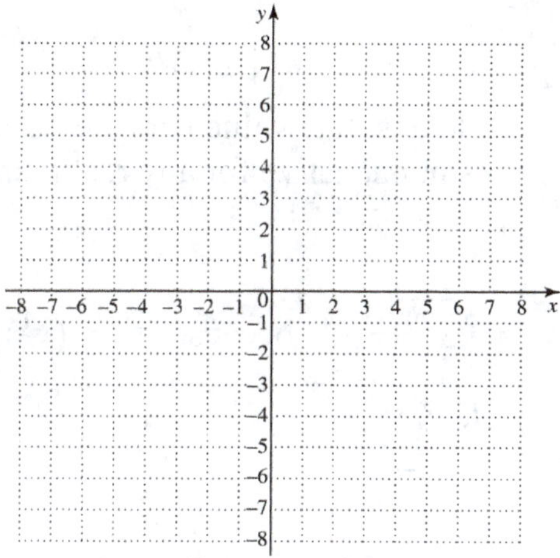

domain: _______________________

range: _______________________

Objective 4   Find *n*th roots of *n*th powers.

Review these examples for Objective 4:

4.   Find each square root. In part (d), m is a real number.

a. $\sqrt{33^2}$

$\sqrt{33^2} = |33| = 33$

d. $\sqrt{(-m)^2}$

$\sqrt{(-m)^2} = |-m| = |m|$

5.   Simplify each root.

a. $\sqrt[4]{(-5)^4}$

n is even. Use absolute value.

$\sqrt[4]{(-5)^4} = |-5| = 5$

b. $\sqrt[5]{(-3)^5}$

n is odd.

$\sqrt[5]{(-3)^5} = -3$

c. $-\sqrt[6]{(-8)^6}$

n is even. Use absolute value.

$-\sqrt[6]{(-8)^6} = -|-8| = -8$

Now Try:

4.   Find each square root. In part (d), n is a real number.

a. $\sqrt{73^2}$

d. $\sqrt{(-n)^2}$

____

5.   Simplify each root.

a. $\sqrt[8]{(-4)^8}$

b. $\sqrt[7]{(-5)^7}$

____

c. $-\sqrt[4]{(-2)^4}$

____

                     267

d. $-\sqrt{r^{12}}$

$-\sqrt{r^{12}} = -\left|r^6\right| = -r^6$

No absolute value bars are needed here since r^6 is nonnegative for any real number value of r.

e. $\sqrt[5]{s^{20}}$

$\sqrt[5]{s^{20}} = s^4$, because $s^{20} = (s^4)^5$.

f. $\sqrt[4]{x^{20}}$

$\sqrt[4]{x^{20}} = \left|x^5\right|$

d. $-\sqrt{x^8}$

____

e. $\sqrt[3]{w^{30}}$

____

f. $\sqrt[6]{x^{30}}$

____

Objective 4 Practice Exercises

For extra help, see Examples 4–5 on page 459 of your text.

Simplify each root.

10. $\sqrt{(-9)^2}$ **10.** _______________

11. $-\sqrt[5]{x^5}$ **11.** _______________

12. $-\sqrt[4]{x^{16}}$ **12.** _______________

Objective 5   Use a calculator to find roots.

Review these examples for Objective 5:

6. Use a calculator to approximate each radical to three decimal places.

a. $\sqrt{12}$

Use the square root key on a calculator.
$\sqrt{12} \approx 3.464$

b. $-\sqrt{596}$

$-\sqrt{596} \approx -24.413$

c. $\sqrt[3]{61}$

$\sqrt[3]{61} \approx 3.936$

d. $\sqrt[4]{6902}$

$\sqrt[4]{6902} \approx 9.115$

Now Try:

6. Use a calculator to approximate each radical to three decimal places.

a. $\sqrt{14}$

____

b. $-\sqrt{678}$

____

c. $\sqrt[3]{431}$

____

d. $\sqrt[4]{8142}$

____

7. The minimum speed at which a driver is traveling when a car skids to a stop can be estimated by the formula $s = \sqrt{30\,fd}$, where s is the speed of the car in miles per hour when the brakes are applied, f is a constant drag factor, and d is the length of the skid marks in feet. Suppose that a car skids to a stop leaving a skid of 90 feet. Assume that the drag factor is 0.7. How fast was the driver going? Give your answer to the nearest tenth.

Start with the given formula. Substitute for f and d, and use a calculator.

$$s = \sqrt{30\,fd}$$

$$s = \sqrt{30(0.7)(90)}$$

$$s \approx 43.5$$

The driver was traveling at about 43.5 miles per hour at the start of the skid.

7. Use the formula at the left to find out how fast a driver was traveling if the skid mark is 20 feet and the drag factor is 0.9. Give your answer to the nearest tenth.

____

Objective 5 Practice Exercises

For extra help, see Examples 6–7 on page 460 of your text.

Use a calculator to find a decimal approximation for each radical. Give the answer to the nearest thousandth.

13. $\sqrt[3]{701}$

13. _______________

14. $-\sqrt{990}$

14. _______________

15. The time t in seconds for one complete swing of a simple pendulum, where L is the length of the pendulum in feet is $t = 2\pi\sqrt{\dfrac{L}{32}}$. Find the time of a complete swing of a 4-ft pendulum to the nearest tenth of a second.

15. _______________

Chapter 7 ROOTS, RADICALS, AND ROOT FUNCTIONS

7.2   Rational Exponents

Learning Objectives
1   Use exponential notation for nth roots.
2   Define and use expressions of the form $a^{m/n}$.
3   Convert between radicals and rational exponents.
4   Use the rules for exponents with rational exponents.

Key Terms

Use the vocabulary terms listed below to complete each statement in exercises 1–3.

product rule for exponents   **quotient rule for exponents**

power rule for exponents

1. $\left(x^2 y^3\right)^4 = x^8 y^{12}$ is an example of the _______________________________.

2. $w^5 w^3 = w^8$ is an example of the _______________________________.

3. $\dfrac{z^6}{z^4} = z^2$ is an example of the _______________________________.

Objective 1   Use exponential notation for nth roots.

Review these examples for Objective 1:

1.   Evaluate each exponential.

   a. $8^{1/3}$

$8^{1/3} = \sqrt[3]{8} = 2$

   b. $25^{1/2}$

$25^{1/2} = \sqrt{25} = 5$

   c. $-81^{1/4}$

$-81^{1/4} = -\sqrt[4]{81} = -3$

   d. $(-81)^{1/4}$

$(-81)^{1/4} = \sqrt[4]{-81}$ is not a real number.

   e. $(-8)^{1/3}$

$(-8)^{1/3} = \sqrt[3]{-8} = -2$

Now Try:

1.   Evaluate each exponential.

   a. $216^{1/3}$

   ____

   b. $121^{1/2}$

   ____

   c. $-1024^{1/10}$

   ____

   d. $(-1024)^{1/10}$

   ____

   e. $(-243)^{1/5}$

   ____

f. $\left(\dfrac{1}{125}\right)^{1/3}$

$$\left(\dfrac{1}{125}\right)^{1/3} = \sqrt[3]{\dfrac{1}{125}} = \dfrac{1}{5}$$

f. $\left(\dfrac{1}{16}\right)^{1/4}$

____

Objective 1 Practice Exercises

For extra help, see Example 1 on page 465 of your text.

Evaluate each exponential.

1. $-256^{1/4}$

1. _______________

2. $16^{1/2}$

2. _______________

3. $(-3375)^{1/3}$

3. _______________

Objective 2   Define and use expressions of the form $a^{m/n}$.

Review these examples for Objective 2:
2.   Evaluate each polynomial.

a. $100^{3/2}$

$$100^{3/2} = (100^{1/2})^3 = 10^3 = 1000$$

b. $64^{2/3}$

$$64^{2/3} = (64^{1/3})^2 = 4^2 = 16$$

c. $-729^{5/6}$

$$-729^{5/6} = -(729^{1/6})^5 = -(3)^5 = -243$$

d. $(-125)^{2/3}$

$$(-125)^{2/3} = [(-125)^{1/3}]^2 = (-5)^2 = 25$$

e. $(-729)^{5/6}$

$(-729)^{5/6} = [(-729)^{1/6}]^5$ is not a real number
since $(-729)^{1/6}$ is not a real number.

Now Try:
2.   Evaluate each polynomial.

a. $27^{2/3}$

____

b. $16^{3/4}$

____

c. $-36^{3/2}$

____

d. $(-27)^{2/3}$

____

e. $(-36)^{3/2}$

____

3. Evaluate each exponential.

  a. $625^{-3/4}$

$$625^{-3/4} = \frac{1}{625^{3/4}} = \frac{1}{(625^{1/4})^3} = \frac{1}{\left(\sqrt[4]{625}\right)^3}$$

$$= \frac{1}{5^3} = \frac{1}{125}$$

  b. $36^{-3/2}$

$$36^{-3/2} = \frac{1}{36^{3/2}} = \frac{1}{\left(\sqrt{36}\right)^3} = \frac{1}{6^3} = \frac{1}{216}$$

  c. $\left(\dfrac{81}{16}\right)^{-3/4}$

$$\left(\frac{81}{16}\right)^{-3/4} = \left(\frac{16}{81}\right)^{3/4} = \left(\sqrt[4]{\frac{16}{81}}\right)^3 = \left(\frac{2}{3}\right)^3 = \frac{8}{27}$$

3. Evaluate each exponential.

  a. $32^{-2/5}$

    ____

  b. $125^{-4/3}$

    ____

  c. $\left(\dfrac{27}{64}\right)^{-2/3}$

    ____

Objective 2 Practice Exercises

For extra help, see Examples 2–3 on pages 465–466 of your text.

Evaluate each exponential.

4. $-81^{5/4}$

    4. ____________

5. $36^{5/2}$

    5. ____________

6. $\left(\dfrac{125}{27}\right)^{-2/3}$

    6. ____________

Objective 3   **Convert between radicals and rational exponents.**

Review these examples for Objective 3:

4. Write each radical as an exponential. Assume that all variables represent positive real numbers. Use the definition that takes the root first.

  a. $17^{1/2}$

$$17^{1/2} = \sqrt{17}$$

  b. $10^{5/6}$

$$10^{5/6} = \left(\sqrt[6]{10}\right)^5$$

Now Try:

4. Write each radical as an exponential. Assume that all variables represent positive real numbers. Use the definition that takes the root first.

  a. $23^{1/3}$

    ____

  b. $21^{3/4}$

    ____

c. $8x^{3/4}$

$$8x^{3/4} = 8\left(\sqrt[4]{x}\right)^3$$

d. $2x^{2/5} - (4x)^{5/6}$

$$2x^{2/5} - (4x)^{5/6} = 2\left(\sqrt[5]{x}\right)^2 - \left(\sqrt[6]{4x}\right)^5$$

e. $x^{-4/5}$

$$x^{-4/5} = \frac{1}{x^{4/5}} = \frac{1}{\left(\sqrt[4]{x}\right)^5}$$

f. $\left(x^3 + y^2\right)^{1/5}$

$$\left(x^3 + y^2\right)^{1/5} = \sqrt[5]{x^3 + y^2}$$

5. Write each radical as an exponential and simplify. Assume that all variables represent positive real numbers.

a. $\sqrt{14}$

$$\sqrt{14} = 14^{1/2}$$

b. $\sqrt{7^4}$

$$\sqrt{7^4} = 7^{4/2} = 7^2 = 49$$

c. $\sqrt[3]{3^6}$

$$\sqrt[3]{3^6} = 3^{6/3} = 3^2 = 9$$

d. $\sqrt[7]{w^7}$

$$\sqrt[7]{w^7} = w^{7/7} = w^1 = w$$

c. $5x^{5/4}$

____

d. $(2x)^{4/3} - 3x^{2/5}$

____

e. $x^{-3/2}$

____

f. $\left(x^2 - y^2\right)^{1/4}$

____

5. Write each radical as an exponential and simplify. Assume that all variables represent positive real numbers.

a. $\sqrt{22}$

____

b. $\sqrt{10^4}$

____

c. $\sqrt[5]{3^{10}}$

____

d. $\sqrt[8]{m^8}$

____

Objective 3 Practice Exercises

For extra help, see Examples 4–5 on pages 467–468 of your text.

Write with radicals. Assume that all variables represent positive real numbers.

7. $4y^{2/5} + (5x)^{1/5}$

7. _______________

8. $\left(2x^4 - 3y^2\right)^{-4/3}$ **8.** _______________

Simplify the radical by rewriting it with a rational exponent. Write answer in radical form if necessary. Assume that variables represent positive real numbers.

9. $\sqrt[8]{a^2}$ **9.** _______________

Objective 4   Use the rules for exponents with rational exponents.

Review these examples for Objective 4:

6. Write with only positive exponents. Assume that all variables represent positive real numbers.

a. $13^{4/5} \cdot 13^{1/2}$

$13^{4/5} \cdot 13^{1/2} = 13^{4/5+1/2} = 13^{13/10}$

b. $\dfrac{8^{3/4}}{8^{1/4}}$

$\dfrac{8^{3/4}}{8^{1/4}} = 8^{3/4-1/4} = 8^{1/2}$

c. $\dfrac{\left(n^{7/4}w^{1/2}\right)^2}{w^{3/4}}$

$\dfrac{\left(n^{7/4}w^{1/2}\right)^2}{w^{3/4}} = \dfrac{\left(n^{7/4}\right)^2\left(w^{1/2}\right)^2}{w^{3/4}}$

$= \dfrac{n^{7/2}w^1}{w^{3/4}}$

$= n^{7/2}w^{1-3/4}$

$= n^{7/2}w^{1/4}$

Now Try:

5. Write with only positive exponents. Assume that all variables represent positive real numbers.

a. $5^{3/4} \cdot 5^{7/4}$

____

b. $\dfrac{a^{4/5}}{a^{2/3}}$

____

c. $\dfrac{\left(x^{1/3}y^{2/3}\right)^6}{y^{1/2}}$

____

d. $\left(\dfrac{c^6 x^3}{c^{-2} x^{1/2}}\right)^{-3/4}$

$$\left(\dfrac{c^6 x^{1/2}}{c^{-2} x^3}\right)^{-3/4} = \left(c^{6-(-2)} x^{1/2-3}\right)^{-3/4}$$

$$= \left(c^8 x^{-5/2}\right)^{-3/4}$$

$$= \left(c^8\right)^{-3/4} \left(x^{-5/2}\right)^{-3/4}$$

$$= c^{-6} x^{15/8}$$

$$= \dfrac{x^{15/8}}{c^6}$$

e. $a^{3/4}\left(a^{2/3} - a^{1/2}\right)$

$$a^{3/4}\left(a^{2/3} - a^{1/2}\right) = a^{3/4} a^{2/3} - a^{3/4} a^{1/2}$$

$$= a^{3/4+2/3} - a^{3/4+1/2}$$

$$= a^{17/12} - a^{5/4}$$

7. Write all radicals as exponentials, and then apply the rules for rational exponents. Leave answers in exponential form. Assume that all variables represent positive real numbers.

a. $\sqrt[4]{x^3} \cdot \sqrt[5]{x}$

$$\sqrt[4]{x^3} \cdot \sqrt[5]{x} = x^{3/4} \cdot x^{1/5}$$

$$= x^{3/4+1/5}$$

$$= x^{15/20+4/20}$$

$$= x^{19/20}$$

b. $\dfrac{\sqrt[3]{y^5}}{\sqrt{y^3}}$

$$\dfrac{\sqrt[3]{y^5}}{\sqrt{y^3}} = \dfrac{y^{5/3}}{y^{3/2}} = y^{5/3-3/2} = y^{1/6}$$

c. $\sqrt{\sqrt[4]{y^3}}$

$$\sqrt{\sqrt[4]{y^3}} = \sqrt{y^{3/4}} = \left(y^{3/4}\right)^{1/2} = y^{3/8}$$

d. $\left(\dfrac{x^{-1} y^{2/3}}{x^{1/3} y^{1/2}}\right)^{-3/2}$

____

e. $r^{1/2}\left(r^{2/3} - r^{8/3}\right)$

____

7. Write all radicals as exponentials, and then apply the rules for rational exponents. Leave answers in exponential form. Assume that all variables represent positive real numbers.

a. $\sqrt[6]{x^3} \cdot \sqrt[3]{x^2}$

____

b. $\dfrac{\sqrt[4]{y^5}}{\sqrt[3]{y^2}}$

____

c. $\sqrt[3]{\sqrt[4]{x^3}}$

____

Objective 4 Practice Exercises

For extra help, see Examples 6–7 on pages 468–470 of your text.

Use the rules of exponents to simplify each expression. Write all answers with positive exponents. Assume that variables represent positive real numbers.

10. $y^{7/3} \cdot y^{-4/3}$

10. ________________

11. $\dfrac{a^{2/3} \cdot a^{-1/3}}{\left(a^{-1/6}\right)^3}$

11. ________________

12. $\dfrac{\left(x^{-3}y^2\right)^{2/3}}{\left(x^2 y^{-5}\right)^{2/5}}$

12. ________________

Chapter 7 ROOTS, RADICALS, AND ROOT FUNCTIONS

7.3 Simplifying Radicals, the Distance Formula, and Circles

Learning Objectives	
1	Use the product rule for radicals.
2	Use the quotient rule for radicals.
3	Simplify radicals
4	Simplify products and quotients of radicals with different indexes.
5	Use the Pythagorean theorem.
6	Use the distance formula.
7	Find an equation of a circle given its center and radius.

Key Terms

Use the vocabulary terms listed below to complete each statement in exercises 1−6.

       index   **radicand**   **hypotenuse**   **legs**

       radius   **circle**   **center**

1. In a right triangle, the side opposite the right angle is called the

   ________________________.

2. In the expression $\sqrt[4]{x^2}$, the "4" is the ____________________ and x^2 is
   the ____________________________.

3. In a right triangle, the sides that form the right angle are called the

   ________________________.

4. A(n) ____________________________ is the set of all points in a plane that lie a
   fixed distance from a fixed point.

5. A fixed point such that every point on a circle is a fixed distance from it is the

   ________________________.

6. The distance from the center of a circle to a point on the circle is called the

   ________________________.

Objective 1   Use the product rule for radicals.

Review these examples for Objective 1:

1. Multiply. Assume that all variables represent positive real numbers.

   a. $\sqrt{13}\cdot\sqrt{5}$

   $$\sqrt{13}\cdot\sqrt{5}=\sqrt{13\cdot 5}=\sqrt{65}$$

Now Try:

1. Multiply. Assume that all variables represent positive real numbers.

   a. $\sqrt{2}\cdot\sqrt{7}$

   ____

b. $\sqrt{5}\cdot\sqrt{2r}$

$$\sqrt{5}\cdot\sqrt{2r} = \sqrt{5\cdot 2r} = \sqrt{10r}$$

c. $\sqrt{5x}\cdot\sqrt{2yz}$

$$\sqrt{5x}\cdot\sqrt{2yz} = \sqrt{10xyz}$$

2. Multiply. Assume that all variables represent positive real numbers.

   a. $\sqrt[4]{2}\cdot\sqrt[4]{2x}$

   $$\sqrt[4]{2}\cdot\sqrt[4]{2x} = \sqrt[4]{2\cdot 2x} = \sqrt[4]{4x}$$

   b. $\sqrt[3]{8x}\cdot\sqrt[3]{2y^2}$

   $$\sqrt[3]{8x}\cdot\sqrt[3]{2y^2} = \sqrt[3]{8x\cdot 2y^2} = \sqrt[3]{16xy^2}$$

   c. $\sqrt[5]{6r^2}\cdot\sqrt[5]{4r^2}$

   $$\sqrt[5]{6r^2}\cdot\sqrt[5]{4r^2} = \sqrt[5]{6r^2\cdot 4r^2} = \sqrt[5]{24r^4}$$

   d. $\sqrt[5]{2}\cdot\sqrt[4]{6}$

   $\sqrt[5]{2}\cdot\sqrt[4]{6}$ cannot be simplified using the product rule for radicals, because the indexes (5 and 4) are different.

b. $\sqrt{5x}\cdot\sqrt{7}$

____

c. $\sqrt{3}\cdot\sqrt{11mn}$

____

2. Multiply. Assume that all variables represent positive real numbers.

   a. $\sqrt[3]{3}\cdot\sqrt[3]{7}$

   ____

   b. $\sqrt[3]{7x}\cdot\sqrt[3]{5y}$

   ____

   c. $\sqrt[5]{4w}\cdot\sqrt[5]{2w^3}$

   ____

   d. $\sqrt{3}\cdot\sqrt[3]{64}$

   ____

Objective 1 Practice Exercises

For extra help, see Examples 1–2 on page 473 of your text.

Multiply. Assume that variables represent positive real numbers.

1. $\sqrt{7x}\cdot\sqrt{6t}$

1. ______________

2. $\sqrt[5]{6r^2t^3}\cdot\sqrt[5]{4r^2t}$

2. ______________

3. $\sqrt{3}\cdot\sqrt[3]{7}$

3. ______________

Objective 2   Use the quotient rule for radicals.

Review these examples for Objective 2:	**Now Try:**

Review these examples for Objective 2:

3. Simplify. Assume that all variables represent positive real numbers.

a. $\sqrt{\dfrac{64}{9}}$

$$\sqrt{\dfrac{64}{9}} = \dfrac{\sqrt{64}}{\sqrt{9}} = \dfrac{8}{3}$$

b. $\sqrt{\dfrac{5}{16}}$

$$\sqrt{\dfrac{5}{16}} = \dfrac{\sqrt{5}}{\sqrt{16}} = \dfrac{\sqrt{5}}{4}$$

c. $\sqrt[3]{-\dfrac{27}{8}}$

$$\sqrt[3]{-\dfrac{27}{8}} = \dfrac{\sqrt[3]{-27}}{\sqrt[3]{8}} = \dfrac{-3}{2} = -\dfrac{3}{2}$$

d. $\sqrt[5]{-\dfrac{a^3}{243}}$

$$\sqrt[5]{-\dfrac{a^3}{243}} = \dfrac{\sqrt[5]{a^3}}{\sqrt[5]{-243}} = \dfrac{\sqrt[5]{a^3}}{-3} = -\dfrac{\sqrt[5]{a^3}}{3}$$

e. $\sqrt{\dfrac{z^4}{36}}$

$$\sqrt{\dfrac{z^4}{36}} = \dfrac{\sqrt{z^4}}{\sqrt{36}} = \dfrac{z^2}{6}$$

Now Try:

3. Simplify. Assume that all variables represent positive real numbers.

a. $\sqrt{\dfrac{36}{49}}$

____

b. $\sqrt{\dfrac{13}{81}}$

____

c. $\sqrt[3]{-\dfrac{343}{125}}$

____

d. $\sqrt[3]{-\dfrac{a^6}{125}}$

____

e. $\sqrt[4]{\dfrac{m}{81}}$

____

Objective 2 Practice Exercises

For extra help, see Example 3 on page 474 of your text.

Simplify each radical. Assume that variables represent positive real numbers.

4. $\sqrt[3]{\dfrac{27}{8}}$

4. _______________

5. $\sqrt[5]{\dfrac{7x}{32}}$

5. _______________

6. $\sqrt[3]{-\dfrac{x^9}{216}}$ **6.** ______________

Objective 3   Simplify radicals.

Review these examples for Objective 3:

4. Simplify.

a. $\sqrt{90}$

$$\sqrt{90} = \sqrt{9 \cdot 10}$$
$$= \sqrt{9} \cdot \sqrt{10}$$
$$= 3\sqrt{10}$$

b. $\sqrt{288}$

$$\sqrt{288} = \sqrt{144 \cdot 2}$$
$$= \sqrt{144} \cdot \sqrt{2}$$
$$= 12\sqrt{2}$$

c. $\sqrt{35}$

No perfect square (other than 1) divides into 35, so $\sqrt{35}$ cannot be simplified further.

d. $\sqrt[3]{81}$

$$\sqrt[3]{81} = \sqrt[3]{27 \cdot 3} = \sqrt[3]{27} \cdot \sqrt[3]{3} = 3\sqrt[3]{3}$$

e. $-\sqrt[4]{3125}$

$$-\sqrt[4]{3125} = -\sqrt[4]{5^5} = \sqrt[4]{5^4 \cdot 5}$$
$$= -\sqrt[4]{5^4} \cdot \sqrt[4]{5}$$
$$= -5\sqrt[4]{5}$$

5. Simplify. Assume that all variables represent positive real numbers.

a. $\sqrt{81x^3}$

$$\sqrt{81x^3} = \sqrt{9^2 \cdot x^2 \cdot x} = 9x\sqrt{x}$$

Now Try:

4. Simplify.

a. $\sqrt{84}$

____

b. $\sqrt{162}$

____

c. $\sqrt{95}$

____

d. $\sqrt[3]{256}$

____

e. $-\sqrt[5]{512}$

____

5. Simplify. Assume that all variables represent positive real numbers.

a. $\sqrt{100y^3}$

____

          Copyright © 2020 Pearson Education, Inc.

b. $\sqrt{56x^7y^6}$

$$\sqrt{56x^7y^6} = \sqrt{4\cdot14\cdot\left(x^3\right)^2\cdot x\cdot\left(y^3\right)^2}$$
$$= 2x^3y^3\sqrt{14x}$$

c. $\sqrt[3]{-270b^4c^8}$

$$\sqrt[3]{-270b^4c^8} = \sqrt[3]{\left(-27b^3c^6\right)\left(10bc^2\right)}$$
$$= \sqrt[3]{-27b^3c^6}\cdot\sqrt[3]{10bc^2}$$
$$= -3bc^2\sqrt[3]{10bc^2}$$

d. $-\sqrt[6]{448a^7b^7}$

$$-\sqrt[6]{448a^7b^7} = -\sqrt[6]{\left(64a^6b^6\right)\left(7ab\right)}$$
$$= -\sqrt[6]{64a^6b^6}\cdot\sqrt[6]{7ab}$$
$$= -2ab\sqrt[6]{7ab}$$

6. Simplify. Assume that all variables represent positive real numbers.

a. $\sqrt[24]{5^4}$

$$\sqrt[24]{5^4} = \left(5^4\right)^{1/24} = 5^{4/24} = 5^{1/6} = \sqrt[6]{5}$$

b. $\sqrt[12]{x^8}$

$$\sqrt[12]{x^8} = \left(x^8\right)^{1/12} = x^{8/12} = x^{2/3} = \sqrt[3]{x^2}$$

c. $\sqrt[6]{x^{15}}$

$$\sqrt[6]{x^{15}} = \left(x^{15}\right)^{1/6} = x^{5/2} = \sqrt{x^5} = \sqrt{x^4}\cdot\sqrt{x}$$
$$= x^2\sqrt{x}$$

b. $\sqrt{48m^5r^9}$

____

c. $\sqrt[3]{-32n^7t^5}$

____

d. $-\sqrt[4]{405x^3y^9}$

____

6. Simplify. Assume that all variables represent positive real numbers.

a. $\sqrt[12]{11^9}$

____

b. $\sqrt[30]{z^{24}}$

____

c. $\sqrt[6]{w^{21}}$

____

Objective 3 Practice Exercises

For extra help, see Examples 4–6 on pages 475–476 of your text.

Simplify each radical. Assume that variables represent positive real numbers.

7. $\sqrt[42]{x^{28}}$

7. _______________

8. $\sqrt{8x^3 y^6 z^{11}}$ 8. _______________

9. $\sqrt[3]{1250a^5 b^7}$ 9. _______________

Objective 4   Simplify products and quotients of radicals with different indexes.

Review this example for Objective 4:

7. Simplify $\sqrt{3} \cdot \sqrt[5]{6}$.

Because the different indexes, 2 and 5, have least common multiple index of 10, we use rational exponents to write each radical as a tenth root.

$$\sqrt{3} = 3^{1/2} = 3^{5/10} = \sqrt[10]{3^5} = \sqrt[10]{243}$$

$$\sqrt[5]{6} = 6^{1/5} = 6^{2/10} = \sqrt[10]{6^2} = \sqrt[10]{36}$$

$$\sqrt{3} \cdot \sqrt[5]{6} = \sqrt[10]{243} \cdot \sqrt[10]{36}$$

$$= \sqrt[10]{243 \cdot 36}$$

$$= \sqrt[10]{8748}$$

Now Try:

7. Simplify $\sqrt[3]{3} \cdot \sqrt[6]{7}$.

____

Objective 4 Practice Exercises

For extra help, see Example 7 on page 477 of your text.

Simplify each radical. Assume that variables represent positive real numbers.

10. $\sqrt{r} \cdot \sqrt[3]{r}$ 10. _______________

11. $\sqrt[4]{2} \cdot \sqrt[8]{7}$ 11. _______________

12. $\sqrt{3} \cdot \sqrt[5]{64}$ 12. _______________

             Copyright © 2020 Pearson Education, Inc.

Objective 5   Use the Pythagorean theorem.

Review this example for Objective 5:

8. Use the Pythagorean theorem to find the length of the unknown side of the triangle.

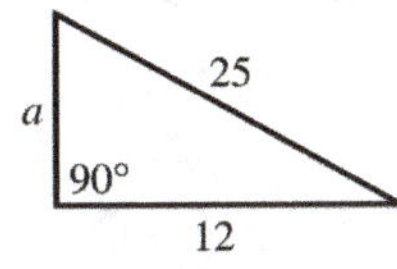

$$a^2 + b^2 = c^2$$
$$12^2 + b^2 = 25^2$$
$$144 + b^2 = 625$$
$$b^2 = 481$$
$$b = \sqrt{481}$$

The length of the side is $\sqrt{481}$.

Now Try:

8. Use the Pythagorean theorem to find the length of the unknown side of the triangle.

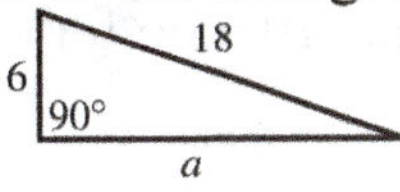

____

Objective 5 Practice Exercises

For extra help, see Example 8 on page 478 of your text.

Find the unknown length in each right triangle. Simplify the answer if necessary.

13.

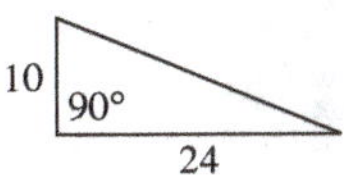

13. _______________

14.

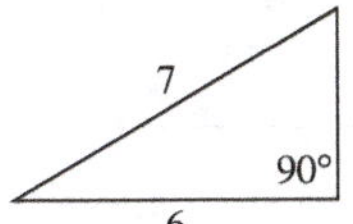

14. _______________

15.

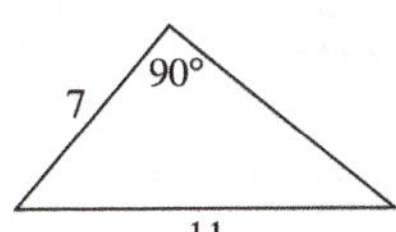

15. _______________

Objective 6   Use the distance formula.

Review this example for Objective 6:

9. Find the distance between the points $(2, -2)$ and $(-6, 1)$.

Use the distance formula. Let $(x_1, y_1) = (2, -2)$ and $(x_2, y_2) = (-6, 1)$.

$$d = \sqrt{(x_2 - x_1)^2 + (y_2 - y_1)^2}$$
$$= \sqrt{(-6 - 2)^2 + [1 - (-2)]^2}$$
$$= \sqrt{(-8)^2 + 3^2}$$
$$= \sqrt{64 + 9}$$
$$= \sqrt{73}$$

Now Try:

9. Find the distance between the points $(-1, -2)$ and $(-4, 3)$.

____

Objective 6 Practice Exercises

For extra help, see Example 9 on page 479 of your text.

Find the distance between each pair of points.

16. $(3, 4)$ and $(-1, -2)$

16. _________________

17. $(-2, -3)$ and $(-5, 1)$

17. _________________

18. $(4, 2)$ and $(3, -1)$

18. _________________

Name: _______________________   Date: _______________________
Instructor: _________________   Section: ____________________

Objective 7   Find an equation of a circle given its center and radius.

Review these examples for Objective 7:

10. Find an equation of the circle with center $(0, 0)$ and radius 2, and graph it.

If the point (x, y) is on the circle, then the distance from (x, y) to the center $(0, 0)$ is 2.

$$\sqrt{(x_2 - x_1)^2 + (y_2 - y_1)^2} = d$$

$$\sqrt{(x - 0)^2 + (y - 0)^2} = 2$$

$$\left(\sqrt{x^2 + y^2}\right)^2 = 2^2$$

$$x^2 + y^2 = 4$$

The equation of this circle is $x^2 + y^2 = 4$.

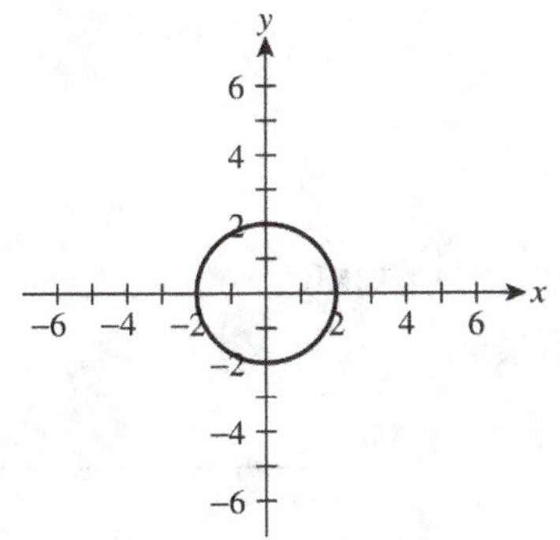

11. Find an equation of the circle with center $(-3, 2)$ and radius 3, and graph it.

$$\sqrt{(x_2 - x_1)^2 + (y_2 - y_1)^2} = d$$

$$\sqrt{(x - (-3))^2 + (y - 2)^2} = 3$$

$$\left(\sqrt{(x + 3)^2 + (y - 2)^2}\right)^2 = 3^2$$

$$(x + 3)^2 + (y - 2)^2 = 9$$

To graph the circle, plot the center $(-3, 2)$, then move three units right, left, up, and down from the center, plotting the points $(0, 2)$, $(-3, 5)$, $(-6, 2)$, and $(-3, -1)$. Draw a smooth curve through the points.

Now Try:

10. Find an equation of the circle with and center $(0, 0)$ and radius 5, and graph it.

____

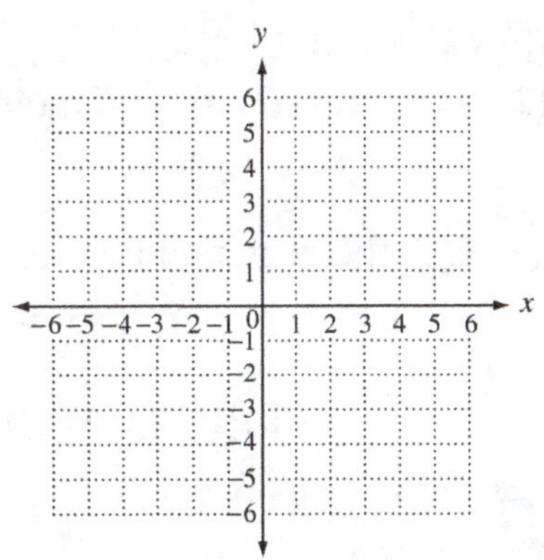

11. Find an equation of the circle with center $(-5, 4)$ and radius 4, and graph it.

____

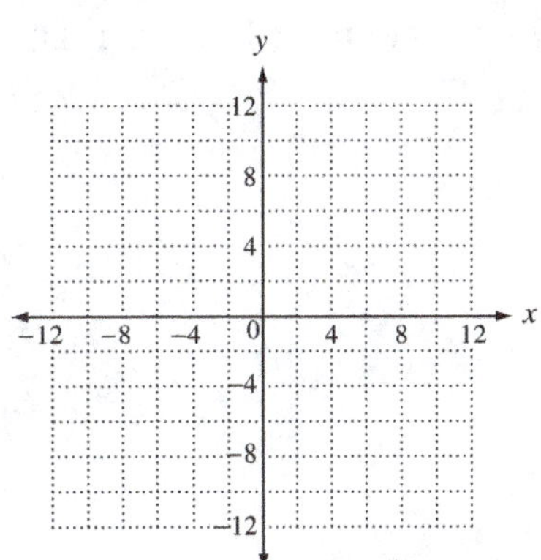

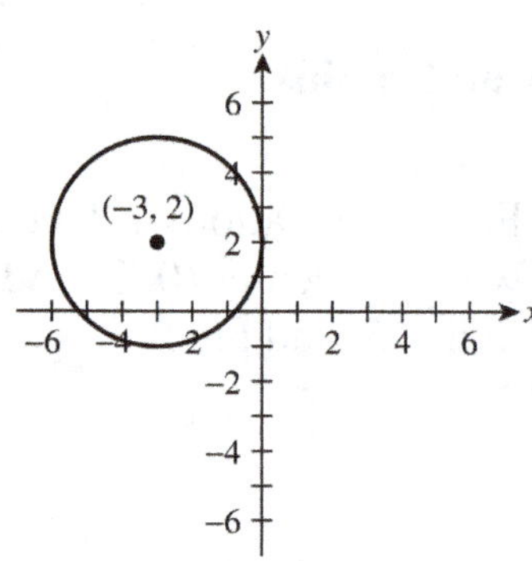

12. Write an equation of a circle with center $(-2,-4)$ and radius $2\sqrt{5}$.

Use the center-radius form.

$$(x-h)^2 + (y-k)^2 = r^2$$

$$[x-(-2)]^2 + [y-(-4)]^2 = \left(2\sqrt{5}\right)^2$$

$$(x+2)^2 + (y+4)^2 = 20$$

12. Write an equation of a circle with center $(3,-1)$ and radius $\sqrt{6}$.

____

Objective 7 Practice Exercises

For extra help, see Examples 10–12 on pages 480–481 of your text.

Find the equation of a circle satisfying the given conditions.

19.   center: $(3,-4)$; radius: 5

19. ______________

20.   center: $(-2,-2)$; radius: 3

20. ______________

21.   center: $(0, 3)$; radius: $\sqrt{2}$

21. ______________

Chapter 7 ROOTS, RADICALS, AND ROOT FUNCTIONS

7.4   Adding and Subtracting Radical Expressions

Learning Objectives
1   Simplify radical expressions involving addition and subtraction.

Key Terms

Use the vocabulary terms listed below to complete each statement in exercises 1–2.

like radicals   unlike radicals

1.   The expressions $2\sqrt{2}$ and $6\sqrt[3]{2}$ are _________________________________.

2.   The expressions $2\sqrt{2}$ and $7\sqrt{2}$ are _________________________________.

Objective 1   Simplify radical expressions involving addition and subtraction.

Review these examples for Objective 1:

1. Add or subtract to simplify each radical expression.

a. $5\sqrt{3}+7\sqrt{3}$

$$5\sqrt{3}+7\sqrt{3}=(5+7)\sqrt{3}$$
$$=12\sqrt{3}$$

b. $3\sqrt{2}-6\sqrt{2}$

$$3\sqrt{2}-6\sqrt{2}=(3-6)\sqrt{2}$$
$$=-3\sqrt{2}$$

c. $3\sqrt{13}+5\sqrt{52}$

$$3\sqrt{13}+5\sqrt{52}=3\sqrt{13}+5\sqrt{4}\sqrt{13}$$
$$=3\sqrt{13}+5\cdot2\sqrt{13}$$
$$=3\sqrt{13}+10\sqrt{13}$$
$$=(3+10)\sqrt{13}$$
$$=13\sqrt{13}$$

Now Try:

1. Add or subtract to simplify each radical expression.

a. $6\sqrt{5}+2\sqrt{5}$

____

b. $4\sqrt{3}-8\sqrt{3}$

____

c. $3\sqrt{54}-5\sqrt{24}$

____

d. $\sqrt{48x} - \sqrt{12x}, \; x \geq 0$

$$\sqrt{48x} - \sqrt{12x} = \sqrt{16} \cdot \sqrt{3x} - \sqrt{4} \cdot \sqrt{3x}$$
$$= 4\sqrt{3x} - 2\sqrt{3x}$$
$$= (4-2)\sqrt{3x}$$
$$= 2\sqrt{3x}$$

e. $7\sqrt{3} - 6\sqrt{21}$

The radicands differ and are already simplified, so this expression cannot be simplified further.

2. Simplify. Assume that all variables represent positive real numbers.

a. $7\sqrt[4]{32} - 9\sqrt[4]{2}$

$$7\sqrt[4]{32} - 9\sqrt[4]{2} = 7\sqrt[4]{16} \cdot \sqrt[4]{2} - 9\sqrt[4]{2}$$
$$= 7 \cdot 2 \cdot \sqrt[4]{2} - 9\sqrt[4]{2}$$
$$= 14\sqrt[4]{2} - 9\sqrt[4]{2}$$
$$= (14-9)\sqrt[4]{2}$$
$$= 5\sqrt[4]{2}$$

b. $6\sqrt[3]{27x^5 r} + 2x\sqrt[3]{x^2 r}$

$$6\sqrt[3]{27x^5 r} + 2x\sqrt[3]{x^2 r}$$
$$= 6 \cdot \sqrt[3]{27x^3} \cdot \sqrt[3]{x^2 r} + 2x\sqrt[3]{x^2 r}$$
$$= 18x\sqrt[3]{x^2 r} + 2x\sqrt[3]{x^2 r}$$
$$= (18x + 2x)\sqrt[3]{x^2 r}$$
$$= 20x\sqrt[3]{x^2 r}$$

d. $3\sqrt{18z} + 2\sqrt{8z}, \; z \geq 0$

____

e. $3\sqrt{7} + 2\sqrt{6}$

____

2. Simplify. Assume that all variables represent positive real numbers.

a. $7\sqrt[3]{54} - 6\sqrt[3]{128}$

____

b. $\sqrt[4]{32y^2 z^5} + 3z\sqrt[4]{2y^2 z}$

____

c. $3\sqrt{40x^5} + 5\sqrt[3]{48x^5}$

$3\sqrt{40x^5} + 5\sqrt[3]{48x^5}$

$= 3\cdot\sqrt{4x^4 \cdot 10x} + 5\sqrt[3]{8x^3 \cdot 6x^2}$

$= 3\cdot\sqrt{4x^4} \cdot \sqrt{10x} + 5\sqrt[3]{8x^3} \cdot \sqrt[3]{6x^2}$

$= 3\cdot 2x^2 \cdot \sqrt{10x} + 5\cdot 2x \cdot \sqrt[3]{6x^2}$

$= 6x^2\sqrt{10x} + 10x\sqrt[3]{6x^2}$

c. $2\sqrt[3]{54x^7} + 2\sqrt{27x^7}$

____

3. Simplify. Assume that all variables represent positive real numbers.

a. $\dfrac{\sqrt{32}}{3} + \dfrac{\sqrt{8}}{\sqrt{18}}$

$\dfrac{\sqrt{32}}{3} + \dfrac{\sqrt{8}}{\sqrt{18}} = \dfrac{\sqrt{16\cdot 2}}{3} + \dfrac{\sqrt{4\cdot 2}}{\sqrt{9\cdot 2}}$

$= \dfrac{4\sqrt{2}}{3} + \dfrac{2\sqrt{2}}{3\sqrt{2}}$

$= \dfrac{4\sqrt{2}}{3} + \dfrac{2}{3}$

$= \dfrac{4\sqrt{2}+2}{3}$

b. $\sqrt[3]{\dfrac{81}{y^6}} + 5\sqrt[3]{\dfrac{27}{y^3}}$

$\sqrt[3]{\dfrac{81}{y^6}} + 5\sqrt[3]{\dfrac{27}{y^3}} = \dfrac{\sqrt[3]{81}}{\sqrt[3]{y^6}} + 5\dfrac{\sqrt[3]{27}}{\sqrt[3]{y^3}}$

$= \dfrac{3\sqrt[3]{3}}{y^2} + 5\left(\dfrac{3}{y}\right)$

$= \dfrac{3\sqrt[3]{3}}{y^2} + \dfrac{15}{y}$

$= \dfrac{3\sqrt[3]{3}+15y}{y^2}$

3. Simplify. Assume that all variables represent positive real numbers.

a. $\sqrt{\dfrac{10}{18}} + \dfrac{\sqrt{15}}{\sqrt{27}}$

____

b. $\sqrt[3]{\dfrac{216}{w^6}} + \sqrt{\dfrac{121}{w^4}}$

____

Objective 1 Practice Exercises

For extra help, see Examples 1–3 on pages 487–489 of your text.

Add or subtract. Assume that all variables represent positive real numbers.

1. $\sqrt{100x} - \sqrt{9x} + \sqrt{25x}$

1. _________________

2. $2\sqrt[3]{16r} + \sqrt[3]{54r} - \sqrt[3]{16r}$

2. _________________

3. $\sqrt[3]{\dfrac{y^7}{125}} + y^2\sqrt[3]{\dfrac{y}{27}}$

3. _________________

Chapter 7 ROOTS, RADICALS, AND ROOT FUNCTIONS

7.5   Multiplying and Dividing Radical Expressions

Learning Objectives
1   Multiply radical expressions.
2   Rationalize denominators with one radical term.
3   Rationalize denominators with binomials involving radicals.
4   Write radical quotients in lowest terms.

Key Terms

Use the vocabulary terms listed below to complete each statement in exercises 1–2.

rationalizing the denominator   conjugate

1.   The ___________________________ of $a + b$ is $a - b$.

2.   The process of removing radicals from the denominator so that the denominator contains only rational quantities is called _________________________________.

Objective 1   Multiply radical expressions.

Review these examples for Objective 1:

2.   Multiply, using the FOIL method.

a. $\left(\sqrt{6} + \sqrt{5}\right)\left(\sqrt{6} - \sqrt{5}\right)$

This is the difference of squares.

$$\left(\sqrt{6} + \sqrt{5}\right)\left(\sqrt{6} - \sqrt{5}\right) = \left(\sqrt{6}\right)^2 - \left(\sqrt{5}\right)^2$$
$$= 6 - 5$$
$$= 1$$

b. $\left(5 - \sqrt{3}\right)\left(\sqrt{2} + \sqrt{5}\right)$

$$\left(5 - \sqrt{3}\right)\left(\sqrt{2} + \sqrt{5}\right)$$
$$= 5 \cdot \sqrt{2} + 5 \cdot \sqrt{5} - \sqrt{3} \cdot \sqrt{2} - \sqrt{3} \cdot \sqrt{5}$$
$$= 5\sqrt{2} + 5\sqrt{5} - \sqrt{6} - \sqrt{15}$$

Now Try:

2.   Multiply, using the FOIL method.

a. $\left(\sqrt{14} - \sqrt{2}\right)\left(\sqrt{14} + \sqrt{2}\right)$

____

b. $\left(3 - \sqrt{2}\right)\left(2 + \sqrt{7}\right)$

____

c. $\left(\sqrt{11}-6\right)^2$

$$\left(\sqrt{11}-6\right)^2 = \left(\sqrt{11}-6\right)\left(\sqrt{11}-6\right)$$
$$= \sqrt{11}\cdot\sqrt{11}-6\sqrt{11}-6\sqrt{11}+6\cdot 6$$
$$= 11-12\sqrt{11}+36$$
$$= 47-12\sqrt{11}$$

c. $\left(3-\sqrt{2}\right)^2$

____

Objective 1 Practice Exercises

For extra help, see Examples 1–2 on pages 492–493 of your text.

Multiply each product, then simplify. Assume that variables represent positive real numbers.

1. $\left(\sqrt{5}+\sqrt{6}\right)\left(\sqrt{2}-4\right)$

1. ___________________

2. $\left(\sqrt{2}-\sqrt{12}\right)^2$

2. ___________________

3. $\left(2+\sqrt[3]{5}\right)\left(2-\sqrt[3]{5}\right)$

3. ___________________

                     Copyright © 2020 Pearson Education, Inc.

Objective 2   Rationalize denominators with one radical term.

Review these examples for Objective 2: | **Now Try:**

3.   Rationalize the denominator.

$$\frac{4}{\sqrt{3}}$$

$$\frac{4}{\sqrt{3}} = \frac{4 \cdot \sqrt{3}}{\sqrt{3} \cdot \sqrt{3}} = \frac{4\sqrt{3}}{3}$$

4.   Simplify the radical.

$$-\sqrt{\frac{27}{98}}$$

$$-\sqrt{\frac{27}{98}} = -\frac{\sqrt{27}}{\sqrt{98}}$$

$$= -\frac{\sqrt{9 \cdot 3}}{\sqrt{49 \cdot 2}}$$

$$= -\frac{3\sqrt{3}}{7\sqrt{2}}$$

$$= -\frac{3\sqrt{3} \cdot \sqrt{2}}{7\sqrt{2} \cdot \sqrt{2}}$$

$$= -\frac{3\sqrt{6}}{7 \cdot 2}$$

$$= -\frac{3\sqrt{6}}{14}$$

5.   Simplify.

$$\sqrt[3]{\frac{16}{9}}$$

$$\sqrt[3]{\frac{16}{9}} = \frac{\sqrt[3]{8 \cdot 2}}{\sqrt[3]{9}} = \frac{2\sqrt[3]{2}}{\sqrt[3]{9}}$$

$$= \frac{2\sqrt[3]{2} \cdot \sqrt[3]{3}}{\sqrt[3]{9} \cdot \sqrt[3]{3}}$$

$$= \frac{2\sqrt[3]{6}}{\sqrt[3]{27}}$$

$$= \frac{2\sqrt[3]{6}}{3}$$

Now Try:

3.   Rationalize the denominator.

$$\frac{2}{\sqrt{15}}$$

____

4.   Simplify the radical.

$$-\sqrt{\frac{45}{32}}$$

____

5.   Simplify.

$$\sqrt[3]{\frac{8}{100}}$$

____

Objective 2 Practice Exercises

For extra help, see Examples 3–5 on pages 494–496 of your text.

Simplify. Assume that variables represent positive real numbers.

4. $\sqrt{\dfrac{5a^2b^3}{6}}$

4. _________________

5. $\sqrt{\dfrac{7y^2}{12b}}$

5. _________________

6. $\sqrt[3]{\dfrac{5}{49x}}$

6. _________________

                Copyright © 2020 Pearson Education, Inc.

Objective 3   Rationalize denominators with binomials involving radicals.

Review this example for Objective 3:	**Now Try:**
6. Rationalize the denominator.	**6.** Rationalize the denominator.

6. Rationalize the denominator.

$$\frac{5}{5+\sqrt{2}}$$

$$\frac{5}{5+\sqrt{2}} = \frac{5\left(5-\sqrt{2}\right)}{\left(5+\sqrt{2}\right)\left(5-\sqrt{2}\right)}$$

$$= \frac{5\left(5-\sqrt{2}\right)}{25-2}$$

$$= \frac{5\left(5-\sqrt{2}\right)}{23}$$

Now Try:

6. Rationalize the denominator.

$$\frac{2}{\sqrt{3}-2}$$

Objective 3 Practice Exercises

For extra help, see Example 6 on pages 497–498 of your text.

Rationalize each denominator. Write quotients in lowest terms. Assume that variables represent positive real numbers.

7. $\dfrac{4}{\sqrt{3}+2}$

7. _______________

8. $\dfrac{5}{\sqrt{3}-\sqrt{10}}$

8. _______________

9. $\dfrac{\sqrt{6}+2}{\sqrt{2}-4}$

9. _______________

Objective 4   Write radical quotients in lowest terms.

Review this example for Objective 4:

7. Write the quotient in lowest terms.

$$\frac{72\sqrt{2}-16\sqrt{7}}{24}$$

$$\frac{72\sqrt{2}-16\sqrt{7}}{24}=\frac{8\left(9\sqrt{2}-2\sqrt{7}\right)}{24}$$

$$=\frac{9\sqrt{2}-2\sqrt{7}}{3}$$

Now Try:

7. Write the quotient in lowest terms.

$$\frac{9+6\sqrt{15}}{12}$$

____

Objective 4 Practice Exercises

For extra help, see Example 7 on page 498 of your text.

Write each quotient in lowest terms. Assume that variables represent positive real numbers.

10. $\dfrac{7-\sqrt{98}}{14}$

10. _______________

11. $\dfrac{16-12\sqrt{72}}{24}$

11. _______________

12. $\dfrac{2x-\sqrt{8x^2}}{4x}$

12. _______________

Chapter 7 ROOTS, RADICALS, AND ROOT FUNCTIONS

7.6   Solving Equations with Radicals

Learning Objectives
1   Solve radical equations using the power rule.
2   Solve radical equations that require additional steps.
3   Solve radical equations with indexes greater than 2.
4   Use the power rule to solve a formula for a specified variable.

Key Terms

Use the vocabulary terms listed below to complete each statement in exercises 1–2.

radical equation   **extraneous solution**

1.   A(n) ________________________________ is a potential solution to an equation that does not satisfy the equation.

2.   An equation with a variable in the radicand is a(n) ________________________________.

Objective 1   Solve radical equations using the power rule.

Review these examples for Objective 1:

Now Try:

1.   Solve $\sqrt{3w+4}=7$.

$$\left(\sqrt{3w+4}\right)^2 = 7^2$$
$$3w+4 = 49$$
$$3w = 45$$
$$w = 15$$

Check $\sqrt{3w+4} = 7$
$$\sqrt{3(15)+4} \overset{?}{=} 7$$
$$\sqrt{49} \overset{?}{=} 7$$
$$7 = 7 \quad \text{True}$$

Since 15 satisfies the original equation, the solution set is {15}.

1.   Solve $\sqrt{7x-6}=8$.

____

2.   Solve $\sqrt{12p+1}+7=0$.

Step 1 $\sqrt{12p+1}=-7$

Step 2 $\left(\sqrt{12p+1}\right)^2 =(-7)^2$

2.   Solve $\sqrt{4x-19}+5=0$.

____

Step 3 $12p + 1 = 49$

$12p = 48$

$p = 4$

Step 4 Check $\sqrt{12p + 1} + 7 = 0$

$\sqrt{12(4) + 1} + 7 \overset{?}{=} 0$

$\sqrt{49} + 7 \overset{?}{=} 0$

$14 = 0$ False

The false result shows that the proposed solution 4 is not a solution of the original equation. It is extraneous. The solution set is $\varnothing$.

Objective 1 Practice Exercises

For extra help, see Examples 1–2 on pages 503–504 of your text.

Solve each equation.

1. $\sqrt{4x - 19} = 5$ 1. _________________

2. $\sqrt{12p + 1} + 7 = 0$ 2. _________________

3. $\sqrt{4x - 3} = 7$ 3. _________________

Objective 2   Solve radical equations that require additional steps.

Review these examples for Objective 2:

Now Try:

3.   Solve $\sqrt{x+3} = x-3$.

3.   Solve $\sqrt{x+11} = x-1$.

Step 1 The radical is isolated on the left side of the equation.

Step 2 Square each side.

$$\left(\sqrt{x+3}\right)^2 = (x-3)^2$$
$$x+3 = x^2 - 6x + 9$$

Step 3 Write the equation in standard form and solve.

$$0 = x^2 - 7x + 6$$
$$0 = (x-1)(x-6)$$
$$x-1=0 \quad \text{or} \quad x-6=0$$
$$x=1 \quad \text{or} \quad x=6$$

Step 4 Check each proposed solution in the original equation.

$$\sqrt{x+3} = x-3 \qquad\qquad \sqrt{x+3} = x-3$$
$$\sqrt{1+3} \stackrel{?}{=} 1-3 \qquad \sqrt{6+3} \stackrel{?}{=} 6-3$$
$$\sqrt{4} \stackrel{?}{=} -2 \qquad\qquad \sqrt{9} \stackrel{?}{=} 3$$
$$2 = -2 \quad \text{False} \qquad 3 = 3 \quad \text{True}$$

The solution set is $\{6\}$. The other proposed solution, 1, is extraneous.

4.   Solve $\sqrt{x^2 + 7x - 14} = x+2$.

4.   Solve $\sqrt{x^2 + 8x - 17} = x+1$.

$$\left(\sqrt{x^2 + 7x - 14}\right)^2 = (x+2)^2$$
$$x^2 + 7x - 14 = x^2 + 4x + 4$$
$$3x = 18$$
$$x = 6$$

Check $\sqrt{x^2 + 7x - 14} = x+2$

$$\sqrt{6^2 + 7(6) - 14} \stackrel{?}{=} 6+2$$
$$\sqrt{64} \stackrel{?}{=} 8$$
$$8 = 8 \quad \text{True}$$

The solution set is $\{6\}$.

      299

5. Solve $\sqrt{3x} - 4 = \sqrt{x-2}$.

$$\left(\sqrt{3x} - 4\right)^2 = \left(\sqrt{x-2}\right)^2$$
$$3x - 8\sqrt{3x} + 16 = x - 2$$
$$-8\sqrt{3x} = -2x - 18$$
$$\left(-8\sqrt{3x}\right)^2 = (-2x - 18)^2$$
$$192x = 4x^2 + 72x + 324$$
$$0 = 4x^2 - 120x + 324$$
$$0 = 4\left(x^2 - 30x + 81\right)$$
$$0 = 4(x - 3)(x - 27)$$
$$x - 3 = 0 \quad \text{or} \quad x - 27 = 0$$
$$x = 3 \quad \text{or} \qquad x = 27$$

Check
$$\sqrt{3x} - 4 = \sqrt{x-2} \qquad\qquad \sqrt{3x} - 4 = \sqrt{x-2}$$
$$\sqrt{3(3)} - 4 \overset{?}{=} \sqrt{3-2} \qquad \sqrt{3(27)} - 4 \overset{?}{=} \sqrt{27-2}$$
$$3 - 4 \overset{?}{=} \sqrt{1} \qquad\qquad 9 - 4 \overset{?}{=} \sqrt{25}$$
$$-1 = 1 \quad \text{False} \qquad\qquad 5 = 5 \quad \text{True}$$

The proposed solution, 27, is valid, but 3 is extraneous and must be rejected. The solution set is $\{27\}$.

5. Solve $\sqrt{3x + 4} = \sqrt{9x} - 2$.

____

Objective 2 Practice Exercises

For extra help, see Examples 3–5 on pages 505–506 of your text.

Solve each equation.

4. $\sqrt{k+10} + \sqrt{2k+19} = 2$

4. _________________

5. $\sqrt{x-5} = x - 5$

5. _________________

6. $\sqrt{x^2-3x-15}=x-3$ **6.** _____________

Objective 3   Solve radical equations with indexes greater than 2.

Review this example for Objective 3:	Now Try:

Review this example for Objective 3:

6.   Solve $\sqrt[3]{5r-6}=\sqrt[3]{3r+4}$.

$$\left(\sqrt[3]{5r-6}\right)^3=\left(\sqrt[3]{3r+4}\right)^3$$
$$5r-6=3r+4$$
$$2r=10$$
$$r=5$$

Check $\sqrt[3]{5r-6}=\sqrt[3]{3r+4}$
$$\sqrt[3]{5(5)-6}\overset{?}{=}\sqrt[3]{3(5)+4}$$
$$\sqrt[3]{19}=\sqrt[3]{19}\quad\text{True}$$

The solution set is $\{5\}$.

Now Try:

6.   Solve $\sqrt[4]{8x+5}=\sqrt[4]{7x+7}$.

____

Objective 3 Practice Exercises

For extra help, see Example 6 on page 507 of your text.

Solve each equation.

7. $\sqrt[3]{2a-63}+5=0$ **7.** _____________

8. $\sqrt[5]{5a+1}-\sqrt[5]{2a-11}=0$ **8.** _____________

9. $\sqrt[4]{8x+5}=\sqrt[4]{7x+7}$ **9.** _____________

Objective 4 Use the power rule to solve a formula for a specified variable.

Review this example for Objective 4:	**Now Try:**
7. Solve the formula $d = \sqrt{\dfrac{H}{1.6n}}$ for n.	7. Solve the formula $r = \sqrt{\dfrac{3v}{\pi h}}$ for h.

$$d^2 = \left(\sqrt{\frac{H}{1.6n}}\right)^2$$

$$d^2 = \frac{H}{1.6n}$$

$$1.6d^2 n = H$$

$$n = \frac{H}{1.6d^2}$$

Objective 4 Practice Exercises

For extra help, see Example 7 on page 507 of your text.

Solve each equation for the indicated variable.

10. $Z = \sqrt{\dfrac{L}{C}}$, for L 10. __________

11. $f = \dfrac{1}{2\pi\sqrt{LC}}$, for C 11. __________

12. $N = \dfrac{1}{2\pi}\sqrt{\dfrac{a}{r}}$, for r 12. __________

Chapter 7 ROOTS, RADICALS, AND ROOT FUNCTIONS

7.7   Complex Numbers

Learning Objectives
1   Simplify numbers of the form $\sqrt{-b}$, where $b > 0$.
2   Identify subsets of the complex numbers.
3   Add and subtract complex numbers.
4   Multiply complex numbers.
5   Divide complex numbers.
6   Simplify powers of i.

Key Terms

Use the vocabulary terms listed below to complete each statement in exercises 1–7.

complex number   **real part**   **imaginary part**

pure imaginary number   **standard form (of a complex number)**

nonreal complex number   **complex conjugate**

1. A ________________________________ is a number that can be written in the form $a + bi$, where a and b are real numbers.

2. The ________________________________ of $a + bi$ is $a - bi$.

3. The ________________________________ of $a + bi$ is bi.

4. The ________________________________ of $a + bi$ is a.

5. A complex number is in ________________________________ if it is written in the form $a + bi$.

6. A complex number $a + bi$ with $a = 0$ and $b \neq 0$ is called a

   ________________________________.

7. A complex number $a + bi$ with $b \neq 0$ is called a ________________________________.

Objective 1   Simplify numbers of the form $\sqrt{-b}$, where $b > 0$.

Review these examples for Objective 1:

1. Write each number as a product of a real number and i.

   a. $\sqrt{-36}$

   $\sqrt{-36} = i\sqrt{36} = 6i$

Now Try:

1. Write each number as a product of a real number and i.

   a. $\sqrt{-16}$

   ____

b. $-\sqrt{-121}$

$$-\sqrt{-121} = -i\sqrt{121} = -11i$$

c. $\sqrt{-3}$

$$\sqrt{-3} = i\sqrt{3}$$

d. $\sqrt{-75}$

$$\sqrt{-75} = i\sqrt{25 \cdot 3} = 5i\sqrt{3}$$

2. Multiply.

$$\sqrt{-6} \cdot \sqrt{-7}$$

$$\begin{aligned}
\sqrt{-6} \cdot \sqrt{-7} &= i\sqrt{6} \cdot i\sqrt{7} \\
&= i^2\sqrt{6 \cdot 7} \\
&= (-1)\sqrt{42} \\
&= -\sqrt{42}
\end{aligned}$$

3. Divide.

a. $\dfrac{\sqrt{-125}}{\sqrt{-5}}$

$$\begin{aligned}
\frac{\sqrt{-125}}{\sqrt{-5}} &= \frac{i\sqrt{125}}{i\sqrt{5}} \\
&= \sqrt{\frac{125}{5}} \\
&= \sqrt{25} \\
&= 5
\end{aligned}$$

b. $\dfrac{\sqrt{-28}}{\sqrt{7}}$

$$\begin{aligned}
\frac{\sqrt{-28}}{\sqrt{7}} &= \frac{i\sqrt{28}}{\sqrt{7}} \\
&= i\sqrt{\frac{28}{7}} \\
&= i\sqrt{4} \\
&= 2i
\end{aligned}$$

b. $-\sqrt{-144}$

____

c. $\sqrt{-11}$

____

d. $\sqrt{-128}$

____

2. Multiply.

$$\sqrt{-5} \cdot \sqrt{-6}$$

____

3. Divide.

a. $\dfrac{\sqrt{-200}}{\sqrt{-8}}$

____

b. $\dfrac{\sqrt{-80}}{\sqrt{5}}$

____

Objective 1 Practice Exercises

For extra help, see Examples 1–3 on pages 510–511 of your text.

Write the number as a product of a real number and i. Simplify all radical expressions.

1. $-\sqrt{-162}$ 1. _______________

Multiply or divide as indicated

2. $\sqrt{-5}\cdot\sqrt{-3}\cdot\sqrt{-7}$ 2. _______________

3. $\dfrac{\sqrt{-42}\cdot\sqrt{-6}}{\sqrt{-7}}$ 3. _______________

Objective 2   Identify subsets of the complex numbers.

For extra help, see page 512 of your text.

Classify each of the following complex numbers as real *or* imaginary.

4. $\sqrt{5}$ 4. _______________

5. $\sqrt{3}-i\sqrt{5}$ 5. _______________

6. $i\sqrt{7}$ 6. _______________

Objective 3   Add and subtract complex numbers.

Review these examples for Objective 3:

4. Add.

$$(2+9i)+(10-3i)$$

$$(2+9i)+(10-3i)=(2+10)+(9-3)i$$
$$=12+6i$$

Now Try:

4. Add.

$$(4-7i)+(6-2i)$$

____

      305

5. Subtract.

$(7-9i)-(-5-6i)$

$(7-9i)-(-5-6i)=[7-(-5)]+[-9-(-6)]i$
$=(7+5)+(-9+6)i$
$=12-3i$

5. Subtract.

$(12+2i)-(-12-2i)$

____

Objective 3 Practice Exercises

For extra help, see Examples 4–5 on page 513 of your text.

Add or subtract as indicated. Write answers in standard form.

7. $(-7-2i)-(-3-3i)$

7. _______________

8. $4i-(9+5i)+(2+3i)$

8. _______________

9. $(7-9i)-(5-6i)$

9. _______________

Objective 4 Multiply complex numbers.

Review these examples for Objective 4:

6. Multiply.

a. $6i(2-7i)$

$6i(2-7i)=6i(2)+6i(-7i)$
$=12i-42i^2$
$=12i-42(-1)$
$=42+12i$

b. $(3+2i)(5-i)$

$(3+2i)(5-i)=3(5)+3(-i)+2i(5)+2i(-i)$
$=15-3i+10i-2i^2$
$=15+7i-2(-1)$
$=15+7i+2$
$=17+7i$

c. $(1+6i)(2+5i)$

$(1+6i)(2+5i)=2(1)+1(5i)+6i(2)+(6i)(5i)$
$=2+5i+12i+30i^2$
$=2+17i+30(-1)$
$=-28+17i$

Now Try:

6. Multiply.

a. $2i(4+7i)$

____

b. $(12+5i)(1-i)$

____

c. $(1+3i)(2-5i)$

____

Name: Date:
Instructor: Section:

Objective 4 Practice Exercises

For extra help, see Example 6 on pages 513–514 of your text.

Multiply.

10. $(2-5i)(2+5i)$ 10. _______________

11. $(1+3i)^2$ 11. _______________

12. $(12+2i)(-1+i)$ 12. _______________

Objective 5   Divide complex numbers.

Review this example for Objective 5:   **Now Try:**
7.   Find the quotient.   **7.**   Find the quotient.

$$\frac{6-i}{2-3i}$$ $$\frac{4+i}{5-2i}$$

Multiply the numerator and denominator
by $2 + 3i$, the conjugate of the denominator.

$$\frac{6-i}{2-3i} = \frac{(6-i)(2+3i)}{(2-3i)(2+3i)}$$

$$= \frac{12+18i-2i-3i^2}{2^2+3^2}$$

$$= \frac{12+16i-3(-1)}{4+9}$$

$$= \frac{15+16i}{13}, \text{ or } \frac{15}{13}+\frac{16}{13}i$$

Objective 5 Practice Exercises

For extra help, see Example 7 on pages 514–515 of your text.

Write each quotient in the form a + bi.

13. $\dfrac{3-2i}{2+i}$ 13. _______________

14. $\dfrac{5+2i}{9-4i}$

14. ____________________

15. $\dfrac{6-i}{2-3i}$

15. ____________________

Objective 6   Simplify powers of i.

Review these examples for Objective 6:

8. Find each power of i.

   a. i^{100}

$$i^{100} = (i^4)^{25} = 1^{25} = 1$$

   b. i^{27}

$$i^{27} = i^{24} \cdot i^3 = (i^4)^6 \cdot (-i) = 1^6 \cdot (-i) = -i$$

   c. i^{-3}

$$i^{-3} = \frac{1}{i^3} = \frac{1}{i^3} \cdot \frac{i}{i} = \frac{i}{i^4} = \frac{i}{1} = i$$

   d. i^{49}

$$i^{49} = i^{48} \cdot i = (i^4)^{12} \cdot i = 1^{12} \cdot i = i$$

Now Try:

8. Find each power of i.

   a. i^{48}

____

   b. i^{77}

____

   c. i^{-5}

____

   d. i^{55}

____

Objective 6 Practice Exercises

For extra help, see Example 8 on page 515 of your text.

Find each power of i.

16. i^{14}

16. ____________________

17. i^{113}

17. ____________________

18. i^{-21}

18. ____________________

Chapter 8   QUADRATIC EQUATIONS, INEQUALITIES, AND FUNCTIONS

8.1   The Square Root Property and Completing the Square

Learning Objectives
1   Review the zero-factor property.
2   Learn the square root property.
3   Solve quadratic equations of the form $(ax + b)^2 = c$ by extending the square root property.
4   Solve quadratic equations by completing the square.
5   Solve quadratic equations with nonreal complex solutions.

Key Terms

Use the vocabulary terms listed below to complete each statement in exercises 1–4.

> **quadratic equation**   **zero-factor property**
>
> **completing the square**   **square root property**

1. The _______________________ says that, if k is positive and $a^2 = k$, then $a = \pm\sqrt{k}$.

2. Use the process called _______________________ in order to rewrite an equation so it can be solved using the square root property.

3. An equation that can be written in the form $ax^2 + bx + c = 0$ is a

   _______________________.

4. The _______________________ states that if a product equals 0, then at least one of the factors of the product also equals zero.

Objective 1   Review the zero-factor property.

Review this example for Objective 1:

1. Solve $2x^2 + 5x - 3 = 0$.

   Use the zero factor property.
   $$2x^2 + 5x - 3 = 0$$
   $$(2x - 1)(x + 3) = 0$$
   $$2x - 1 = 0 \quad \text{or} \quad x + 3 = 0$$
   $$2x = 1 \quad \text{or} \quad x = -3$$
   $$x = \frac{1}{2}$$

   The solution set is $\left\{-3, \frac{1}{2}\right\}$.

Now Try:

1. Solve $3x^2 - 5x - 28 = 0$.

   ____

Objective 1 Practice Exercises

For extra help, see Example 1 on page 532 of your text.

Solve each equation by factoring.

1. $15s^2 - 2 = s$ 1. ________________

2. $z^2 = 6z - 9$ 2. ________________

3. $16m^2 - 64 = 0$ 3. ________________

Objective 2   Learn the square root property.

Review these examples for Objective 2:

2. Solve the equation.

   a. $x^2 = 7$

$$x^2 = 7$$
$$x = \sqrt{7} \ \text{ or } \ x = -\sqrt{7}$$

The solution set is $\left\{\sqrt{7}, -\sqrt{7}\right\}$, or $\left\{\pm\sqrt{7}\right\}$.

   b. $5p^2 - 100 = 0$

$$5p^2 - 100 = 0$$
$$5p^2 = 100$$
$$p^2 = 20$$
$$p = \sqrt{20} \ \text{ or } \ p = -\sqrt{20}$$
$$p = 2\sqrt{5} \ \text{ or } \ p = -2\sqrt{5}$$

The solution set is $\left\{2\sqrt{5}, -2\sqrt{5}\right\}$, or $\left\{\pm 2\sqrt{5}\right\}$.

3. Use Galileo's formula to determine how long it will take a penny dropped from the 86th floor Observatory deck of the Empire State Building to reach the ground. The deck is 1050 feet above the ground. Round your answer to the nearest tenth.

Galileo's formula is $d = 16t^2$, where d is the distance in feet that an object falls, and t is the

Now Try:

2. Solve the equation.

   a. $r^2 = 13$

            ____

   b. $3x^2 - 54 = 0$

            ____

3. A child dropped a ball from a hotel balcony that is 113 ft above the ground. Use Galileo's formula to determine how long it takes for the ball to reach the ground. Round your answer to the nearest tenth.

            ____

time in seconds.

$$d = 16t^2$$

$$1050 = 16t^2$$

$$65.625 = t^2$$

$$t = \sqrt{65.625} \quad \text{or} \quad t = -\sqrt{65.625}$$

Time cannot be negative, so we discard $t = -\sqrt{65.625}$. Using a calculator, $\sqrt{65.625} \approx 8.1$, so $t \approx 8.1$. The penny would fall to the ground in about 8.1 seconds.

Objective 2 Practice Exercises

For extra help, see Examples 2–3 on pages 533–534 of your text.

Solve each equation by using the square root property. Express all radicals in simplest form.

4. $r^2 = 30$

4. ____________________

5. $x^2 - 98 = 0$

5. ____________________

6. $3d^2 - 750 = 0$

6. ____________________

Objective 3 **Solve equations of the form $(ax+b)^2 = k$ by extending the square root property.**

Review these examples for Objective 3:

4. Solve $(x-2)^2 = 25$

$$(x-2)^2 = 25$$

$$x - 2 = \sqrt{25} \quad \text{or} \quad x - 2 = -\sqrt{25}$$

$$x - 2 = 5 \quad \text{or} \quad x - 2 = -5$$

$$x = 7 \quad \text{or} \quad x = -3$$

The solution set is $\{-3, 7\}$.

Now Try:

4. Solve $(x-1)^2 = 16$

____

Name: Date:
Instructor: Section:

5. Solve $(3x+4)^2 = 40$.

$$(3x+4)^2 = 40$$
$$3x+4 = \sqrt{40} \qquad \text{or} \quad 3x+4 = -\sqrt{40}$$
$$3x = -4+\sqrt{40} \quad \text{or} \qquad 3x = -4-\sqrt{40}$$
$$x = \frac{-4+2\sqrt{10}}{3} \quad \text{or} \qquad x = \frac{-4-2\sqrt{10}}{3}$$

Check

$$(3x+4)^2 = 40$$
$$\left[3\left(\frac{-4-2\sqrt{10}}{3}\right)+4\right]^2 \overset{?}{=} 40$$
$$\left(-4-2\sqrt{10}+4\right)^2 \overset{?}{=} 40$$
$$\left(-2\sqrt{10}\right)^2 \overset{?}{=} 40$$
$$40 = 40$$

The check for the second solution is similar.

The solution sets is $\left\{\dfrac{-4-2\sqrt{10}}{3}, \dfrac{-4+2\sqrt{10}}{3}\right\}$.

5. Solve $(2x+5)^2 = 32$.

____

Objective 3 Practice Exercises

For extra help, see Examples 4–5 on pages 534–535 of your text.

Solve each equation by using the square root property. Express all radicals in simplest form.

7. $(y+2)^2 = 16$

7. _____________

8. $(q-4)^2 = 7$

8. _____________

9. $(3f+4)^2 = 32$

9. _____________

Objective 4   Solve quadratic equations by completing the square.

Review these examples for Objective 4: | **Now Try:**

6.   Solve $x^2 - 12x + 24 = 0$.   **6.**   Solve $x^2 - 6x + 1 = 0$.

$$x^2 - 12x + 24 = 0$$

$$x^2 - 12x = -24$$

Take half of the coefficient of the first-degree term, $-12x$, and square the result.

$$\left[\tfrac{1}{2}(-12)\right]^2 = (-6)^2 = 36$$

Add 36 to each side.

$$x^2 - 12x + 36 = -24 + 36$$

$$(x - 6)^2 = 12$$

$$x - 6 = \sqrt{12} \quad \text{or} \quad x - 6 = -\sqrt{12}$$

$$x = 6 + 2\sqrt{3} \quad \text{or} \quad x = 6 - 2\sqrt{3}$$

A check indicates the solution set is

$$\{6 - 2\sqrt{3}, \ 6 + 2\sqrt{3}\}.$$

7.   Solve $x^2 + 5x + 2 = 0$.   **7.**   Solve $x^2 - 11x + 8 = 0$.

Since the coefficient of the second-degree term is 1, begin with Step 2.

Step 2 $x^2 + 5x = -2$

Step 3   Take half the coefficient of the first-degree term and square the result.

$$\left[\tfrac{1}{2}(5)\right]^2 = \left(\tfrac{5}{2}\right)^2 = \tfrac{25}{4}$$

$$x^2 + 5x + \tfrac{25}{4} = -2 + \tfrac{25}{4}$$

$$\left(x + \tfrac{5}{2}\right)^2 = \tfrac{17}{4}$$

Step 4

$$x + \tfrac{5}{2} = \sqrt{\tfrac{17}{4}} \qquad \text{or} \quad x + \tfrac{5}{2} = -\sqrt{\tfrac{17}{4}}$$

$$x + \tfrac{5}{2} = \tfrac{\sqrt{17}}{2} \qquad \text{or} \quad x + \tfrac{5}{2} = -\tfrac{\sqrt{17}}{2}$$

$$x = -\tfrac{5}{2} + \tfrac{\sqrt{17}}{2} \quad \text{or} \qquad x = -\tfrac{5}{2} - \tfrac{\sqrt{17}}{2}$$

A check shows the solution set is

$$\left\{-\tfrac{5}{2} - \tfrac{\sqrt{17}}{2}, \ -\tfrac{5}{2} + \tfrac{\sqrt{17}}{2}\right\}.$$

8. Solve $3x^2 - 6x - 2 = 0$.

$$x^2 - 2x - \frac{2}{3} = 0 \qquad \text{Step 1}$$

$$x^2 - 2x = \frac{2}{3} \qquad \text{Step 2}$$

$$\left[\frac{1}{2}(-2)\right]^2 = (-1)^2 = 1 \qquad \text{Step 3}$$

$$x^2 - 2x + 1 = \frac{2}{3} + 1$$

$$(x-1)^2 = \frac{5}{3}$$

$$x - 1 = \sqrt{\frac{5}{3}} \qquad \text{or} \quad x - 1 = -\sqrt{\frac{5}{3}} \qquad \text{Step 4}$$

$$x = 1 + \sqrt{\frac{5}{3}} \qquad \text{or} \qquad x = 1 - \sqrt{\frac{5}{3}}$$

$$x = 1 + \frac{\sqrt{15}}{3} \qquad \text{or} \qquad x = 1 - \frac{\sqrt{15}}{3}$$

$$x = \frac{3 + \sqrt{15}}{3} \qquad \text{or} \qquad x = \frac{3 - \sqrt{15}}{3}$$

The solution set is $\left\{\dfrac{3 - \sqrt{15}}{3}, \dfrac{3 + \sqrt{15}}{3}\right\}$.

8. Solve $2p^2 + 6p - 1 = 0$.

____

Objective 4 Practice Exercises

For extra help, see Examples 6–8 on pages 536–538 of your text.

Solve each equation by completing the square.

10. $x^2 - 4x = 2$

10. ______________

11. $x^2 - 9x + 8 = 0$

11. ______________

12. $6q^2 + 4q = 1$

12. ______________

Name: Date:

Instructor: Section:

Objective 5   Solve quadratic equations with nonreal complex solutions.

Review these examples for Objective 5: | **Now Try:**

9. Solve each equation.

   a. $x^2 = -48$

$$x^2 = -48$$
$$x = \sqrt{-48} \quad \text{or} \quad x = -\sqrt{-48}$$
$$x = 4i\sqrt{3} \quad \text{or} \quad x = -4i\sqrt{3}$$

The solution set is $\{-4i\sqrt{3},\ 4i\sqrt{3}\}$.

   b. $(x-3)^2 = -25$

$$(x-3)^2 = -25$$
$$x - 3 = \sqrt{-25} \quad \text{or} \quad x - 3 = -\sqrt{-25}$$
$$x - 3 = 5i \quad \text{or} \quad x - 3 = -5i$$
$$x = 3 + 5i \quad \text{or} \quad x = 3 - 5i$$

The solution set is $\{3 - 5i,\ 3 + 5i\}$.

   c. $x^2 + 4x + 9 = 0$

$$x^2 + 4x + 9 = 0$$
$$x^2 + 4x = -9$$
$$x^2 + 4x + 4 = -9 + 4$$
$$(x+2)^2 = -5$$
$$x + 2 = \pm i\sqrt{5}$$
$$x = -2 \pm i\sqrt{5}$$

The solution set is $\{-2 - i\sqrt{5}, -2 + i\sqrt{5}\}$.

Now Try:

9. Solve each equation.

   a. $y^2 = -32$

   ____

   b. $(x+2)^2 = -49$

   ____

   c. $x^2 + 6x + 13 = 0$

   ____

Objective 5 Practice Exercises

For extra help, see Example 9 on page 538 of your text.

Find the complex solutions of each equation.

13. $(10m - 5)^2 + 9 = 0$

13. ______________

14. $(m + 1)^2 = -36$

14. ______________

15. $(x - 1)^2 + 2 = 0$

15. ______________

Chapter 8   QUADRATIC EQUATIONS, INEQUALITIES, AND FUNCTIONS

8.2   The Quadratic Formula

Learning Objectives
1   Derive the quadratic formula.
2   Solve quadratic equations using the quadratic formula.
3   Use the discriminant to determine the number and type of solutions.

Key Terms

Use the vocabulary terms listed below to complete each statement in exercises 1–2.

quadratic formula   **discriminant**

1.   The expression under the radical in the quadratic formula is called the

_______________________________.

2.   The formula $x = \dfrac{-b \pm \sqrt{b^2 - 4ac}}{2a}$ is called the _______________________________.

Objective 1   Derive the quadratic formula.

For extra help, see page 542 of your text.

Objective 2   Solve quadratic equations using the quadratic formula.

Review these examples for Objective 2:

1.   Solve $5x^2 - 13x - 6 = 0$.

Use the quadratic formula with $a = 5$, $b = -13$, and $c = -6$.

$$x = \frac{-b \pm \sqrt{b^2 - 4ac}}{2a}$$

$$x = \frac{-(-13) \pm \sqrt{(-13)^2 - 4(5)(-6)}}{2(5)}$$

$$x = \frac{13 \pm \sqrt{169 + 120}}{10}$$

$$x = \frac{13 \pm \sqrt{289}}{10}$$

$$x = \frac{13 \pm 17}{10}$$

There are two solutions.

$$x = \frac{13 + 17}{10} = 3 \quad \text{or} \quad x = \frac{13 - 17}{10} = \frac{-4}{10} = -\frac{2}{5}$$

The solution set is $\left\{ -\dfrac{2}{5},\ 3 \right\}$.

Now Try:

1.   Solve $6x^2 - 17x + 12 = 0$.

____

2. Solve $16x^2 + 8x + 1 = 0$.

Use the quadratic formula with
$a = 16$, $b = 8$, $c = 1$

$$x = \frac{-b \pm \sqrt{b^2 - 4ac}}{2a}$$

$$x = \frac{-8 \pm \sqrt{(8)^2 - 4(16)(1)}}{2(16)}$$

$$x = \frac{-8 + \sqrt{64 - 64}}{32} = \frac{8 \pm 0}{32}$$

$$x = -\frac{1}{4}$$

There is one *distinct* solution. The solution set is $\left\{ -\frac{1}{4} \right\}$.

2. Solve $25x^2 + 30x + 9 = 0$.

____

3. Solve $4x^2 = -4x + 1$.

First write the equation in standard form as $4x^2 + 4x - 1 = 0$.
$a = 4$, $b = 4$, $c = -1$

$$x = \frac{-b \pm \sqrt{b^2 - 4ac}}{2a}$$

$$x = \frac{-4 \pm \sqrt{(4)^2 - 4(4)(-1)}}{2(4)}$$

$$x = \frac{-4 \pm \sqrt{16 + 16}}{8} = \frac{-4 \pm \sqrt{32}}{8}$$

$$x = \frac{-4 \pm 4\sqrt{2}}{8}$$

$$x = \frac{4(-1 \pm \sqrt{2})}{4(2)}$$

$$x = \frac{-1 \pm \sqrt{2}}{2}$$

The solution set is $\left\{ \dfrac{-1 - \sqrt{2}}{2}, \ \dfrac{-1 + \sqrt{2}}{2} \right\}$.

3. Solve $2x^2 = 2x + 3$.

____

4. Solve $(5x - 2)(x + 2) = -9$.

$$(5x - 2)(x + 2) = -9$$
$$5x^2 + 8x - 4 = -9$$
$$5x^2 + 8x + 5 = 0$$

From the standard form, we identify $a = 5$, $b = 8$, and $c = 5$.

4. Solve $(2x - 6)(x + 1) = -16$.

____

$$x = \frac{-b \pm \sqrt{b^2 - 4ac}}{2a}$$

$$x = \frac{-8 \pm \sqrt{(8)^2 - 4(5)(5)}}{2(5)}$$

$$x = \frac{-8 \pm \sqrt{-36}}{10} = \frac{-8 \pm 6i}{10}$$

$$x = \frac{2(-4 \pm 3i)}{2(5)}$$

$$x = \frac{-4 \pm 3i}{5} = -\frac{4}{5} \pm \frac{3}{5}i$$

The solution set is $\left\{-\frac{4}{5} - \frac{3}{5}i, -\frac{4}{5} + \frac{3}{5}i\right\}$.

Objective 2 Practice Exercises

For extra help, see Examples 1–4 on pages 543–545 of your text.

Use the quadratic formula to solve each equation. (All solutions for these equations are real numbers.)

1. $(z+2)^2 = 2(5z-2)$ 1. _______________

2. $5k^2 + 4k - 2 = 0$ 2. _______________

3. $34 - 10x = -x^2$ 3. _______________

Objective 3   Use the discriminant to determine the number and type of solutions.

Review these examples for Objective 3:	**Now Try:**

Review these examples for Objective 3:

5. Find the discriminant. Use it to predict the number and type of solutions for each equation. Then tell whether the equation can be solved by factoring or whether the quadratic formula should be used.

a. $3x^2 + x - 2 = 0$

First identify the values of a, b, and c.
$\quad a = 3$, $b = 1$, and $c = -2$.
Then find the discriminant.
$$b^2 - 4ac = 1^2 - 4(3)(-2)$$
$$= 1 + 24$$
$$= 25, \text{ or } 5^2$$

Since a, b, and c are integers and the discriminant 25 is a perfect square, there will be two rational solutions. The equation can be solved by factoring.

b. $16x^2 + 25 = 40x$

Write in standard form: $16x^2 - 40x + 25 = 0$.
$a = 16$, $b = -40$, and $c = 25$
$$b^2 - 4ac = (-40)^2 - 4(16)(25)$$
$$= 1600 - 1600$$
$$= 0$$

Because the discriminant is 0, this quadratic equation will have one distinct rational solution. The equation can be solved by factoring.

c. $5y^2 - 5y + 2 = 0$

$a = 5$, $b = -5$, and $c = 2$
$$b^2 - 4ac = (-5)^2 - 4(5)(2)$$
$$= 25 - 40$$
$$= -15$$

Because the discriminant is negative and a, b, and c are integers, this quadratic equation will have two nonreal complex solutions. The quadratic equation should be used to solve it.

Now Try:

5. Find the discriminant. Use it to predict the number and type of solutions for each equation. Then tell whether the equation can be solved by factoring or whether the quadratic formula should be used.

a. $10x^2 + 21x + 9 = 0$

____

b. $25x^2 + 9 = 30x$

____

c. $2y^2 + 4y + 8 = 0$

____

Objective 3 Practice Exercises

For extra help, see Example 5 on pages 546–547 of your text.

Use the discriminant to determine whether the solutions for each equation are

    A. *two rational numbers*   B. *one rational number,*
    C. *two irrational numbers*   D. *two imaginary numbers.*

Do not actually solve.

4. $m^2 - 4m + 4 = 0$ 4. __________________

5. $z^2 + 6z + 3 = 0$ 5. __________________

6. $16x^2 - 12x + 9 = 0$ 6. __________________

Chapter 8 QUADRATIC EQUATIONS, INEQUALITIES, AND FUNCTIONS

8.3 Equations That Lead to Quadratic Methods

Learning Objectives
1   Solve rational equations that lead to quadratic equations.
2   Solve applied problems involving quadratic equations.
3   Solve radical equations that lead to quadratic equations.
4   Solve equations that are quadratic in form

Key Terms

Use the vocabulary terms listed below to complete each statement in exercises 1–2.

quadratic in form   **standard form**

1. A quadratic equation written in the form $ax^2 + bx + c = 0$, $a \neq 0$ is written in

    _______________________________.

2. A nonquadratic equation that can be written as a quadratic equation is called

    _______________________________.

Objective 1   Solve rational equations that lead to quadratic equations.

Review this example for Objective 1:

1. Solve $5 + \dfrac{6}{m+1} = \dfrac{14}{m}$.

Multiply each side by the least common denominator, $m(m + 1)$. The domain must be restricted to $m \neq 0$, $m \neq -1$.

$$5 + \frac{6}{m+1} = \frac{14}{m}$$

$$m(m+1)\left(5 + \frac{6}{m+1}\right) = m(m+1)\frac{14}{m}$$

$$m(m+1)(5) + m(m+1)\frac{6}{m+1} = m(m+1)\frac{14}{m}$$

$$5m^2 + 5m + 6m = 14m + 14$$

$$5m^2 + 11m = 14m + 14$$

$$5m^2 - 3m - 14 = 0$$

$$(5m + 7)(m - 2) = 0$$

$$5m + 7 = 0 \quad \text{or} \quad m - 2 = 0$$

$$m = -\frac{7}{5} \quad \text{or} \quad m = 2$$

The solution set is $\left\{-\dfrac{7}{5},\, 2\right\}$.

Now Try:

1. Solve $4 - \dfrac{8}{x-1} = -\dfrac{35}{x}$.

    ____

Name: Date:
Instructor: Section:

Objective 1 Practice Exercises

For extra help, see Example 1 on page 549 of your text.

Solve each equation. Check your solutions.

1. $\dfrac{5}{x} + \dfrac{1}{2x+7} = -\dfrac{2}{3}$

1. _______________

2. $\dfrac{2m}{m-5} + \dfrac{7}{m+1} = 0$

2. _______________

3. $1 + \dfrac{49}{2x} = \dfrac{15}{x+1}$

3. _______________

Objective 2   Solve applied problems involving quadratic equations.

Review these examples for Objective 2:

2. Amy rows her boat 6 miles upstream and then returns in $2\dfrac{6}{7}$ hours. The speed of the current is 2 miles per hour. How fast can she row?

Step 1 Read the problem carefully.

Step 2 Assign a variable. Let $x =$ the rate that Amy rows in still water. The current slows Amy when she is going upstream, so Amy's rate going upstream is her rate in still water less the rate of the current, or $x - 2$. Similarly, the current makes Amy row faster when she is going downstream,

Now Try:

2. Mike can row 3 miles per hour in still water. It takes him 3 hours and 36 minutes to row 3 miles upstream and return. Find the speed of the current.

____

        Copyright © 2020 Pearson Education, Inc.

so her downstream rate is $x + 2$.
Complete a table. Recall that $d = rt$.

	d	r	t
Upstream	6	$x - 2$	$\dfrac{6}{x-2}$
Downstream	6	$x + 2$	$\dfrac{6}{x+2}$

Step 3 Write an equation. The time upstream plus the time downstream equals the total time, $2\dfrac{6}{7}$ hours, or $\dfrac{20}{7}$ hours.

$$\frac{6}{x-2} + \frac{6}{x+2} = \frac{20}{7}$$

Step 4 Solve the equation. The LCD is $7(x-2)(x+2)$.

$$7(x-2)(x+2)\left(\frac{6}{x-2} + \frac{6}{x+2}\right)$$
$$= 7(x-2)(x+2)\left(\frac{20}{7}\right)$$

$$7(x+2)6 + 7(x-2)6 = (x-2)(x+2)20$$

$$42(x+2) + 42(x-2) = 20(x^2 - 4)$$

$$42x + 84 + 42x - 84 = 20x^2 - 80$$

$$20x^2 - 84x - 80 = 0$$

$$4(5x^2 - 21x - 20) = 0$$

$$4(5x + 4)(x - 5) = 0$$

$$5x + 4 = 0 \quad \text{or} \quad x - 5 = 0$$

$$x = -\frac{4}{5} \quad \text{or} \quad x = 5$$

Step 5 State the answer. The rate cannot be $-\dfrac{4}{5}$ mph, so the answer is Amy rows at 5 mph.

Step 6 Check that this value satisfies the original equation.

3. It takes two painters working together $4\dfrac{4}{5}$ hours to paint a house. If each worked alone, one of them could do the job in 4 hours less than the other. How long would it take each painter to complete the job alone?

Step 1 Read the problem carefully. There will be two answers.

3. Working together, Tom and Huck painted a fence in 8 hours. If each worked alone, Tom could paint the fence 12 hours faster than Huck could. How long would it take each to paint the fence alone?

Step 2 Assign a variable. Let $x =$ the number of hours for the slower painter to complete the job alone. Then the faster painter could do the entire job in $x - 4$ hours. The faster painter's rate is $\dfrac{1}{x-4}$ and the slower painter's rate is $\dfrac{1}{x}$.

Together, they do the job in $4\dfrac{4}{5}$ hours.

Complete a table.

	Rate	Time working together	Fractional part of the job done
Slower painter	$\dfrac{1}{x}$	$\dfrac{24}{5}$	$\dfrac{1}{x}\left(\dfrac{24}{5}\right)$
Faster painter	$\dfrac{1}{x-4}$	$\dfrac{24}{5}$	$\dfrac{1}{x-4}\left(\dfrac{24}{5}\right)$

Step 3 Write an equation. The sum of the two fractional parts is 1.

$$\frac{24}{5x} + \frac{24}{5(x-4)} = 1$$

Step 4 Solve the equation. The LCD is $5x(x-4)$.

$$5x(x-4)\left(\frac{24}{5x} + \frac{24}{5(x-4)}\right) = 5x(x-4)(1)$$

$$24(x-4) + 24x = 5x(x-4)$$

$$24x - 96 + 24x = 5x^2 - 20x$$

$$48x - 96 = 5x^2 - 20x$$

$$5x^2 - 68x + 96 = 0$$

$$(5x-8)(x-12) = 0$$

$$5x - 8 = 0 \quad \text{or} \quad x - 12 = 0$$

$$x = \frac{8}{5} \quad \text{or} \quad x = 12$$

Step 5 State the answer.

If the slower painter can do the job in $\dfrac{8}{5}$ hours (or 1.6 hours), then the faster painter's time to do the job is $1.6 - 4 = -2.4$ hr, which cannot represent the time for the slower painter.
If the slower painter can do the job in 12 hours, then the faster painter's time is $12 - 4 = 8$ hours.

Step 6 Check that these results satisfy the original problem.

Name: Date:
Instructor: Section:

Objective 2 Practice Exercises

For extra help, see Examples 2–3 on pages 550–552 of your text.

Solve each problem. Round answers to the nearest tenth, if necessary.

4. Two pipes together can fill a large tank in 10 hours.   4. pipe 1 ___________
    One of the pipes, used alone, takes 15 hours longer
    than the other to fill the tank. How long would each   pipe 2 ___________
    pipe used alone take to fill the tank?

5. A jet plane traveling at a constant speed goes 1200   5. ___________
    miles with the wind, then turns around and travels
    for 1000 miles against the wind. If the speed of the
    wind is 50 miles per hour and the total flight takes 4
    hours, find the speed of the plane.

6. A man rode a bicycle for 12 miles and then hiked an   6. bike___________
    additional 8 miles. The total time for the trip was 5
    hours. If his rate when he was riding the bicycle was   hike ___________
    10 miles per hour faster than his rate walking, what
    was each rate?

                    325

Objective 3   Solve radical equations that lead to quadratic equations.

Review these examples for Objective 3:	**Now Try:**

Review these examples for Objective 3:

4. Solve each equation.

    a. $y = \sqrt{y+42}$

Start by squaring each side.

$$y = \sqrt{y+42}$$

$$y^2 = \left(\sqrt{y+42}\right)^2$$

$$y^2 = y + 42$$

$$y^2 - y - 42 = 0$$

$$(y+6)(y-7) = 0$$

$$y + 6 = 0 \quad \text{or} \quad y - 7 = 0$$

$$y = -6 \quad \text{or} \quad y = 7$$

We must check all proposed solutions in the original equation because squaring each side of an equation can introduce extraneous solutions.

Check

$$y = \sqrt{y+42} \qquad \qquad y = \sqrt{y+42}$$

$$-6 \overset{?}{=} \sqrt{-6+42} \qquad 7 \overset{?}{=} \sqrt{7+42}$$

$$-6 \overset{?}{=} \sqrt{36} \qquad \qquad 7 \overset{?}{=} \sqrt{49}$$

$$-6 = 6 \quad \text{False} \qquad 7 = 7 \quad \text{True}$$

The solution set is $\{7\}$.

    b. $\sqrt{x} + 2 = x$

$$\sqrt{x} = x - 2$$

$$\left(\sqrt{x}\right)^2 = (x-2)^2$$

$$x = x^2 - 4x + 4$$

$$0 = x^2 - 5x + 4$$

$$0 = (x-1)(x-4)$$

$$x - 1 = 0 \quad \text{or} \quad x - 4 = 0$$

$$x = 1 \quad \text{or} \quad x = 4$$

Check each proposed solution in the original equation.

Now Try:

4. Solve each equation.

    a. $x = \sqrt{x+2}$

            ____

    b. $\sqrt{2x} + 4 = x$

            ____

$$\sqrt{x}+2=x \qquad\qquad \sqrt{x}+2=x$$

$$\sqrt{1}+2\overset{?}{=}1 \qquad\qquad \sqrt{4}+2\overset{?}{=}4$$

$$1+2\overset{?}{=}1 \qquad\qquad 2+2\overset{?}{=}4$$

$$3=1 \quad \text{False} \qquad\qquad 4=4 \quad \text{True}$$

The solution set is $\{4\}$.

Objective 3 Practice Exercises

For extra help, see Example 4 on page 552 of your text.

Solve each equation. Check your solutions.

7. $\sqrt{7y-10}=y$ 7. ______________

8. $x=\sqrt{\dfrac{x+3}{2}}$ 8. ______________

9. $\sqrt{4x}+3=x$ 9. ______________

Objective 4   Solve equations that are quadratic in form.

Review these examples for Objective 4:

6. Solve the equation.

$$c^4 - 20c^2 + 64 = 0$$

Write this equation in quadratic form by substituting u for c^2.

$$u^2 - 20u + 64 = 0$$
$$(u - 4)(u - 16) = 0$$

$$u - 4 = 0 \quad \text{or} \quad u - 16 = 0$$
$$u = 4 \quad \text{or} \quad u = 16$$
$$c^2 = 4 \quad \text{or} \quad c^2 = 16$$
$$c = \pm 2 \quad \text{or} \quad c = \pm 4$$

The solution set is $\{-4, -2, 2, 4\}$.

7. Solve each equation.

a. $(m + 5)^2 + 6(m + 5) + 8 = 0$

Step 1 Substitute u for $m + 5$.

$$(m + 5)^2 + 6(m + 5) + 8 = 0$$
$$u^2 + 6u + 8 = 0$$

Step 2 $(u + 4)(u + 2) = 0$

$$u + 4 = 0 \quad \text{or} \quad u + 2 = 0$$
$$u = -4 \quad \text{or} \quad u = -2$$

Step 3 $m + 5 = -4 \quad \text{or} \quad m + 5 = -2$

Step 4 $m = -9 \quad \text{or} \quad m = -7$

Step 5 Check that the solution set of the original equation is $\{-9, -7\}$.

b. $x^{4/3} - 20x^{2/3} + 36 = 0$

Step 1 Substitute u for $x^{2/3}$.

$$x^{4/3} - 20x^{2/3} + 36 = 0$$
$$u^2 - 20u + 36 = 0$$

Step 2 $(u - 2)(u - 18) = 0$

$$u - 2 = 0 \quad \text{or} \quad u - 18 = 0$$
$$u = 2 \quad \text{or} \quad u = 18$$

Now Try:

6. Solve the equation.

$$x^4 - 5x^2 + 4 = 0$$

____

7. Solve each equation.

a. $(x - 5)^2 + 2(x - 5) - 35 = 0$

____

b. $x^{2/3} - 2x^{1/3} = 3$

____

Name: Date:
Instructor: Section:

$Step\ 3$ $\quad x^{2/3} = 2$ or $x^{2/3} = 18$

$Step\ 4$ $\left(x^{2/3}\right)^{3/2} = (2)^{3/2}$ or $\left(x^{2/3}\right)^{3/2} = (18)^{3/2}$

$\qquad\qquad x = 2\sqrt{2}$ or $\qquad x = 54\sqrt{2}$

$Step\ 5$ Check that the solution set of the original equation is $\{2\sqrt{2},\ 54\sqrt{2}\}$.

Objective 4 Practice Exercises

For extra help, see Examples 5–7 on pages 553–556 of your text.

Solve each equation. Check your solutions.

10. $\quad 4t^4 = 21t^2 - 5$

10. ___________________

11. $\quad p^{4/3} - 12p^{2/3} + 27 = 0$

11. ___________________

12. $\quad \left(t^2 - 3t\right)^2 = 14\left(t^2 - 3t\right) - 40$

12. ___________________

    329

Chapter 8   QUADRATIC EQUATIONS, INEQUALITIES, AND FUNCTIONS

8.4   Formulas and Further Applications

Learning Objectives
1   Solve formulas involving squares and square roots for specified variables.
2   Solve applied problems using the Pythagorean theorem.
3   Solve applied problems using area formulas.
4   Solve applied problems using quadratic functions as models.

Key Terms

Use the vocabulary terms listed below to complete each statement in exercises 1–2.

quadratic function   Pythagorean theorem

1.   A function defined by $f(x) = ax^2 + bx + c$, for real numbers a, b, and c, with $a \neq 0$, is a ________________________________.

2.   The ________________________________ states that the sum of the squares of the lengths of the legs of a right triangle equals the square of the length of the hypotenuse.

Objective 1   Solve formulas involving squares and square roots for specified variables.

Review these examples for Objective 1:

1.   Solve each formula for the given variable.

a. $y = \dfrac{1}{2}gt^2$ for t

The goal is to isolate t on one side.

$$y = \frac{1}{2}gt^2$$

$$2y = gt^2$$

$$\frac{2y}{g} = t^2$$

$$t = \pm\sqrt{\frac{2y}{g}}$$

$$t = \pm\frac{\sqrt{2y}}{\sqrt{g}} \cdot \frac{\sqrt{g}}{\sqrt{g}}$$

$$t = \pm\frac{\sqrt{2yg}}{g}$$

Now Try:

1.   Solve each formula for the given variable.

a. $F = \dfrac{mx}{t^2}$ for t

____

                Copyright © 2020 Pearson Education, Inc.

b. $D = \sqrt{kh}$ for k

The goal is to isolate k on one side.

$$D = \sqrt{kh}$$

$$D^2 = kh$$

$$\frac{D^2}{h} = k$$

b. $p = \dfrac{y}{\sqrt{6z}}$ for z

____

2. Solve $rk^2 - 3k = -s$ for k.

Write the equation in standard form and then use the quadratic formula to solve for k.

$$rk^2 - 3k = -s$$

$$rk^2 - 3k + s = 0$$

Let $a = r$, $b = -3$, and $c = s$.

$$k = \frac{-(-3) \pm \sqrt{(-3)^2 - 4(r)(s)}}{2r}$$

$$k = \frac{3 \pm \sqrt{9 - 4rs}}{2r}$$

The solutions are $\dfrac{3 + \sqrt{9 - 4rs}}{2r}$ and

$\dfrac{3 - \sqrt{9 - 4rs}}{2r}$.

2. Solve $p^2 q^2 + pkq = k^2$ for q.

____

Objective 1 Practice Exercises

For extra help, see Examples 1–2 on pages 561–562 of your text.

Solve each equation for the indicated variable. (Leave ± in your answers.)

1. $F = \dfrac{kl}{\sqrt{d}}$ for d

1. ___________

2. $p = \sqrt{\dfrac{kl}{g}}$ for k

2. ___________

3. $b^2 a^2 + 2bca = c^2$ for a

3. ___________

Objective 2   Solve applied problems using the Pythagorean theorem.

Review this example for Objective 2:

3. A 13-foot ladder is leaning against a building. The distance from the bottom of the ladder to the building is 2 feet more than twice the distance from the top of the ladder to the ground. How far is the bottom of the ladder from the building?

Step 1 Read the problem carefully.

Step 2 Assign a variable. Let x = the distance from the top of the ladder to the ground.
Then $2x + 2$ = the distance from the bottom of the ladder to the building. Draw a picture to represent the problem.

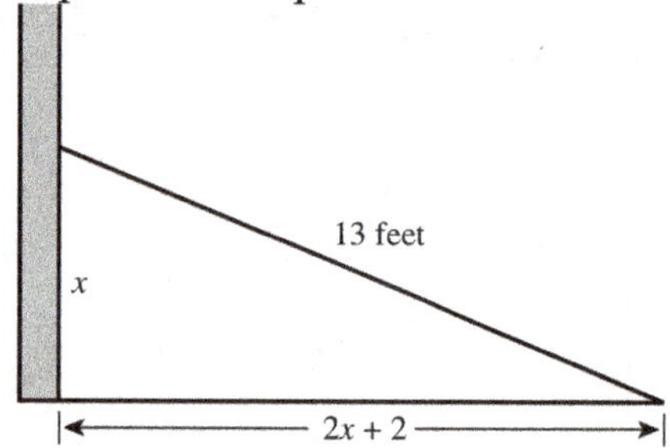

Step 3 Write an equation. Use the Pythagorean theorem.

$$a^2 + b^2 = c^2$$

$$x^2 + (2x+2)^2 = 13^2$$

Step 4 Solve.

$$x^2 + 4x^2 + 8x + 4 = 169$$

$$5x^2 + 8x - 165 = 0$$

$$(5x + 33)(x - 5) = 0$$

$$5x + 33 = 0 \quad \text{or} \quad x - 5 = 0$$

$$x = -\frac{33}{5} \quad \text{or} \quad x = 5$$

Step 5 State the answer. Length cannot be negative, so discard the negative solution. The distance from the top of the ladder to the ground is 5 feet. However, we are asked to find the distance from the bottom of the ladder to the building. This distance is $2(5) + 2 = 12$ feet.

Step 6 Check. Since $5^2 + 12^2 = 13^2$, the answer is correct.

Now Try:

3. Two cars left an intersection at the same time, one heading south, the other heading east. Sometime later, the car traveling south had gone 18 miles farther than the car headed east. At that time they were 90 miles apart. How far had each car traveled?

south   _____________

east   _____________

Objective 2 Practice Exercises

For extra help, see Example 3 on page 562 of your text.

Solve each problem.

4. A child flying a kite has let out 45 feet of string to
    the kite. The distance from the kite to the ground is 9
    feet more than the distance from the child to a point
    directly below the kite. How high up is the kite?

    4. ___________________

5. A ladder is leaning against a building so that the top
    is 8 feet above the ground. The length of the ladder
    is 2 feet less than twice the distance of the bottom of
    the ladder from the building. Find the length of the
    ladder.

    5. ___________________

6. Two cars left an intersection at the same time, one
    heading north, the other heading west. Later they
    were exactly 95 miles apart. The car headed west
    had gone 38 miles less than twice as far as the car
    headed north. How far had each car traveled?

    6. north ___________

       west ___________

Objective 3   Solve applied problems using area formulas.

Review this example for Objective 3:

4. A fish pond is 3 feet by 4 feet. How wide a strip of concrete can be poured around the pond if there is enough concrete for 44 square feet?

Step 1 Read the problem carefully.

Step 2 Assign a variable. Let x = the width of the strip of concrete. Then $2x + 3$ = the width of the fish pond with the two strips of concrete and $2x + 4$ = the length of the fish pond with the two strips of concrete.

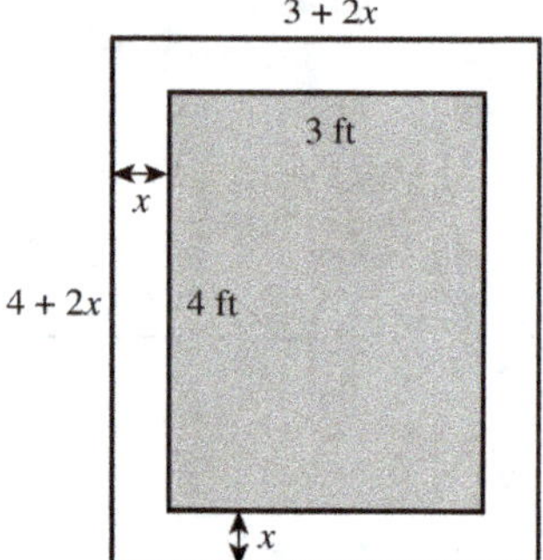

Step 3 Write an equation. The area of the strip is 44 sq ft and the area of the fish pond is $3(4) = 12$ sq ft, so the total area of the outer rectangle is $44 + 12 = 56$ sq ft.

$$(3+2x)(4+2x)=56$$

Step 4 Solve.

$$(3+2x)(4+2x)=56$$
$$12+14x+4x^2=56$$
$$4x^2+14x-44=0$$
$$2(2x^2+7x-22)=0$$
$$2(2x+11)(x-2)=0$$
$$2x+11=0 \quad \text{or} \quad x-2=0$$
$$x=-\frac{11}{2} \quad \text{or} \quad x=2$$

Step 5 State the answer. Since length cannot be negative, we disregard the negative solution. The concrete strip should be 2 ft wide.

Step 6 Check. If $x = 2$, then the area of the large rectangle is $(3+2\cdot2)(4+2\cdot2)=7(8)=56$ sq ft. The area of the fish pond is $3(4) = 12$ sq ft, so the area of the concrete strip is $56-12=44$ sq ft.

Now Try:

4. A picture 9 inches by 12 inches is to be mounted on a piece of mat board so that there is an even width of mat all around the picture. How wide will the matted border be if the area of the mounted picture is 238 square inches?

____

Name: Date:
Instructor: Section:

Objective 3 Practice Exercises

For extra help, see Example 4 on page 563 of your text.

Solve each problem.

7. A rug is to fit in a room so that a border of even
    width is left on all four sides. If the room is
    16 feet by 20 feet and the area of the rug is 165
    square feet, how wide to the nearest tenth of a foot
    will the border be?

7. ___________________

8. A rectangular garden has an area of 12 feet by 5 feet.
    A gravel path of equal width is to be built around the
    garden. How wide can the path be if there is enough
    gravel for 138 square feet?

8. ___________________

9. A doghouse 2 feet by 4 feet is to be built with a
    cement path around it of equal width on all sides.
    The area available for the doghouse and path is 120
    square feet. How wide will the path be?

9. ___________________

Objective 4   Solve applied problems using quadratic functions as models.

Review this example for Objective 4:	**Now Try:**

5.   A certain projectile is located at a distance of $d(t) = 3t^2 - 6t + 1$ feet from its starting point after t seconds. How many seconds will it take the projectile to travel 10 feet?

Let $d = 10$ in the formula and solve for t.

$$d = 3t^2 - 6t + 1$$
$$10 = 3t^2 - 6t + 1$$
$$9 = 3t^2 - 6t$$
$$3 = t^2 - 2t$$
$$3 + 1 = t^2 - 2t + 1$$
$$4 = (t-1)^2$$
$$\sqrt{4} = t - 1 \quad \text{or} \quad -\sqrt{4} = t - 1$$
$$2 = t - 1 \quad \text{or} \quad -2 = t - 1$$
$$3 = t \quad\quad \text{or} \quad\quad -1 = t$$

Since t represents time, we reject the negative solution. It will take 3 seconds for the projectile to travel 10 feet.

5.   A baseball is thrown upward from a building 20 m high with a velocity of 15 m/sec. Its distance from the ground after t seconds is modeled by the function
$$f(t) = -4.9t^2 + 15t + 20.$$
When will the ball hit the ground? Round your answer to the nearest tenth.

____

For extra help, see Examples 5–6 on pages 564–565 of your text.

Solve each problem. Round answers to the nearest tenth.

10.   A population of microorganisms grows according to the function $p(x) = 100 + 0.2x + 0.5x^2$, where x is given in hours. How many hours does it take to reach a population of 250 microorganisms?

10. ________________

11. An object is thrown downward from a tower 280 feet   **11.** _________________
high. The distance the object has fallen at time t in
seconds is given by $s(t) = 16t^2 + 68t$. How long will
it take the object to fall 100 feet?

12. A widget manufacturer estimates that her monthly   **12.** _________________
revenue can be modeled by the function
$R(x) = -0.006x^2 + 32x - 10{,}000$. What is the
minimum number of items that must be sold for the
revenue to equal \$30,000?

Chapter 8   QUADRATIC EQUATIONS, INEQUALITIES, AND FUNCTIONS

8.5   Graphs of Quadratic Functions

> **Learning Objectives**
> 1   Graph a quadratic function.
> 2   Graph parabolas with horizontal and vertical shifts.
> 3   Use the coefficient of x^2 to predict the shape and direction in which a parabola opens.
> 4   Find a quadratic function to model data.

Key Terms

Use the vocabulary terms listed below to complete each statement in exercises 1−4.

parabola   vertex   axis   quadratic function

1.   The vertical (or horizontal) line through the vertex of a vertical (or horizontal) parabola is its _______________________.

2.   The point on a parabola that has the least y-value (if the parabola opens up) or the greatest y-value (if the parabola opens down) is called the _______________ of the parabola.

3.   A function defined by $f(x) = ax^2 + bx + c$, for real numbers a, b, and c, with $a \neq 0$, is a _______________________________.

4.   The graph of a quadratic function is a _________________________________.

Objective 1   Graph a quadratic function.

For extra help, see pages 570–571 of your text.

Objective 2   Graph parabolas with horizontal and vertical shifts.

Review these examples for Objective 2:

1.   Graph $g(x) = x^2 + 2$. Give the vertex, axis, domain, and range.

The graph of $g(x)$ has the same shape as that of $f(x) = x^2$ but shifted 2 units up with vertex (0, 2). Every function value is 2 more than the corresponding function value of $f(x) = x^2$.

Now Try:

1.   Graph $f(x) = x^2 - 1$. Give the vertex, axis, domain, and range.

Vertex   _____________

Axis   _____________

Domain   _____________

Range   _____________

x	$f(x)=x^2$	$g(x)=x^2+2$
-2	4	6
-1	1	3
0	0	2
1	1	3
2	4	6

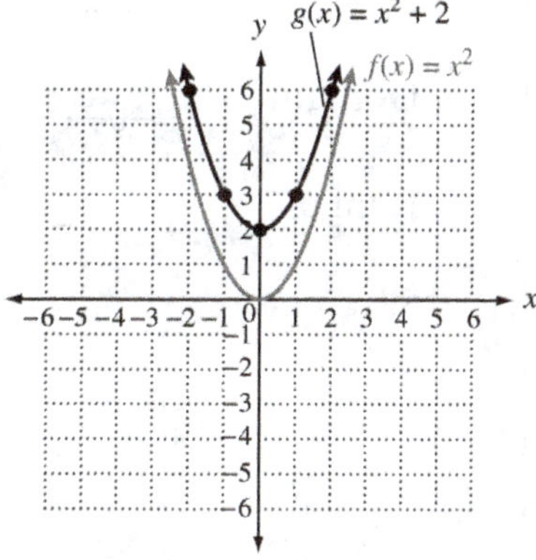

The vertex is $(0, 2)$. The axis is $x = 0$. The domain is $(-\infty, \infty)$. The range is $[2, \infty)$.

2. Graph $g(x)=(x-1)^2$. Give the vertex, axis, domain, and range.

The graph of $g(x)$ has the same shape as that of $f(x)=x^2$ but shifted 1 unit right with vertex $(1, 0)$.

x	$f(x)=x^2$	$g(x)=(x-1)^2$
-2	4	9
-1	1	4
0	0	1
1	1	0
2	4	1
3	9	4

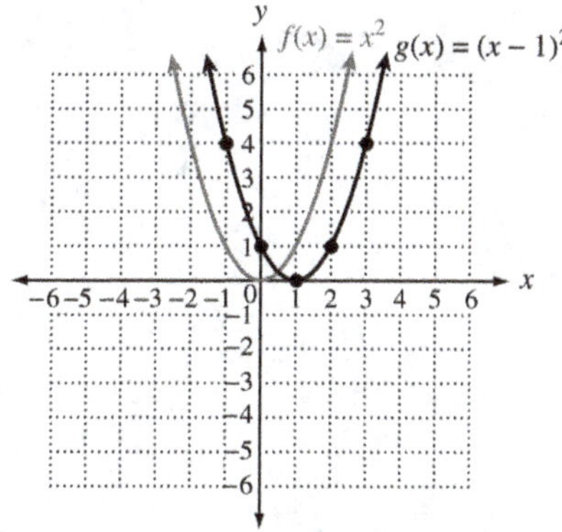

The vertex is $(1, 0)$. The axis is $x = 1$. The domain is $(-\infty, \infty)$. The range is $[0, \infty)$.

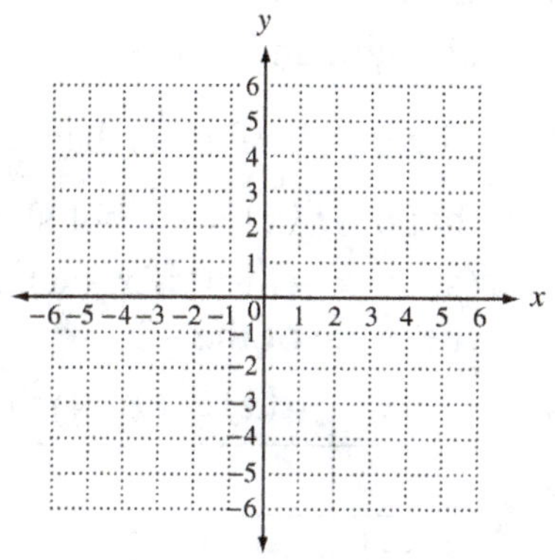

2. Graph $f(x)=(x-3)^2$. Give the vertex, axis, domain, and range.

Vertex   ________________

Axis   ________________

Domain   ________________

Range   ________________

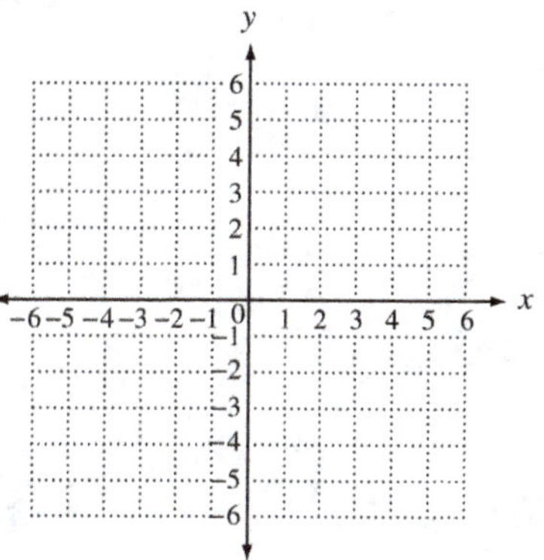

3. Graph $g(x) = (x-1)^2 - 2$. Give the vertex, axis, domain, and range.

The graph of $g(x)$ has the same shape as that of $f(x) = x^2$ but shifted 1 unit right (since $x - 1 = 0$ if $x = 1$) and 2 units down (because of the -2).

x	$g(x) = (x-1)^2 - 2$
-2	7
-1	2
0	-1
1	-2
2	-1
3	2
4	7

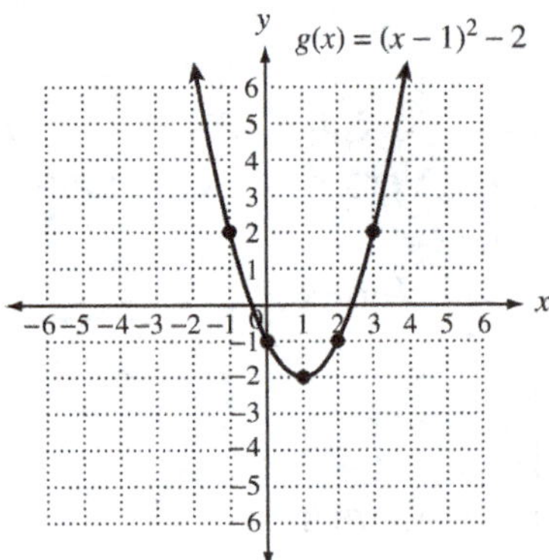

The vertex is $(1, -2)$. The axis is $x = 1$. The domain is $(-\infty, \infty)$. The range is $[-2, \infty)$.

3. Graph $f(x) = (x+2)^2 - 1$. Give the vertex, axis, domain, and range.

Vertex   ____________

Axis   ____________

Domain   ____________

Range   ____________

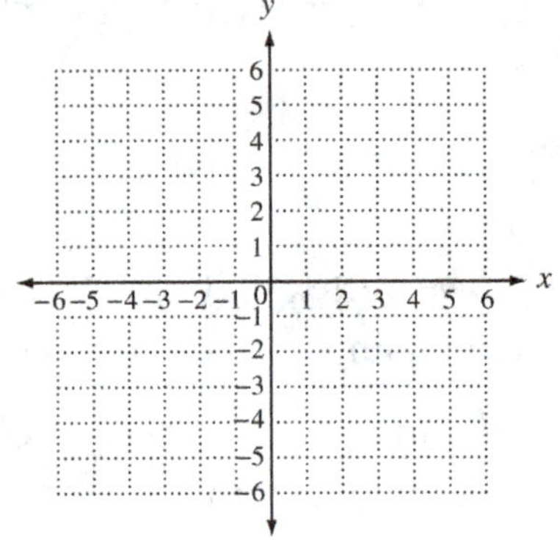

Name: Date:

Instructor: Section:

Objective 2 Practice Exercises

For extra help, see Examples 1–3 on pages 571–573 of your text.

Sketch the graph of each parabola. Give the vertex, axis, domain, and range.

1. $f(x) = x^2 - 4$

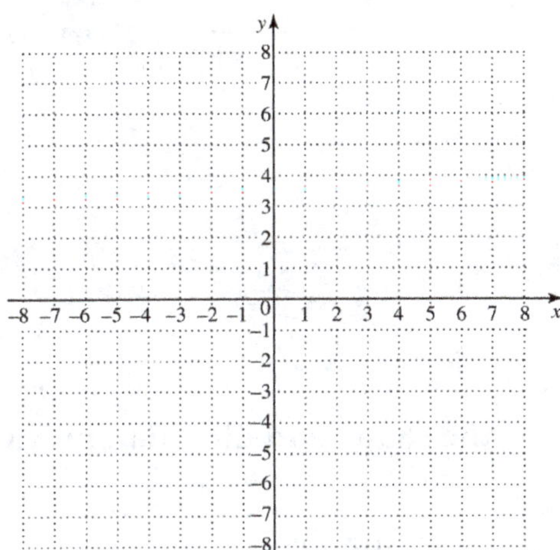

1. vertex________________

   axis________________

   domain____________

   range ______________

2. $f(x) = (x - 3)^2$

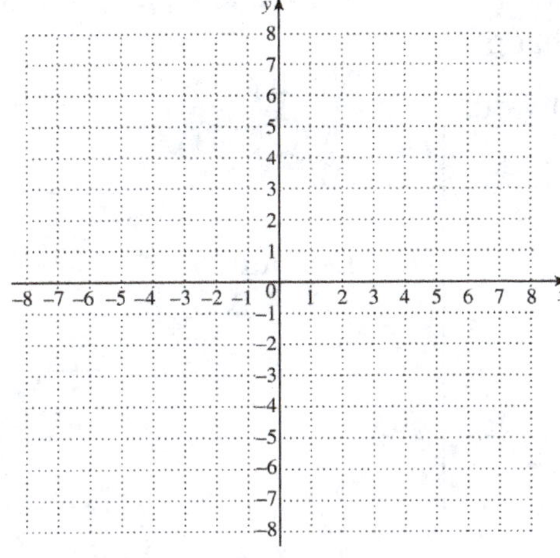

2. vertex________________

   axis________________

   domain______________

   range ______________

3. $f(x) = (x - 3)^2 - 1$

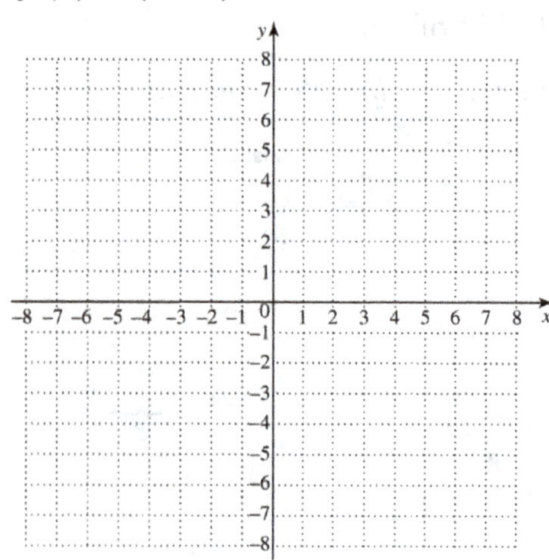

3. vertex _______________

axis _______________

domain _______________

range _______________

Objective 3   **Use the coefficient of x^2 to predict the shape and direction in which a parabola opens.**

Review these examples for Objective 3:

4.   Graph $g(x) = -2x^2$. Give the vertex, axis, domain, and range.

The graph of $g(x)$ has the same shape as that of $f(x) = x^2$ but is narrower and opens downward.

x	$g(x) = -2x^2$
-2	-8
-1	-2
0	0
1	-2
2	-8

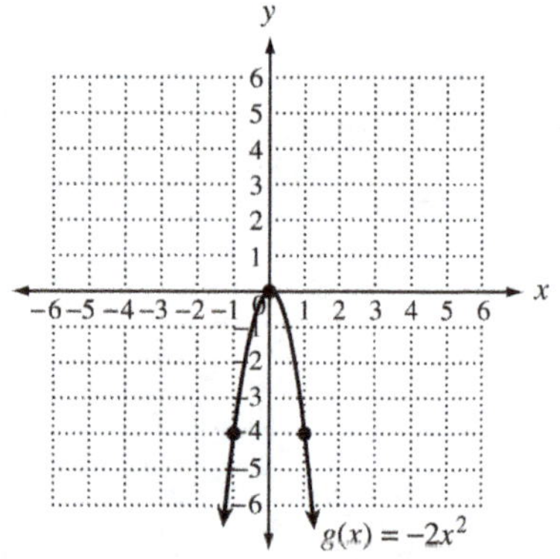

The vertex is $(0, 0)$. The axis is $x = 0$. The domain is $(-\infty, \infty)$. The range is $(-\infty, 0]$.

Now Try:

4.   Graph $g(x) = -\dfrac{1}{4}x^2$. Give the vertex, axis, domain, and range.

Vertex   _______________

Axis   _______________

Domain   _______________

Range   _______________

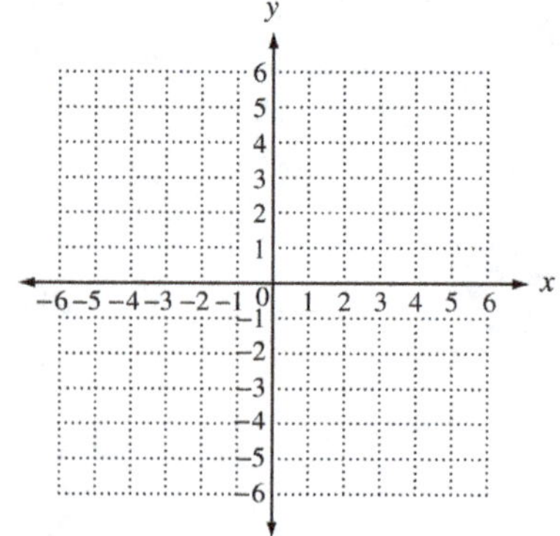

              Copyright © 2020 Pearson Education, Inc.

5. Graph $g(x) = -\dfrac{1}{2}(x+1)^2 - 2$. Give the vertex, axis, domain, and range.

The parabola opens down because $a < 0$ and is wider than the graph of $f(x) = x^2$. The parabola has vertex $(-1,-2)$.

x	$g(x) = -\dfrac{1}{2}(x+1)^2 - 2$
-3	-4
-2	-2.5
-1	-2
0	-2.5
1	-4

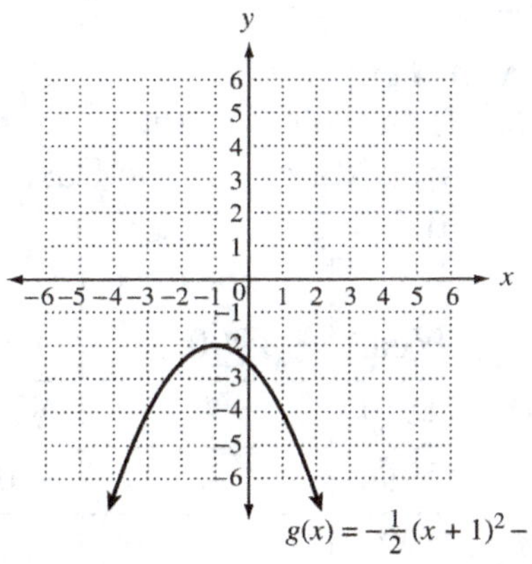

The vertex is $(-1,-2)$. The axis is $x = -1$. The domain is $(-\infty, \infty)$. The range is $(-\infty, -2]$.

5. Graph $f(x) = 3(x-1)^2 + 1$. Give the vertex, axis, domain, and range.

Vertex ______________

Axis ______________

Domain ______________

Range ______________

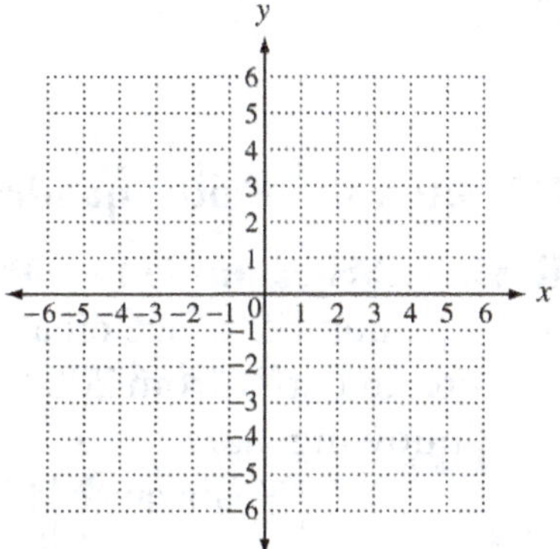

Objective 3 Practice Exercises

For extra help, see Examples 4–5 on pages 573–574 of your text.

For each quadratic function, tell whether the graph opens up or down and whether the graph is wider, narrower, or the same shape as the graph of $f(x) = x^2$. Then give the vertex, domain, and range.

4. $f(x) = -\dfrac{4}{3}x^2 - 1$

4. ________________

vertex ____________

domain ___________

range ____________

5. $f(x) = -2(x+1)^2$

5. _____________

     vertex __________

     domain __________

     range __________

6. $f(x) = \frac{5}{4}(x-1)^2 + 7$

6. _____________

     vertex __________

     domain __________

     range __________

Objective 4   Find a quadratic function to model data.

Review this example for Objective 4:

6. The number of ice cream cones sold by an ice cream parlor from 2003–2009 is shown in the following table.

Year	Years since 2003, x	Number of cones sold
2003	0	1775
2004	1	4194
2005	2	5063
2006	3	5161
2007	4	4663
2008	5	4639
2009	6	3710

Use the ordered pairs (x, number of cones sold) to make a scatter diagram of the data. Determine a quadratic function that models these data by using a system of equations. Use the ordered pairs (0, 1775), (3, 5161), and (6, 3710). Round the values of a, b, and c in your model to the nearest tenth, as necessary.

Now Try:

6. The table lists the average price of a Major League Baseball ticket.

Year	Years since 1990, x	Price
1991	1	$9.14
1994	4	$10.60
1997	7	$12.49
2000	10	$16.81
2004	14	$19.82
2010	20	$26.74

Use the ordered pairs (x, price) to make a scatter diagram of the data. Determine a quadratic function that models these data by using a system of equations. Use the ordered pairs (1, 9.14), (10, 16.81), and (20, 26.74). Round the values of a, b, and c in your model to the nearest hundredth, as necessary.

____

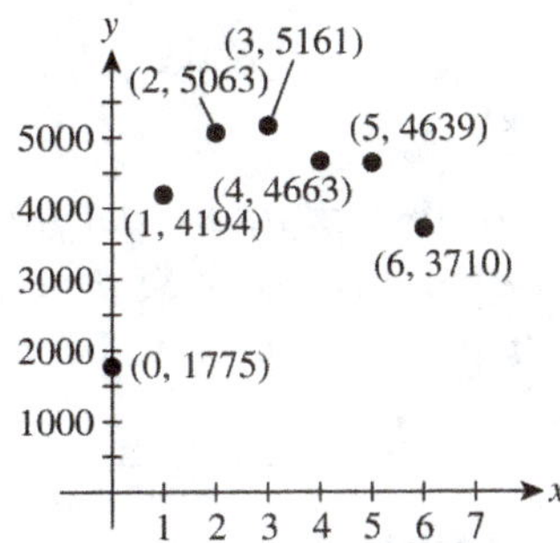

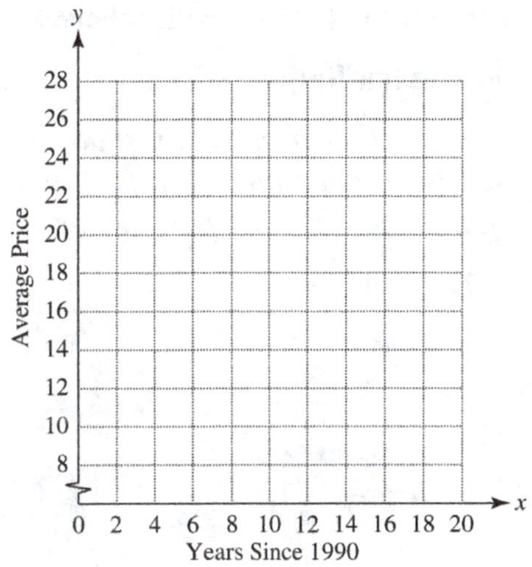

It appears that the parabola opens down, so the coefficient a is negative.

Using the chosen ordered pairs, we substitute the x- and y-values into the quadratic form

$y = ax^2 + bx + c$ to obtain the three equations.

$$a(0)^2 + b(0) + c = 1775 \quad (1)$$

$$a(3)^2 + b(3) + c = 5161 \quad (2)$$

$$a(6)^2 + b(6) + c = 3710 \quad (3)$$

Equation (1) simplifies to $c = 1775$, so substitute 1775 for c in equation (2) and (3).

$$9a + 3b + 1775 = 5161 \quad (2)$$

$$36a + 6b + 1775 = 3710 \quad (3)$$

Subtract 1775 from each side of both equations.

$$9a + 3b = 3386 \quad (2)$$

$$36a + 6b = 1935 \quad (3)$$

Solve by elimination. Multiply equation (2) by –2 and add to equation (3).

$$-18a - 6b = -6772 \quad \text{Multiply (2) by} -2.$$

$$\underline{36a + 6b = 1935 \quad\quad (3)}$$

$$18a = -4837$$

$$a \approx -268.7 \quad \text{Round to the tenth.}$$

Substitute this value for a into equation (2) and solve for b.

$$9a + 3b = 3386 \quad (2)$$

$$9(-268.7) + 3b = 3386$$

$$-2418.3 + 3b = 3386$$

$$3b = 5804.3$$

$$b \approx 1934.8$$

Therefore, the model

is $y = -268.7x^2 + 1934.8x + 1775$.

                345

Objective 4 Practice Exercises

For extra help, see Example 6 on pages 574–575 of your text.

Tell whether a linear or quadratic function would be a more appropriate model for each set of graphed data. If linear, tell whether the slope should be positive or negative. If quadratic, tell whether the coefficient a of x^2 should be positive or negative.

7. 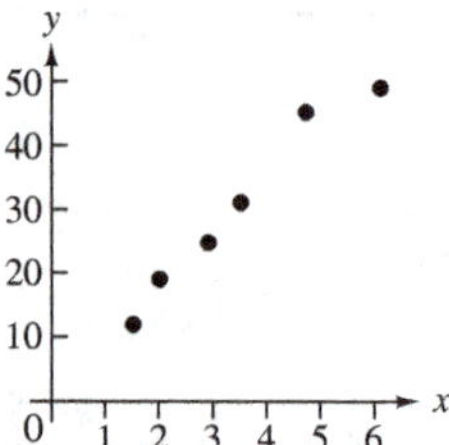

7. _________________

8. 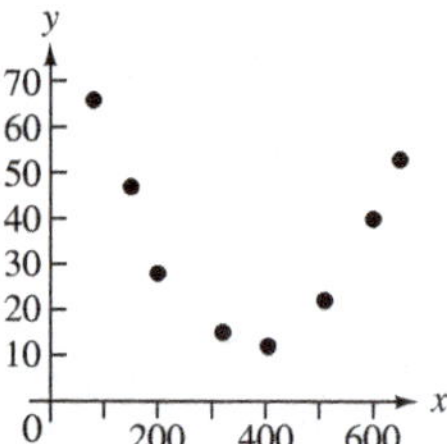

8. _________________

Solve the problem.

9. The number of publicly traded companies filing for bankruptcy for selected years between 1990 and 2000 are shown in the table, with 0 representing 1990, 2 representing 1992, etc.

Year	Number of Bankruptcies
0	115
2	91
4	70
6	84
8	120
10	176

Use the ordered pairs to make a scatter diagram of the data.

Use the ordered pairs (0, 115), (4, 70), and (8, 120) to find a function that models the data. Round the values of a, b, and c to three decimal places, if necessary.

9.

____

Source: Lial, Margaret L., John Hornsby, Terry McGinnis, *Intermediate Algebra* Eighth Edition. Boston: Pearson Education, 2006.

Chapter 8 QUADRATIC EQUATIONS, INEQUALITIES, AND FUNCTIONS

8.6   More about Parabolas and Their Applications

Learning Objectives

1	Find the vertex of a vertical parabola.
2	Graph a quadratic function.
3	Use the discriminant to find the number of x-intercepts of a parabola with a vertical axis.
4	Use quadratic functions to solve problems involving maximum or minimum value.
5	Graph parabolas with horizontal axes.

Key Terms

Use the vocabulary terms listed below to complete each statement in exercises 1−2.

discriminant   **vertex**

1.   The________________________________ of a quadratic function is found by using the formula $b^2 - 4ac$.

2.   The maximum or minimum value of a quadratic function occurs at the ________________________________ of its graph.

Objective 1   Find the vertex of a vertical parabola.

Review these examples for Objective 1:

1.   Find the vertex of the graph of
$$f(x) = x^2 + 6x + 10.$$

We can express $x^2 + 6x + 10$ in the form $(x - h)^2 + k$ by completing the square on $x^2 + 6x$. Because we want to keep $f(x)$ alone on one side of the equation, we add and subtract the appropriate number on just one side.

$$f(x) = x^2 + 6x + 10 \qquad \left[\tfrac{1}{2}(6)\right]^2 = 9$$

$$f(x) = (x^2 + 6x + 9 - 9) + 10$$

$$f(x) = (x^2 + 6x + 9) + 10 - 9$$

$$f(x) = (x + 3)^2 + 1$$

The vertex of the parabola is (−3, 1).

Now Try:

1.   Find the vertex of the graph of
$$f(x) = x^2 - 6x + 4.$$

____

2. Find the vertex of the graph of $f(x) = 3x^2 + 6x + 10$.

Because the x^2-term has a coefficient other than 1, we factor that coefficient out of the first two terms before completing the square.

$$f(x) = 3x^2 + 6x + 10$$

$$f(x) = 3(x^2 + 2x) + 10 \qquad \left[\frac{1}{2}(2)\right]^2 = 1$$

$$f(x) = 3(x^2 + 2x + 1 - 1) + 10$$

$$f(x) = 3(x^2 + 2x + 1) + 3(-1) + 10$$

$$f(x) = 3(x + 1)^2 + 7$$

The vertex is $(-1, 7)$.

3. Use the vertex formula to find the vertex of the graph of $f(x) = -4x^2 + 5x + 3$.

The x-coordinate of the vertex of the parabola is given by $\frac{-b}{2a}$.

$$\frac{-b}{2a} = \frac{-5}{2(-4)} = \frac{5}{8}$$

The y-coordinate is $f\left(\frac{-b}{2a}\right) = f\left(\frac{5}{8}\right)$.

$$f\left(\frac{5}{8}\right) = -4\left(\frac{5}{8}\right)^2 + 5\left(\frac{5}{8}\right) + 3 = \frac{73}{16}$$

The vertex is $\left(\frac{5}{8}, \frac{73}{16}\right)$.

2. Find the vertex of the graph of $f(x) = -2x^2 + 4x - 1$.

____

3. Use the vertex formula to find the vertex of the graph of $f(x) = 2x^2 - 6x + 5$.

____

Objective 1 Practice Exercises

For extra help, see Examples 1–3 on pages 579–581 of your text.

Find the vertex of each parabola.

1. $f(x) = x^2 - 2x + 4$

1. _______________

2. $f(x) = 5x^2 - 4x + 1$

2. _______________

3. $f(x) = -\frac{1}{4}x^2 - 3x - 9$

3. _______________

Objective 2   Graph a quadratic function.

Review this example for Objective 2:

4. Graph the quadratic function defined by $f(x) = x^2 - 3x + 2$. Give the vertex, axis, domain, and range.

Step 1 From the equation, $a = 1$, so the graph opens up.

Step 2 The x-coordinate of the vertex is $\frac{3}{2}$. The y-coordinate of the vertex is

$$f\left(\frac{3}{2}\right) = \left(\frac{3}{2}\right)^2 - 3\left(\frac{3}{2}\right) + 2 = -\frac{1}{4}.$$ The vertex is $\left(\frac{3}{2}, -\frac{1}{4}\right).$

Step 3 Find any intercepts. Since the vertex is in quadrant IV and the graph opens up, there will be two x-intercepts. Let $f(x) = 0$ and solve.

$$x^2 - 3x + 2 = 0$$
$$(x-1)(x-2) = 0$$
$$x - 1 = 0 \quad \text{or} \quad x - 2 = 0$$
$$x = 1 \quad \text{or} \quad x = 2$$

The x-intercepts are $(1, 0)$ and $(2, 0)$. Find the y-intercept by evaluating $f(0)$.

$$f(0) = 0^2 - 3(0) + 2$$

The y-intercept is $(0, 2)$.

Step 4 Plot the points found so far and additional points as needed using symmetry about the axis, $x = \frac{3}{2}$.

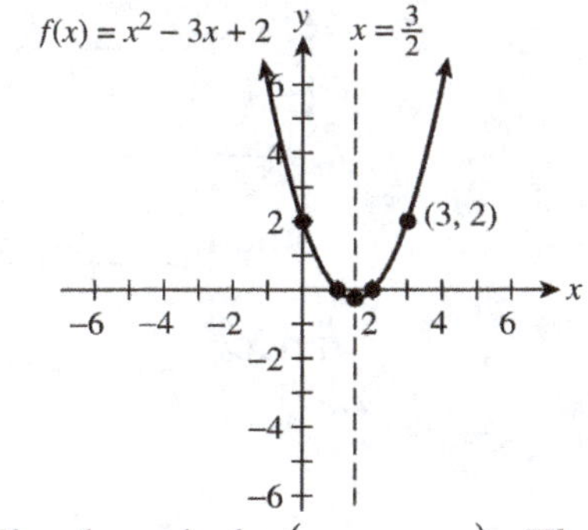

The domain is $(-\infty, \infty)$. The range is $\left[-\frac{1}{4}, \infty\right).$

Now Try:

4. Graph the quadratic function defined by $f(x) = x^2 + 4x + 5$. Give the vertex, axis, domain, and range.

Vertex   ____________

Axis   ____________

Domain   ____________

Range   ____________

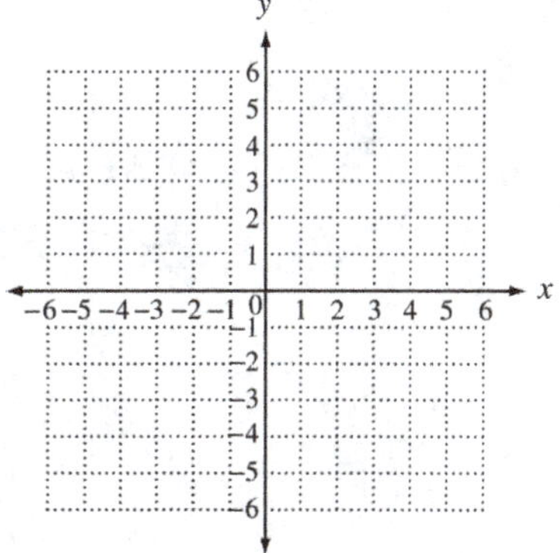

Objective 2 Practice Exercises

For extra help, see Example 4 on pages 581–582 of your text.

Sketch the graph of each parabola. Give the vertex, axis, domain, and range.

4. $f(x) = -x^2 + 8x - 10$

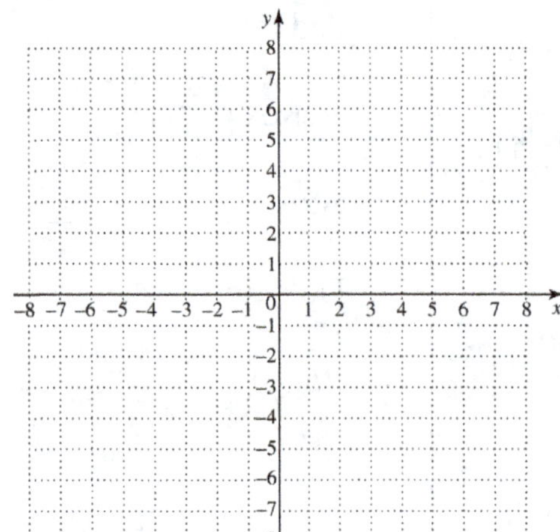

4. vertex _______________

   axis _______________

   domain _______________

   range _______________

5. $f(x) = 3x^2 + 6x + 2$

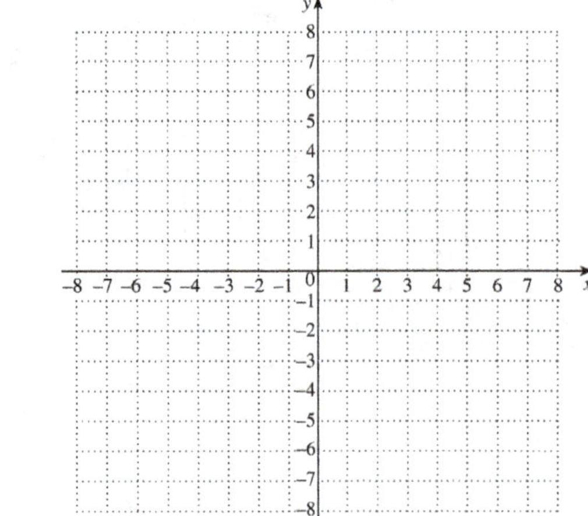

5. vertex _______________

   axis _______________

   domain _______________

   range _______________

6. $f(x) = -2x^2 + 4x + 1$

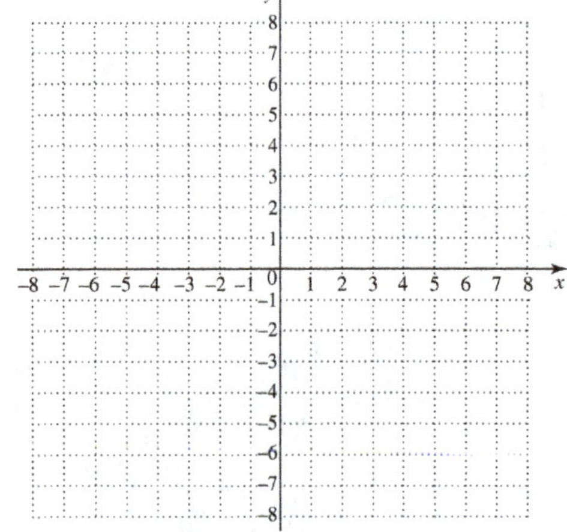

6. vertex _______________

   axis _______________

   domain _______________

   range _______________

              Copyright © 2020 Pearson Education, Inc.

Objective 3   Use the discriminant to find the number of x-intercepts of a parabola with a vertical axis.

Review this example for Objective 3:

5. Use the discriminant to determine the number of x-intercepts of the graph of the quadratic function.

$$f(x) = 4x^2 + 12x + 9$$

$$b^2 - 4ac = 12^2 - 4(4)(9) = 0$$

Since the discriminant is zero, the graph has only one x-intercept, its vertex.

Now Try:

5. Use the discriminant to determine the number of x-intercepts of the graph of the quadratic function.

$$f(x) = 9x^2 - 24x + 16$$

____

Objective 3 Practice Exercises

For extra help, see Example 5 on page 583 of your text.

Use the discriminant to determine the number of x-intercepts of the graph of each function.

7. $f(x) = 2x^2 - 3x + 2$

7. ______________

8. $f(x) = -3x^2 - x + 5$

8. ______________

9. $f(x) = 3x^2 - 6x + 3$

9. ______________

Objective 4   Use quadratic functions to solve problems involving maximum or minimum value.

Review these examples for Objective 4:

6. A farmer has 1000 yards of fencing to enclose a rectangular field. What is the largest area that the farmer can enclose? What are the dimensions of the field when the area is maximized?

If the length of the field is represented by l and the width of the field is represented by w, the perimeter is given by $2l + 2w = 1000$ or $l + w = 500$ or $w = 500 - l$.

The area is given by $A = lw$ or

$$A(l) = l(500 - l) = 500l - l^2 = -l^2 + 500l.$$

Now Try:

6. A farmer has 1000 yards of fencing to enclose a rectangular field next to a building. What is the largest area that the farmer can enclose? What are the dimensions of the field when the area is maximized?

____

This is a quadratic equation, so its maximum occurs at the vertex of its graph. The

x-coordinate is given by $\dfrac{-b}{2a} = \dfrac{-500}{2(-1)} = 250.$

The y-coordinate is

$A(250) = -250^2 + 500(250) = 62,500.$

If $l = 250$, then $w = 500 - 250 = 250$. Therefore, the maximum area is 62,500 sq yd when the length of the field is 250 yd and the width is 250 yd.

7. An object is launched directly upward at 64 feet per second from a platform 80 feet high. Its height above the ground is given by

$s(t) = -16t^2 + 64t + 80,$ where t is the number of seconds after launch. What will be the object's maximum height? When will it reach this height?

For this function, $a = -16$, $b = 64$, and $c = 80$.

The vertex formula gives $t = \dfrac{-b}{2a} = \dfrac{-64}{2(-16)} = 2.$

This indicates that the maximum height is reached at 2 seconds. Now calculate $s(2)$ to find the maximum height.

$s(2) = -16(2)^2 + 64(2) + 80 = 144$

Thus, it takes 2 seconds to reach the maximum height of 144 feet above the ground.

7. An object is launched directly upward at 48 feet per second from a platform 250 feet high. Its height above the ground is given by

$s(t) = -16t^2 + 48t + 250$

where t is the number of seconds after launch. What will be the object's maximum height? When will it reach this height?

____

Objective 4 Practice Exercises

For extra help, see Examples 6–7 on pages 584–585 of your text.

Solve each problem.

10. Jean sells ceramic pots. She has weekly costs of

$C(x) = x^2 - 100x + 2700,$ where x is the number of pots she sells each week. How many pots should she sell to minimize her costs? What is the minimum cost?

10. units _______________

cost _______________

11. The length and width of a rectangle have a sum of 48. What width will produce the maximum area?

11. ______________________

12. A projectile is fired upward so that its distance (in feet) above the ground t seconds after firing is given by $s(t) = -16t^2 + 80t + 156$. Find the maximum height it reaches and the number of seconds it takes to reach that height.

12. height   ______________

time______________

Objective 5   Graph parabolas with horizontal axes.

Review this example for Objective 5:	**Now Try:**

9. Graph $x = -y^2 + 6y - 9$. Give the vertex, axis, domain, and range.

We must complete the square in order to write the equation in $x = (y - k)^2 + h$ form.

$$x = -(y^2 - 6y) - 9$$

$$x = -(y^2 - 6y + 9 - 9) - 9$$

$$x = -(y^2 - 6y + 9) - 1(-9) - 9$$

$$x = -(y - 3)^2$$

The vertex is (0, 3). The axis is $y = 3$.

x	y
-4	1
0	3
-4	5

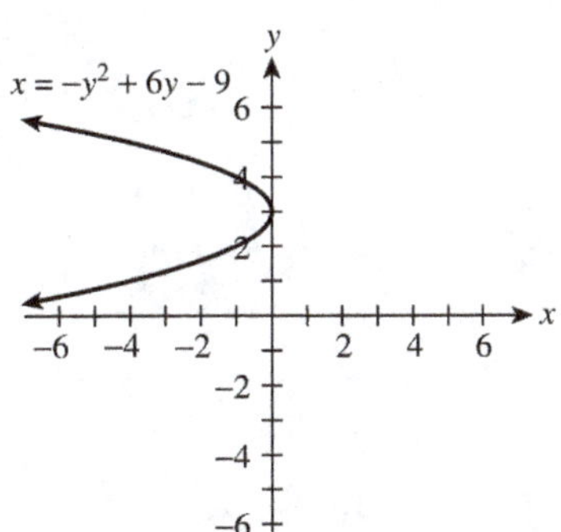

domain: $(-\infty, 0]$

range: $(-\infty, \infty)$

9. Graph $x = -y^2 + 4y - 4$. Give the vertex, axis, domain, and range.

Vertex   ______________

Axis   ______________

Domain   ______________

Range   ______________

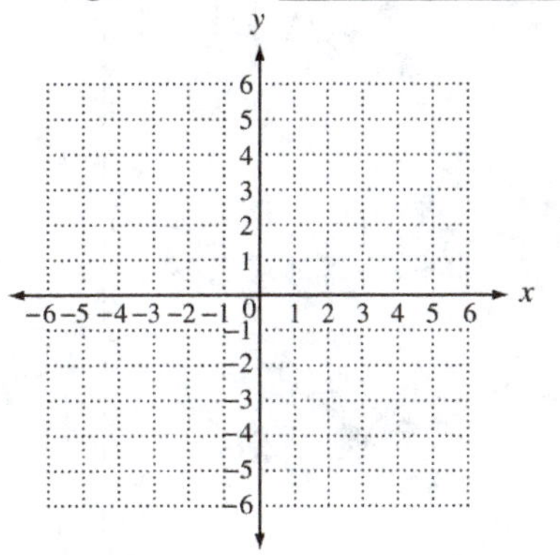

Name: Date:
Instructor: Section:

Objective 5 Practice Exercises

For extra help, see Examples 8–9 on pages 585–586 of your text.

Sketch the graph of each parabola. Give the vertex, axis, domain, and range.

13. $x = -y^2 + 2$

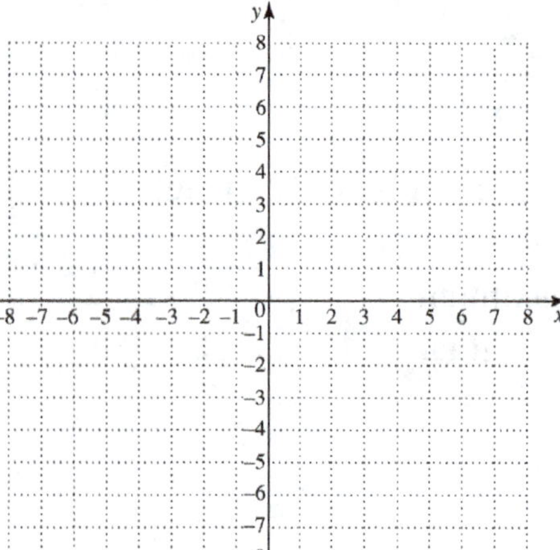

13. vertex __________

axis __________

domain __________

range __________

14. $x = y^2 - 3$

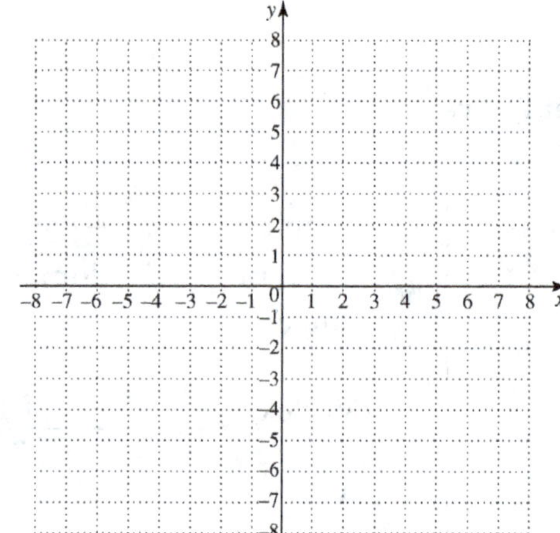

14. vertex __________

axis __________

domain __________

range __________

15. $x = -y^2 - 6y - 10$

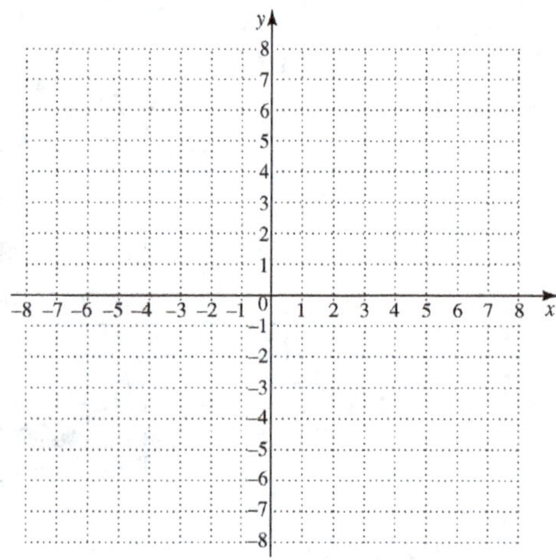

15. vertex __________

axis __________

domain __________

range __________

Chapter 8 QUADRATIC EQUATIONS, INEQUALITIES, AND FUNCTIONS

8.7   Polynomial and Rational Inequalities

Learning Objectives
1   Solve quadratic inequalities.
2   Solve polynomial inequalities of degree 3 or greater.
3   Solve rational inequalities.

Key Terms

Use the vocabulary terms listed below to complete each statement in exercises 1–2.

> **quadratic inequality**   **rational inequality**

1.   An inequality that involves a rational expression is a _______________________.

2.   An inequality that can be written in the form $ax^2 + bx + c < 0$ or $ax^2 + bx + c > 0$, where a, b, and c are real numbers with $a \neq 0$ is called a

_______________________________.

Objective 1   Solve quadratic inequalities.

Review these examples for Objective 1:

1.   Solve each inequality.

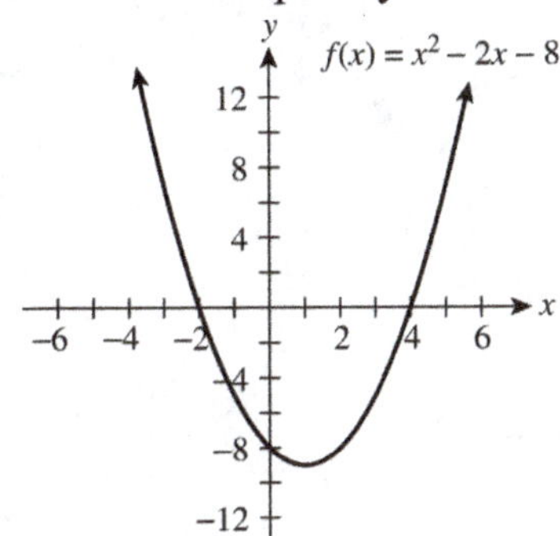

a. $x^2 - 2x - 8 > 0$

From the graph, we see that the y-values are greater than 0 when the x-values are less than –2 or greater than 4. Therefore, the solution set of

$x^2 - 2x - 8 > 0$ is $(-\infty, -2) \cup (4, \infty)$.

b. $x^2 - 2x - 8 < 0$

From the graph, we see that the y-values are less than 0 when the x-values are greater than –2 and less than 4. Therefore, the solution set of

$x^2 - 2x - 8 < 0$ is $(-2, 4)$.

Now Try:

1.   Solve each inequality.

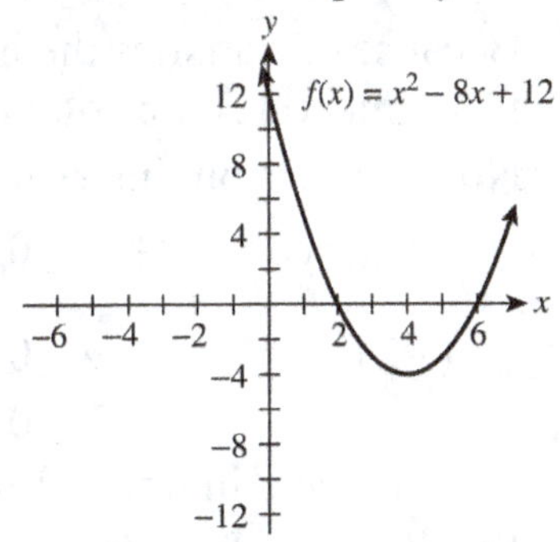

a. $x^2 - 8x + 12 > 0$

____

b. $x^2 - 8x + 12 < 0$

____

2. Solve and graph the solution set of $x^2 + 5x + 4 \geq 0$.

Solve the quadratic equation by factoring.
$$(x+1)(x+4) = 0$$
$$x+1 = 0 \quad \text{or} \quad x+4 = 0$$
$$x = -1 \quad \text{or} \quad x = -4$$

The numbers –4 and –1 divide a number line into intervals A, B, and C, as shown below.

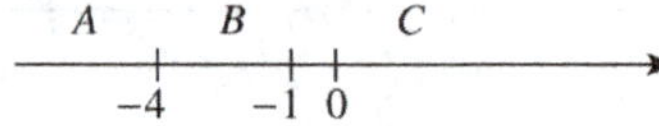

Since the numbers –4 and –1 are the only numbers that make the quadratic expression $x^2 + 5x + 4$ equal to 0, all other numbers make the expression either positive or negative. If one number in an interval satisfies the inequality, then all the numbers in that interval will satisfy the inequality.

Choose any number in interval A as a test number; we will choose –5.
$$x^2 + 5x + 4 \geq 0$$
$$(-5)^2 + 5(-5) + 4 \overset{?}{\geq} 0$$
$$4 \geq 0 \quad \text{True}$$

Because –5 satisfies the inequality, all numbers from interval A are solutions.

Now try –2 from interval B.
$$x^2 + 5x + 4 \geq 0$$
$$(-2)^2 + 5(-2) + 4 \overset{?}{\geq} 0$$
$$-2 \geq 0 \quad \text{False}$$

The numbers in interval B are not solutions.
Finally, try 0 from interval C.
$$x^2 + 5x + 4 \geq 0$$
$$0^2 + 5(0) + 4 \overset{?}{\geq} 0$$
$$4 \geq 0 \quad \text{True}$$

Because 0 satisfies the inequality, all numbers from interval C are solutions.

Because the inequality is greater than or equal to zero, we include the endpoints of the intervals in the solution set. Thus, the solution set is $(-\infty, -4] \cup [-1, \infty)$.

2. Solve and graph the solution set of $x^2 - x - 2 < 0$.

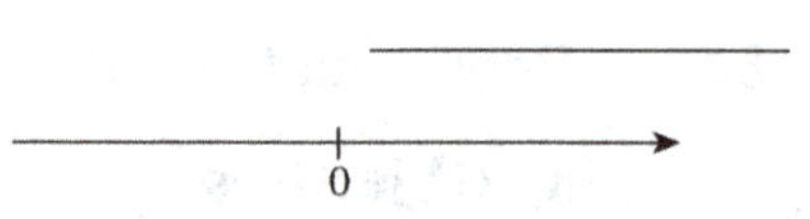

4. Solve each inequality.

    a. $(2k+5)^2 \geq -1$

Because $(2k+5)^2$ is never negative, it is always greater than -1. The solution set is $(-\infty,\ \infty)$.

    b. $(2k+5)^2 \leq -1$

Because $(2k+5)^2$ is never negative, there is no solution. The solution set is $\varnothing$.

4. Solve each inequality.

    a. $(4m+1)^2 \geq -3$

____

    b. $(4m+1)^2 \leq -3$

____

Objective 1 Practice Exercises

For extra help, see Examples 1–4 on pages 591–594 of your text.

Solve each inequality, and graph the solution set.

1. $a^2 - a - 2 \leq 0$

1. _______________________

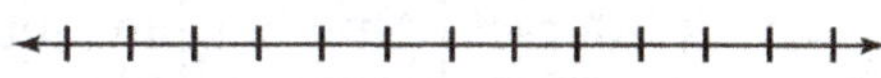

2. $8k^2 + 10k > 3$

2. _______________________

3. $(3x-2)^2 < -1$

3. _______________________

Objective 2   Solve polynomial inequalities of degree 3 or greater.

Review this example for Objective 2: | **Now Try:**

5. Solve and graph the solution set of
$(x+1)(x-2)(x+4)\le 0.$

Set the factored polynomial equal to 0, then use the zero-factor property.

$x+1=0$ or $x-2=0$ or $x+4=0$

$x=-1$ or $x=2$ or $x=-4$

Locate -4, -1, and 2 on a number line to determine the intervals A, B, C, and D.

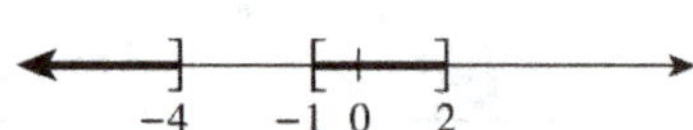

Substitute a test number from each interval in the original inequality to determine which intervals satisfy the inequality.

Interval	Test Number	Test of inequality	True or False?
A	-5	$-28\le 0$	T
B	-2	$8\le 0$	F
C	0	$-8\le 0$	T
D	5	$162\le 0$	F

The numbers in intervals A and C are in the solution set. The three endpoints are included in the solution set since the inequality symbol, $\le$, includes equality. Thus, the solution set is $(-\infty,-4]\cup[-1,\ 2].$

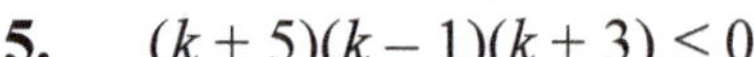

Now Try:

5. Solve and graph the solution set of $(2x-1)(2x+3)(3x+1)\le 0.$

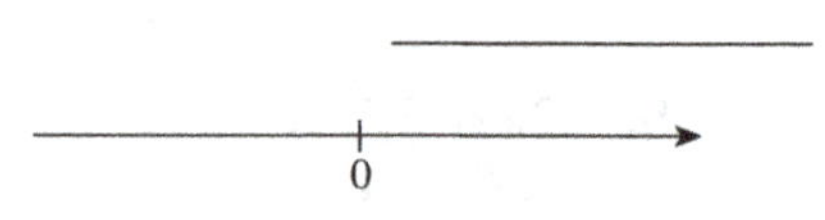

Objective 2 Practice Exercises

For extra help, see Example 5 on pages 594–595 of your text.

Solve each inequality, and graph the solution set.

4. $(y+2)(y-1)(y-2)<0$

4. ______________________

5. $(k+5)(k-1)(k+3)\le 0$

5. ______________________

          Copyright © 2020 Pearson Education, Inc.

6. $(x-1)(x-3)(x+2) \geq 0$

6. ________________________

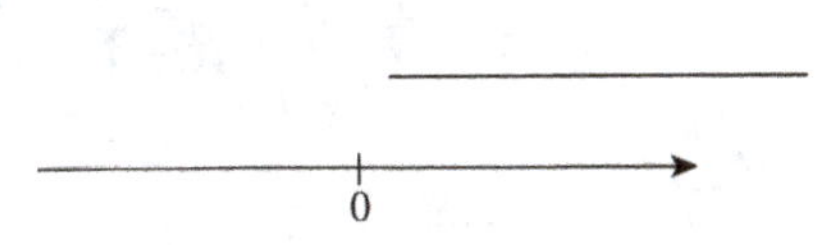

Objective 3   Solve rational inequalities.

Review these examples for Objective 3:

6. Solve and graph the solution set of $\dfrac{7}{x-1} < 1$.

Write the inequality so that 0 is on one side.

$$\dfrac{7}{x-1} - 1 < 0$$

$$\dfrac{7}{x-1} - \dfrac{x-1}{x-1} < 0 \qquad \text{The LCD is } x-1.$$

$$\dfrac{7-x+1}{x-1} < 0$$

$$\dfrac{8-x}{x-1} < 0$$

The sign of $\dfrac{8-x}{x-1}$ will change from positive to negative or negative to positive only at those numbers that make the numerator or denominator 0. These two numbers, 1 and 8, divide a number line into three intervals.

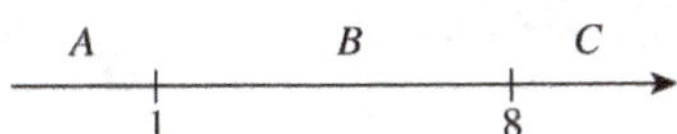

Test a number in each interval using the original inequality.

Interval	Test Number	Test of inequality	True or False?
A	0	$-7 < 1$	T
B	2	$7 < 1$	F
C	10	$\dfrac{7}{9} < 1$	T

The solution set is $(-\infty,\ 1) \cup (8,\ \infty)$. This interval does not include 1 because it would make the denominator of the original inequality 0. The number 8 is not included because the inequality symbol, $<$, does not include equality.

Now Try:

6. Solve and graph the solution set of $\dfrac{y}{y+1} > 3$.

____

7. Solve and graph the solution set of $\dfrac{x+1}{x-5} \geq 3$.

Write the inequality so that 0 is on one side.

$$\frac{x+1}{x-5} - 3 \geq 0$$

$$\frac{x+1}{x-5} - \frac{3(x-5)}{x-5} \geq 0$$

$$\frac{x+1}{x-5} - \frac{3x-15}{x-5} \geq 0$$

$$\frac{x+1-3x+15}{x-5} \geq 0$$

$$\frac{-2x+16}{x-5} \geq 0$$

The sign of $\dfrac{-2x+16}{x-5}$ will change from positive to negative or negative to positive only at those numbers that make the numerator or denominator 0. These two numbers, 5 and 8, divide a number line into three intervals.

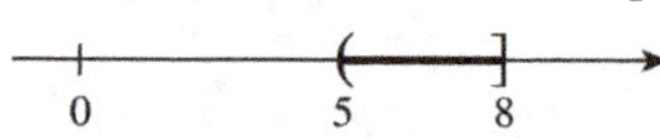

Test a number in each interval using original inequality.

Interval	Test Number	Test of inequality	True or False?
A	0	$-\dfrac{1}{5} \geq 3$	F
B	6	$7 \geq 3$	T
C	10	$\dfrac{11}{5} \geq 3$	F

The solution set is $(5,\ 8]$. This interval does not include 5 because it would make the denominator of the original inequality 0. The number 8 is included because the inequality symbol, $\geq$, does includes equality.

7. Solve and graph the solution set of $\dfrac{z+2}{z-3} \leq 2$.

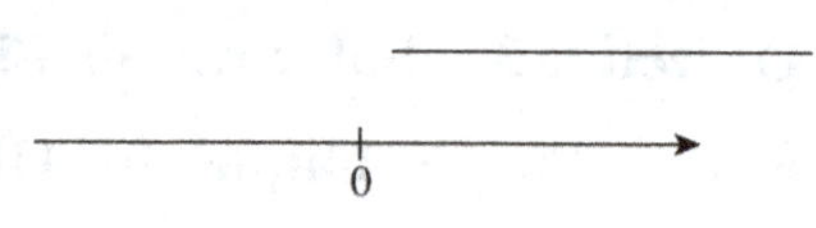

  Copyright © 2020 Pearson Education, Inc.

Objective 3 Practice Exercises

For extra help, see Examples 6–7 on pages 595–597 of your text.

Solve each inequality, and graph the solution set.

7. $\dfrac{7}{x-1} \le 1$

7. ______________________

8. $\dfrac{2p-1}{3p+1} \le 1$

8. ______________________

9. $\dfrac{5}{x-3} \le -1$

9. ______________________

Chapter 9 INVERSE, EXPONENTIAL, AND LOGARITHMIC FUNCTIONS

9.1 Inverse Functions

Learning Objectives
1   Decide whether a function is one-to-one and, if it is, find its inverse.
2   Use the horizontal line test to determine whether a function is one-to-one.
3   Find the equation of the inverse of a function.
4   Graph f^{-1} from the graph of f.

Key Terms

Use the vocabulary terms listed below to complete each statement in exercises 1−2.

> **one-to-one function**   **inverse of a function** f

1. A function in which each x-value corresponds to just one y-value and each y-value corresponds to just one x-value is a(n) _______________________________.

2. If f is a one-to-one function, the ____ is the set of all ordered pairs of the form (y, x) where (x, y) belongs to f.

Objective 1 Decide whether a function is one-to-one and, if it is, find its inverse.

Review these examples for Objective 1:

1. Find the inverse of each function that is one-to-one.

   a. $G = \{(-3,-1), (-2, 0), (-1, 1), (0, 2)\}$

   Every x-value in G corresponds to only one y-value, and every y-value corresponds to only one x-value, so G is a one-to-one function.
   The inverse function is found by interchanging the x- and y-values in each ordered pair.

   $$G^{-1} = \{(-1,-3), (0,-2), (1,-1), (2, 0)\}$$

   b. $F = \{(2, 1), (-1, 1), (0, 0), (1, 1)\}$

   Every x-value in F corresponds to only one y-value. However, the y-value 1 corresponds to two x-values, so F is not a one-to-one function.

Now Try:

1. Find the inverse of each function that is one-to-one.

   a. $G = \{(3, 2), (-3,-2), (2, 3), (-2,-3)\}$

   ____

   b. $F = \{(2, 4), (-1, 1), (0, 0), (1, 1), (2, 6)\}$

   ____

c.

State	Number of National Parks
AK	8
AZ	3
CA	8
CO	4
FL	3
HI	2
UT	5

c.

State	Number of representatives
AK	1
AZ	8
CA	53
FL	25
NY	29
DE	1

Let N be the function defined in the table, with the states forming the domain and the number of national parks forming the range. Then, N is not one-to-one, because two different states have the same number of national parks.

____

Objective 1 Practice Exercises

For extra help, see Example 1 on page 613 of your text.

If the function is one-to-one, find its inverse.

1.   {(–3,–1), (–2, 2), (–1, 3), (0, 4)}   **1.** ________________

2.   {(1, 0), (2, 0), (3, 5), (4, 1)}   **2.** ________________

3.   {(0, 0), (1, 1), (–1,–1), (2, 2), (–2,–2)}   **3.** ________________

Name: Date:
Instructor: Section:

Objective 2   **Use the horizontal line test to determine whether a function is one-to-one.**

Review these examples for Objective 2:

2. Use the horizontal line test to determine whether each graph is the graph of a one-to-one function.

a.

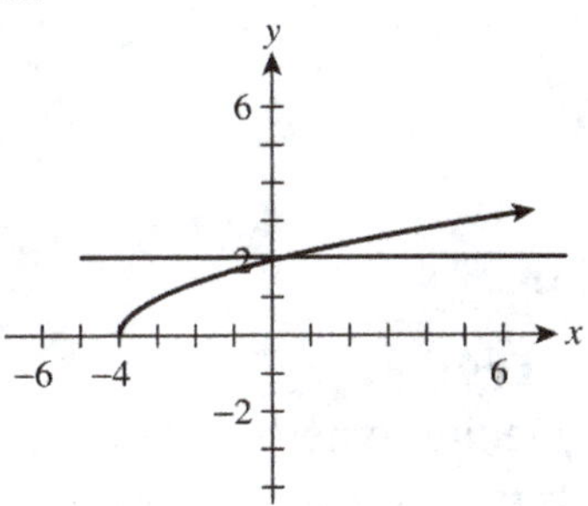

Every horizontal line will intersect the graph in exactly one point. The function is one-to-one.

b.

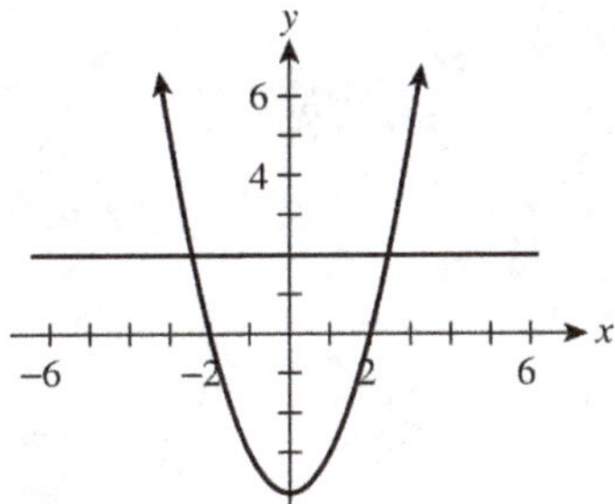

Because a horizontal line intersects the graph in more than one point, the function is not one-to-one.

Now Try:

2. Use the horizontal line test to determine whether each graph is the graph of a one-to-one function.

a.

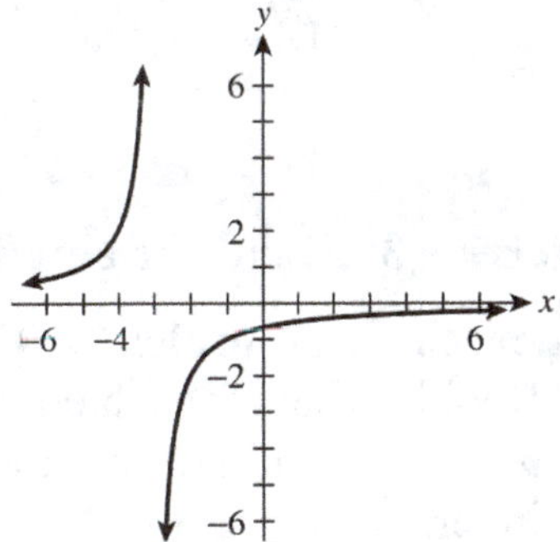

____

b.

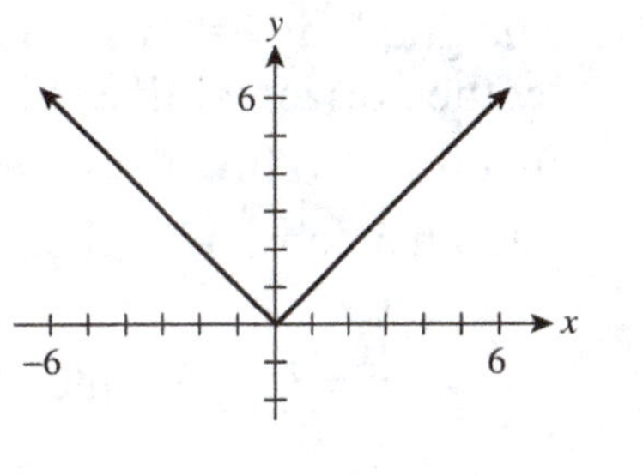

____

Objective 2 Practice Exercises

For extra help, see Example 2 on page 614 of your text.

Use the horizontal line test to determine whether each function is one-to-one.

4.

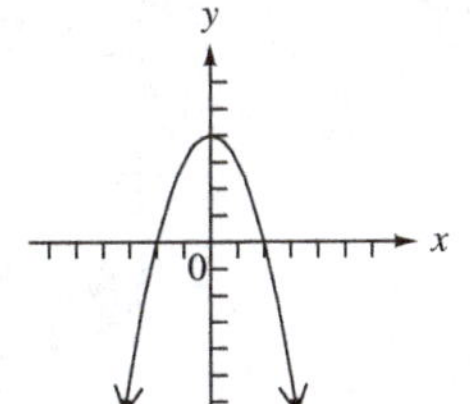

4. _______________

5.

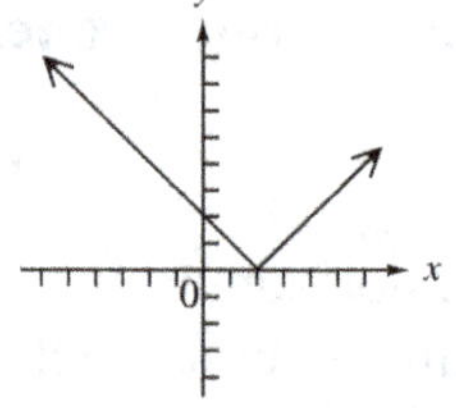

5. ______________

6.

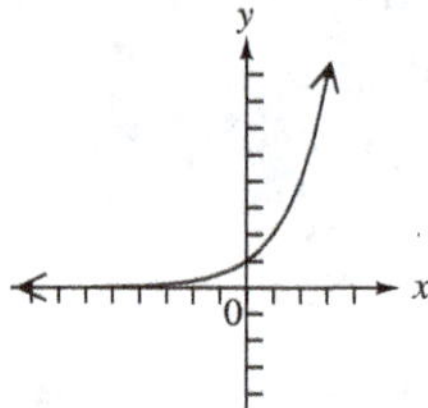

6. ______________

Objective 3   Find the equation of the inverse of a function.

Review these examples for Objective 3:

3. Decide whether each equation represents a one-to-one function. If so, find the equation for the inverse.

a. $f(x) = 3x - 5$

The graph of $y = 3x - 5$ is a nonvertical line, so by the horizontal line test, f is a one-to-one function. To find the inverse, let $y = f(x)$, interchange x and y, then solve for y.

$$y = 3x - 5$$

$$x = 3y - 5 \quad \text{Interchange } x \text{ and } y.$$

$$x + 5 = 3y$$

$$\frac{x + 5}{3} = y$$

$$f^{-1}(x) = \frac{x + 5}{3} = \frac{x}{3} + \frac{5}{3}$$

$$f^{-1}(x) = \frac{1}{3}x + \frac{5}{3}$$

b. $f(x) = 2x^2 + 3$

The graph of $y = 2x^2 + 3$ is a vertical parabola, so by the horizontal line test, f is not a one-to-one function and does not have an inverse.

Now Try:

3. Decide whether each equation represents a one-to-one function. If so, find the equation for the inverse.

a. $f(x) = 4x - 1$

____

b. $f(x) = -\frac{3}{2}x^2$

____

c. $f(x) = x^3 + 1$ **c.** $f(x) = 2x^3 - 3$

The graph of $y = x^3 + 1$ is a cubing function.
The function is one-to-one and has an inverse.

$$y = x^3 + 1$$

$$x = y^3 + 1 \quad \text{Interchange } x \text{ and } y.$$

$$x - 1 = y^3$$

$$\sqrt[3]{x - 1} = y$$

$$f^{-1}(x) = \sqrt[3]{x - 1}$$

____

Objective 3 Practice Exercises

For extra help, see Examples 3–4 on pages 614–616 of your text.

If the function is one-to-one, find its inverse.

7. $f(x) = 2x - 5$ 7. _________________

8. $f(x) = x^3 - 1$ 8. _________________

9. $f(x) = x^2 - 1$ 9. _________________

Objective 4   Graph f^{-1} from the graph of f.

Review this example for Objective 4:

5.   Use the given graph to graph the inverse of f.

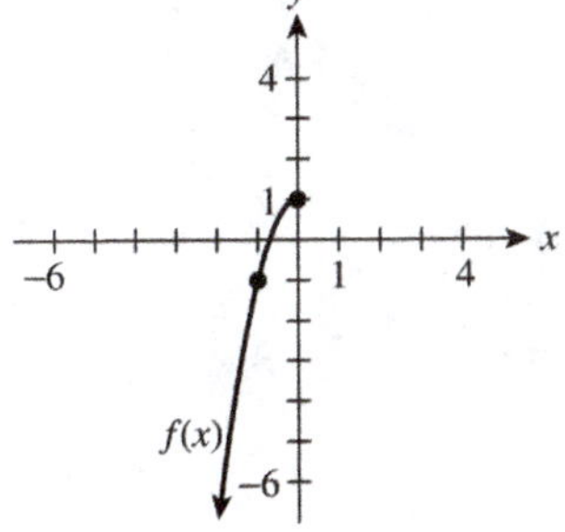

Now Try:

5.   Use the given graph to graph the inverse of f.

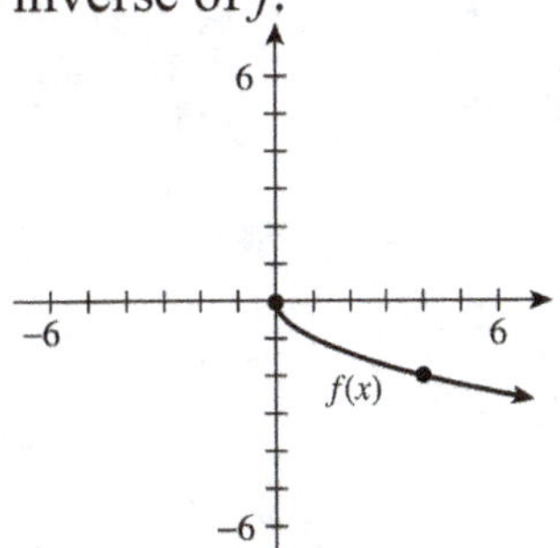

We can find the graph of f^{-1} from the graph of f
by locating the mirror image of each point in f
with respect to the line $y = x$.

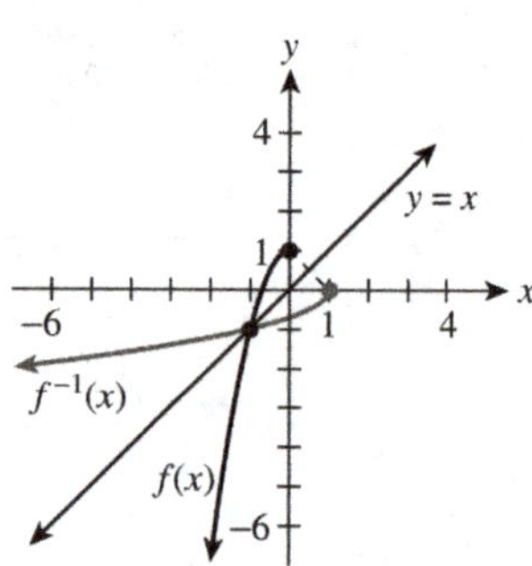

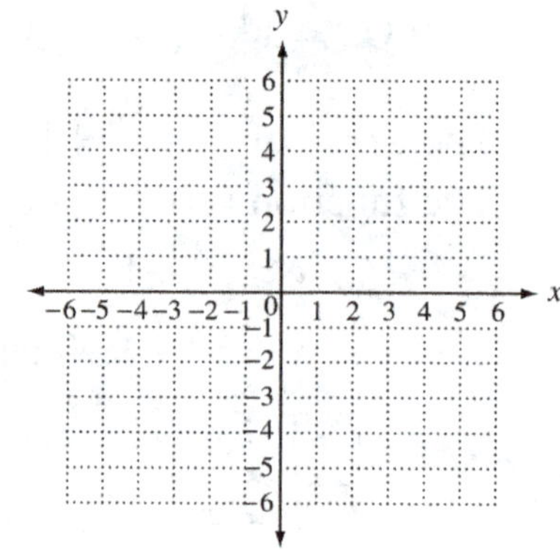

Objective 4 Practice Exercises

For extra help, see Example 5 on page 617 of your text.

If the function is one-to-one, graph the function f and its inverse f^{-1} on the same set of axes.

10.

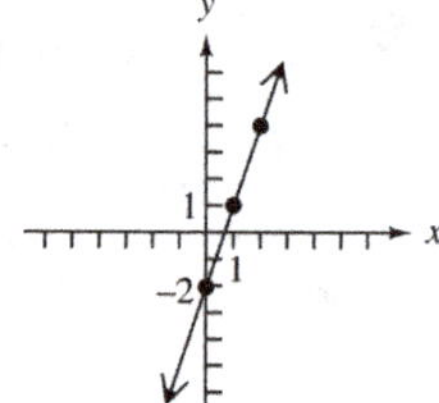

10. _______________

11.

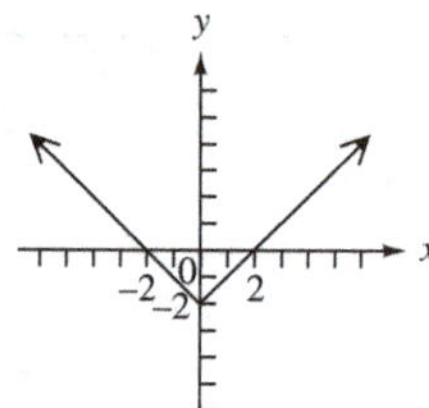

11. _______________

12.

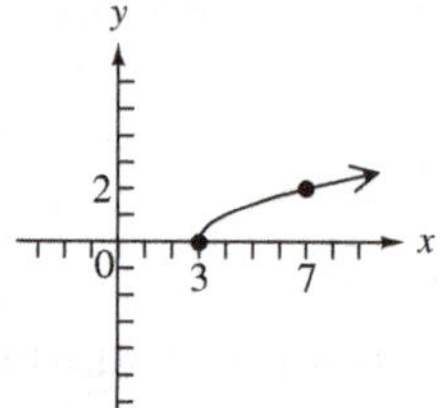

12. _______________

         Copyright © 2020 Pearson Education, Inc.

Chapter 9   INVERSE, EXPONENTIAL, AND LOGARITHMIC FUNCTIONS

9.2   Exponential Functions

Learning Objectives

1. Evaluate exponential expression using a calculator.
2. Define and graph exponential functions.
3. Solve exponential equations of the form $a^x = a^k$ for x.
4. Use exponential functions in applications involving growth or decay.

Key Terms

Use the vocabulary terms listed below to complete each statement in exercises 1–2.

exponential equation   **inverse**

1. If f is a one-to-one function, then the ________________________ of f is the set of all ordered pairs formed by interchanging the coordinates of the ordered pairs of f.

2. An equation that has a variable as an exponent, is an ________________________.

Objective 1   Evaluate exponential expression using a calculator.

Review these examples for Objective 1:

1. Use a calculator to approximate each exponential expression to three decimal places.

     a. $3^{1.8}$

     $3^{1.8} \approx 7.225$

     b. $3^{-1.4}$

     $3^{-1.4} \approx 0.215$

     c. $3^{1/4}$

     $3^{1/4} \approx 1.316$

Now Try:

1. Use a calculator to approximate each exponential expression to three decimal places.

     a. $3^{1.9}$

     ____

     b. $3^{-1.6}$

     ____

     c. $3^{1/5}$

     ____

Name: _______________________   Date: _______________________

Instructor: _______________________   Section: _______________________

Objective 1 Practice Exercises

For extra help, see Example 1 on page 621 of your text.

Use a calculator to find an approximation to three decimal places for each exponential expression.

1. $3^{1.2}$

1. _______________________

2. $3^{-1.2}$

2. _______________________

3. $3^{1/3}$

3. _______________________

Objective 2   Define and graph exponential functions.

Review these examples for Objective 2:

2. Graph $f(x) = 6^x$.

   Create a table of values, then plot the points and draw a smooth curve through them.

   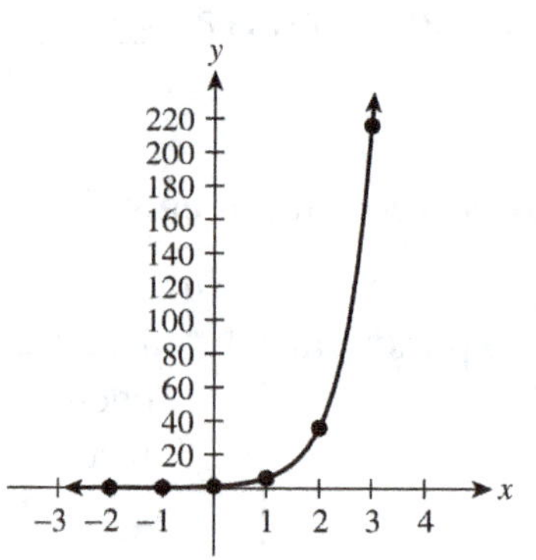

x	$f(x) = 6^x$
-2	$\dfrac{1}{36}$
-1	$\dfrac{1}{6}$
0	1
1	6
2	36
3	216

3. Graph $f(x) = \left(\dfrac{1}{6}\right)^x$.

   Create a table of values, then plot the points and draw a smooth curve through them.

   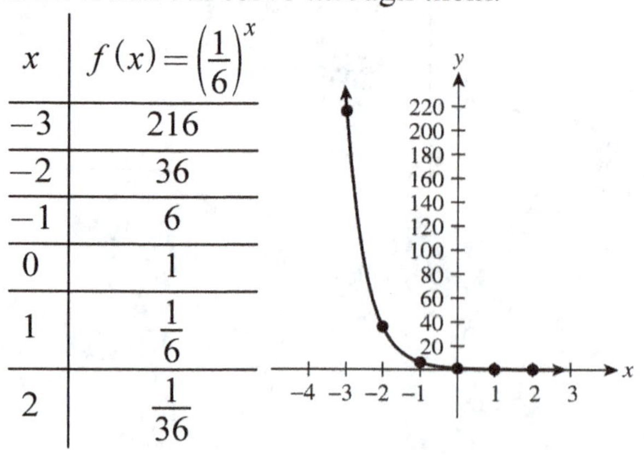

x	$f(x) = \left(\dfrac{1}{6}\right)^x$
-3	216
-2	36
-1	6
0	1
1	$\dfrac{1}{6}$
2	$\dfrac{1}{36}$

Now Try:

2. Graph $f(x) = 3^x$.

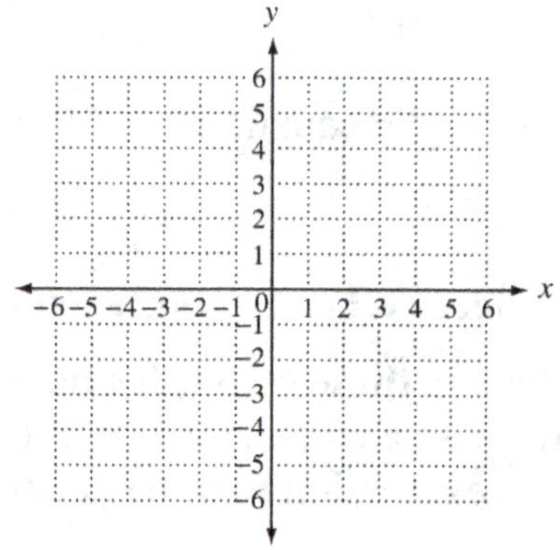

3. Graph $f(x) = \left(\dfrac{1}{3}\right)^x$.

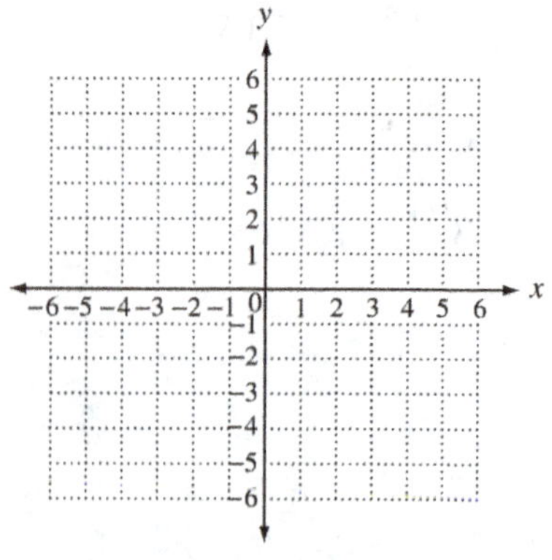

4. Graph $f(x)=3^{2x-1}$.

Create a table of values, then plot the points and draw a smooth curve through them.

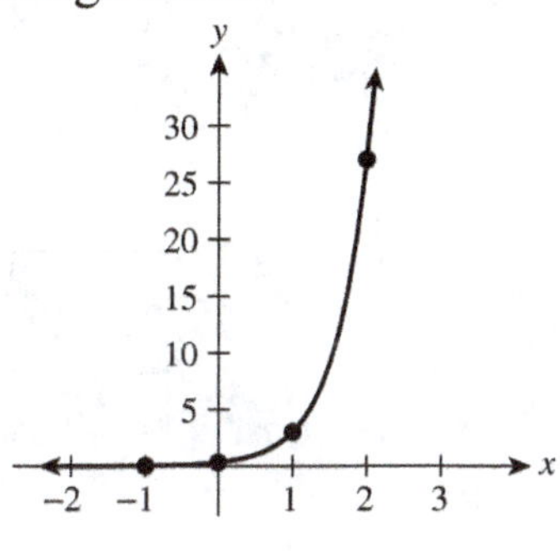

x	$2x-1$	$f(x)=3^{2x-1}$
-1	-3	$\dfrac{1}{27}$
0	-1	$\dfrac{1}{3}$
1	1	3
2	3	27

4. Graph $f(x)=2^{1-x}$.

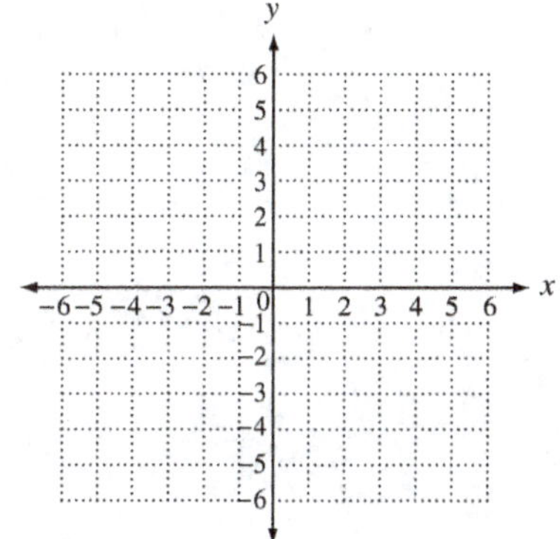

Objective 2 Practice Exercises

For extra help, see Examples 2–4 on pages 622–623 of your text.

Graph each exponential function.

4. $f(x)=2^{-x}$

4.

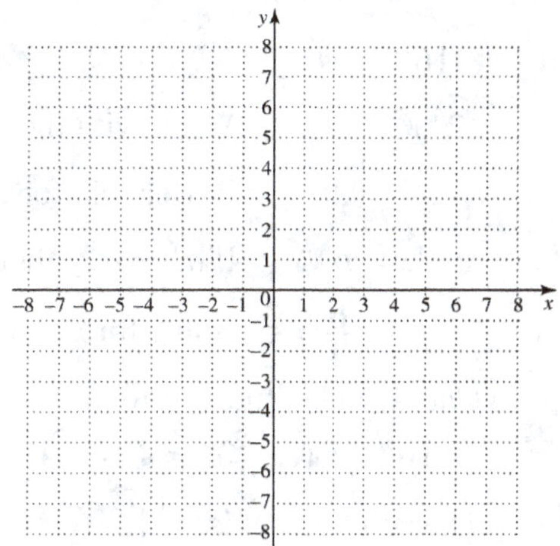

5. $f(x)=\left(\dfrac{1}{8}\right)^x$

5.

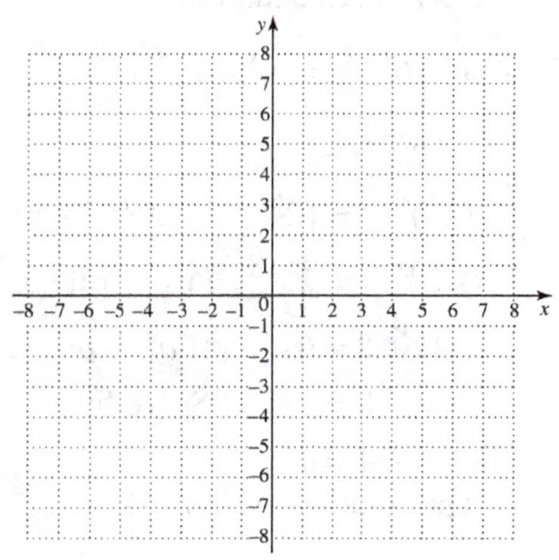

6. $f(x) = 4^{2x-3}$

6.

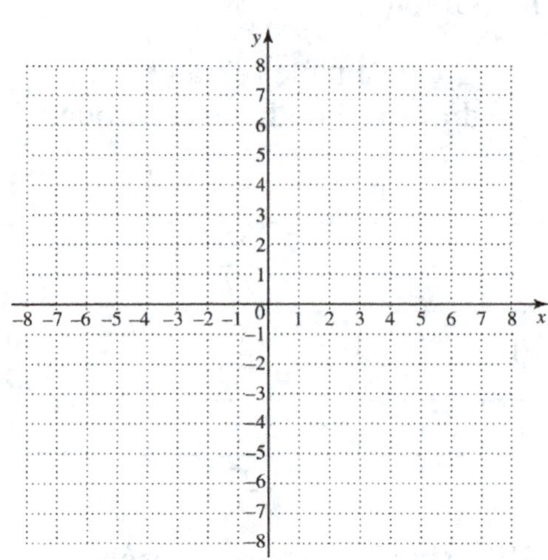

Objective 3 Solve exponential equations of the form $a^x = a^k$ for x.

Review these examples for Objective 3:

5. Solve the equation $16^x = 64$.

$$16^x = 64$$

$$(2^4)^x = 2^6 \qquad \text{Write with the same base.}$$

$$2^{4x} = 2^6 \qquad \text{Power rule for exponents}$$

$$4x = 6 \qquad \text{If } a^x = a^y, \text{ then } x = y.$$

$$x = \frac{6}{4} = \frac{3}{2} \quad \text{Solve for } x; \text{ simplify.}$$

Check: Substitute 3/2 for x.

$$16^{3/2} = (16^{1/2})^3 = 4^3 = 64$$

The solution set is $\left\{\frac{3}{2}\right\}$.

6. Solve each equation.

a. $16^{x-2} = 64^x$

$$16^{x-2} = 64^x$$

$$(2^4)^{x-2} = (2^6)^x \quad \text{Write with the same base.}$$

$$2^{4x-8} = 2^{6x} \quad \text{Power rule for exponents}$$

$$4x - 8 = 6x \qquad \text{If } a^x = a^y, \text{ then } x = y.$$

$$-8 = 2x \qquad \text{Solve for } x.$$

$$-4 = x$$

The solution set is $\{-4\}$.

Now Try:

5. Solve the equation $25^x = 125$.

____

6. Solve each equation.

a. $4^{x-1} = 8^x$

____

b. $4^x = \dfrac{1}{64}$ **b.** $3^x = \dfrac{1}{243}$

$$4^x = \frac{1}{64}$$

$$4^x = \frac{1}{4^3} \qquad 64 = 4^3$$ ______________

$4^x = 4^{-3}$ Write with the same base.

$x = -3$ Set exponents equal.

The solution set is $\{-3\}$.

c. $\left(\dfrac{2}{5}\right)^x = \dfrac{125}{8}$ **c.** $\left(\dfrac{3}{2}\right)^x = \dfrac{16}{81}$

$$\left(\frac{2}{5}\right)^x = \frac{125}{8}$$

$$\left(\frac{2}{5}\right)^x = \left(\frac{8}{125}\right)^{-1}$$ ______________

$\left(\dfrac{2}{5}\right)^x = \left[\left(\dfrac{2}{5}\right)^3\right]^{-1}$ Write with the same base.

$\left(\dfrac{2}{5}\right)^x = \left(\dfrac{2}{5}\right)^{-3}$ Power rule for exponents

$x = -3$ Set exponents equal.

The solution set is $\{-3\}$.

Objective 3 Practice Exercises

For extra help, see Examples 5–6 on pages 624–625 of your text.

Solve each equation.

7. $25^{1-t} = 5$ **7.** ______________

8. $8^{2x+1} = 4^{4x}$ **8.** ______________

9. $\left(\dfrac{3}{4}\right)^x = \dfrac{16}{9}$ **9.** ______________

Objective 4   Use exponential functions in applications involving growth or decay.

Review these examples for Objective 4:

7. Suppose the number of bacteria present in a certain culture after t minutes is given by the equation $Q(t) = 2500(2^{0.05t})$,

How many bacteria were present after 20 minutes?

Start with the given function. Replace t with 20.

$$Q(t) = 2500(2^{0.05t})$$

$$Q(20) = 2500(2^{0.05 \times 20})$$

$$Q(20) = 2500(2^1)$$

$$Q(20) = 5000$$

There were 5000 bacteria present after 20 minutes.

8. The amount of radioactive material in a sample is given by the function $A(t) = 90\left(\dfrac{1}{2}\right)^{t/18}$, where $A(t)$ is the amount present, in grams, t days after the initial measurement.

How many grams will be present after 3 days? Round to the nearest hundredth.

Start with the given function. Replace t with 3.

$$A(t) = 90\left(\frac{1}{2}\right)^{t/18}$$

$$A(3) = 90\left(\frac{1}{2}\right)^{3/18}$$

$$A(3) \approx 80.18$$

After 3 days, there were about 80.18 grams in the sample.

Now Try:

7. The population of Evergreen Park is now 16,000. The population t years from now is given by the formula

$$P = 16,000(2^{t/10}).$$

Using the model, what will be the population 40 years from now?

____

8. An industrial city in Ohio has found that its population is declining according to the equation $y = 70,000(2)^{-0.01x}$, where x is the time in years from 1910.

According to the model, what will the city's population be in the year 2020?

____

Objective 4 Practice Exercises

For extra help, see Examples 7–8 on pages 626–627 of your text.

Solve each problem.

10. The population of Canadian geese that spend the summer at Gemini Lake each year has been growing according to the function $f(x) = 56(2)^{0.2x}$, where x is the time in years from 1990. Find the number of geese in 2010.

10. 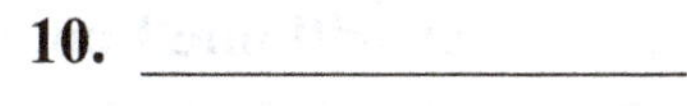 _______________

11. A sample of a radioactive substance with mass in grams decays according to the function $f(x) = 100(10)^{-0.2x}$, where x is the time in hours after the original measurement. Find the mass of the substance after 10 hours.

11.  _______________

12. A culture of a certain kind of bacteria grows according to $f(x) = 7750(x)^{0.75x}$, where x is the number of hours after 12 noon. Find the number of bacteria in the culture at 12 noon.

12.  _______________

Chapter 9   INVERSE, EXPONENTIAL, AND LOGARITHMIC FUNCTIONS

9.3   Logarithmic Functions

Learning Objectives
1   Define a logarithm.
2   Convert between exponential and logarithmic forms, and evaluate logarithms.
3   Solve logarithmic equations of the form $\log_a b = k$ for a, b, or k.
4   Use the definition of logarithm to simplify logarithmic expressions.
5   Define and graph logarithmic functions.
6   Use logarithmic functions in applications involving growth or decay.

Key Terms

Use the vocabulary terms listed below to complete each statement in exercises 1−2.

logarithm   **logarithmic equation**

1.   The ______________________________ of a positive number is the exponent indicating the power to which it is necessary to raise a given number (the base) to give the original number.

2.   An equation with a logarithm in at least one term is a ______________________________.

Objective 1   Define a logarithm.

For extra help, see pages 630–631 of your text.

Objective 2   Convert between exponential and logarithmic forms, and evaluate logarithms.

Review these examples for Objective 2:
1.

   a. Write $5^3 = 125$ in logarithmic form.

   $\log_5 125 = 3$

   b. Write $\log_{16} 4 = \dfrac{1}{2}$ in exponential form.

   $16^{1/2} = 4$

Now Try:
1.

   a. Write $8^2 = 64$ in logarithmic form.

   ____

   b. Write $\log_{16} \dfrac{1}{4} = -\dfrac{1}{2}$ in exponential form.

   ____

                Copyright © 2020 Pearson Education, Inc.

2. Use a calculator to approximate each logarithm to four decimal places.

   a. $\log_2 9$

$\log_2 9 \approx 3.1699$

   b. $\log_7 15$

$\log_7 15 \approx 1.3917$

   c. $\log_{1/3} 20$

$\log_{1/3} 20 \approx -2.7268$

   d. $\log_{10} 17$

$\log_{10} 17 \approx 1.2304$

2. Use a calculator to approximate each logarithm to four decimal places.

   a. $\log_2 6$

____

   b. $\log_4 21$

____

   c. $\log_{1/4} 25$

____

   d. $\log_{10} 12$

____

Objective 2 Practice Exercises

For extra help, see Examples 1–2 on page 631 of your text.

Write in exponential form.

   1. $\log_{10} 0.001 = -3$

1. _______________

Write in logarithmic form.

   2. $2^{-7} = \dfrac{1}{128}$

2. _______________

Use a calculator to approximate the logarithm to four decimal places.

   3. $\log_3 25$

3. _______________

Objective 3   Solve logarithmic equations of the form $\log_a b = k$ for a, b, or k.

Review these examples for Objective 3:

3. Solve each equation.

   a. $\log_{3/2} x = -2$

By definition, $\log_{3/2} x = -2$ is equivalent to

$x = \left(\dfrac{3}{2}\right)^{-2}$, and $\left(\dfrac{3}{2}\right)^{-2} = \left(\dfrac{2}{3}\right)^{2} = \dfrac{4}{9}.$ The solution

set is $\left\{\dfrac{4}{9}\right\}.$

Now Try:

2. Solve each equation.

   a. $\log_4 x = -3$

____

b. $\log_5(3x+1)=2$

$\log_5(3x+1)=2$

$\quad 3x+1=5^2$ Write in exponential form.

$\quad 3x=24$ Apply the exponent; subtract 1.

$\quad x=8$ Divide by 3.

The solution set is $\{8\}$.

c. $\log_x 6 = 2$

$\log_x 6 = 2$

$\quad x^2 = 6$ Write in exponential form.

$\quad x = \pm\sqrt{6}$ Take square root.

Only the principal square root satisfies the equation since the base must be a positive number. The solution set is $\left\{\sqrt{6}\right\}$.

d. $\log_{64}\sqrt[4]{8} = x$

$\log_{64}\sqrt[4]{8} = x$

$\quad 64^x = \sqrt[4]{8}$ Write in exponential form.

$\quad \left(8^2\right)^x = 8^{1/4}$ Write with the same base.

$\quad 8^{2x} = 8^{1/4}$ Power rule for exponents

$\quad 2x = \dfrac{1}{4}$

$\quad x = \dfrac{1}{8}$

The solution set is $\left\{\dfrac{1}{8}\right\}$.

b. $\log_9(2x+1)=2$

____

c. $\log_x 12 = 2$

____

d. $\log_{81}\sqrt[3]{9} = x$

____

Objective 3 Practice Exercises

For extra help, see Example 3 on pages 632–633 of your text.

Solve each equation.

4. $x = \log_{32} 8$

4. ________________

5. $\log_{1/3} r = -4$

5. ________________

6. $\log_a 4 = \dfrac{1}{2}$

6. ______________________

Objective 4   Use the definition of logarithm to simplify logarithmic expressions.

Review these examples for Objective 4:	**Now Try:**

4.   Use special properties to evaluate each expression.

   a. $\log_8 8$

   $\log_8 8 = 1$

   c. $\log_{64} 1$

   $\log_{64} 1 = 0$

   d. $\log_{0.2} 1$

   $\log_{0.2} 1 = 0$

   e. $\log_4 4^{11}$

   $\log_4 4^{11} = 11$

   g. $8^{\log_8 5}$

   $8^{\log_8 5} = 5$

   i. $\log_4 64$

   $\log_4 64 = \log_4 4^3 = 3$

   j. $\log_2 \dfrac{1}{2}$

   $\log_2 \dfrac{1}{2} = \log_2 2^{-1} = -1$

4.   Use special properties to evaluate each expression.

   a. $\log_4 4$

   ____

   c. $\log_{100} 1$

   ____

   d. $\log_{1/3} 1$

   ____

   e. $\log_6 6^3$

   ____

   g. $6^{\log_6 9}$

   ____

   i. $\log_5 125$

   ____

   j. $\log_5 \dfrac{1}{5}$

   ____

Objective 4 Practice Exercises

For extra help, see Example 4 on page 634 of your text.

Use the special properties to evaluate each expression.

7. $\log_{3.4} 1$

7. ______________________

8. $\log_8 8^3$

8. _______________

9. $2^{\log_2 5}$

9. _______________

Objective 5   Define and graph logarithmic functions.

Review these examples for Objective 5:

5. Graph $f(x) = \log_5 x$.

Begin by writing $y = \log_5 x$ in exponential form as $x = 5^y$. Then, create a table of values, plot the points and draw a smooth curve through them.

$x = 5^y$	y
$\dfrac{1}{5}$	-1
1	0
5	1
25	2

Now Try:

5. Graph $f(x) = \log_3 x$.

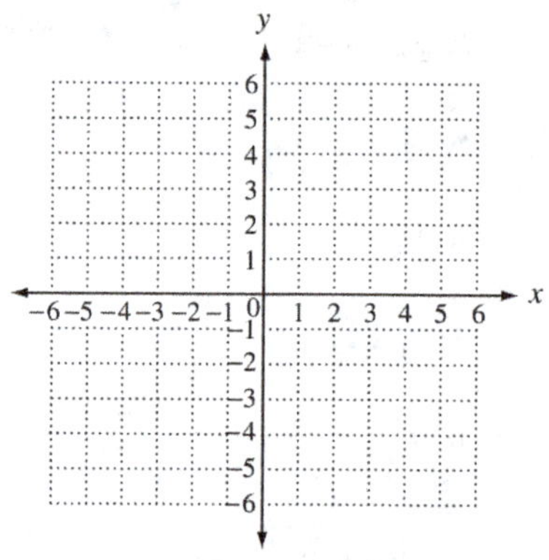

6. Graph $f(x) = \log_{1/3} x$.

Begin by writing $y = \log_{1/3} x$. in exponential form. Then, create a table of values, plot the points and draw a smooth curve through them.

$x = \left(\dfrac{1}{3}\right)^y$	y
$\dfrac{1}{3}$	1
1	0
3	-1
9	-2

6. Graph $f(x) = \log_{1/4} x$.

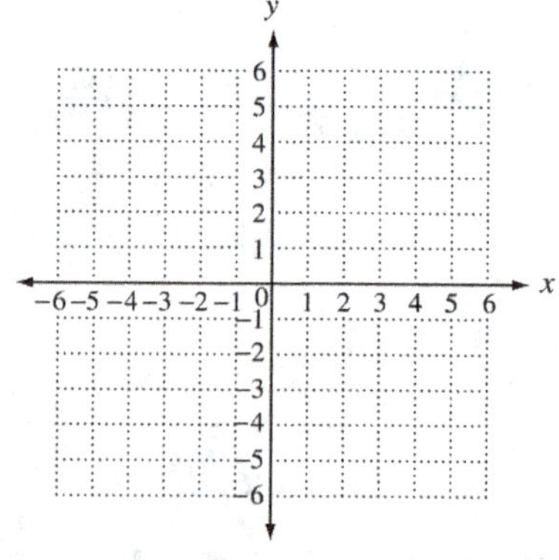

Name: Date:
Instructor: Section:

Objective 5 Practice Exercises

For extra help, see Examples 5–6 on pages 634–635 of your text.

Graph each logarithmic function.

10. $y = \log_9 x$

10.

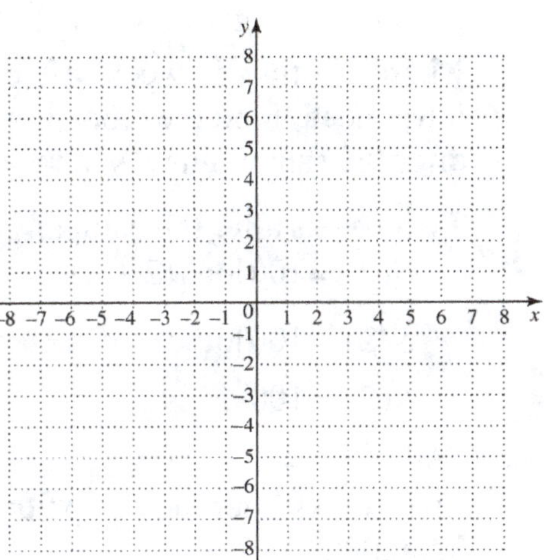

11. $y = \log_{1/4} x$

11.

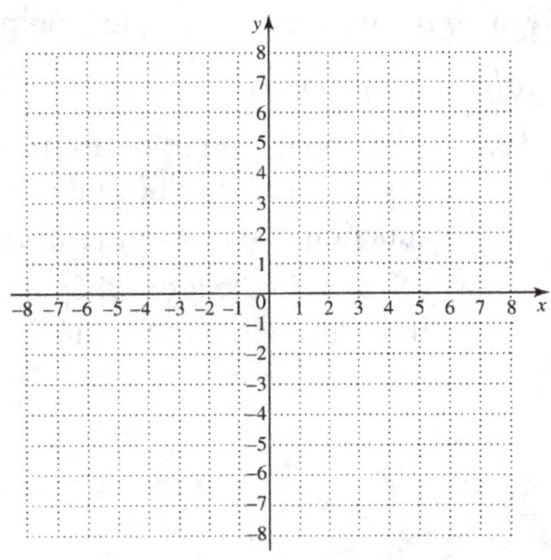

                              381

Objective 6   Use logarithmic functions in applications involving growth or decay.

Review this example for Objective 6:

Now Try:

6. A company analyst has found that total sales in thousands of dollars after a major advertising campaign are given by $S(x) = 100\log_2(x+2)$, where x is time in weeks after the campaign was introduced. Find the amount of sales two weeks after the campaign was introduced.

Two weeks after the campaign, $x = 2$, so we have
$$S(2) = 100\log_2(2+2)$$
$$S(2) = 100\log_2(2)^2$$
$$S(2) = 100\cdot 2$$
$$S(2) = 200$$
Two weeks after the campaign, sales were $200,000.

6. The number of fish in an aquarium is given by the function $f(t) = 8\log_5(2t+5)$, where t is time in months. Find the number of fish present after 10 months.

____

Objective 6 Practice Exercises

For extra help, see Example 7 on page 635 of your text.

Solve each problem.

12. The population of foxes in an area t months after the foxes were introduced there is approximated by the function $F(t) = 500\log_{10}(2t+10)$. Find the number of foxes in the area when the foxes were first introduced into the area.

12. ______________________

13. A population of mites in a laboratory is growing according to the function
$p = 50\log_3(20t+7) - 25\log_9(80t+1)$, where t is the number of days after a study is begun. Find the number of mites present 1 day after the beginning of the study.

13. ______________________

14. Sales (in thousands) of a new product are approximated by

$$S = 125 + 20\log_2(30t + 4) + 30\log_4(35t - 6),\ \text{where}$$

t is the number of years after the product is introduced. Find the total sales 2 years after the product is introduced.

14. ______________________

Chapter 9   INVERSE, EXPONENTIAL, AND LOGARITHMIC FUNCTIONS

9.4   Properties of Logarithms

Learning Objectives
1   Use the product rule for logarithms.
2   Use the quotient rule for logarithms.
3   Use the power rule for logarithms.
4   Use properties to write alternative forms of logarithmic expressions.

Key Terms

Use the vocabulary terms listed below to complete each statement in exercises 1−4.

product rule for logarithms   **quotient rule for logarithms**

power rule for logarithms   **special properties**

1.   The equations $b^{\log_b x} = x,\ x > 0$ and $\log_b b^x = x$ are referred to as

___________________ of logarithms.

2.   The equation $\log_b \frac{x}{y} = \log_b x - \log_b y$ is referred to as the _________________.

3.   The equation $\log_b xy = \log_b x + \log_b y$ is referred to as the _________________.

4.   The equation $\log_b x^r = r \log_b x$ is referred to as the _________________.

Objective 1   Use the product rule for logarithms.

Review these examples for Objective 1:

1.   Use the product rule to rewrite each logarithm. Assume $x > 0$.

a. $\log_4 (6 \cdot 11)$

Use the product rule.
$$\log_4 (6 \cdot 11) = \log_4 6 + \log_4 11$$

b. $\log_3 8 + \log_3 2$

Use the product rule.
$$\log_3 8 + \log_3 2 = \log_3 (8 \cdot 2) = \log_3 16$$

c. $\log_2 (2x)$

$$\log_2 (2x) = \log_2 2 + \log_2 x \quad \text{Product rule}$$
$$= 1 + \log_2 x \quad \log_2 2 = 1$$

Now Try:

1.   Use the product rule to rewrite each logarithm. Assume $x > 0$.

a. $\log_6 (5 \cdot 3)$

____

b. $\log_5 7 + \log_5 3$

____

c. $\log_4 (4x)$

____

d. $\log_3 x^4$ **d.** $\log_5 x^4$

$$\log_3 x^4 = \log_3 (x \cdot x \cdot x \cdot x)$$
$$= \log_3 x + \log_3 x + \log_3 x + \log_3 x$$
$$= 4 \log_3 x$$

Objective 1 Practice Exercises

For extra help, see Example 1 on page 640 of your text.

Use the product rule to express each logarithm as a sum of logarithms.

1. $\log_7 5m$ 1. ___________________

2. $\log_2 6xy$ 2. ___________________

Use the product rule to express the sum as a single logarithm.

3. $\log_4 7 + \log_4 3$ 3. ___________________

Objective 2   Use the quotient rule for logarithms.

Review these examples for Objective 2:   **Now Try:**

2. Use the quotient rule to rewrite each logarithm.   2. Use the quotient rule to rewrite
    Assume $x > 0$.   each logarithm. Assume $x > 0$.

a. $\log_4 \dfrac{8}{7}$ **a.** $\log_5 \dfrac{4}{9}$

$$\log_4 \frac{8}{7} = \log_4 8 - \log_4 7$$ ___________________

b. $\log_3 8 - \log_3 x$ **b.** $\log_6 x - \log_6 3$

$$\log_3 8 - \log_3 x = \log_3 \frac{8}{x}$$ ___________________

c. $\log_2 \dfrac{16}{11}$ **c.** $\log_4 \dfrac{16}{11}$

$$\log_2 \frac{16}{11} = \log_2 16 - \log_2 11 = 4 - \log_2 11$$ ___________________

d. $\log_5 54 - \log_5 6$ **d.** $\log_7 30 - \log_7 5$

$$\log_5 54 - \log_5 6 = \log_5 \frac{54}{6} = \log_5 9$$ ___________________

Objective 2 Practice Exercises

For extra help, see Example 2 on page 640 of your text.

Use the quotient rule for logarithms to express each logarithm as a difference of logarithms, or as a single number if possible.

4. $\log_2 \dfrac{5}{m}$ 4. _______________

5. $\log_6 \dfrac{k}{3}$ 5. _______________

Use the quotient rule for logarithms to express the difference as a single logarithm.

6. $\log_2 7q^4 - \log_2 5q^2$ 6. _______________

Objective 3   Use the power rule for logarithms.

Review these examples for Objective 3:

3. Use the power rule to rewrite each logarithm. Assume that $b > 0$, $x > 0$, and $b \neq 1$.

a. $\log_4 3^5$

$\log_4 3^5 = 5\log_4 3$

c. $\log_b \sqrt{11}$

$\log_b \sqrt{11} = \log_b 11^{1/2} = \dfrac{1}{2}\log_b 11$

d. $\log_2 \sqrt[5]{x^4}$

$\log_2 \sqrt[5]{x^4} = \log_2 x^{4/5} = \dfrac{4}{5}\log_2 x$

e. $\log_3 \dfrac{1}{x^5}$

$\log_3 \dfrac{1}{x^5} = \log_3 x^{-5} = -5\log_3 x$

Now Try:

3. Use the power rule to rewrite each logarithm. Assume that $b > 0$, $x > 0$, and $b \neq 1$.

a. $\log_6 4^3$

____

c. $\log_b \sqrt{13}$

____

d. $\log_3 \sqrt[4]{x^3}$

____

e. $\log_3 \dfrac{1}{x^7}$

____

Objective 3 Practice Exercises

For extra help, see Example 3 on page 642 of your text.

Use the power rule for logarithms to rewrite each logarithm or as a single number if possible.

7. $\log_m 2^7$ 7. _______________

8. $\log_3 \sqrt[3]{5}$ 8. _______________

9. $3^{\log_3 \sqrt[3]{7}}$ 9. _______________

Objective 4   Use properties to write alternative forms of logarithmic expressions.

Review these examples for Objective 4:

4. Use the properties of logarithms to rewrite each expression if possible. Assume that all variables represent positive real numbers.

a. $\log_5 6x^3$

$$\log_5 6x^3 = \log_5 6 + \log_5 x^3$$
$$= \log_5 6 + 3\log_5 x$$

b. $\log_b \sqrt{\dfrac{5}{x}}$

$$\log_b \sqrt{\dfrac{5}{x}} = \log_b \left(\dfrac{5}{x}\right)^{1/2}$$
$$= \dfrac{1}{2}\log_b \dfrac{5}{x}$$
$$= \dfrac{1}{2}(\log_b 5 - \log_b x)$$

d. $3\log_b x - \left(2\log_b y + \dfrac{1}{2}\log_b z\right)$

$$3\log_b x - \left(2\log_b y + \dfrac{1}{2}\log_b z\right)$$
$$= \log_b x^3 - (\log_b y^2 + \log_b z^{1/2})$$
$$= \log_b x^3 - \log_b y^2 \sqrt{z}$$
$$= \log_b \dfrac{x^3}{y^2 \sqrt{z}}$$

Now Try:

4. Use the properties of logarithms to rewrite each expression if possible. Assume that all variables represent positive real numbers.

a. $\log_6 36x^3$

____

b. $\log_b \sqrt{\dfrac{x}{3}}$

____

d. $\log_b x + 4\log_b y - \log_b z$

____

e. $2\log_3 x + \log_3(x-1) - \dfrac{1}{2}\log_3(x+1)$

$2\log_3 x + \log_3(x-1) - \dfrac{1}{2}\log_3(x+1)$

$= \log_3 x^2 + \log_3(x-1) - \log_3(x+1)^{1/2}$

$= \log_3\left(x^2(x-1)\right) - \log_3\sqrt{x+1}$

$= \log_3 \dfrac{x^3 - x^2}{\sqrt{x+1}}$

f. $\log_2(2x+3y)$

$\log_2(2x+3y)$ cannot be rewritten using the properties of logarithms.

5. Given that $\log_2 9 \approx 3.1699$ and $\log_2 11 \approx 3.4594,$ use properties of logarithms to evaluate each expression.

a. $\log_2 99$

$\log_2 99 = \log_2(9 \cdot 11)$

$\qquad = \log_2 9 + \log_2 11$

$\qquad = 3.1699 + 3.4594$

$\qquad = 6.6293$

b. $\log_2 \dfrac{1}{9}$

$\log_2 \dfrac{1}{9} = \log_2 1 - \log_2 9$

$\qquad = 0 - 3.1699$

$\qquad = -3.1699$

c. $\log_2 121$

$\log_2 121 = \log_2 11^2$

$\qquad = 2\log_2 11$

$\qquad = 2(3.4594)$

$\qquad = 6.9188$

e.

$2\log_2 x + \log_2(x-1) - \dfrac{1}{3}\log_2(x^2+1)$

____

f. $\log_3(3x-y)$

____

5. Given that $\log_2 9 \approx 3.1699$ and $\log_2 11 \approx 3.4594,$ use properties of logarithms to evaluate each expression.

a. $\log_2 198$

____

b. $\log_2 \dfrac{1}{11}$

____

c. $\log_2 729$

____

6. Decide whether each statement is *true* or *false*.

    a. $\log_2 32 - \log_2 16 = \log_2 16$

Evaluate each side.
Left side:
$$\log_2 32 - \log_2 16 = \log_2 2^5 - \log_2 2^4$$
$$= 5 - 4 \log_2 2$$
$$= 1$$
Right side:
$$\log_2 16 = \log_2 2^4 = 4$$
The statement is false because $1 \neq 4$.

    b. $\log_3(\log_4 64) = \dfrac{\log_{12} 144}{\log_6 36}$

Evaluate each side.
Left side:
$$\log_3(\log_4 64) = \log_3(\log_4 4^3)$$
$$= \log_3 3$$
$$= 1$$
Right side:
$$\frac{\log_{12} 144}{\log_6 36} = \frac{\log_{12} 12^2}{\log_6 6^2} = \frac{2}{2} = 1$$
The statement is true because $1 = 1$.

6. Decide whether each statement is *true* or *false*.

    a. $\log_3 27 + \log_3 9 = \log_5 5$

    ____

    b. $(\log_2 8)(\log_2 4) = \log_2 32$

    ____

For extra help, see Examples 4–6 on pages 642–644 of your text.

Use the properties of logarithms to express the sum or difference of logarithms as a single logarithm, or as a single number if possible.

10. $\quad \log_4 10y + \log_4 3y - \log_4 6y^3$

10. ________________

Given that $\log_2 6 \approx 2.5850$ *and* $\log_2 12 \approx 3.5850$, *use properties of logarithms to evaluate the expression.*

11. $\quad \log_2 72$

11. ________________

Decide whether the statement is true *or* false.

12. $\quad \log_2 4p^3 = 6 + 3\log_2 p$

12. ________________

Chapter 9   INVERSE, EXPONENTIAL, AND LOGARITHMIC FUNCTIONS

9.5   Common and Natural Logarithms

Learning Objectives
1   Evaluate common logarithms using a calculator.
2   Use common logarithms in applications.
3   Evaluate natural logarithms using a calculator.
4   Use natural logarithms in applications.
5   Use the change-of-base rule.

Key Terms

Use the vocabulary terms listed below to complete each statement in exercises 1−2.

common logarithm   **natural logarithm**

1.   A logarithm to the base e is a ____.

2.   A logarithm to the base 10 is a ____.

Objective 1   Evaluate common logarithms using a calculator.

Review this example for Objective 1:

1. Using a calculator, evaluate the logarithm to four decimal places.

$$\log 436.2$$

$$\log 436.2 \approx 2.6397$$

Now Try:

1. Using a calculator, evaluate the logarithm to four decimal places.

$$\log 983.5$$

____

Objective 1 Practice Exercises

For extra help, see Example 1 on pages 646–647 of your text.

Use a calculator to find each logarithm. Give an approximation to four decimal places.

1.   log 57.23   1. _________________

2.   log 0.0914   2. _________________

3.   log 87,123   3. _________________

Objective 2   Use common logarithms in applications.

Review these examples for Objective 2:

2. Wetlands are classified as bogs, fens, marshes, and swamps, on the basis of pH values. A pH value between 6.0 and 7.5 indicates that the wetland is a "rich fen." When the pH is between 3.0 and 6.0, the wetland is a "poor fen," and if the pH falls to 3.0 or less, it is a "bog." Suppose that the hydronium ion concentration of a sample of water from a wetland is 5.4×10^{-4}. Find the pH value for the water and determine how the wetland should be classified.

$$\text{pH} = -\log\left(5.4 \times 10^{-4}\right) \quad \text{Definition of pH}$$

$$= -\left(\log 5.4 + \log 10^{-4}\right) \quad \text{Product rule}$$

$$= -\left(0.7324 - 4\right) \quad \begin{array}{l}\text{Use a calculator to}\\\text{find log 5.4.}\end{array}$$

$$= 3.2676$$

Since the pH is between 3.0 and 6.0, the wetland is a poor fen.

3. Find the hydronium ion concentration of a solution with pH 5.4.

$$\text{pH} = -\log\left[H_3O^+\right] \quad \text{Definition of pH}$$

$$5.4 = -\log\left[H_3O^+\right]$$

$$\log\left[H_3O^+\right] = -5.4 \quad \text{Multiply by } -1.$$

$$H_3O^+ = 10^{-5.4} \quad \begin{array}{l}\text{Write in}\\\text{exponential form.}\end{array}$$

$$\approx 4.0 \times 10^{-6} \quad \text{Use a calculator.}$$

4. Find the decibel level to the nearest whole number of the sound with intensity I of $5.012 \times 10^{10} I_0$.

$$D = 10\log\left(\frac{I}{I_0}\right) = 10\log\left(\frac{5.012 \times 10^{10} I_0}{I_0}\right)$$

$$= 10\log\left(5.012 \times 10^{10}\right)$$

$$\approx 107 \text{ db}$$

Now Try:

2. Suppose that the hydronium ion concentration of a sample of water from a wetland is 6.2×10^{-8}. Find the pH value for the water and determine how the wetland should be classified.

____

3. Find the hydronium ion concentration of a solution with pH 3.6.

____

4. Find the decibel level to the nearest whole number of the sound with intensity I of $3.16 \times 10^8 I_0$.

____

Objective 2 Practice Exercises

For extra help, see Examples 2–4 on pages 647–648 of your text.

Solve each problem.

4. Find the pH of a solution with the given hydronium ion concentration. Round the answer to the nearest tenth.

   a. 4.3×10^{-9} **b.** 2.8×10^{-6}

4. a.______________

   b.______________

5. Find the decibel level to the nearest whole number of the sound with intensity I of $2.5 \times 10^{13} I_0$.

5. ______________

6. Find the hydronium ion concentration of a solution with the given pH value.

   a. 5.2   **b.** 1.3

6. a.______________

   b.______________

Objective 3   Evaluate natural logarithms using a calculator.

Review this example for Objective 3:	Now Try:
5. Using a calculator, evaluate the logarithm to four decimal places.	5. Using a calculator, evaluate the logarithm to four decimal places.
$\ln 436.2$	$\ln 98$
$\ln 436.2 \approx 6.0781$	______________

Objective 3 Practice Exercises

For extra help, see Example 5 on page 649 of your text.

Find each natural logarithm. Give an approximation to four decimal places.

7. $\ln 76.3$

7. ______________

8. $\ln 0.102$

8. ______________

9. $\ln 50$

9. ______________

Objective 4 Use natural logarithms in applications.

Review this example for Objective 4:	**Now Try:**

6. The time t in years for an investment increasing at a rate of r percent (in decimal form) to double is given by
$$t = \frac{\ln 2}{\ln(1+r)}.$$
This is called the doubling time. Find the doubling time to the nearest tenth for an investment at 4%.

$4\% = 0.04$, so $t = \dfrac{\ln 2}{\ln(1+0.04)} = \dfrac{\ln 2}{\ln 1.04} \approx 17.7$

The doubling time for the investment is about 17.7 years.

6. Use the formula at the left to find the doubling time to the nearest tenth for an investment at 6%.

____

Objective 4 Practice Exercises

For extra help, see Example 6 on page 650 of your text.

The time t in years for an amount increasing at a rate of r (in decimal form) to double (the doubling time) is given by $t = \dfrac{\ln 2}{\ln(1+r)}$. *Find the doubling time for an investment at the interest rate. Round to the nearest whole number.*

10. 3% **10.** ___________________

The half-life of a radioactive substance is the time it takes for half of the material to decay. The amount A in pounds of substance remaining after t years is given by $\ln\dfrac{A}{C} = -\dfrac{t}{h}\ln 2$, *where C is the initial amount in pounds, and h is its half-life in years. Use the formula to solve the following problems. Round to the nearest whole number.*

11. The half-life of radium-226 is 1620 years. How long, to the nearest year, will it take for 100 pounds to decay to 25 pounds? **11.** ___________________

Newton's Law of Cooling describes the cooling of a warmer object to the cooler temperature of the surrounding environment. The formula can be given as

$$t = \frac{1}{k}\ln\frac{T_s - T_1}{T_s - T_2},$$ *where t is the elapsed time, T_1 is the initial temperature measurement of the object, T_2 is the second temperature measurement of the object, and T_s is the temperature of the surrounding environment. Use this formula to solve the problem. Round to the nearest tenth.*

12.   A corpse was discovered in a motel room at midnight and its temperature was 80°F. The temperature in the room was 60°F. Assuming that the person's temperature at the time of death was 98.6° F and using $k = 0.1438$, determine t and the time of death.

12. t _______________

time _____________

Objective 5   Use the change-of-base rule.

Review this example for Objective 5:

7.   Use the change-of-base rule to approximate $\log_7 28$ to four decimal places.

$$\log_7 28 = \frac{\log 28}{\log 7} = 1.7124$$

Now Try:

7.   Use the change-of-base rule to approximate $\log_5 180$ to four decimal places.

____

Objective 5 Practice Exercises

For extra help, see Example 7 on page 651 of your text.

Use the change-of-base rule to find each logarithm. Give approximations to four decimal places.

13. $\log_{16} 27$

13. _______________

14. $\log_6 0.25$

14. _______________

15. $\log_{1/2} 5$

15. _______________

Chapter 9   INVERSE, EXPONENTIAL, AND LOGARITHMIC FUNCTIONS

9.6   Exponential and Logarithmic Equations; Further Applications

Learning Objectives
1   Solve equations involving variables in the exponents.
2   Solve equations involving logarithms.
3   Solve applications of compound interest.
4   Solve applications involving base e exponential growth and decay.

Key Terms

Use the vocabulary terms listed below to complete each statement in exercises 1–2.

compound interest   **continuous compounding**

1. The formula for ________________________ is $A = Pe^{rt}$.

2. The formula for ________________________ is $A = P\left(1 + \frac{r}{n}\right)^{nt}$.

Objective 1   Solve equations involving variables in the exponents.

Review these examples for Objective 1:

1. Solve $4^x = 30$. Approximate the solution to three decimal places.

$$4^x = 30$$

$$\log 4^x = \log 30 \quad \text{If } x = y, \text{ and } x > 0, \ y > 0, \text{ then } \log_b x = \log_b y.$$

$$x \log 4 = \log 30 \quad \text{Power rule}$$

$$x = \frac{\log 30}{\log 4} \quad \text{Divide by } \log 4.$$

$$x \approx 2.453 \quad \text{Use a calculator.}$$

Check

$$4^x = 4^{2.453} \approx 30$$

The solution set is $\{2.453\}$.

Now Try:

1. Solve $3^x = 15$. Approximate the solution to three decimal places.

____

2. Solve $e^{0.005x} = 9$. Approximate the solution to three decimal places.

$$e^{0.005x} = 9$$

$$\ln e^{0.005x} = \ln 9 \qquad \text{If } x = y, \text{ and } x > 0,$$
$$\phantom{\ln e^{0.005x} = \ln 9 \quad} y > 0, \text{ then } \ln x = \ln y.$$

$$0.005x \ln e = \ln 9 \qquad \text{Power rule}$$

$$0.005x = \ln 9 \qquad \ln e = 1$$

$$x = \frac{\ln 9}{0.005} \qquad \text{Divide by 0.005.}$$

$$x \approx 439.445 \text{ Use a calculator.}$$

The solution set is $\{439.445\}$.

2. Solve $e^{0.4x} = 15$. Approximate the solution to three decimal places.

____

Objective 1 Practice Exercises

For extra help, see Examples 1–3 on pages 654–655 of your text.

Solve each equation. Give solutions to three decimal places.

1. $25^{x+2} = 125^{3-x}$

1. _______________

2. $4^{x-1} = 3^{2x}$

2. _______________

3. $e^{-0.45x} = 7$

3. _______________

        Copyright © 2020 Pearson Education, Inc.

Objective 2   Solve equations involving logarithms.

Review these examples for Objective 2:

5.   Solve $\log_3 (x-1)^3 = 5$. Give the exact solution.

$$\log_3 (x-1)^3 = 5$$

$$(x-1)^3 = 3^5 \qquad \text{Write in exponential form.}$$

$$(x-1)^3 = 243$$

$$x-1 = \sqrt[3]{243} \qquad \text{Take the cube root of each side.}$$

$$x-1 = 3\sqrt[3]{9} \qquad \text{Simplify the cube root.}$$

$$x = 1 + 3\sqrt[3]{9} \quad \text{Add 1.}$$

Check:

$$\log_3 (x-1)^3 = 5$$

$$\log_3 \left(1 + 3\sqrt[3]{9} - 1\right)^3 \overset{?}{=} 5$$

$$\log_3 \left(\sqrt[3]{243}\right)^3 \overset{?}{=} 5$$

$$\log_3 (243) \overset{?}{=} 5$$

$$\log_3 3^5 \overset{?}{=} 5$$

$$5 = 5$$

The solution set is $\left\{1 + 3\sqrt[3]{9}\right\}$.

6.   Solve $\log_3 (5x+42) - \log_3 x = \log_3 26$.

$$\log_3 (5x+42) - \log_3 x = \log_3 26.$$

$$\log_3 \frac{5x+42}{x} = \log_3 26$$

$$\frac{5x+42}{x} = 26 \qquad \begin{array}{l}\text{If } \log_b x = \log_b y \\ \text{then } x = y.\end{array}$$

$$5x+42 = 26x \quad \text{Multiply by } x.$$

$$42 = 21x \quad \text{Subtract } 5x.$$

$$2 = x \qquad \text{Divide by 21.}$$

Now Try:

5.   Solve $\log_6 (x+1)^3 = 2$. Give the exact solution.

____

6.   Solve
$$\log_6 (2x+7) - \log_6 x = \log_6 16.$$

____

Check:
$$\log_3 (5x + 42) - \log_3 x = \log_3 26$$
$$\log_3 (5 \cdot 2 + 42) - \log_3 2 \overset{?}{=} \log_3 26$$
$$\log_3 52 - \log_3 2 \overset{?}{=} \log_3 26$$
$$\log_3 \tfrac{52}{2} \overset{?}{=} \log_3 26$$
$$\log_3 26 = \log_3 26$$

The solution set is $\{2\}$.

7. Solve $\log_2 (x + 7) + \log_2 (x + 3) = \log_2 77$.

$$\log_2 (x + 7) + \log_2 (x + 3) = \log_2 77$$
$$\log_2 [(x + 7)(x + 3)] = \log_2 77$$

Product rule

$$(x + 7)(x + 3) = 77$$

If $\log_b x = \log_b y$
then $x = y$.

$$x^2 + 10x + 21 = 77 \quad \text{Multiply.}$$
$$x^2 + 10x - 56 = 0 \quad \text{Subtract 77.}$$
$$(x - 4)(x + 14) = 0 \quad \text{Factor.}$$
$$x - 4 = 0 \quad \text{or} \quad x + 14 = 0$$
$$x = 4 \qquad\qquad x = -14$$

The value -14 must be rejected since it leads to the logarithm of a negative number in the original equation.
A check shows that the only solution is 4.
The solution set is $\{4\}$.

7. Solve.
$$\log_4 (4x - 3) + \log_4 x = \log_4 (2x - 1)$$

____

Objective 2 Practice Exercises

For extra help, see Examples 4–7 on pages 655–657 of your text.

Solve each equation. Give exact solution.

4. $\log(-a) + \log 4 = \log(2a + 5)$

4. _______________

                  Copyright © 2020 Pearson Education, Inc.

5. $\log_3(x^2-10)-\log_3 x=1$ 5. ________________

6. $\ln(x+4)+\ln(x-2)=\ln 7$ 6. ________________

Objective 3   Solve applications of compound interest.

Review these examples for Objective 3:

8. How much money will be in an account at the end of 5 years if $5000 is deposited at 4% compounded monthly?

Because interest is compounded monthly, $n=12$. The other given values are $P=5000$, $r=0.04$, and $t=5$.

$$A=P\left(1+\frac{r}{n}\right)^{nt}$$

$$A=5000\left(1+\frac{0.04}{12}\right)^{12\cdot 5}$$

$$A=5000(1.0033)^{60}$$

$$A=6104.98$$

There will be $6104.98 in the account at the end of 5 years.

9. Approximate the time it would take for money deposited in an account paying 5% interest compounded quarterly to double. Round to the nearest hundredth.

We want the number of years t for P dollars to grow to $2P$ dollars at a rate of 5% per year. In the compound interest formula, we substitute $2P$ for A, and let $r=0.05$ and $n=4$.

Now Try:

8. How much money will be in an account at the end of 5 years if $10,000 is deposited at 4% compounded quarterly?

____

9. Approximate the time it would take for money deposited in an account paying 5% interest compounded monthly to double. Round to the nearest hundredth.

____

$$2P = P\left(1 + \frac{0.05}{4}\right)^{4t}$$

$$2 = 1.0125^{4t}$$

$$\log 2 = \log 1.025^{4t}$$

$$\log 2 = 4t \log 1.0125$$

$$t = \frac{\log 2}{4 \log 1.0125}$$

$$t \approx 13.95$$

It will take about 13.95 years for the investment to double.

10. Suppose that $5000 is invested at 4% interest for 3 years.

a. How much will the investment be worth if it is compounded continuously?

$$A = Pe^{rt}$$

$$A = 5000e^{0.04 \cdot 3}$$

$$A = 5000e^{0.12}$$

$$A = 5637.48$$

The investment will be worth $5637.48.

b. Approximate the amount of time it would take for the investment to double. Round to the nearest tenth.

Find the value of t that will cause A to be $2(\$5000) = \$10,000$.

$$A = Pe^{rt}$$

$$10,000 = 5000e^{0.04t}$$

$$2 = e^{0.04t} \qquad \text{Divide by 5000.}$$

$$\ln 2 = \ln e^{0.04t} \qquad \begin{array}{l}\text{If } x = y,\\ \text{then } \ln x = \ln y.\end{array}$$

$$\ln 2 = 0.04t \qquad \ln e^{k} = k$$

$$\frac{\ln 2}{0.04} = t \qquad \text{Divide by 0.04.}$$

$$t \approx 17.3$$

It will take about 17.3 years for the amount to double.

10. Suppose that $5000 is invested at 2% interest for 3 years.

a. How much will the investment be worth if it is compounded continuously?

____

b. Approximate the amount of time it would take for the investment to double. Round to the nearest tenth.

____

Objective 3 Practice Exercises

For extra help, see Examples 8–10 on pages 658–659 of your text.

Solve each problem.

7. How much will be in an account after 10 years if $25,000 is invested at 8% compounded quarterly? Round to the nearest cent.

    7. ________________

8. How much will be in an account after 5 years if $10,000 is invested at 4.5% compounded continuously? Round to the nearest cent.

    8. ________________

9. How long will it take an investment to double if it is placed in an account paying 9% interest compounded continuously? Round to the nearest tenth.

    9. ________________

Objective 4   Solve applications involving base e exponential growth and decay.

Review these examples for Objective 4:

11. A sample of 500 g of lead-210 decays according to the function $y = y_0 e^{-0.032t}$, where t is the time in years, y is the amount of the sample at time t, and y_0 is the initial amount present at $t = 0$.

    a. How much lead will be left in the sample after 20 years? Round to the nearest tenth of a gram.

    Let $t = 20$ and $y_0 = 500$.

$$y = 500e^{-0.032 \cdot 20} \approx 263.6$$

    There will be about 263.6 grams after 20 years.

Now Try:

11. Cesium-137, a radioactive isotope used in radiation therapy, decays according to the function $y = y_0 e^{-0.0231t}$, where t is the time in years and y_0 is the initial amount present at $t = 0$.

    a. If an initial sample contains 36 mg of cesium-137, how much cesium-137 will be left in the sample after 50 years? Round to the nearest tenth.

    ____

b. Approximate the half-life of lead-210 to the nearest tenth.

Let $y = \frac{1}{2}(500) = 250$.

$$250 = 500e^{-0.032t}$$

$$0.5 = e^{-0.032t}$$

$$\ln 0.5 = \ln e^{-0.032t}$$

$$\ln 0.5 = -0.032t$$

$$t = \frac{\ln 0.5}{-0.032} \approx 21.7$$

The half-life of lead-210 is about 21.7 years.

b. Approximate the half-life of cesium-137 to the nearest tenth.

____

Objective 4 Practice Exercises

For extra help, see Example 11 on page 660 of your text.

Solve each problem.

10. Radioactive strontium decays according to the function $y = y_0 e^{-0.0239t}$, where t is the time in years. If an initial sample contains $y_0 = 15$ g of radioactive strontium, how many grams will be present after 25 years? Round to the nearest hundredth of a gram.

10. _____________

11. How long will it take the initial sample of strontium in exercise 19 to decay to half of its original amount?

11. _____________

12. The concentration of a drug in a person's system decreases according to the function $C(t) = 2e^{-0.2t}$, where $C(t)$ is given in mg and t is in hours. How much of the drug will be in the person's system after one hour? Approximate answer to the nearest hundredth.

12. _____________

                Copyright © 2020 Pearson Education, Inc.

Chapter 10 NONLINEAR FUNCTIONS, CONIC SECTIONS, AND NONLINEAR SYSTEMS

10.1   Additional Graphs of Functions

Learning Objectives

1   Recognize the graphs of the absolute values, reciprocal, and square root functions, and graph their translations.
2   Recognize and graph step functions.

Key Terms

Use the vocabulary terms listed below to complete each statement in exercises 1–3.

asymptotes   **greatest integer function**   **step function**

1. A _________________________________ is a function that looks like a series of steps.

2. The function defined by $f(x) = [\![x]\!]$ is called the _________________________.

3. Lines that a graph approaches without actually touching are called

    _________________________.

Objective 1   Recognize the graphs of the absolute value, reciprocal, and square root functions and graph their translations.

Review these examples for Objective 1:

1. Graph $f(x) = |x + 4|$. Give the domain and range.

    The graph of $f(x) = |x + 4|$ is found by shifting the graph of $y = |x|$ four units to the left.

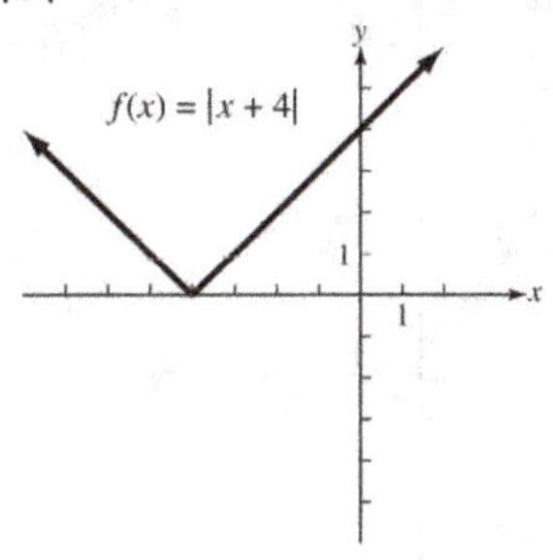

x	y
-6	2
-5	1
-4	0
-3	1
-2	2

The domain is $(-\infty, \infty)$. The range is $[0, \infty)$.

Now Try:

1. Graph $f(x) = \sqrt{x + 3}$. Give the domain and range.

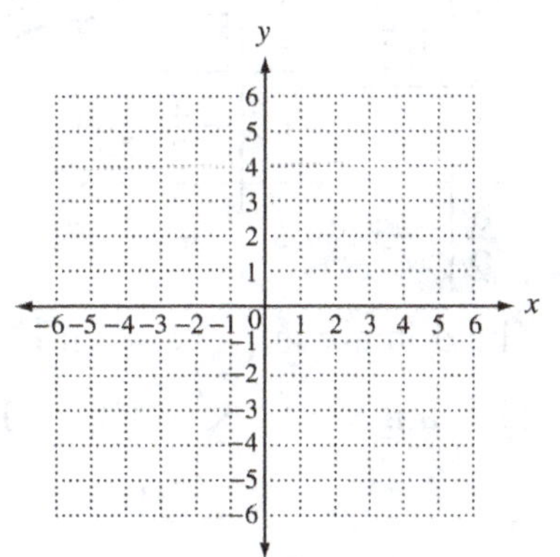

2. Graph $f(x) = \sqrt{x} - 2$. Give the domain and range.

The graph of $y = \sqrt{x} - 2$ is obtained by shifting the graph of $y = \sqrt{x}$ two units down.

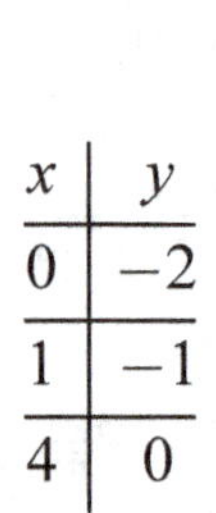

x	y
0	-2
1	-1
4	0

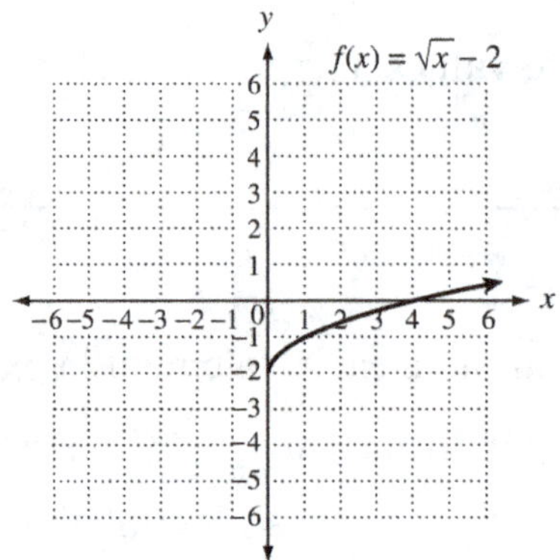

The domain is $[0, \infty)$. The range is $[-2, \infty)$.

3. Graph $f(x) = \dfrac{1}{x+2} - 1$. Give the domain and range.

The graph of $y = \dfrac{1}{x+2} - 1$ is obtained by shifting the graph of $y = \dfrac{1}{x}$ two units to the left and then one unit down.

x	y	x	y
-6	$-\dfrac{5}{4}$	$-\dfrac{3}{2}$	1
-5	$-\dfrac{4}{3}$	-1	0
-4	$-\dfrac{3}{2}$	0	$-\dfrac{1}{2}$
-3	-2	1	$-\dfrac{2}{3}$
$-\dfrac{5}{2}$	-3	2	$-\dfrac{3}{4}$

The domain is $(-\infty, -2) \cup (-2, \infty)$.
The range is $(-\infty, -1) \cup (-1, \infty)$.

2. Graph $f(x) = |x| - 1$. Give the domain and range.

____

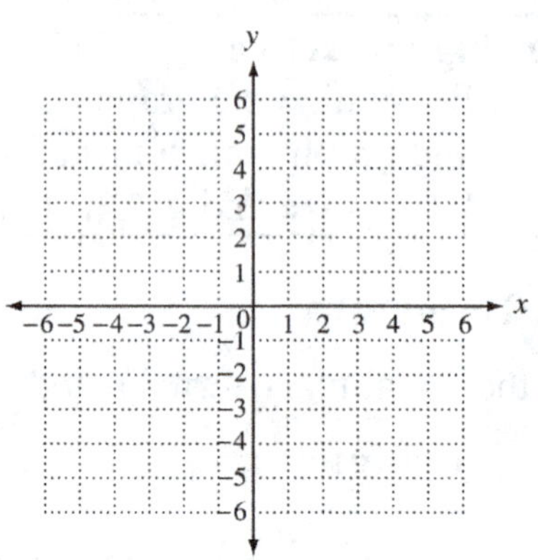

3. Graph $f(x) = \dfrac{1}{x-1} - 2$. Give the domain and range.

____

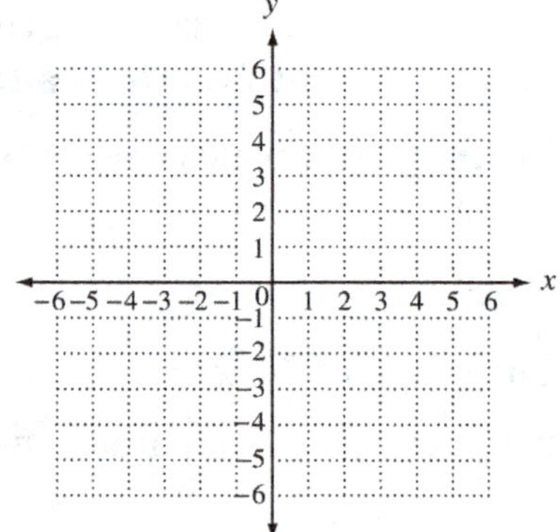

Objective 1 Practice Exercises

For extra help, see Examples 1–3 on pages 679–680 of your text.

Graph each function. Give the domain and range.

1. $f(x) = |x - 2| + 3$

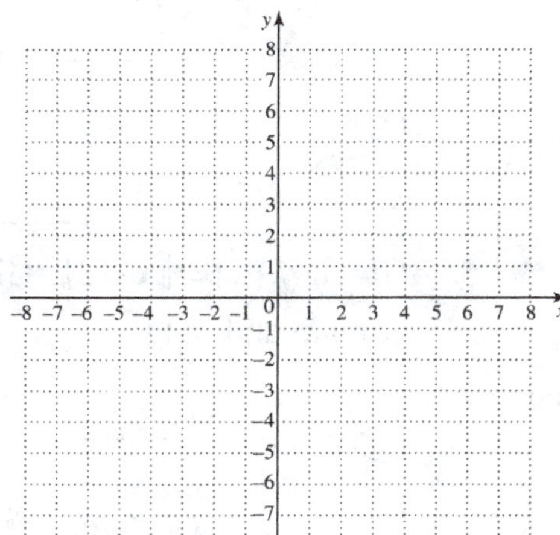

1. domain ___________

  range___________

2. $f(x) = \sqrt{x} + 3$

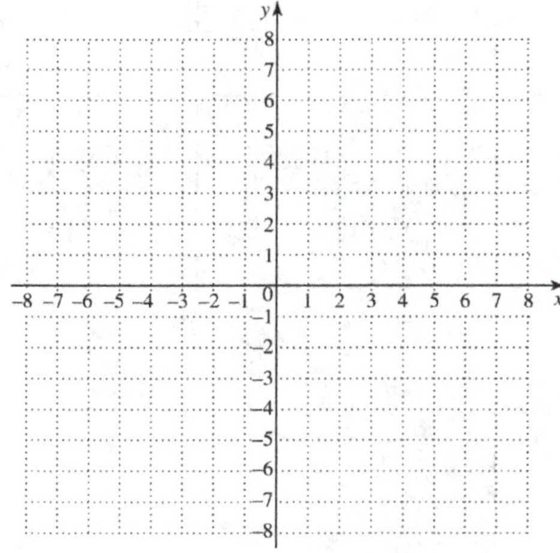

2. domain ___________

  range___________

3. $f(x) = \dfrac{1}{x - 3}$

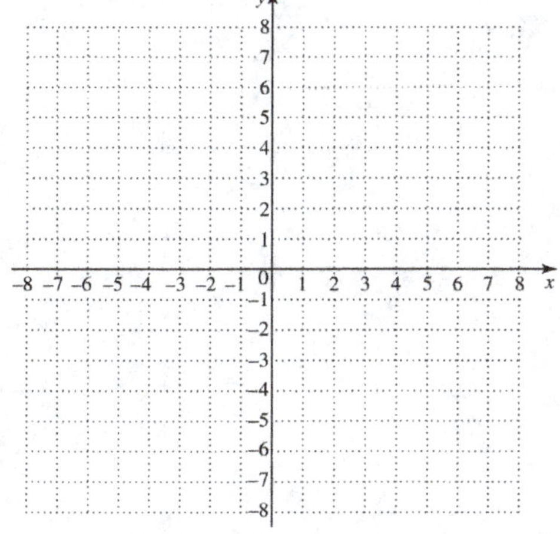

3. domain ___________

  range___________

Objective 2   Recognize and graph step functions.

Review these examples for Objective 2:

4.   Evaluate each expression.

   d. $[\![9.76]\!]$

   $$[\![9.76]\!]=9$$

   e. $[\![-3.5]\!]$

   $$[\![-3.5]\!]=-4$$

5.   Graph $f(x)=[\![x]\!]+2$. Give the domain and range.

The graph of $y=[\![x]\!]+2$ is obtained by shifting the graph of $y=[\![x]\!]$ two units up.

If $-3\le x<-2$, then $[\![x]\!]+2=-3+2=-1$.

If $-2\le x<-1$, then $[\![x]\!]+2=-2+2=0$.

If $-1\le x<0$, then $[\![x]\!]+2=-1+2=1$.

If $0\le x<1$, then $[\![x]\!]+2=0+2=2$, etc.

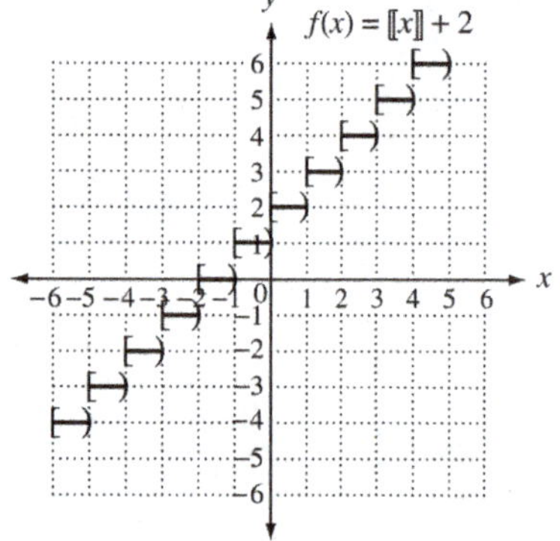

The domain is $(-\infty,\ \infty)$. The range is $\{...,-2,-1,\ 0,\ 1,\ 2,...\}$ (the set of integers)

Now Try:

4.   Evaluate each expression.

   d. $[\![1.76]\!]$

   ____

   e. $[\![-10.01]\!]$

   ____

5.   Graph $f(x)=[\![x-2]\!]$. Give the domain and range.

   ____

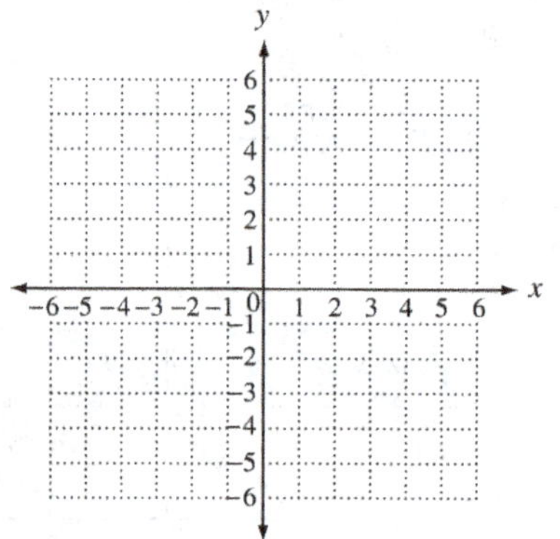

6. In 2011, U.S. first class letter postage is 46¢ for the first ounce and 18¢ for each additional ounce. Let $p(x)$ = the cost of sending a letter that weighs x ounces. Graph the function over the interval (0, 6].

For x in the interval (0, 1], $y = \$0.46$.
For x in the interval (1, 2],
   $y = \$0.46 + \$0.18 = \$0.64$.
For x in the interval (2, 3],
   $y = \$0.64 + \$0.18 = \$0.82$.
For x in the interval (3, 4],
   $y = \$0.82 + \$0.18 = \$1.00$.
For x in the interval (4, 5],
   $y = \$1.00 + \$0.18 = \$1.18$.
For x in the interval (5, 6],
   $y = \$1.18 + \$0.18 = \$1.36$.

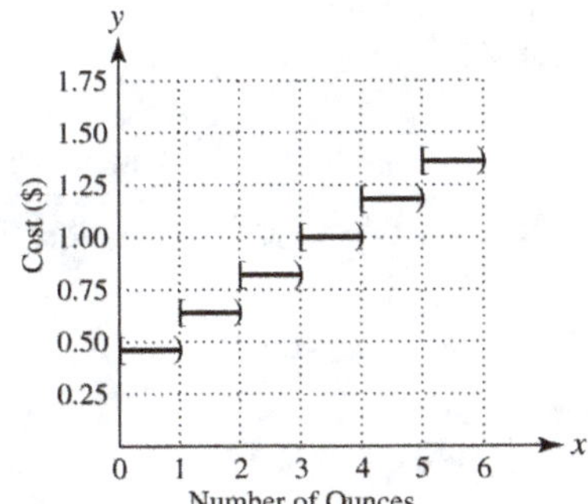

6. At the end of December, Mary's dad lent her $500 to buy an iPad. She agreed to repay the loan at $50 per month on the first of each month, starting in January. If $f(x)$ represents the amount to be repaid in month x, graph the function over its entire domain.

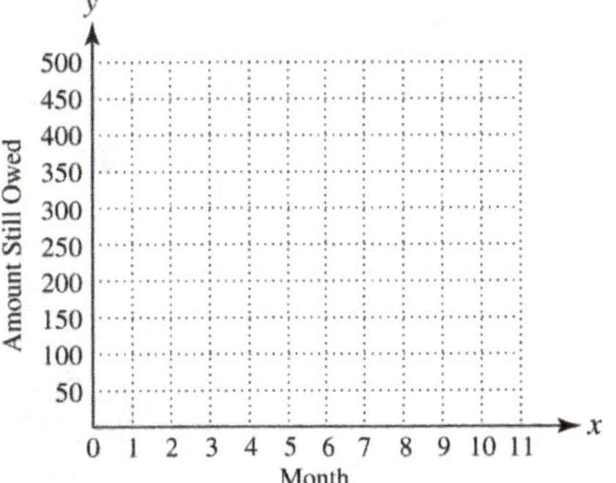

Objective 2 Practice Exercises

For extra help, see Examples 4–6 on pages 680–681 of your text.

Graph each function.

4. $f(x) = [\![x + 2]\!] - 3$

4.

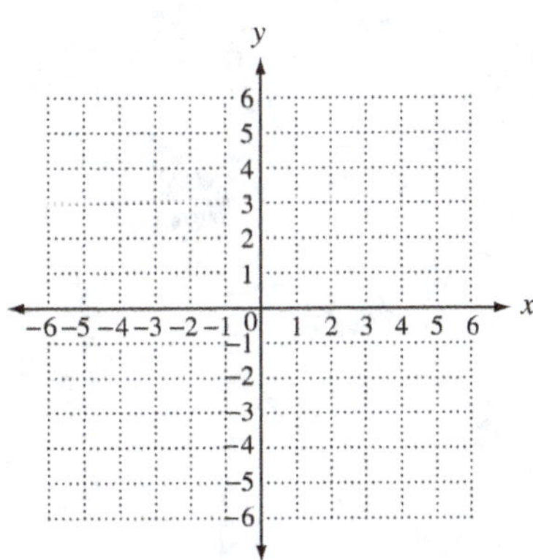

5. The cost of parking a car in an hourly parking lot is \$6.00 for the first hour and \$4.00 for each additional hour or fraction of an hour. Graph the function f that models the cost of parking a car for x hours over the interval $(0, 6]$.

5.

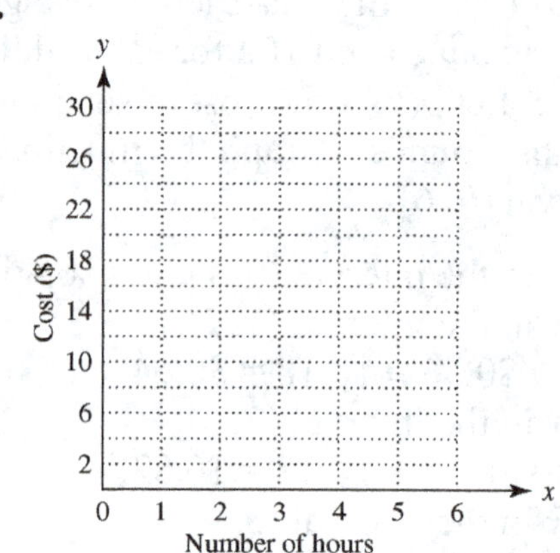

Chapter 10 NONLINEAR FUNCTIONS, CONIC SECTIONS, AND NONLINEAR SYSTEMS

10.2   Circles Revisited and Ellipses

Learning Objectives	
1	Graph circles.
2	Write an equation of a circle given its center and radius.
3	Determine the center and radius of a circle given its equation.
4	Recognize the equation of an ellipse.
5	Graph ellipses.

Key Terms

Use the vocabulary terms listed below to complete each statement in exercises 1−8.

conic sections	**circle**	**center (of a circle)**	**radius**
center-radius form	**ellipse**	**foci**	**center (of an ellipse)**

1. A(n) _________________________ is the set of all points in a plane that lie a fixed distance from a fixed point.

2. A(n) _________________________ is the set of all points in a plane the sum of whose distances from two fixed points is constant.

3. Figures that result from the intersection of an infinite cone with a plane are called __________________________________.

4. A fixed point such that every point on a circle is a fixed distance from it is the __________________________.

5. The distance from the center of a circle to a point on the circle is called the __________________________.

6. The __________________ of the equation of a circle with center (h, k) and radius r is $(x-h)^2 + (y-k)^2 = r^2$.

7. __________________ are fixed points used to determine the points that form a parabola, and ellipse, or a hyperbola.

8. The __________________ of the ellipse defined by $\dfrac{(x-3)^2}{9} + \dfrac{(y+2)^2}{16} = 1$ is $(3, -2)$.

Objective 1   Graph circles.

Review these examples for Objective 1:	**Now Try:**

Review these examples for Objective 1:

1. Find the center and radius of each circle. Then graph the circle.

   a. $x^2 + y^2 = 4$

$$(x-0)^2 + (y-0)^2 = 2^2$$
Here, $h = 0$, and $k = 0$.
The center is $(0, 0)$ and the radius is 2.

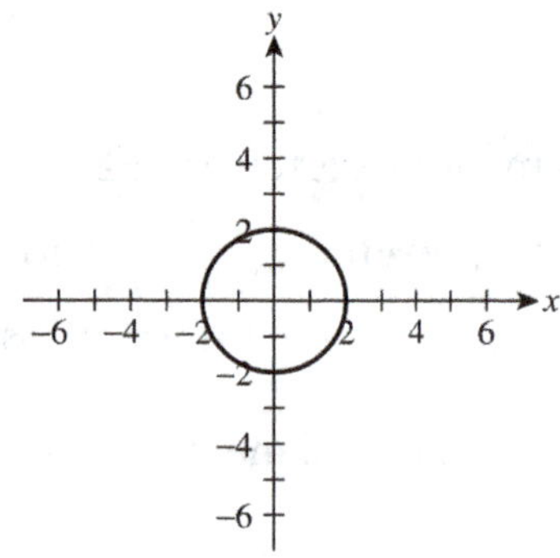

   b. $(x+3)^2 + (y-2)^2 = 9$

$$[x-(-3)]^2 + (y-2)^2 = 3^2$$
The center (h, k) is $(-3, 2)$ and the radius is 3.

To graph the circle, plot the center $(-3, 2)$, then move three units right, left, up, and down from the center, plotting the points $(0, 2)$, $(-3, 5)$, $(-6, 2)$, and $(-3, -1)$. Draw a smooth curve through the points.

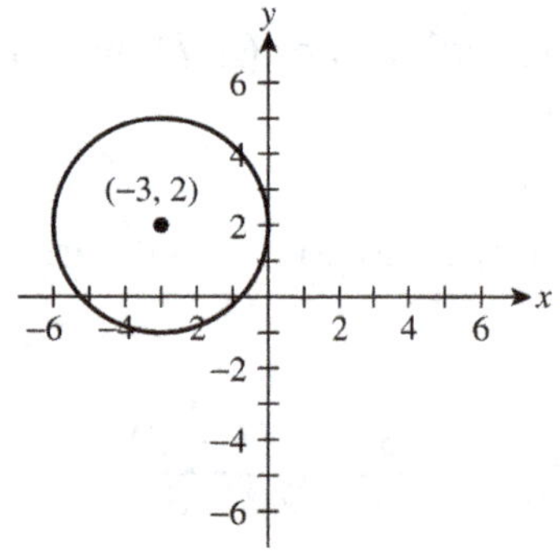

Now Try:

1. Find the center and radius of each circle. Then graph the circle.

   a. $x^2 + y^2 = 25$

     Center: _____________

     Radius: _____________

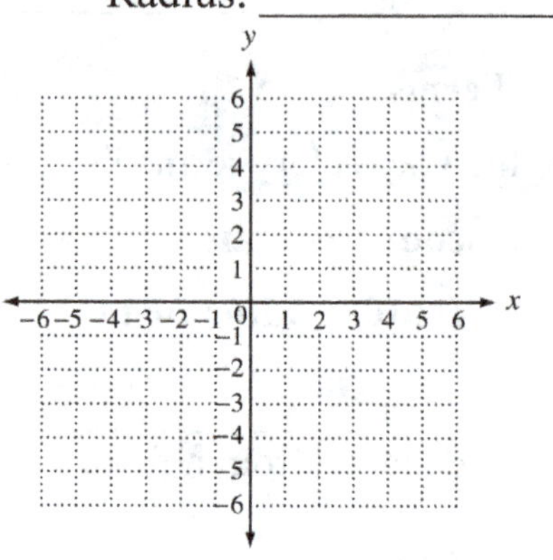

   b. $(x+5)^2 + (y-4)^2 = 16$

     Center: _____________

     Radius: _____________

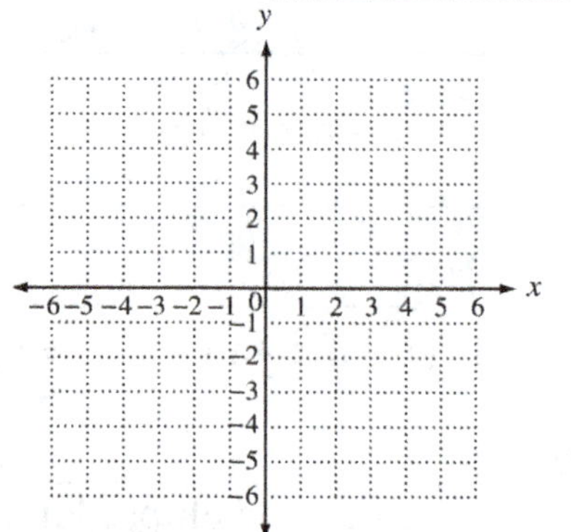

Objective 1 Practice Exercises

For extra help, see Example 1 on page 685 of your text.

Find the center and radius of each circle. Then graph the circle.

1. $(x-1)^2 + (y-4)^2 = 4$

1. Center: ______________

Radius: ______________

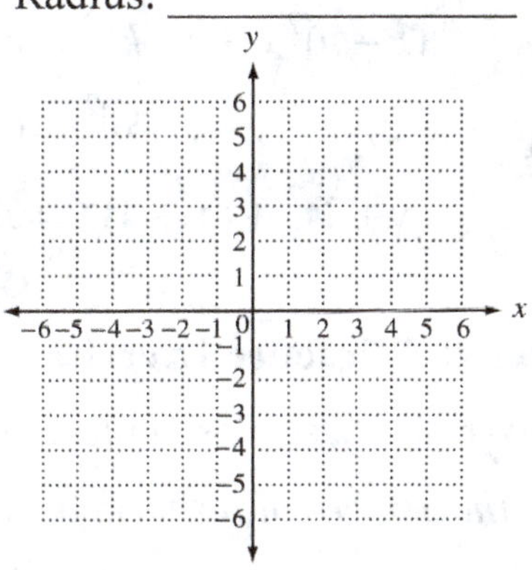

2. $(x-2)^2 + (y+3)^2 = 25$

2. Center: ______________

Radius: ______________

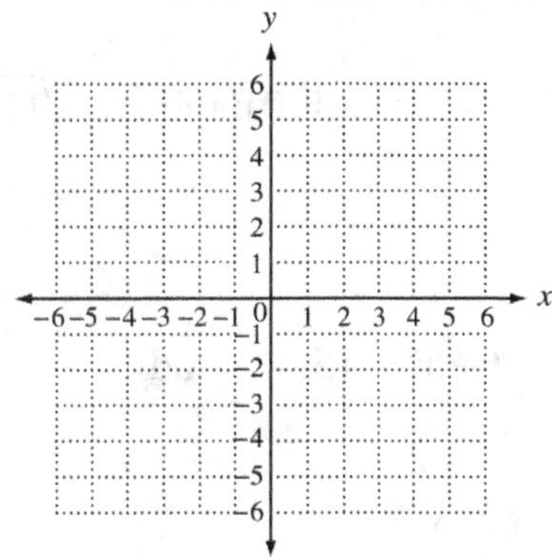

3. $x^2 + (y-5)^2 = 9$

3. Center: ______________

Radius: ______________

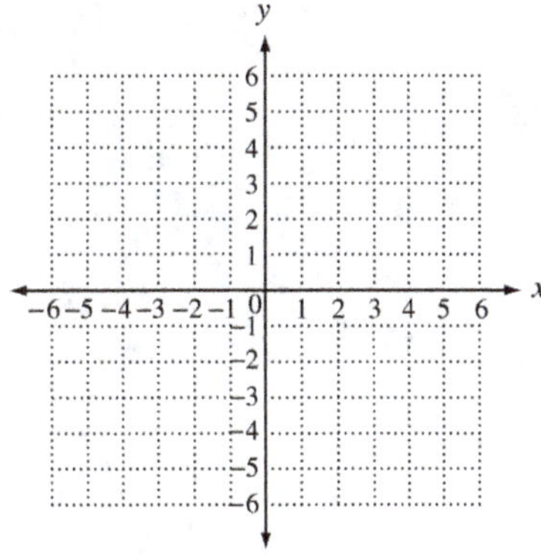

Objective 2   Write an equation of a circle given its center and radius.

Review this example for Objective 2:

2. Find an equation of the circle with center $(-2,-4)$ and radius $2\sqrt{5}$.

Here $(h, k) = (-2, -4)$ and $r = 2\sqrt{5}$.

$$(x - h)^2 + (y - k)^2 = r^2$$
$$[x - (-2)]^2 + [y - (-4)]^2 = (2\sqrt{5})^2$$
$$(x + 2)^2 + (y + 4)^2 = 20$$

Now Try:

2. Find an equation of the circle with center $(0, 3)$ and radius $\sqrt{2}$.

____

Objective 2 Practice Exercises

For extra help, see Example 2 on page 686 of your text.

Write the center-radius form of each circle described.

4.   Center: $(-1, 3)$; radius: $\sqrt{7}$

4. ______________

5.   Center: $(4, 0)$; radius: $\sqrt{11}$

5. ______________

6.   Center: $(5,-2)$; radius: $\sqrt{18}$

6. ______________

Objective 3   Determine the center and radius of a circle given its equation.

Review this example for Objective 3:

3. Find the center and radius of the circle.

$$x^2 + y^2 + 8x + 4y - 29 = 0$$

To find the center and radius, complete the squares on x and y.

$$x^2 + y^2 + 8x + 4y - 29 = 0$$
$$x^2 + y^2 + 8x + 4y = 29$$
$$\left(x^2 + 8x \quad\right) + \left(y^2 + 4y \quad\right) = 29$$
$$\left[\tfrac{1}{2}(8)\right]^2 = 16 \quad \left[\tfrac{1}{2}(4)\right]^2 = 4$$
$$(x^2 + 8x + 16) + (y^2 + 4y + 4) = 29 + 16 + 4$$
$$(x+4)^2 + (y+2)^2 = 49$$
$$\left[x-(-4)\right]^2 + \left[y-(-2)\right]^2 = 7^2$$

The center is $(-4, -2)$ and the radius is 7.

Now Try:

3. Find the center and radius of the circle.

$$x^2 + y^2 - 8x - 2y + 15 = 0$$

Center: _______________

Radius: _______________

Objective 3 Practice Exercises

For extra help, see Example 3 on page 687 of your text.

Find the center and radius of each circle.

7. $x^2 + y^2 - 4x + 8y + 11 = 0$

7.

Center: _______________

Radius: _______________

8. $x^2 + y^2 - 2x - 6y - 15 = 0$

8.

Center: _______________

Radius: _______________

9. $3x^2 + 3y^2 + 12y + 30x = 21$

9.

Center: _______________

Radius: _______________

      413

Objective 4   Recognize the equation of an ellipse.

For extra help, see pages 687–688 of your text.

Objective 5   Graph ellipses.

Review these examples for Objective 5:   **Now Try:**

4.   Graph $\dfrac{x^2}{25}+\dfrac{y^2}{9}=1$.

$a^2 = 25$, so $a = 5$, and
the x-intercepts are $(5, 0)$ and $(-5, 0)$.
$b^2 = 9$, so $b = 3$, and
the y-intercepts are $(0, 3)$ and $(0, -3)$.

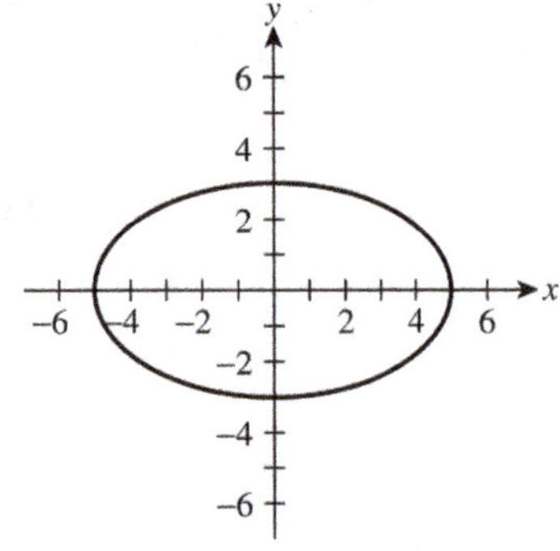

4.   Graph $\dfrac{x^2}{16}+\dfrac{y^2}{9}=1$.

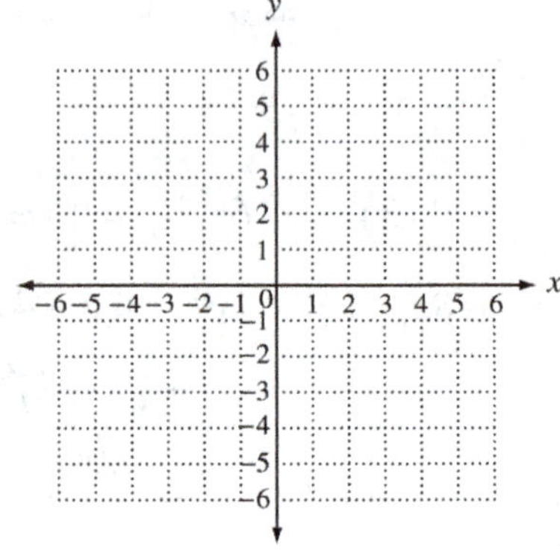

5.   Graph $\dfrac{(x-1)^2}{9}+\dfrac{(y+2)^2}{4}=1$.

The center is $(1, -2)$, $a = 3$, and $b = 2$. The
ellipse passes through the points $(4, -2)$,
$(1, -4)$, $(-2, -2)$, and $(1, 0)$.

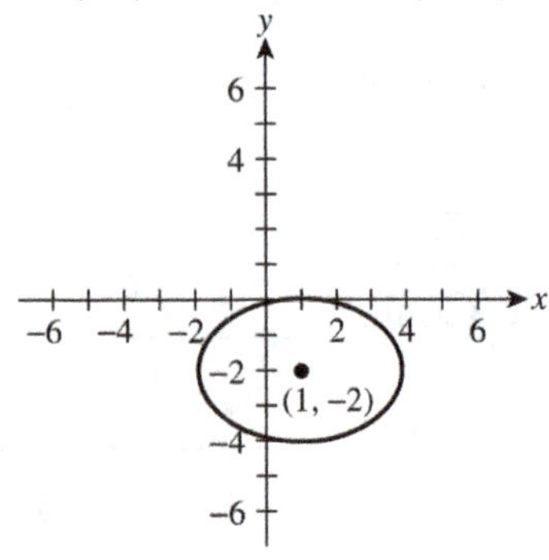

5.   Graph $\dfrac{(x+1)^2}{4}+\dfrac{(y-2)^2}{9}=1$.

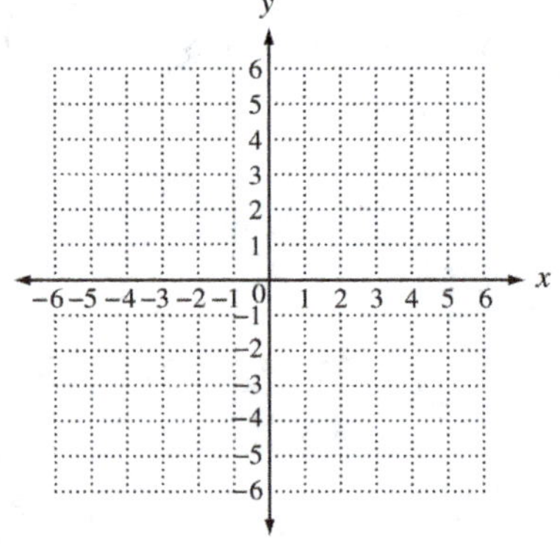

Objective 5 Practice Exercises

For extra help, see Examples 4–5 on pages 688–689 of your text.

Graph each ellipse.

10. $\dfrac{x^2}{36}+\dfrac{y^2}{9}=1$

10.

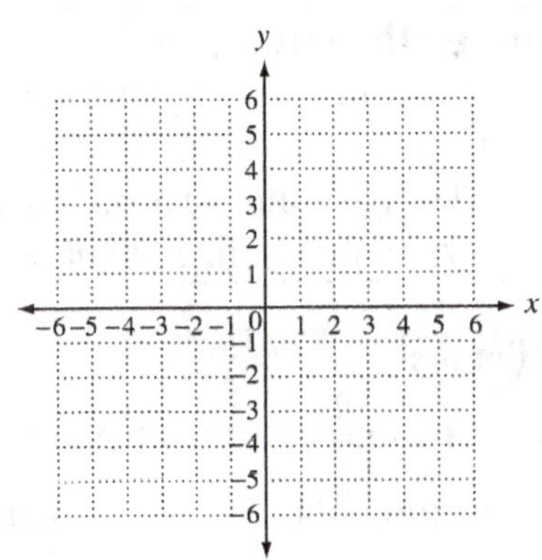

11. $\dfrac{(x+2)^2}{4}+\dfrac{(y-3)^2}{25}=1$

11.

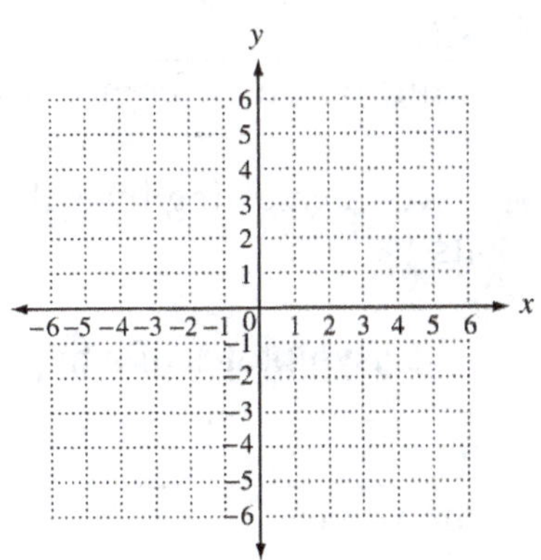

12. $\dfrac{(x+3)^2}{9}+\dfrac{(y+3)^2}{4}=1$

12.

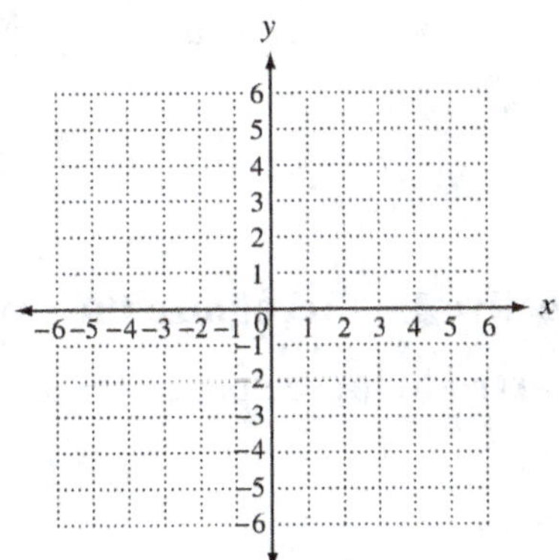

Chapter 10 NONLINEAR FUNCTIONS, CONIC SECTIONS, AND NONLINEAR SYSTEMS

10.3   Hyperbolas and Functions Defined by Radicals

Learning Objectives
1   Recognize the equation of a hyperbola.
2   Graph hyperbolas using asymptotes.
3   Identify conic sections by their equations.
4   Graph generalized square root functions.

Key Terms

Use the vocabulary terms listed below to complete each statement in exercises 1−5.

hyperbola   **transverse axis**   **asymptotes**

fundamental rectangle   **generalized square root function**

1.   A(n) _____________________ is the set of all points in a plane such that the absolute value of the difference of the distances from two fixed points is constant.

2.   Two intersecting lines that the branches of a hyperbola approach, but never reach, are its _____________________.

3.   The asymptotes of a hyperbola are the extended diagonals of its _____________________.

4.   The _____________________ of a hyperbola with x-intercepts $(a, 0)$ and $(-a, 0)$ lies on the x-axis.

5.   For an algebraic expression in x defined by u, with $u \geq 0$, a function of the form $f(x) = \sqrt{u}$, is a _____________________.

Objective 1   Recognize the equation of a hyperbola.

For extra help, see page 694 of your text.

Objective 2   Graph hyperbolas using asymptotes.

Review this example for Objective 2:

1. Graph $\dfrac{x^2}{9} - \dfrac{y^2}{16} = 1$.

Step 1: $a = 3$, $b = 4$. The *x*-intercepts are $(3, 0)$ and $(-3, 0)$.

Step 2: The four points $(3, 4)$, $(3, -4)$, $(-3, -4)$, and $(-3, 4)$ are the vertices of the fundamental rectangle.

Step 3: The equations of the asymptotes are $y = \pm \frac{4}{3}x$.

Step 4: Sketch the graph.

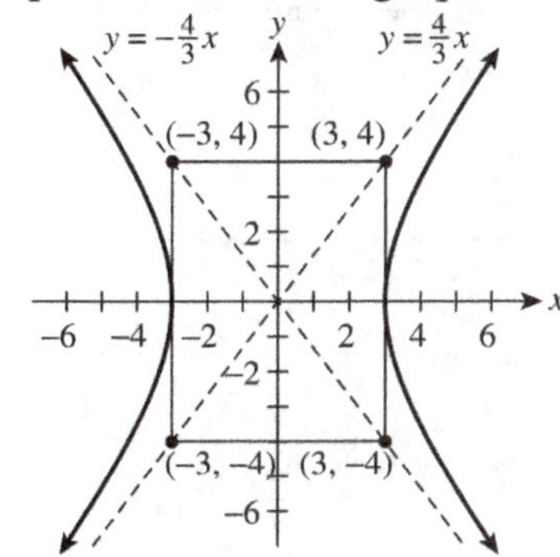

Now Try:

1. Graph $\dfrac{x^2}{16} - \dfrac{y^2}{4} = 1$.

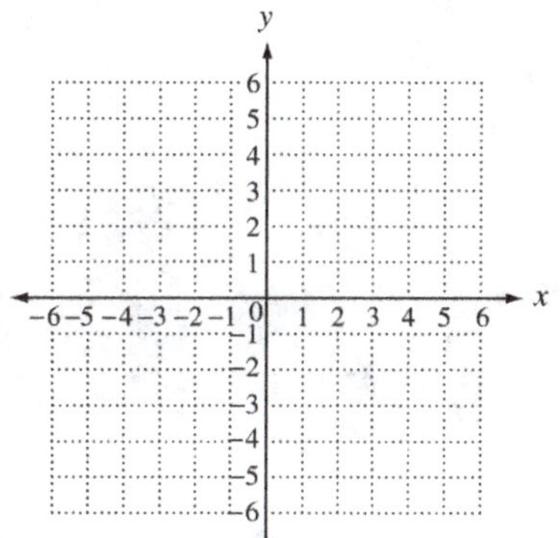

Objective 2 Practice Exercises

For extra help, see Examples 1–2 on pages 695–696 of your text.

Graph each hyperbola.

1. $\dfrac{x^2}{36} - \dfrac{y^2}{49} = 1$

1.

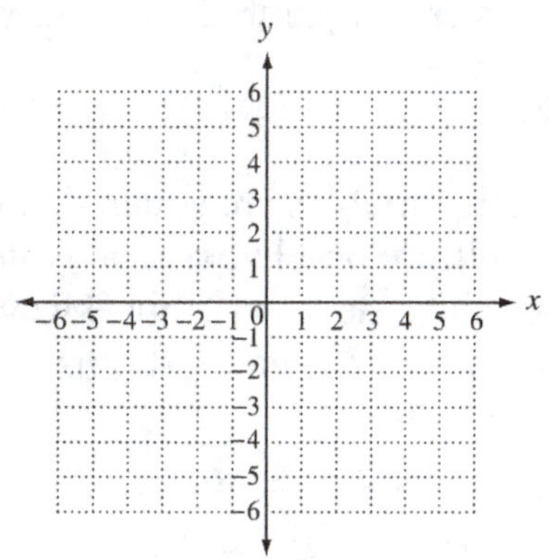

                                    417

2. $\dfrac{y^2}{4} - \dfrac{x^2}{25} = 1$

2.

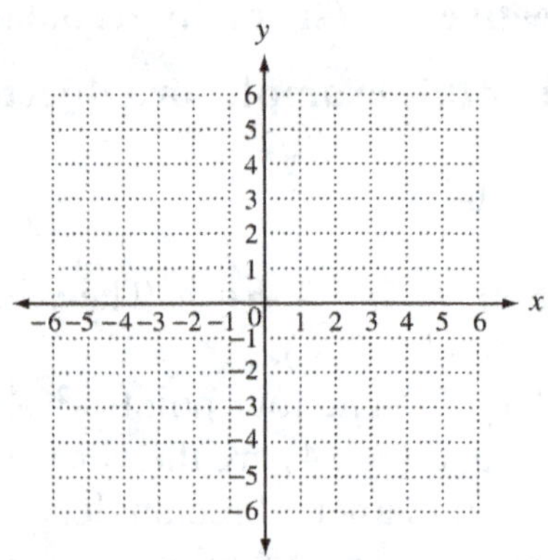

3. $\dfrac{x^2}{16} - y^2 = 1$

3.

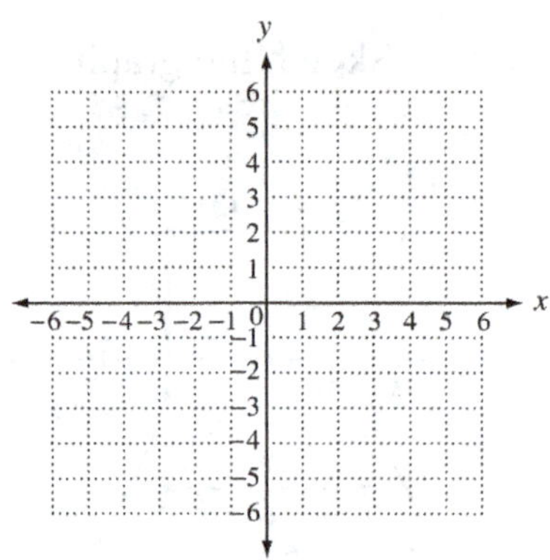

Objective 3   Identify conic sections using their equations.

Review these examples for Objective 3:

3. Identify the graph of each equation.

  a. $3x^2 + y = 36$

Only one of the variables is squared, so this is the vertical parabola $y = -3x^2 + 36$.

  b. $4x^2 = 36 - 4y^2$

Both variables are squared, so the graph is either an ellipse or a hyperbola. Note that a circle is a special case of an ellipse. Rewrite the equation so that the x^2- and y^2-terms are on one side of the equation.

$$4x^2 + 4y^2 = 36$$
$$x^2 + y^2 = 9$$

The graph of this equation is a circle with center $(0, 0)$ and radius 3.

Now Try:

3. Identify the graph of each equation.

  a. $5x^2 - 6y^2 = 30$

    ____

  b. $4y^2 = 12 - 3x^2$

    ____

c. $4x^2 = 36 + 4y^2$ **c.** $2x - 9 = y^2$

Again, both variables are squared, so the graph is
either an ellipse or a hyperbola. Rewrite the   _____________
equation so that the x^2- and y^2-terms are on one
side of the equation.

$$4x^2 - 4y^2 = 36$$

$$\frac{x^2}{9} - \frac{y^2}{9} = 1$$

The graph of this equation is a hyperbola.

Objective 3 Practice Exercises

For extra help, see Example 3 on page 696 of your text.

Identify the graph of each equation as a parabola, circle, ellipse, or hyperbola.

4. $2x^2 + y^2 = 16$ **4.** _____________________

5. $25y^2 + 100 = 4x^2$ **5.** _____________________

6. $5x^2 = 25 - 5y^2$ **6.** _____________________

Objective 4 Graph generalized square root functions.

Review these examples for Objective 4: | **Now Try:**

4. Graph $f(x) = \sqrt{36 - x^2}$. Give the domain and range.

$$f(x) = \sqrt{36 - x^2}$$
$$y = \sqrt{36 - x^2}$$
$$y^2 = 36 - x^2$$
$$x^2 + y^2 = 36$$

This is the graph of a circle with center at $(0, 0)$ and radius 6. Since $f(x)$ represents a principal square root in the original equation, $f(x)$ must be nonnegative. This restricts the graph to the upper half of the circle.

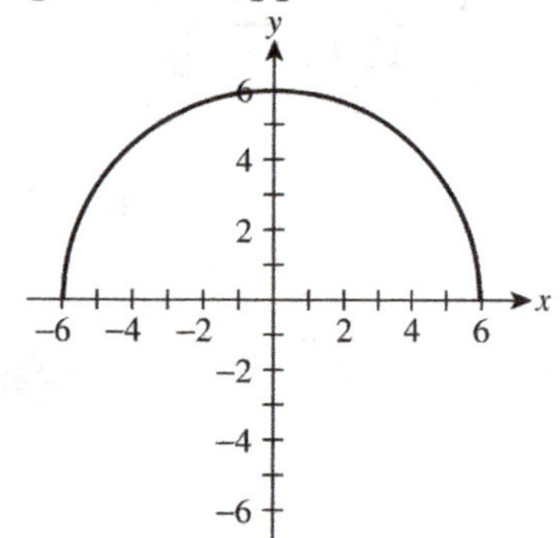

Domain: $[-6, 6]$; range: $[0, 6]$.

Now Try:

4. Graph $f(x) = \sqrt{9 - x^2}$. Give the domain and range.

Domain: _____________

Range: _____________

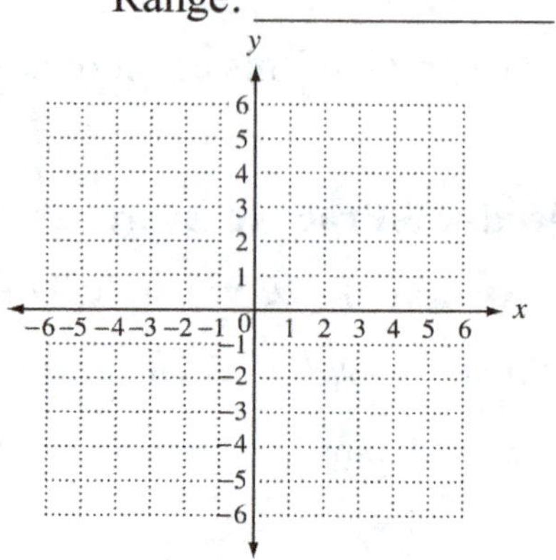

                Copyright © 2020 Pearson Education, Inc.

Name: ________________________   Date: ________________________

Instructor: ________________________   Section: ________________________

5. Graph $f(x) = -3\sqrt{1 - \dfrac{x^2}{4}}$. Give the domain and range.

$$f(x) = -3\sqrt{1 - \frac{x^2}{4}}$$

$$y^2 = 9\left(1 - \frac{x^2}{4}\right)$$

$$\frac{y^2}{9} = 1 - \frac{x^2}{4}$$

$$\frac{x^2}{4} + \frac{y^2}{9} = 1$$

This is the equation of an ellipse with x-intercepts $(2, 0)$ and $(-2, 0)$ and y-intercepts $(0, 3)$ and $(0, -3)$.

Since the original equation has a negative square root, y must be nonpositive, restricting the graph to the lower half of the ellipse.

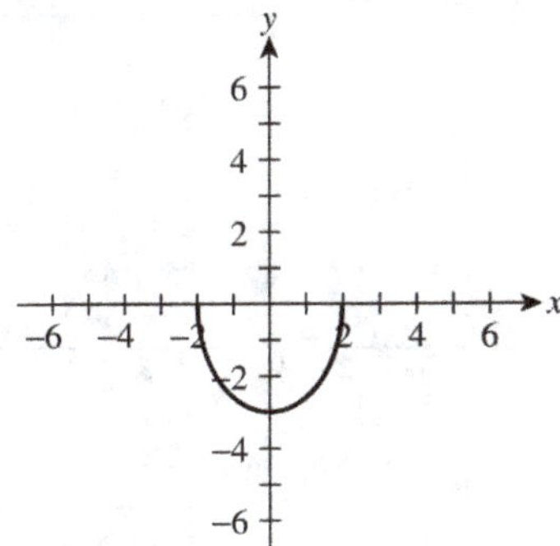

Domain: $[-2, 2]$; range: $[0, -3]$.

5. Graph $f(x) = \sqrt{9 - 9x^2}$. Give the domain and range.

Domain: ________________________

Range: ________________________

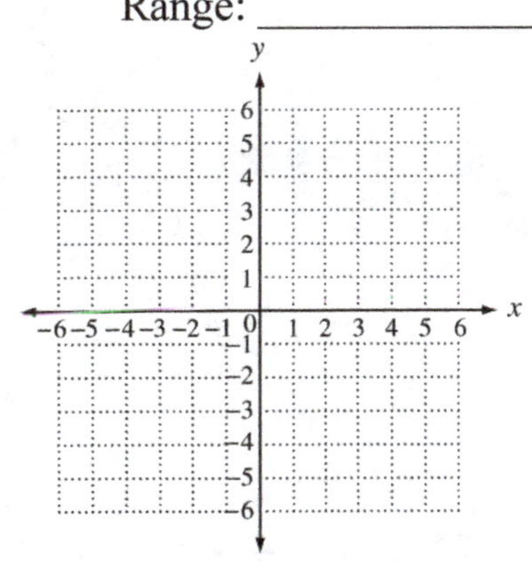

Objective 4 Practice Exercises

For extra help, see Examples 4–5 on pages 698–699 of your text.

Graph each generalized square root function. Give the domain and range.

7. $f(x) = -\sqrt{16 - x^2}$

7. Domain: _______________

Range: _______________

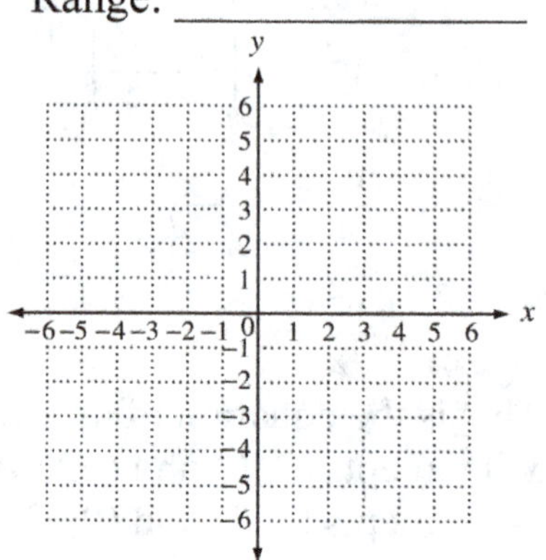

8. $f(x) = -\sqrt{1 + x}$

8. Domain: _______________

Range: _______________

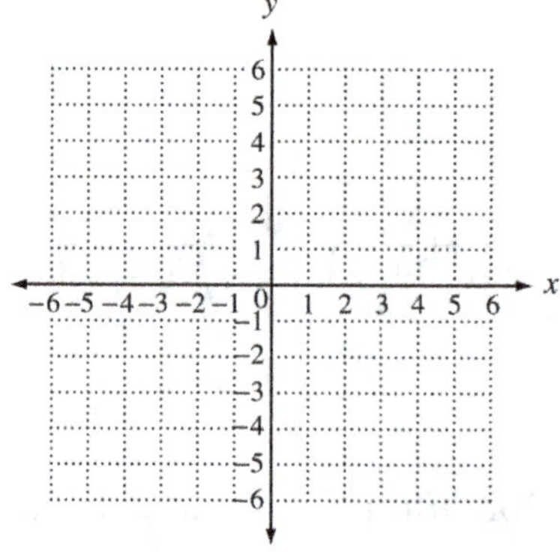

9. $f(x) = -3\sqrt{1 + \dfrac{x^2}{25}}$

9. Domain: _______________

Range: _______________

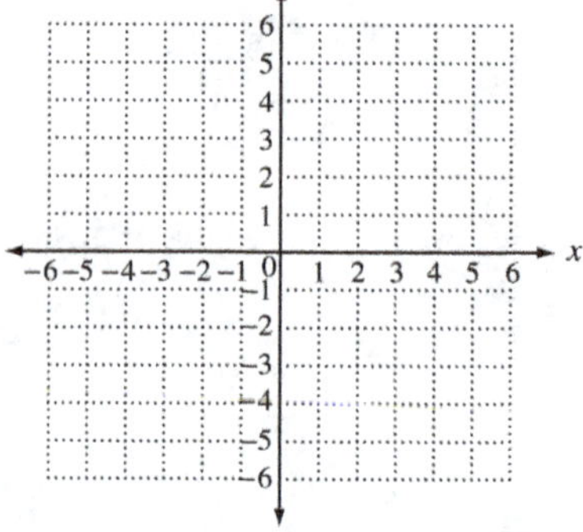

Chapter 10 NONLINEAR FUNCTIONS, CONIC SECTIONS, AND NONLINEAR SYSTEMS

10.4   Nonlinear Systems of Equations

Learning Objectives
1   Solve a nonlinear system using substitution.
2   Solve a nonlinear system with two second-degree equations using elimination.
3   Solve a nonlinear system that requires a combination of methods.

Key Terms

Use the vocabulary terms listed below to complete each statement in exercises 1–2.

nonlinear equation   **nonlinear system of equations**

1.   An equation in which some terms have more than one variable or a variable of degree 2 or greater is a ________________________________.

2.   A system with at least one nonlinear equation is a ____________________.

Objective 1   Solve a nonlinear system using substitution.

Review these examples for Objective 1:

1.   Solve the system.

$$x^2 + 3y^2 = 3 \quad (1)$$

$$x + y = -1 \quad (2)$$

The graph of (1) is an ellipse and the graph of (2) is a line, so the graphs could intersect in zero, one, or two points.

Solve (2) for x, then substitute that expression into (1) and solve for y.

$$x + y = -1 \quad (2)$$

$$x = -y - 1$$

$$x^2 + 3y^2 = 3 \quad (1)$$

$$(-y - 1)^2 + 3y^2 = 3$$

$$y^2 + 2y + 1 + 3y^2 = 3$$

$$4y^2 + 2y - 2 = 0$$

$$2(2y - 1)(y + 1) = 0$$

$$(2y - 1)(y + 1) = 0$$

$$2y - 1 = 0 \quad \text{or} \quad y + 1 = 0$$

$$y = \tfrac{1}{2} \qquad\qquad y = -1$$

Now Try:

1.   Solve the system.

$$x^2 + y^2 = 17$$

$$x + y = -3$$

____

To solve for x, substitute $y = \frac{1}{2}$ in (2): $x = -\frac{3}{2}$.

To solve for x, substitute $y = -1$ in (1): $x = 0$.

The solution set is $\left\{\left(-\frac{3}{2}, \frac{1}{2}\right), (0, -1)\right\}$.

2. Solve the system. $$xy = -10 \quad (1)$$ $$2x - y = 9 \quad (2)$$	**2.** Solve the system. $$xy = 1$$ $$x + y = 2$$

The graph of (1) is a hyperbola and the graph of (2) is a line. There may be zero, one, or two points of intersection. Since neither equation has a squared term, solve either equation for one of the variables and then substitute the result into the other equation.

Solving (1) for y gives $y = -\frac{10}{x}$. Substituting into (2) gives

$$2x - \left(-\frac{10}{x}\right) = 9$$

$$2x + \frac{10}{x} = 9$$

$$x\left[2x + \frac{10}{x}\right] = 9x$$

$$2x^2 + 10 = 9x$$

$$2x^2 - 9x + 10 = 0$$

$$(2x - 5)(x - 2) = 0$$

$$2x - 5 = 0 \quad \text{or} \quad x - 2 = 0$$

$$x = \frac{5}{2} \qquad\qquad x = 2$$

To solve for y, substitute $x = \frac{5}{2}$ in (1): $y = -4$.

To solve for y, substitute $x = 2$ in (1): $y = -5$.

The solution set is $\left\{(2, -5), \left(\frac{5}{2}, -4\right)\right\}$.

Objective 1 Practice Exercises

For extra help, see Examples 1–2 on pages 702–704 of your text.

Solve each system by the substitution method.

1. $\quad 2x^2 - y^2 = -1$ $\qquad 2x + y = 7$	**1.** _________

2. $xy = -6$

       $x + y = 1$

2. _______________

3. $xy = 24$

       $y = 2x + 2$

3. _______________

Objective 2   **Solve a nonlinear system with two second-degree equations using elimination.**

Review this example for Objective 2:

3. Solve the system.

$$3x^2 + y^2 = 35 \quad (1)$$
$$2x^2 - y^2 = 15 \quad (2)$$

The graph of (1) is an ellipse and the graph of (2) is a hyperbola. There may be zero, one, or two points of intersection. Adding the two equations will eliminate y.

$$3x^2 + y^2 = 35 \quad (1)$$
$$\underline{2x^2 - y^2 = 15} \quad (2)$$
$$5x^2 = 50$$
$$x^2 = 10$$
$$x = \pm\sqrt{10}$$

Substitute the values for x in (1) and solve for y.

For $x = \sqrt{10}$, $3(\sqrt{10})^2 + y^2 = 35$
$$30 + y^2 = 35$$
$$y^2 = 5$$
$$y = \pm\sqrt{5}$$

For $x = -\sqrt{10}$, $3(-\sqrt{10})^2 + y^2 = 35$
$$30 + y^2 = 35$$
$$y^2 = 5$$
$$y = \pm\sqrt{5}$$

The solution set is $\{(\sqrt{10}, -\sqrt{5}), (\sqrt{10}, \sqrt{5}),$
$(-\sqrt{10}, \sqrt{5}), (-\sqrt{10}, -\sqrt{5})\}.$

Now Try:

3. Solve the system.

$$x^2 - 2y = 8$$
$$x^2 + y^2 = 16$$

____

Objective 2 Practice Exercises

For extra help, see Example 3 on pages 704–705 of your text.

Solve each system by the elimination method.

4. $5x^2 - y^2 = 55$

     $2x^2 + y^2 = 57$

4.   ______________

5. $x^2 + 2y^2 = 11$

     $2x^2 - y^2 = 17$

5.   ______________

6. $3x^2 + 2y^2 = 30$

     $2x^2 + y^2 = 17$

6.   ______________

Objective 3   Solve a nonlinear system that requires a combination of methods.

Review this example for Objective 3:

4.   Solve the system.

$$x^2 + 5xy - y^2 = 13 \quad (1)$$
$$x^2 - \qquad y^2 = 3 \quad (2)$$

We will use the elimination method in combination with the substitution method. Multiply Eq (2) by -1 and add to Eq (1).

$$x^2 + 5xy - y^2 = 13 \quad (1)$$
$$-x^2 + \qquad y^2 = -3 \quad (2)$$
$$\overline{\qquad\qquad 5xy = 10}$$
$$y = \frac{2}{x} \quad (3)$$

Now Try:

4.   Solve the system.

$$5x^2 - xy + 5y^2 = 89$$
$$x^2 + \qquad y^2 = 17$$

____

Now substitute $\frac{2}{x}$ for y in (2) and solve for x.

$$x^2 - \left(\frac{2}{x}\right)^2 = 3$$

$$x^2 - \frac{4}{x^2} = 3$$

$$x^4 - 4 = 3x^2$$

$$x^4 - 3x^2 - 4 = 0$$

$$(x-2)(x+2)(x^2+1) = 0$$

Using the zero-factor property, we have
$x = 2$, $x = -2$, $x = i$, $x = -i$.
Substitute each of these values into (3) and solve
for y.
If $x = 2$, then $y = 1$.
If $x = -2$, then $y = -1$.
If $x = i$, then $y = \frac{2}{i} = \frac{2}{i} \cdot \frac{-i}{-i} = \frac{-2i}{-i^2} = -2i$.

If $x = -i$, then $y = \frac{2}{-i} = \frac{2}{-i} \cdot \frac{i}{i} = \frac{2i}{-i^2} = 2i$.

It is important to check all answers in the
original equations because it is possible to
obtain extraneous solutions. The solution
set is $\{(2,\ 1),\ (-2,\ -1),\ (i,\ -2i),\ (-i,\ 2i)\}$.

Objective 3 Practice Exercises

For extra help, see Example 4 on pages 705–706 of your text.

Solve each system.

7. $4x^2 - 2xy + 4y^2 = 64$
          $x^2 + y^2 = 13$

7. _______________

8. $x^2 + 3xy + 2y^2 = 12$
       $-x^2 + 8xy - 2y^2 = 10$

8. _______________

9. $x^2 + 5xy - y^2 = 20$
          $x^2 - 2xy - y^2 = -8$

9. _______________

Chapter 10 NONLINEAR FUNCTIONS, CONIC SECTIONS, AND NONLINEAR SYSTEMS

10.5   Second-Degree Inequalities and Systems of Inequalities

Learning Objectives
1   Graph second-degree inequalities.
2   Graph the solution set of a system of inequalities.

Key Terms

Use the vocabulary terms listed below to complete each statement in exercises 1–2.

second-degree inequality   **system of inequalities**

1.   A _________________________________ consists of two or more inequalities to be solved at the same time.

2.   A(n)____ is an inequality with at least one variable of degree 2 and no variable with degree greater than 2.

Objective 1   Graph second-degree inequalities.

Review these examples for Objective 1:

2.   Graph $y < -x^2 + 3$.

The boundary, $y = -x^2 + 3$, is a parabola that opens down with vertex $(0, 3)$.
Use $(0, 0)$ as a test point.

$$y < -x^2 + 3$$
$$0 \overset{?}{<} -0^2 + 3$$
$$0 < 3 \text{ True}$$

Because the final inequality is a true statement, the points in the region containing $(0, 0)$ satisfy the inequality. The parabola is drawn as a dashed curve since the points on the parabola itself do not satisfy the inequality and the region inside (or below) the parabola is shaded.

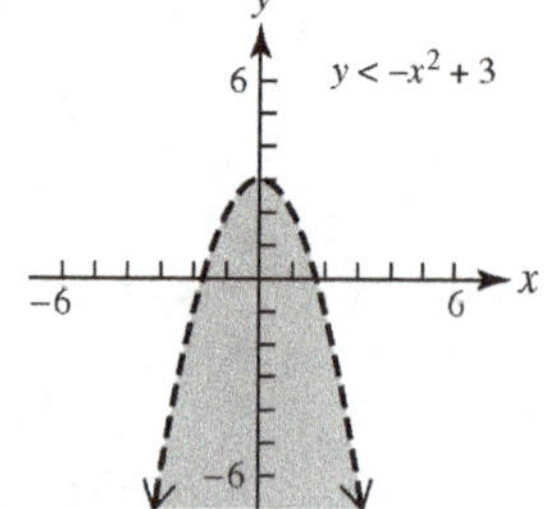

Now Try:

2.   Graph $y \geq x^2 - 4$.

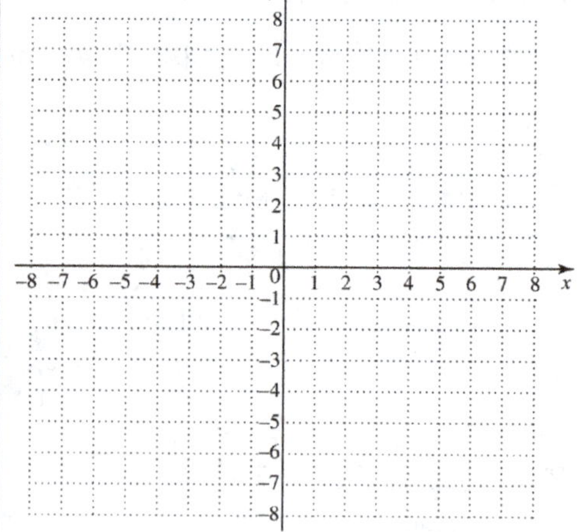

Copyright © 2020 Pearson Education, Inc.

3. Graph $16x^2 < 9y^2 + 144$.

Rewrite the inequality.

$$16x^2 - 9y^2 < 144$$

$$\frac{x^2}{9} - \frac{y^2}{16} < 1$$

The boundary, drawn as a dashed curve, is the following hyperbola. $\dfrac{x^2}{9} - \dfrac{y^2}{16} = 1$

Since the graph is a horizontal hyperbola, the desired region will be either between the branches or the regions to the right of the right branch and to the left of the left branch. Using the test point $(0, 0)$ into the original inequality leads to $0 < 1$, a true statement. So the region between the branches containing $(0, 0)$ is shaded.

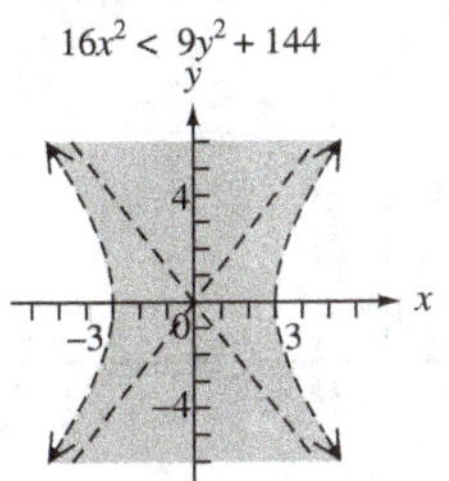

3. Graph $9y^2 - 36x^2 > 144$.

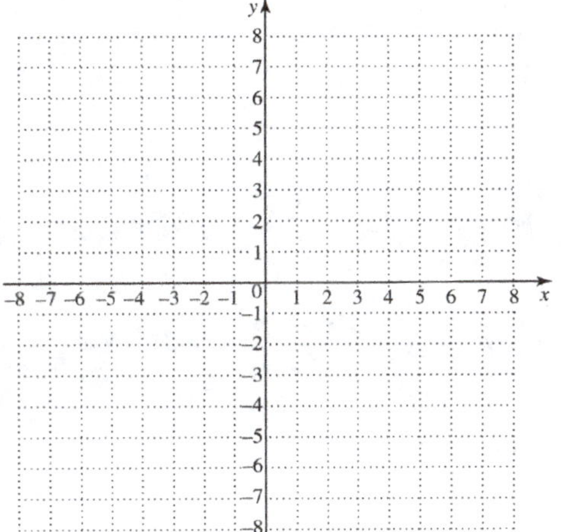

Objective 1 Practice Exercises

For extra help, see Examples 1–3 on pages 709–710 of your text.

Graph each inequality.

1. $x \le 2y^2 + 8y + 9$

1.

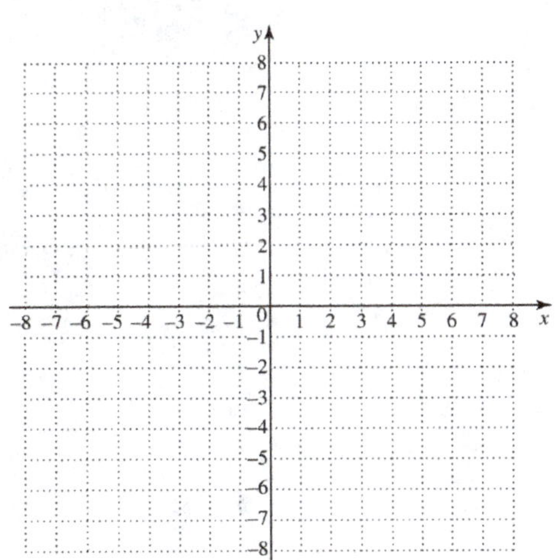

2. $x^2 + 9y^2 > 36$

2.

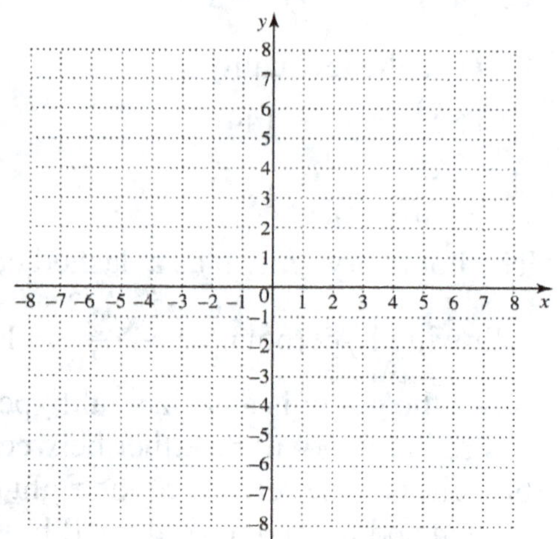

3. $9x^2 - y^2 < 36$

3.

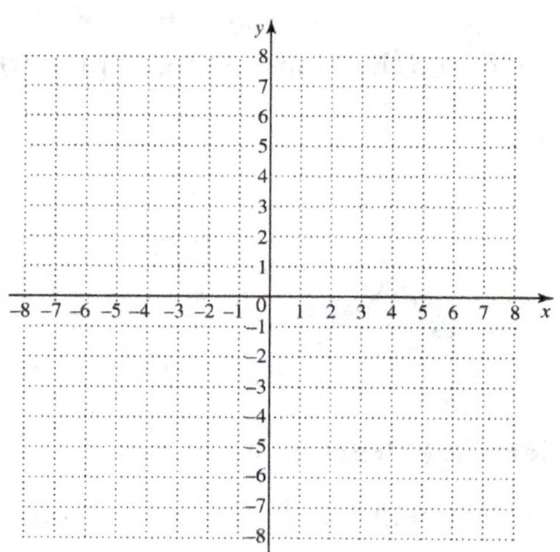

Objective 2   Graph the solution set of a system of inequalities.

Review these examples for Objective 2:

5. Graph the solution set of the system.

$$x^2 + y^2 \le 16$$
$$y > x$$

Begin by graphing $x^2 + y^2 \le 16$. The boundary line is a circle centered at the origin with radius 4. The test point $(0, 0)$ leads to a true statement, so we shade inside the circle.

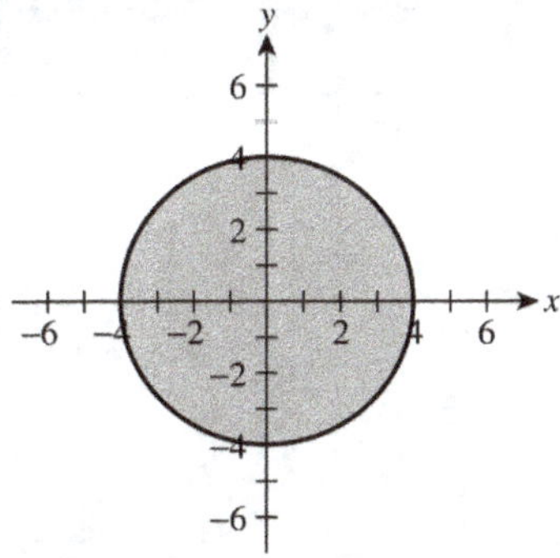

The boundary of the solution set of $y > x$ is a dashed line passing through $(0, 0)$. Using the test point $(0, 1)$ leads to a true statement, so shade above the line.

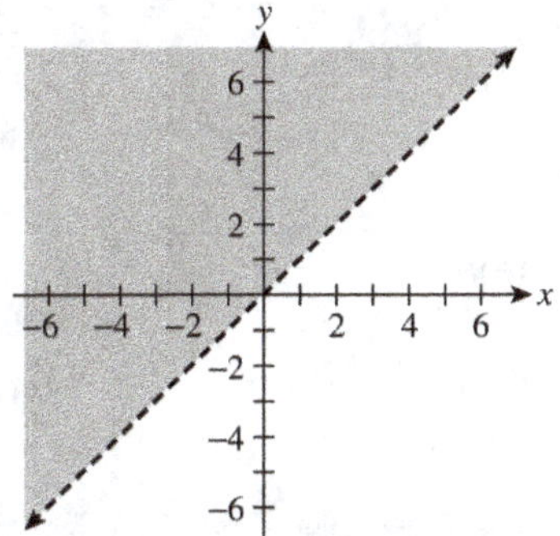

The graph of the solution set of the system is the intersection of the graphs of the two inequalities.

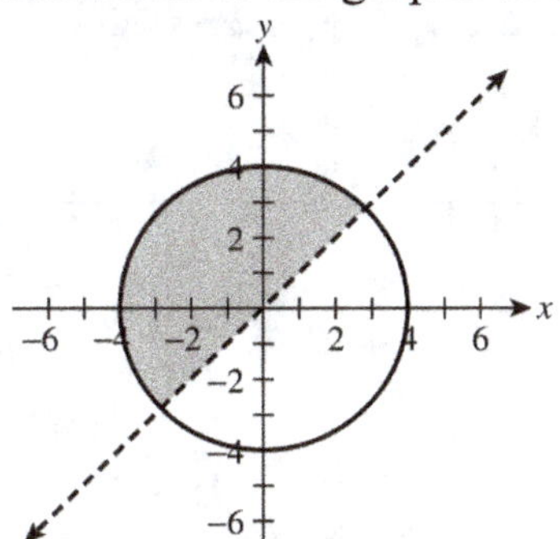

Now Try:

5. Graph the solution set of the system.

$$4y + x^2 < 0$$
$$x \ge 0$$

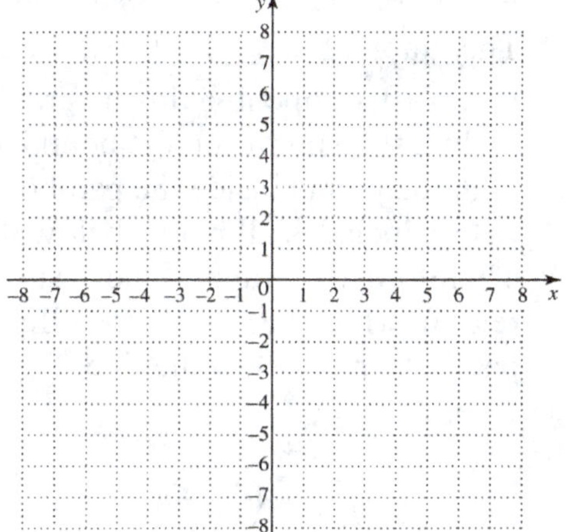

7. Graph the solution set of the system.

$$25x^2 + 9y^2 < 225$$
$$y \le -x^2 + 4$$
$$y < -x$$

The graph of $25x^2 + 9y^2 < 225$ is a dashed ellipse with $a = 3$ and $y = 5$. To satisfy the inequality, a point must lie inside the ellipse.

The graph of $y \le -x^2 + 4$ is a parabola with vertex (0, 4) opening downward. The inequality includes the points on the boundary along with the points inside the parabola. The graph of $y < -x$ includes all points below the line $y = -x$. Therefore, the graph of the system is the shaded region, which lies inside the ellipse and the parabola, and below the line.

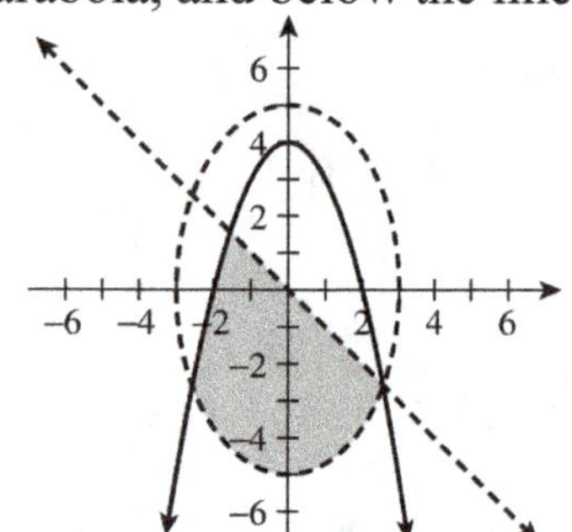

7. Graph the solution set of the system.

$$y \ge (x+2)^2 - 5$$
$$2x + y < -5$$
$$(x+3)^2 + (y-1)^2 < 9$$

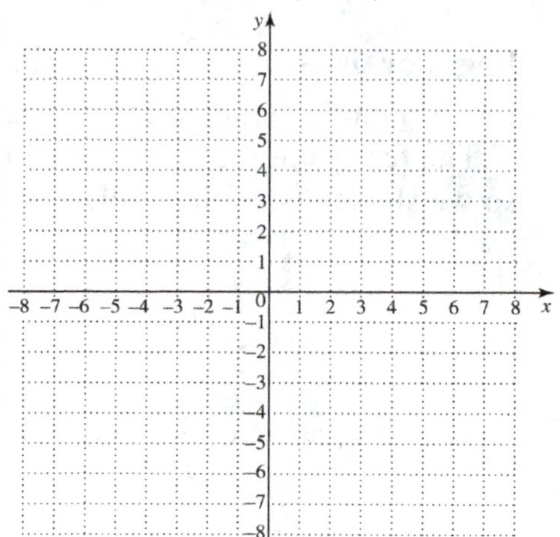

Objective 2 Practice Exercises

For extra help, see Examples 4–7 on pages 710–712 of your text.

Graph each system of inequalities.

4. $-x + y > 2$
   $3x + y > 6$

4.

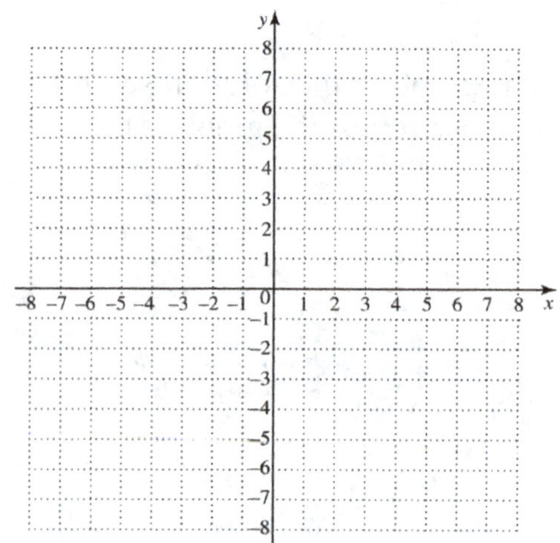

5. $x^2 + y^2 \leq 25$

$3x - 5y > -15$

5.

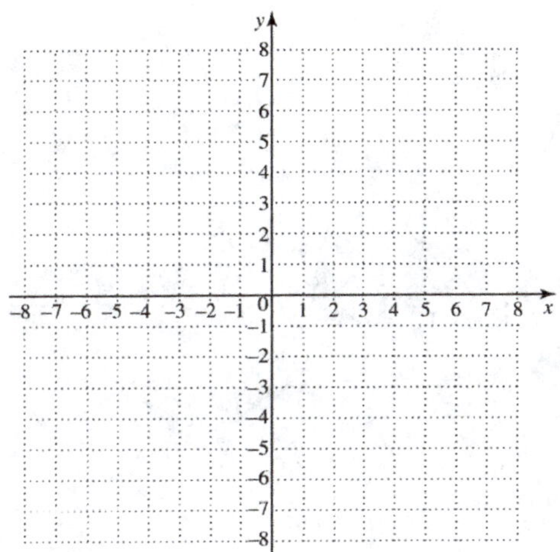

6. $x^2 + 4y^2 \leq 36$

$-5 < x < 2$

$y \geq 0$

6.

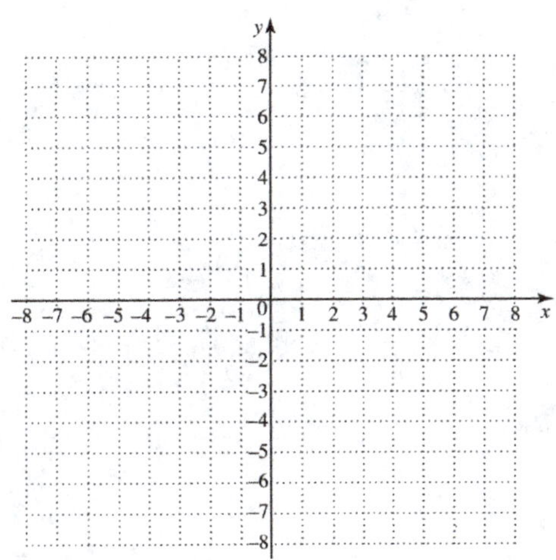

Chapter 11   FURTHER TOPICS IN ALGEBRA

11.1   Sequences and Series

Learning Objectives	
1	Define infinite and finite sequences.
2	Find the terms of a sequence, given the general term.
3	Find the general term of a sequence.
4	Use sequences to solve applied problems.
5	Use summation notation to evaluate a series.
6	Write a series using summation notation.
7	Find the arithmetic mean (average) of a group of numbers.

Key Terms

Use the vocabulary terms listed below to complete each statement in exercises 1–8.

infinite sequence	**finite sequence**	**terms of a sequence**
general term	**series**	**summation notation**
index of summation	**arithmetic mean (average)**	

1. A(n) _________________ is a function whose domain is the set of positive integers.

2. A _________________ is the sum of the terms of a sequence.

3. _________________ is a compact way of writing a series using the general term of the corresponding sequence.

4. When using summation notation $\sum\limits_{i}^{n} f(i)$, the letter i is called the _________________.

5. The expression a_n which defines a sequence, is called the _________________ of the sequence.

6. The _________________ of a group of numbers is given by the formula $\dfrac{\sum\limits_{i=1}^{n} x_i}{n}$.

7. The function values $a_1, a_2, a_3, \ldots$, written in order, are the _________________.

8. A(n) _________________ is a function with domain of the form $\{1, 2, 3, \ldots, n\}$, where n is a positive integer.

                435

Objective 1   Define infinite and finite sequences.

For extra help, see page 726 of your text.

Objective 2   Find the terms of a sequence, given the general term.

Review this example for Objective 1:

1. Given an infinite sequence with $a_n = 3n - 2$, find a_4.

$$a_4 = 3(4) - 2 = 10$$

Now Try:

1. Given an infinite sequence with $a_n = \dfrac{n+2}{5n}$ find a_6.

____

Objective 2 Practice Exercises

For extra help, see Example 1 on page 726 of your text.

Write out the first five terms of each sequence.

1. $a_n = (-1)^n$

1. _________________

2. $a_n = \dfrac{1+n}{n}$

2. _________________

Find the indicated term for the sequence.

3. $a_n = \dfrac{3n-2}{5n+2};\ a_8$

3. _________________

Objective 3   Find the general term of a sequence.

Review this example for Objective 3:

2. Find an expression for the general term a_n of the sequence $1,\ \dfrac{1}{2},\ \dfrac{1}{3},\ \dfrac{1}{4},\ \dfrac{1}{5},\ ....$

    Notice that the denominators are consecutive integers, so the general term is $a_n = \dfrac{1}{n}$.

Now Try:

2. Find an expression for the general term a_n of the sequence $2, -4,\ 8, -16,\ 32,\ ....$

____

               Copyright © 2020 Pearson Education, Inc.

Objective 3 Practice Exercises

For extra help, see Example 2 on page 727 of your text.

Find a general term a_n for the given terms of each sequence.

4.   3, 5, 7, 9, 11, ...   4. ________________

5. $\sqrt{3},\ 3,\ 3\sqrt{3},\ 9,\ 9\sqrt{3},\ ...$ 5. ________________

6. $-\dfrac{1}{5},\ \dfrac{1}{10},\ -\dfrac{1}{15},\ \dfrac{1}{20},\ -\dfrac{1}{25},\ ...$ 6. ________________

Objective 4   **Use sequences to solve applied problems.**

Review this example for Objective 4:

3.   Suppose a copier loses $\frac{1}{4}$ of its value each year; that is, at the end of any given year, the value is $\frac{3}{4}$ of its value at the beginning of the year. If a new copier costs $4800, what is its value at the end of 4 years?

At the end of the first year, the value of the copier is $\frac{3}{4}(\$4800) = \$3600$.

At the end of the second year, the value of the copier is $\frac{3}{4}(\$3600) = \$2700$.

At the end of the third year, the value of the copier is $\frac{3}{4}(\$2700) = \$2025$.

At the end of the fourth year, the value of the copier is $\frac{3}{4}(\$2025) = \$1518.75$.

Now Try:

3.   Carla borrows $6000 and agrees to pay $500 monthly plus interest of 2% on the unpaid balance from the beginning of the first month. Find the payments for the first four months and the remaining debt at the end of that period.

     Month 1 _____________

     Month 2 _____________

     Month 3 _____________

     Month 4 _____________

     Remaining balance ______

Objective 4 Practice Exercises

For extra help, see Example 3 on page 727 of your text.

Solve each applied problem by writing the first few terms of each sequence.

7.   A colony of bacteria doubles in weight every hour. If   **7.** ________________
the colony weighs 2 grams at the beginning of an
experiment, find the weight after 3 hours.

8.   Ms. Burley is offered a new job with a salary of   **8.** ________________
\$35,000 per year and a 2% raise at the end of each
year. Write a sequence showing her salary for each
of the first five years. (Round to the nearest dollar.)

Objective 5   Use summation notation to evaluate a series.

Review this example for Objective 5:

4.   Write out the series as a sum of terms and then find the sum.

$$\sum_{i=1}^{6}(2i-6)$$

$$\sum_{i=1}^{6}(2i-6)$$
$$=[2(1)-6]+[2(2)-6]+[2(3)-6]$$
$$+[2(4)-6]+[2(5)-6]+[2(6)-6]$$
$$=-4+(-2)+0+2+4+6$$
$$=6$$

Now Try:

4.   Write out the series as a sum of terms and then find the sum.

$$\sum_{i=1}^{7}(4i-3)$$

____

Objective 5 Practice Exercises

For extra help, see Example 4 on pages 728–729 of your text.

Write out each series as a sum of terms and then find the sum.

9. $\displaystyle\sum_{i=1}^{4}(2i+3)$ **9.** ________________

10. $\displaystyle\sum_{i=1}^{4}\left(-\frac{1}{2i}\right)$

10. ______________

11. $\displaystyle\sum_{i=1}^{6}(i^2+1)$

11. ______________

Objective 6   Write a series using summation notation.

Review this example for Objective 6:

5. Write out the sum using summation notation.
$$3+9+27+81+243$$

Each term is a power of 3, so $a_n=3^n$. Thus, the sum can be written as $\displaystyle\sum_{i=1}^{5}3^i$.

Now Try:

5. Write out the sum using summation notation.
$$4+8+12+16+20+24$$

____

Objective 6 Practice Exercises

For extra help, see Example 5 on page 729 of your text.

Write each series using summation notation.

12. $1+4+9+16+25$

12. ______________

13. $\dfrac{1}{5}+\dfrac{1}{8}+\dfrac{1}{11}+\dfrac{1}{14}+\dfrac{1}{17}$

13. ______________

14. $1+2x+3x^2+4x^3+5x^4$

14. ______________

Objective 7   Find the arithmetic mean (average) of a group of numbers.

Review this example for Objective 7:

6. The amount of rain that fell in Philadelphia from March 2009 through August 2009 is given in the table.

Month	Amount of Precipitation
March	7.55 in.
April	3.85 in.
May	2.61 in.
June	2.15 in.
July	8.33 in.
August	1.08 in.

Source: Franklin Institute

What was the average amount of rain each month for this six-month period?

$$\overline{x} = \dfrac{\sum\limits_{i=1}^{6} x_i}{6}$$

$$= \dfrac{7.55 + 3.85 + 2.61 + 2.15 + 8.33 + 1.08}{6}$$

$$= \dfrac{25.57}{6} \approx 4.26$$

The average amount of rain was about 4.26 inches per month during the six-month period.

Now Try:

6. The amount of rain that fell in Seattle during October from 2004 through 2010 is given in the table.

Month	Amount of Precipitation
2004	2.81 in.
2005	3.01 in.
2006	1.55 in.
2007	3.32 in.
2008	2.17 in.
2009	5.54 in.
2010	5.24 in.

Source: National Weather Service

What is the average amount of rain that fell during October for the seven-year period?

____

Objective 7 Practice Exercises

For extra help, see Example 6 on page 730 of your text.

Find the arithmetic mean for each collection of numbers.

15.   2, 8, 18, 32, –5, –7

15. ______________

16. $\dfrac{1}{3}, -\dfrac{1}{2}, \dfrac{1}{4}, \dfrac{7}{6}$

16. ______________

Chapter 11   FURTHER TOPICS IN ALGEBRA

11.2   Arithmetic Sequences

Learning Objectives
1   Find the common difference of an arithmetic sequence.
2   Find the general term of an arithmetic sequence.
3   Use an arithmetic sequence in an application.
4   Find any specified term or the number of terms of an arithmetic sequence.
5   Find the sum of a specified number of terms of an arithmetic sequence.

Key Terms

Use the vocabulary terms listed below to complete each statement in exercises 1−2.

arithmetic sequence (arithmetic progression)   **common difference**

1. A(n) ___________________ is a sequence in which each term after the first differs from the preceding term by a constant amount.

2. The ___________________ d is the difference between any two adjacent terms of an arithmetic sequence.

Objective 1   Find the common difference of an arithmetic sequence.

Review these examples for Objective 1:

1. Determine the common difference d for the arithmetic sequence.

    35, 32, 29, 26, …

    Since the sequence is arithmetic, d is the difference between any two adjacent terms. We arbitrarily choose the terms 29 and 26.
    $d = 26 - 29 = -3$

2. Write the first five terms of the arithmetic sequence with first term 9 and common difference 5.

    The second term is found by adding 5 to the first term 9, getting 14.
    For the next term, add 5 to 14, and so on.
    The first five terms are 9, 14, 19, 24, 29.

Now Try:

1. Determine the common difference d for the arithmetic sequence.

    $-13, -11, -9, -7, …$

    ____

2. Write the first five terms of the arithmetic sequence with first term 9 and common difference -5.

    ____

Objective 1 Practice Exercises

For extra help, see Examples 1–2 on page 733 of your text.

Find the common difference d for the arithmetic sequence.

1. $-9, -5, -1, 3, \ldots$

1. _______________

2. $\dfrac{1}{4}, -\dfrac{1}{4}, -\dfrac{3}{4}, -\dfrac{5}{4}, \ldots$

2. _______________

3. Write the first five terms of the arithmetic sequence with first term 6 and common difference –3.

3. _______________

Objective 2   Find the general term of an arithmetic sequence.

Review this example for Objective 2:

3. Determine an expression for the general term of the arithmetic sequence where $a_1 = -3$ and $d = -4$.

Now find a_n.
$$a_n = a_1 + (n-1)d$$
$$a_n = -3 + (n-1)(-4)$$
$$a_n = -3 - 4n + 4$$
$$a_n = -4n + 1$$

Now Try:

3. Determine an expression for the general term of the arithmetic sequence where $a_1 = 35$ and $d = 4$.

____

Objective 2 Practice Exercises

For extra help, see Example 3 on page 734 of your text.

Use the formula for a_n to find the general term of each arithmetic sequence.

4. $a_1 = -7, \ d = -5$

4. _______________

5. $\dfrac{1}{2}, \dfrac{5}{6}, \dfrac{7}{6}, \dfrac{3}{2}, \dfrac{11}{6}, \ldots$

5. _______________

6. $0.8, 0.5, 0.2, \ldots, -29.8, \ldots$

6. _______________

Objective 3   Use an arithmetic sequence in an application.

Review this example for Objective 3:

4. After knee surgery, your trainer tells you to return to your jogging program slowly. He suggests jogging for 12 minutes each day for the first week. Each week thereafter, he suggests that you increase that time by 6 minutes per day. How long will you be jogging in the 6th week?

After n weeks, you will be jogging
$$a_n = 12 + 6(n-1) \text{ minutes.}$$
To find how long you will be jogging in the 6th week, find a_6.
$$a_6 = 12 + 6(6-1) = 42$$
You will be jogging 42 minutes in the 6th week.

Now Try:

4. Mary accepts a teaching job that pays \$32,530 the first year with a guarantee of a \$1,030 raise each year thereafter. What will her salary be in her 10th year on the job?

____

Objective 3 Practice Exercises

For extra help, see Example 4 on page 734 of your text.

Solve each problem.

7. Ben's father has started a savings fund for Ben's college education. He makes an initial deposit of \$5000 and each month contributes an additional \$100. How much money will be in the account after 60 months? (Disregard any interest.)

7. _______________

8. Lisa is starting a swimming program. She plans to swim 50 laps per day the first week and add 5 laps per day each week until she is swimming 125 laps. In which week will this occur?

8. _______________

9. The starting salary at a supermarket is \$8.75 per hour. Every six months an employee receives a raise of \$0.75 per hour. What will the employee's salary be after two years?

9. _______________

                    443

Objective 4   Find any specified term or the number of terms of an arithmetic sequence.

Review these examples for Objective 4:

5. Evaluate the indicated term for each arithmetic sequence.

   a. $a_1 = 28,\ d = 10,\ a_{12}$

$$a_n = a_1 + (n-1)d$$
$$a_{12} = 28 + (12-1)(10)$$
$$a_{12} = 138$$

   b. $a_4 = 24,\ a_{10} = -6,\ a_{20}$

Any term can be found if a_1 and d are known. Use the formula for a_n.

$$a_4 = a_1 + (4-1)d \quad\bigg|\quad a_{10} = a_1 + (10-1)d$$
$$a_4 = a_1 + 3d \quad\qquad\bigg|\quad a_{10} = a_1 + 9d$$
$$24 = a_1 + 3d \quad\qquad\bigg|\quad -6 = a_1 + 9d$$

This gives a system of two equations in two variables.

$$24 = a_1 + 3d \quad (1)$$
$$-6 = a_1 + 9d \quad (2)$$

Multiply (2) by -1, then add the resulting equation to (1) to eliminate a_1.

$$24 = a_1 + 3d$$
$$6 = -a_1 - 9d$$
$$30 = -6d$$
$$-5 = d$$

Now find a_1.

$$24 = a_1 + 3(-5)$$
$$24 = a_1 - 15$$
$$39 = a_1$$

Thus, $a_n = 39 - 5(n-1) = 44 - 5n$.

$$a_{20} = 44 - 5(20) = -56$$

Now Try:

5. Evaluate the indicated term for each arithmetic sequence.

   a. $a_1 = 9,\ d = 6,\ a_{15}$

____

   b. $a_3 = 7,\ a_{10} = -14,\ a_{20}$

____

6. Find the number of terms in the arithmetic sequence 7, 10, 13, …, 55.

$a_1 = 7, \quad a_n = 55, \quad d = 10 - 7 = 3$

$a_n = a_1 + (n-1)d$

$55 = 7 + (n-1)3$

$55 = 7 + 3n - 3$

$55 = 4 + 3n$

$51 = 3n$

$17 = n$

There are 17 terms in the sequence.

6. Find the number of terms in the arithmetic sequence

$3, \ \dfrac{9}{2}, \ 6, \ \dfrac{15}{2}, \ 9, \ ..., \ \dfrac{51}{2}.$

____

Objective 4 Practice Exercises

For extra help, see Examples 5–6 on pages 735–736 of your text.

Find the indicated term for each arithmetic sequence.

10. $a_1 = -5, \ d = 9; \ a_{12}$

10. ______________

11. $a_1 = -\dfrac{1}{2}, \ d = \dfrac{3}{2}; \ a_{15}$

11. ______________

12. Find the number of terms in the arithmetic sequence 19, 26, 33, …, 96.

12. ______________

Objective 5   Find the sum of a specified number of terms of an arithmetic sequence.

Review these examples for Objective 5:

7. Find the sum of the first eight terms of the sequence in which $a_n = 7 - 2n$.

Begin by evaluating a_1 and a_8.

$a_1 = 7 - 2(1) = 5$

$a_8 = 7 - 2(8) = -9$

Now find the sum using $a_1 = 5$, $a_8 = -9$, and $n = 8$.

$S_n = \dfrac{n}{2}\left(a_1 + a_n\right)$

$S_8 = \dfrac{8}{2}\left(5 + (-9)\right) = 4(-4) = -16$

Now Try:

7. Find the sum of the first six terms of the sequence in which $a_n = 3n - 4$.

____

                     445

8. Find the sum of the first 15 terms of the arithmetic sequence having first term 5 and common difference -4.

$$a_1 = 5, \ d = -4, \ n = 15$$

$$S_n = \frac{n}{2}\left[2a_1 + (n-1)d\right]$$

$$S_{15} = \frac{15}{2}\left[2(5) + (15-1)(-4)\right]$$

$$= \frac{15}{2}(-46)$$

$$= -345$$

8. Find the sum of the first 13 terms of the arithmetic sequence having first term 2 and common difference 10.

____

9. Evaluate $\sum\limits_{i=1}^{15}(3i+2)$.

Begin by evaluating a_1 and a_{15}.

$$a_1 = 3(1)+2 = 5$$

$$a_{15} = 3(15)+2 = 47$$

Now find the sum using $a_1 = 5$, $a_{15} = 47$, and $n = 15$.

$$S_n = \frac{n}{2}\left(a_1 + a_n\right)$$

$$S_{15} = \frac{15}{2}\left[5+47\right] = \frac{15}{2}(52) = 390$$

9. Evaluate $\sum\limits_{i=1}^{10}(7i-2)$.

____

Objective 5 Practice Exercises

For extra help, see Examples 7–9 on pages 736–738 of your text.

Solve each problem.

13. Find S_{10} for the arithmetic sequence defined by $a_n = 2 - 3n$.

13. ______________

14. Find S_8 for the arithmetic sequence with $a_1 = 3$ and $d = 2$.

14. ______________

15. Evaluate $\sum\limits_{i=1}^{10}(2i+3)$.

15. ______________

Chapter 11   FURTHER TOPICS IN ALGEBRA

11.3   Geometric Sequences

Learning Objectives
1   Find the common ratio of a geometric sequence.
2   Find the general term of a geometric sequence.
3   Find any specified term of a geometric sequence.
4   Find the sum of a specified number of terms of a geometric sequence.
5   Apply the formula for the future value of an ordinary annuity.
6   Find the sum of an infinite number of terms of a certain geometric sequence.

Key Terms

Use the vocabulary terms listed below to complete each statement in exercises 1−7.

geometric sequence (geometric progression)   **common ratio**

annuity   **ordinary annuity**   **payment period**

future value of an annuity   **term of an annuity**

1. A(n) ___________________ r is the constant multiplier between adjacent terms in a geometric sequence.

2. A(n) ___________________ is a sequence of equal payments made at equal periods of time.

3. A(n) ___________________ is a sequence in which each term after the first is a constant multiple of the preceding term.

4. The ___________________ is the sum of the compound amounts of all the payments, compounded to the end of the term.

5. If the payments for an annuity are made at the end of the period, and if the frequency of payments is the same as the frequency of compounding, the annuity is called a(n) ___________________.

6. The time from the beginning of the first payment period to the end of the last is called the ___________________.

7. The time between payments of an annuity is called the

    ___________________.

Objective 1   Find the common ratio of a geometric sequence.

Review this example for Objective 1:

1. Determine the common ratio r for the geometric sequence.

   $4, -8, 16, -32, 64, \ldots$

   To find r, choose any two successive terms and divide the second one by the first. We choose the third and fourth terms of the sequence.

   $$r = \frac{a_4}{a_3} = \frac{-32}{16} = -2$$

Now Try:

1. Determine the common ratio r for the geometric sequence.

   $3, \dfrac{3}{4}, \dfrac{3}{16}, \dfrac{3}{64}, \ldots$

   ____

Objective 1 Practice Exercises

For extra help, see Example 1 on page 741 of your text.

Find the common ratio for the geometric sequence.

1. $10, \ 10\sqrt{2}, \ 20, \ 20\sqrt{2}, \ 40, \ \ldots$

    1. _______________

2. $-\dfrac{1}{2}, \dfrac{1}{4}, -\dfrac{1}{8}, \dfrac{1}{16}, -\dfrac{1}{32}, \ldots$

    2. _______________

Objective 2   Find the general term of a geometric sequence.

Review this example for Objective 2:

2. Determine an expression for the general term of the sequence.

   $-3, -15, -75, -375, \ldots$

   The first term is $a_1 = -3$ and the common ratio is $r = 5$.

   $$a_n = a_1 r^{n-1} = -3(5)^{n-1}$$

Now Try:

2. Determine an expression for the general term of the sequence.

   $\sqrt{3}, \ 3, \ 3\sqrt{3}, \ 9, \ 9\sqrt{3}, \ 27, \ \ldots$

   ____

Name: Date:
Instructor: Section:

Objective 2 Practice Exercises

For extra help, see Example 2 on page 741 of your text.

Determine an expression for the general term of the geometric sequence.

3.   6, 12, 24, 48, 96, …   3. _______________

4. $-\dfrac{2}{3}, \dfrac{2}{9}, -\dfrac{2}{27}, \dfrac{2}{81}, \dots$ 4. _______________

5. $-\dfrac{4}{5}, -\dfrac{4}{25}, -\dfrac{4}{125}, -\dfrac{4}{625}, \dots$ 5. _______________

Objective 3   Find any specified term of a geometric sequence.

Review these examples for Objective 3:

3.   Evaluate the indicated term for each geometric sequence.

  a. $a_1 = 64$, $r = -\dfrac{1}{2}$; a_6

  $a_n = a_1 r^{n-1}$

  $a_6 = 64\left(-\dfrac{1}{2}\right)^{6-1} = -2$

  b. 4, 12, 36, 108, …; a_9

  $a_n = a_1 r^{n-1}$

  $a_9 = 4(3)^{9-1} = 26{,}244$

4.   Write the first five terms of the geometric sequence whose first term is -3 and whose common ratio is $\dfrac{2}{3}$.

  Use $a_n = a_1 r^{n-1}$ with $a_1 = -3$, $r = \dfrac{2}{3}$, and $n = 1, 2, 3, 4, 5$.

Now Try:

3.   Evaluate the indicated term for each geometric sequence.

  a. $a_1 = -1$, $r = 3$; a_7

  ____

  b. $\dfrac{4}{3}, \dfrac{2}{3}, \dfrac{1}{3}, \dfrac{1}{6}, \dots$; a_{10}

  ____

4.   Write the first five terms of the geometric sequence whose first term is 0.8 and whose common ratio is -5.

  ____

                    449

Objective 3 Practice Exercises

For extra help, see Examples 3–4 on page 742 of your text.

Evaluate the indicated term for each geometric sequence.

6. $a_1 = -4,\ r = 2;\ a_7$

6. _________________

7. $a_1 = \dfrac{2}{3},\ r = -3;\ a_6$

7. _________________

8. Write the first five terms of the geometric sequence whose first term is $\dfrac{1}{2}$ and common ratio 5.

8. _________________

Objective 4   Find the sum of a specified number of terms of a geometric sequence.

Review these examples for Objective 4:

5. Evaluate the sum of the first seven terms of the geometric sequence with first term 36 and common ratio −6.

$$S_n = \frac{a_1\left(1 - r^n\right)}{1 - r}$$

$$S_7 = \frac{36\left(1 - (-6)^7\right)}{1 - (-6)}$$

$$= \frac{36\left(1 - (-279,936)\right)}{7}$$

$$= \frac{36(279,937)}{7}$$

$$= 1,439,676$$

Now Try:

5. Evaluate the sum of the first six terms of the geometric sequence with first term $\frac{2}{3}$ and common ratio 3.

____

6. Evaluate $\displaystyle\sum_{i=1}^{8} 4(3)^i$.

Since the series is in the form $\displaystyle\sum_{i=1}^{n} a \cdot b^i$, it represents the sum of the first n terms of the geometric sequence with $a_1 = a \cdot b^1$ and $r = b$. $a_1 = 4 \cdot (3)^1 = 12$ and $r = 3$.

$$S_n = \frac{a_1\left(1 - r^n\right)}{1 - r}$$

$$S_8 = \frac{12\left(1 - 3^8\right)}{1 - 3}$$

$$= \frac{12(1 - 6561)}{-2}$$

$$= 39,360$$

6. Evaluate $\displaystyle\sum_{i=1}^{5} \frac{1}{2}\left(4^i\right)$.

____

Objective 4 Practice Exercises

For extra help, see Examples 5–6 on page 743 of your text.

9. Evaluate the sum of the first seven terms of the geometric sequence having first term 2 and common ratio 3.

9. _____________

Evaluate each sum.

10. $\displaystyle\sum_{i=1}^{12} 2^i$

10. _____________

11. $\displaystyle\sum_{i=1}^{7} 3(-2)^i$

11. _____________

Objective 5 Apply the formula for the future value of an ordinary annuity.

| **Review these examples for Objective 5:** | **Now Try:** |

Review these examples for Objective 5:

7. Work each problem. Give answers to the nearest cent.

a. Matt decides to start a college fund for his son. He plans on depositing \$1500 at the end of each year into an account paying 4.5% per year, compounded annually. How much will be in the account after 18 years?

The payments form an ordinary annuity with $R = 1500$, $n = 18$, and $i = 0.045$.

$$S = R\left[\frac{(1+i)^n - 1}{i}\right]$$

$$S = 1500\left[\frac{(1+0.045)^{18} - 1}{0.045}\right]$$

$$\approx 40,282.63$$

b. How much will be in the account after 18 years if Matt deposits \$1500 at the end of each year into an account paying 3.5% per year, compounded quarterly?

The payments form an ordinary annuity with $R = 1500$, $n = 4(18)$, and $i = \dfrac{0.035}{4}$.

$$S = R\left[\frac{(1+i)^n - 1}{i}\right]$$

$$S = 1500\left[\frac{\left(1+\dfrac{0.035}{4}\right)^{4\cdot18} - 1}{\dfrac{0.035}{4}}\right]$$

$$\approx 149,566.71$$

Now Try:

7. Work each problem. Give answers to the nearest cent.

a. To save for retirement, Sara decides to deposit \$2000 into an IRA at the end of each year. How much will be in the account at the end of 30 years if it is invested at 10% compounded annually?

____

b. To save for retirement, Sara decides to deposit \$2000 into an IRA at the end of each year. How much will be in the account at the end of 30 years if it is invested at 5% compounded annually?

____

Objective 5 Practice Exercises

For extra help, see Example 7 on page 744 of your text.

Work each problem. Give answers to the nearest cent.

12.   Julio is a professional wrestler who believes that his wrestling career will last ten years. To prepare for retirement, he deposits $14,000 at the end of each year for ten years into an account paying 7% compounded annually. How much will he have on deposit after ten years?

12. __________________

13.   Gordon wants to buy a new car. If he puts $1,100 at the end of each year into an account paying 4.9% compounded annually for 3 years, how much will he have to spend on a car?

13. __________________

Objective 6   **Find the sum of an infinite number of terms of a certain geometric sequence.**

Review these examples for Objective 6:

8. Evaluate the sum of the terms of the infinite geometric sequence with $a_1 = -3$ and $r = \frac{2}{3}$.

$$S = \frac{a_1}{1-r} \quad \text{Infinite sum formula}$$

$$= \frac{-3}{1-\frac{2}{3}}$$

$$= \frac{-3}{\frac{1}{3}}$$

$$= -9$$

Now Try:

8. Evaluate the sum of the terms of the infinite geometric sequence with $a_1 = 5$ and $r = -\frac{1}{5}$.

____

9. Evaluate $\displaystyle\sum_{i=1}^{\infty} 3\left(\tfrac{1}{4}\right)^{i}$.

This is the infinite geometric series with

$a_1 = \dfrac{3}{4}$ and $r = \dfrac{1}{4}$. Since $|r| < 1$, find the sum

using the formula $S = \dfrac{a_1}{1-r}$.

$$S = \frac{\frac{3}{4}}{1-\frac{1}{4}} = 1$$

9. Evaluate $\displaystyle\sum_{i=1}^{\infty} 3^{i}$.

____

Objective 6 Practice Exercises

For extra help, see Examples 8–9 on pages 746–747 of your text.

Evaluate the sum, if possible, of the infinite geometric sequence.

14. $a_1 = 2, \; r = -\dfrac{1}{3}$

14. _____________

15. $\displaystyle\sum_{i=1}^{\infty} -5\left(-\tfrac{5}{3}\right)^{i}$

15. _____________

Chapter 11 FURTHER TOPICS IN ALGEBRA

11.4   The Binomial Theorem

Learning Objectives
1	Expand a binomial raised to a power.
2	Find any specified term of the expansion of a binomial.

Key Terms

Use the vocabulary terms listed below to complete each statement in exercises 1−2.

> **Pascal's triangle**   **binomial theorem (general binomial expansion)**

1. The _________________________ is a formula used to expand a binomial raised to a power.

2. Writing the coefficients of the terms of binomial expansions in a triangular pattern gives _________________________________.

Objective 1   Expand a binomial raised to a power.

Review this example for Objective 1:

1. Evaluate 8!.

$$8! = 8 \cdot 7 \cdot 6 \cdot 5 \cdot 4 \cdot 3 \cdot 2 \cdot 1 = 40,320$$

2. Find the value of each expression.

a. $\dfrac{6!}{4!2!}$

$$\frac{6!}{4!2!} = \frac{6 \cdot 5 \cdot 4 \cdot 3 \cdot 2 \cdot 1}{(4 \cdot 3 \cdot 2 \cdot 1)(2 \cdot 1)}$$
$$= \frac{6 \cdot 5}{2 \cdot 1}$$
$$= 15$$

b. $\dfrac{8!}{4!4!}$

$$\frac{8!}{4!4!} = \frac{8 \cdot 7 \cdot 6 \cdot 5 \cdot 4 \cdot 3 \cdot 2 \cdot 1}{(4 \cdot 3 \cdot 2 \cdot 1)(4 \cdot 3 \cdot 2 \cdot 1)}$$
$$= \frac{8 \cdot 7 \cdot 6 \cdot 5}{4 \cdot 3 \cdot 2 \cdot 1}$$
$$= 70$$

Now Try:

1. Evaluate 10!.

____

2. Find the value of each expression.

a. $\dfrac{9!}{9!0!}$

____

b. $\dfrac{9!}{7!2!}$

____

c. $\dfrac{8!}{8!0!}$

$$\frac{8!}{8!0!} = \frac{8\cdot 7\cdot 6\cdot 5\cdot 4\cdot 3\cdot 2\cdot 1}{(8\cdot 7\cdot 6\cdot 5\cdot 4\cdot 3\cdot 2\cdot 1)(1)}$$

$$= 1$$

d. $\dfrac{8!}{7!1!}$

$$\frac{8!}{7!1!} = \frac{8\cdot 7\cdot 6\cdot 5\cdot 4\cdot 3\cdot 2\cdot 1}{(7\cdot 6\cdot 5\cdot 4\cdot 3\cdot 2\cdot 1)(1)}$$

$$= \frac{8}{1}$$

$$= 8$$

3. Evaluate $_9C_6$.

$$_nC_r = \frac{n!}{r!(n-r)!}$$

$$_9C_6 = \frac{9!}{6!(9-6)!}$$

$$= \frac{9!}{6!3!}$$

$$= \frac{9\cdot 8\cdot 7\cdot 6\cdot 5\cdot 4\cdot 3\cdot 2\cdot 1}{(6\cdot 5\cdot 4\cdot 3\cdot 2\cdot 1)(3\cdot 2\cdot 1)}$$

$$= 84$$

4. Expand $(2x+5y)^4$.

$$(2x+5y)^4$$
$$= (2x)^4 + \frac{4!}{1!3!}(2x)^3(5y) + \frac{4!}{2!2!}(2x)^2(5y)^2$$
$$\quad + \frac{4!}{3!1!}(2x)(5y)^3 + (5y)^4$$
$$= (2x)^4 + 4(2x)^3(5y) + 6(2x)^2(5y)^2$$
$$\quad + 4(2x)(5y)^3 + (5y)^4$$
$$= 16x^4 + 160x^3y + 600x^2y^2$$
$$\quad + 1000xy^3 + 625y^4$$

c. $\dfrac{9!}{8!1!}$

____

d. $\dfrac{9!}{5!4!}$

____

3. Evaluate $_{10}C_8$.

____

4. Expand $(3r-2s)^4$.

5. Expand $\left(\dfrac{1}{2}-a^2\right)^4$.

$$\left(\dfrac{1}{2}-a^2\right)^4$$

$$=\left(\dfrac{1}{2}\right)^4+\dfrac{4!}{1!3!}\left(\dfrac{1}{2}\right)^3\left(-a^2\right)+\dfrac{4!}{2!2!}\left(\dfrac{1}{2}\right)^2\left(-a^2\right)^2$$

$$+\dfrac{4!}{3!1!}\left(\dfrac{1}{2}\right)\left(-a^2\right)^3+\left(-a^2\right)^4$$

$$=\left(\dfrac{1}{2}\right)^4+4\left(\dfrac{1}{2}\right)^3\left(-a^2\right)+6\left(\dfrac{1}{2}\right)^2\left(-a^2\right)^2$$

$$+4\left(\dfrac{1}{2}\right)\left(-a^2\right)^3+\left(-a^2\right)^4$$

$$=a^8-2a^6+\dfrac{3}{2}a^4-\dfrac{1}{2}a^2+\dfrac{1}{16}$$

5. Expand $\left(2-\dfrac{x}{2}\right)^5$.

____

Objective 1 Practice Exercises

For extra help, see Examples 1–5 on pages 751–754 of your text.

Evaluate each expression.

1. $\dfrac{7!}{3!4!}$

1. ______________

2. $_{15}C_3$

2. ______________

Use the binomial theorem to expand the expression.

3. $(2y+3z)^5$

3. ______________

Objective 2   Find any specified term of the expansion of a binomial.

Review this example for Objective 2:	**Now Try:**

6. Find the sixth term of the expansion of $(a-2b)^9$.

Using the general form of the binomial expansion, we let $n = 9$, $x = a$, $y = -2b$, and $r = 6$. In the sixth term, a has an exponent of $9 - (6 - 1) = 4$ and $2b$ has an exponent of $6 - 1 = 5$. Thus, we have

$$\frac{9!}{5!4!}a^4(-2b)^5 = \frac{9\cdot8\cdot7\cdot6}{4\cdot3\cdot2\cdot1}a^4(-2b)^5$$

$$= 126a^4\left(-32b^5\right)$$

$$= -4032a^4b^5$$

6. Find the sixth term of the expansion of $\left(m^3 - 3r\right)^7$.

____

Objective 2 Practice Exercises

For extra help, see Example 6 on page 755 of your text.

Find the indicated term of each binomial expansion.

4.   Third term of $(m-3)^5$

4. _____________

5.   Fourth term of $(4x-3y)^5$

5. _____________

6.   Eighth term of $\left(3k-p^2\right)^9$

6. _____________

Chapter R REVIEW OF THE REAL NUMBER SYSTEM

R.1   Fractions, Decimals, and Percents

Key Terms

1. decimals
2. percent
3. improper fraction
4. numerator
5. proper fraction
6. denominator
7. lowest terms

Objective 1

Now Try

1a. $\dfrac{3}{5}$

1b. $\dfrac{1}{4}$

1c. $\dfrac{8}{25}$

Practice Exercises

1. $\dfrac{7}{25}$

3. $\dfrac{33}{73}$

Objective 2

Now Try

2. $14\dfrac{4}{5}$

3. $\dfrac{86}{7}$

Practice Exercises

5. $\dfrac{122}{9}$

Objective 3

Now Try

4. $\dfrac{1}{10}$

5a. $\dfrac{16}{21}$

5c. $\dfrac{6}{11}$

6a. $\dfrac{3}{4}$

6b. $\dfrac{23}{24}$

6c. $\dfrac{2}{9}$

6d. $1\dfrac{5}{32}$

Practice Exercises

7. $\dfrac{45}{4}$ or $11\dfrac{1}{4}$

9. $\dfrac{119}{24}$ or $4\dfrac{23}{24}$

Objective 4

Now Try

7a. $\dfrac{72}{100}$

7b. $\dfrac{53}{1000}$

7c. $\dfrac{37,058}{10,000}$

Answers

Practice Exercises

11. $\dfrac{1803}{100}$

Objective 5
Now Try

8a. 37.871   8b. 27.282   9a. 251.116

9b. 0.028   9c. 1.43   10a. 513.72

10b. 51,372   10c. 5.1372   10d. 0.051372

Practice Exercises

13. 61.053   15. 0.4292

Objective 6
Now Try

11a. 0.35   11b. $3.\overline{5}$ or 3.555…

Practice Exercises

17. $0.\overline{4}$ or 0.444

Objective 7
Now Try

12a. 0.91   12b. 4.30 or 4.3   12c. 0.06

12d. 43%   12e. 520%   12f. 0.8%

Practice Exercises

19. 3.62   21. 8.4%

Objective 8
Now Try

13a. $\dfrac{3}{10}$ 13b. $1\dfrac{1}{4}$ 14a. 15%

14b. $5\dfrac{5}{9}\%$ exact, or 5.6% rounded

Practice Exercises

23. $\dfrac{139}{250}$

R.2 Basic Concepts from Algebra

Key Terms

1. number line
2. Set-builder notation
3. coordinate
4. set
5. finite set
6. empty set
7. inequality
8. signed numbers
9. infinite set
10. variable
11. absolute value
12. additive inverse
13. equation
14. graph
15. elements

Objective 1

Now Try

1. $\{0, 1, 2, 3, 4, 5\}$
2. $\{x \mid x$ is a natural number less than 6$\}$

Practice Exercises

1. $\{5, 10, 15, \ldots\}$
3. $\{x \mid x$ is a whole number between 7 and 10, inclusive$\}$

Objective 2

Practice Exercises

5. 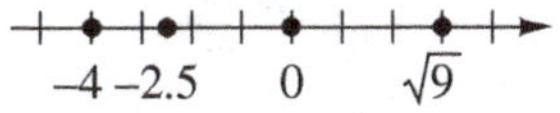

Objective 3

Now Try

3a. $-7, 0, 3$

3b. $-7, -\frac{8}{9}, 0, 0.4, \frac{3}{5}, 2.\overline{15}, 3$

3c. $-\sqrt{3}, \sqrt{7}$

3d. All are real numbers

Practice Exercises

7. (a) 5; (b). 0, 5; (c). –4, 0, 5; (d) $-4, -\frac{1}{3}, 0, \frac{4}{5}, 5$; (e) $-\sqrt{2}, \sqrt{7}$;
   (f) $-4, -\sqrt{2}, -\frac{1}{3}, 0, \frac{4}{5}, \sqrt{7}, 5$
9. false

Objective 4

Practice Exercises

11. $\frac{5}{2}$

Objective 5

Now Try

5a. 11

5b. –10

5c. –10

5d. 9

6. Philadelphia from 2000–2009

Practice Exercises

13. 5

Objective 6
  Now Try
  7. >   8. False
  Practice Exercises
15. false   17. false

R.3   Operations on Real Numbers

Key Terms
  1. product   2. reciprocals   3. difference

  4. quotient   5. sum

Objective 1
  Now Try
  1. −21   2. 3

  Practice Exercises

  1. −12.4   3. $-\dfrac{16}{33}$

Objective 2
  Now Try
  3. −22   4. 13

  Practice Exercises
  5. −3.78

Objective 3
  Now Try
  5. 12

  Practice Exercises
  7. 3   9. 9

Objective 4
  Now Try
6a. 77   6b. −10

  Practice Exercises
11. 3.924

Objective 5
  Now Try
7a. −2   7b. −3   7d. undefined

7e. 0

  Practice Exercises

13. $-\dfrac{1}{5}$ 15. undefined

R.4   Exponents, Roots, and Order of Operations

Key Terms

1. exponential expression

2. base

3. exponent

4. square root

5. algebraic expression

6. factors

Objective 1

  Now Try

1a. 9^3

1b. $\left(\dfrac{3}{8}\right)^4$

1c. $(-10)^4$

1e. z^6

2a. 49

2b. $\dfrac{8}{125}$

2d 256

2e. -32

3a. 625

3b. 625

3c. -625

  Practice Exercises

1. b^7

3. -8

Objective 2

  Now Try

4a. 7

4c. $\dfrac{4}{9}$

4f. -30

4g. not a real number

  Practice Exercises

5. 100

Objective 3

  Now Try

5. 38

6. 34

7. undefined

  Practice Exercises

7. 20

9. -59

Objective 4

  Now Try

8a. -75

8c. 234

  Practice Exercises

11. $-\dfrac{34}{13}$

Answers

R.5   Properties of Real Numbers

Key Terms
  1. coefficient   2. term   3. like terms

  4. combining like terms

Objective 1
  Now Try
  1b. $-12 - 4q$ 1d. $-6r$

  Practice Exercises
  1. $-3z + 21$ 3. 650

Objective 2
  Now Try
  2a. $30q$ 2c. $-4 + 5c$

  Practice Exercises
  5. -2.5

Objective 3
  Practice Exercises

  7. $\dfrac{1}{5}$ 9. $-\dfrac{1}{7}$

Objective 4
  Now Try
  3. $-2x - 16$ 4a. $y - 6$ 4d. $24xy$

  Practice Exercises
11. $-3x + 17$

Chapter 1 LINEAR EQUATIONS, INEQUALITIES, AND APPLICATIONS

1.1   Linear Equation in One Variable

Key Terms
1. equivalent equations
2. linear equation (first-degree) equation in one variable
3. solution set
4. contradiction
5. conditional equation
6. identity
7. solution

Objective 1
Now Try
1a. expression
1b. equation

Practice Exercises
1. equation
3. expression

Objective 2
Practice Exercises
5. nonlinear equation

Objective 3
Now Try
2. {5}

Practice Exercises
7. {−5}
9. {8}

Objective 4
Now Try
3. {4}
4. {6}

Practice Exercises
11. {−8}

Objective 5
Now Try
5. {−5}
6. {2}

Practice Exercises
13. {3}
15. {60}

Objective 6
Now Try
7a. {0}; conditional
7b. ∅; contradiction
7c. {all real numbers}; identity

Practice Exercises
17. contradiction; ∅

1.2 Formulas and Percent

Key Terms
1. formula   2. mathematical model   3. percent

Objective 1
Now Try

1. $L = \dfrac{A}{W}$ 2. $h = \dfrac{3V}{B}$

Practice Exercises

1. $B = \dfrac{3V}{h}$ 3. $y = 2x + 4$

Objective 2
Now Try
5. 60 mph

Practice Exercises
5. 8%

Objective 3
Now Try
6a. 2.6%   6b. 3.5%

7. \$20.625 million or \$20,625,000

Practice Exercises
7. 1100 students   9. 15,000 students

Objective 4
Now Try
8a. about 5.4%   8b. about 20.1%

Practice Exercises
11. 6.5%

Objective 5
Now Try
9. 0.67 m^2

Practice Exercises
13. 1.06 m^2

1.3   Applications of Linear Equations

Key Terms
1. sum, increased by, more than
2. product, double, of
3. quotient, per, ratio
4. difference, less than, decreased by

Objective 1

Practice Exercises

1. $13 - 7x$

3. $\dfrac{4+x}{9}$

Objective 2

Now Try

1a. $3x - 100 = 197$

1b. $\dfrac{x}{x+5} = 23$

Practice Exercises

5. $\dfrac{x}{x-3} = 17$

Objective 3

Now Try

2a. equation; $\{65\}$

2b. expression; $16x + 31$

Practice Exercises

7. expression; $19z - 59$

9. equation; $\{159\}$

Objective 4

Now Try

3. 6 m, 7 m, 9 m

Practice Exercises

11. Bush: 271 votes; Gore: 266 votes

Objective 5

Now Try

5. $250,000

Practice Exercises

13. $5000

15. $6370.50

Objective 6

Now Try

6. 5%: $1600; 7%: $3500

Practice Exercises

17. 7%: $800; 9%: $300

Objective 7

Now Try

7. 8 oz

Practice Exercises

19. 8 lb

1.4   Further Applications of Linear Equations

Key Terms

1. supplementary angles 2. complementary angles

3. vertical angles   4. consecutive integers

Objective 1

Now Try

1. 32 quarters; 18 nickels

Practice Exercises

1. 25 large jars; 55 small jars

3. 1500 general admission; 750 reserved

Objective 2

Now Try

2. plane A: 400 mph; plane B: 360 mph   3. $\frac{5}{6}$ hr or 50 min

Practice Exercises

5. plane speed: 265 mph; wind speed: 35 mph

Objective 3

Now Try

4. 20°, 45°, 115°

Practice Exercises

7. 25°, 55°, 100°   9. 30°, 60°, 90°

Objective 4

Now Try

5. 53, 54, 55

Practice Exercises

11. −2, 0, 2, 4

1.5 Linear Inequalities in One Variable

Key Terms

1. interval notation   2. linear inequality in one variable

3. inequality   4. interval   5. equivalent inequalities

Objective 1
Now Try

1a. $(-1, \infty)$; 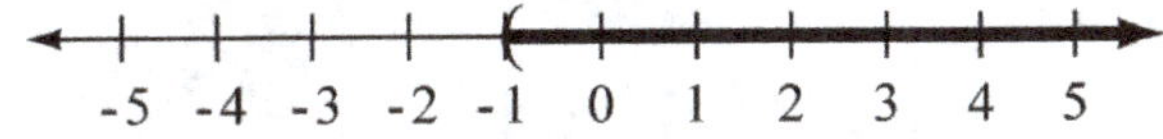

1b. $(-5, -1]$ 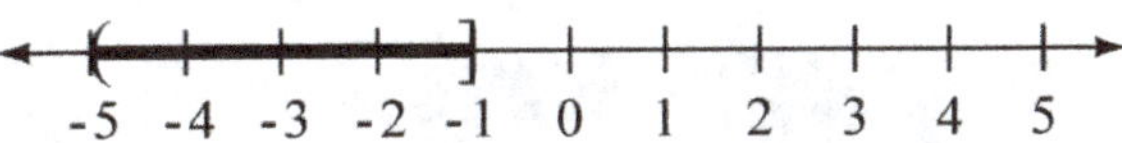

Practice Exercises

1. $(3, \infty)$; 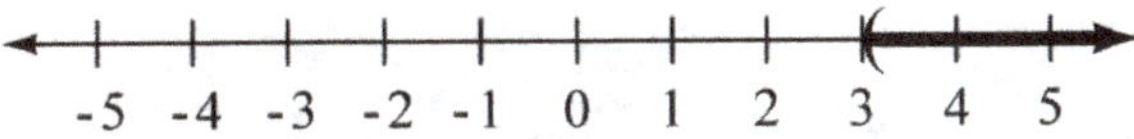

3. $(-3, 2]$; 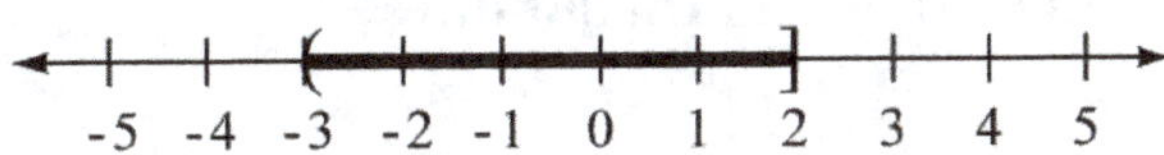

Objective 2
Now Try

2. $(-\infty, -4)$ 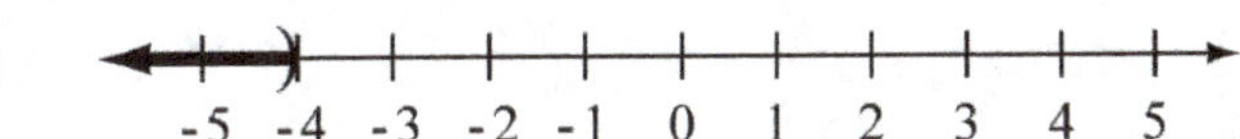

Practice Exercises

5. $(2, \infty)$; 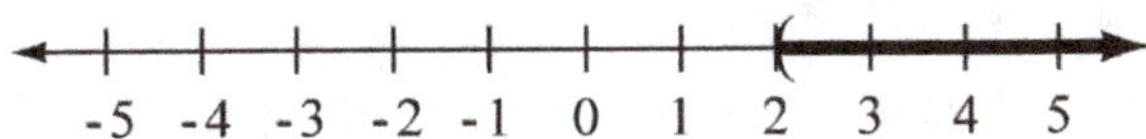

Objective 3
Now Try

4. $(-\infty, -5]$ 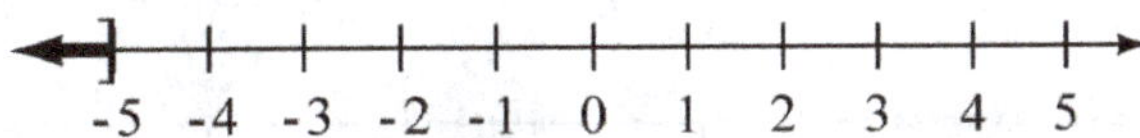

5. $(-\infty, -8]$ 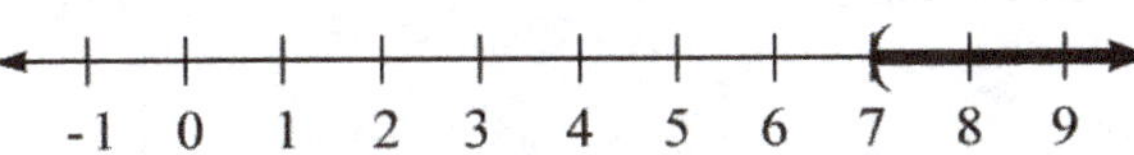

6. $(7, \infty)$ 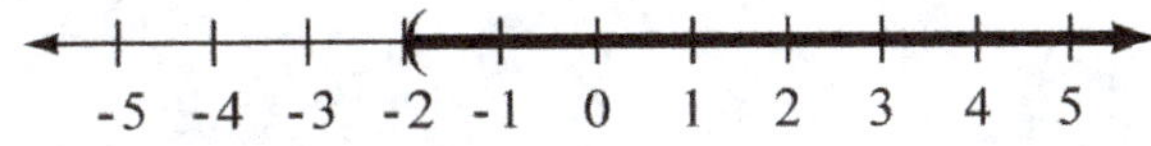

Practice Exercises

7. $(-2, \infty)$; 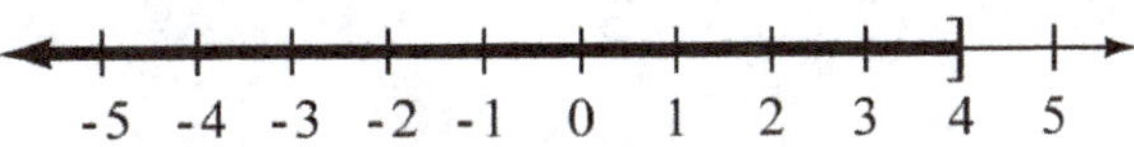

9. $(-\infty, 4]$

Answers

Objective 4
  Now Try

8. $(-4, -2]$

Practice Exercises

11. $[-5, -3)$;

Objective 5
  Now Try
10. 10 feet

  Practice Exercises
13. 89   15. 84

1.6   Set Operations and Compound Inequalities

Key Terms
  1. union   2. compound inequality   3. intersection

Objective 2
  Now Try
  1. $\{30, 50\}$

  Practice Exercises
  1. $\{0, 2, 4\}$ 3. $\{1, 3, 5\}$

Objective 3
  Now Try

2. $[12, 16]$

3. $(-\infty, -3)$

  Practice Exercises

5. $[-4, 3)$

Objective 4
  Now Try
  5. $\{20, 30, 40, 50, 70\}$

  Practice Exercises
  7. $\{0, 1, 2, 3, 4, 5\}$ 9. $\{0, 1, 2, 3, 4, 5\}$

Objective 5
Now Try

6. $(-\infty,\ 2)\cup[5,\ \infty)$ 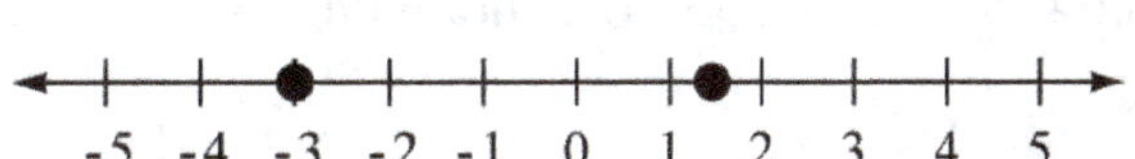

8. $(-\infty,\ \infty)$

Practice Exercises

11. $(-\infty,\infty)$

1.7 Absolute Value Equations and Inequalities

Key Terms
1. absolute value equation   2. absolute value inequality

Objective 2
Now Try

1. $\left\{-3,\ \dfrac{7}{5}\right\}$

Practice Exercises

1. $\left\{-\dfrac{13}{2},\dfrac{7}{2}\right\}$ 3. $\{-2,\ 14\}$

Objective 3
Now Try

2. $(-\infty,-2)\cup(1,\ \infty)$

3. $(-3,\ 2)$

Practice Exercises

5. $(-\infty,-7]\cup[16,\ \infty)$ 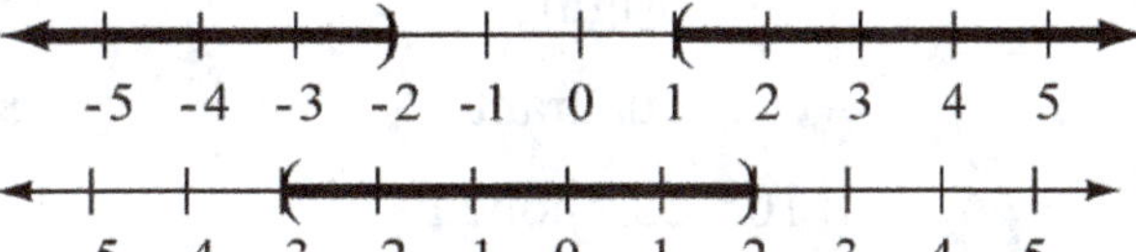

Objective 4
Now Try

5. $\{-4,\ 10\}$ 6a. $(-\infty,-22]\cup[4,\ \infty)$ 6b. $[-22,\ 4]$

Practice Exercises

7. $\{-2,\ 3\}$ 9. $\left\{-\dfrac{5}{4},-\dfrac{1}{4}\right\}$

Objective 5
Now Try
7. $\{-5, -1\}$

Practice Exercises
11. $\left\{-3, \dfrac{11}{3}\right\}$

Objective 6
Now Try

8a. $\varnothing$ 　　　　8b. $\{-5\}$ 　　　　9a. $(-\infty, \infty)$

9b. $\varnothing$ 　　　　9c. $\{4\}$

Practice Exercises
13. $\{-14\}$ 　　　　15. $\varnothing$

Objective 7
Now Try
10. between 16.055 and 17.745 oz, inclusive

Practice Exercises
17. between 16.6465 and 17.1535 oz, inclusive

Chapter 2 LINEAR EQUATIONS, GRAPHS, AND FUNCTIONS

2.1　Linear Equations in Two Variables

Key Terms
1. y-intercept 　　　　2. linear equation in two variables
3. coordinate 　　　　4. origin 　　　　5. x-intercept
6. x-axis 　　　　7. quadrant 　　　　8. ordered pair
9. y-axis 　　　　10. components 　　　　11. plot
12. rectangular (Cartesian) coordinate system
13. graph of an equation 　14. linear equation in two variables

Objective 3
Now Try
1a. $(0, -2)$ 　　　　1b. $(5, 0)$ 　　　　1c. $(-5, -4)$

1d. $(10, 2)$ 　　　　1e.

x	y
0	-2
5	0
-5	-4
10	2

Practice Exercises

1. (a) $(2, 0)$; (b) $\left(4, -\dfrac{5}{2}\right)$; (c) $\left(-\dfrac{2}{5}, 3\right)$; (d) $\left(0, \dfrac{5}{2}\right)$; (e) $\left(\dfrac{2}{5}, 2\right)$

3. $(-4, 4), (0, 4), (6, 4), (-12, 4)$

Objective 5
Now Try

2.
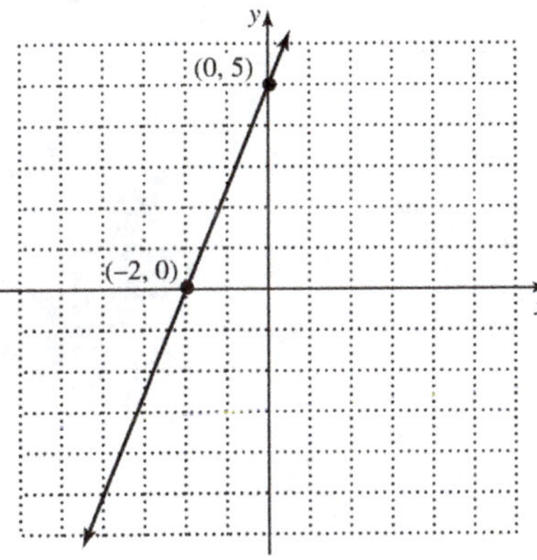

3.
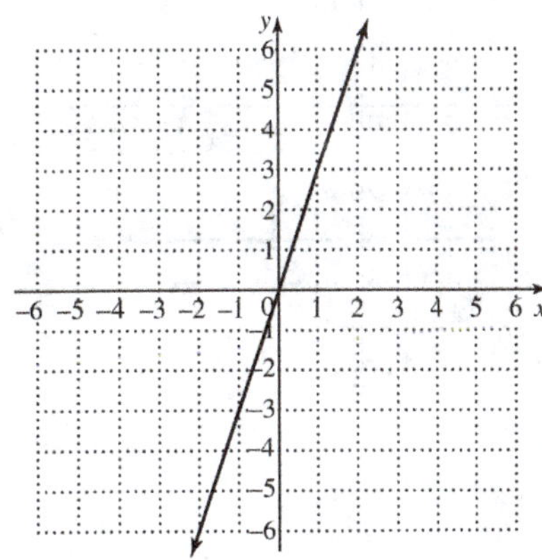

Practice Exercises

5.
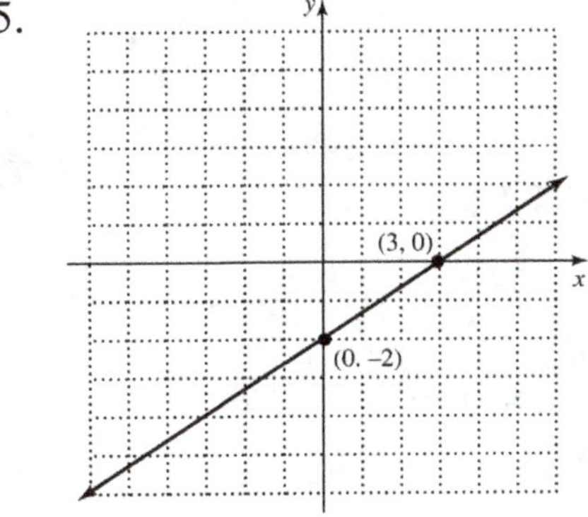

Objective 6
Now Try

4a.
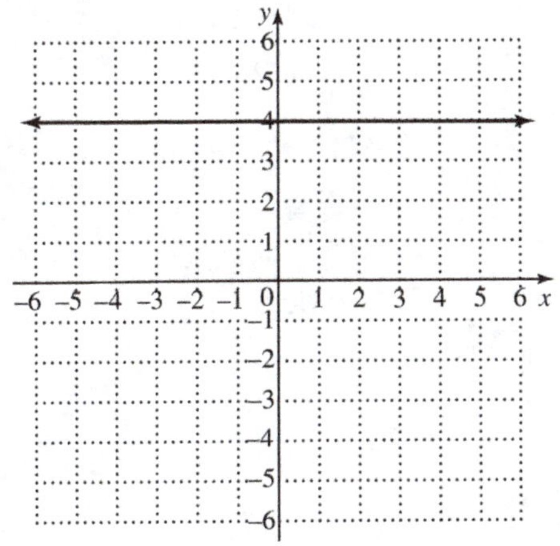

4b.
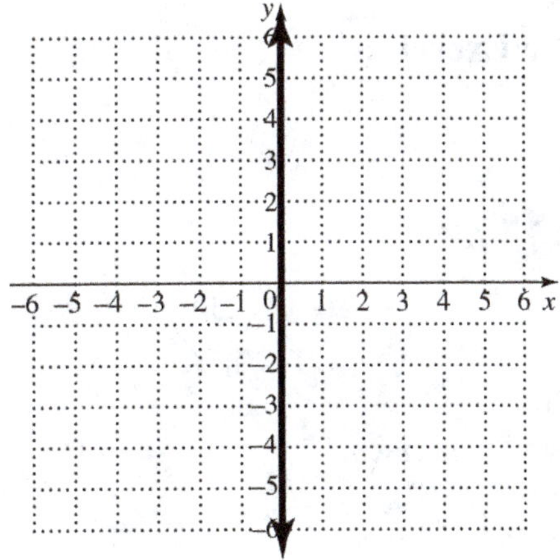

Answers

Practice Exercises

7. $y + 3 = 0$

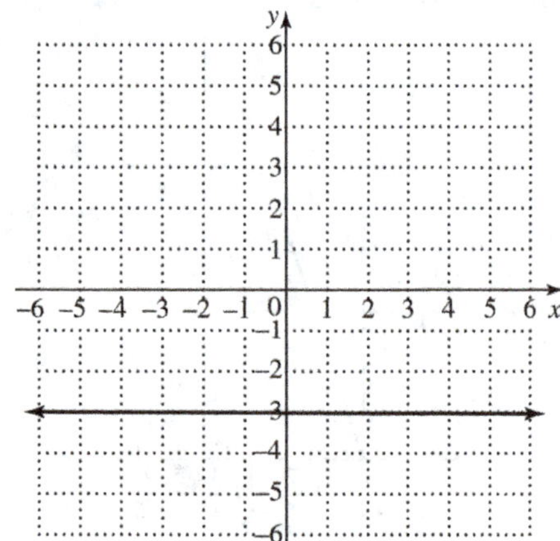

Objective 7
  Now Try
  5. $(5, -2)$

Practice Exercises

9. $\left(\dfrac{5}{2}, -3\right)$

2.2   The Slope of a Line

Key Terms
  1. slope   2. rise   3. run

Objective 1
  Now Try
  1. $-\dfrac{16}{9}$

Practice Exercises

1. -2 3. $-\dfrac{1}{14}$

Objective 2
  Now Try
  2. 7   3a. 0   3b. undefined

  4. $\dfrac{7}{4}$

Practice Exercises

5. $\dfrac{2}{3}$

     Copyright © 2020 Pearson Education, Inc.

Objective 3
Now Try
5.

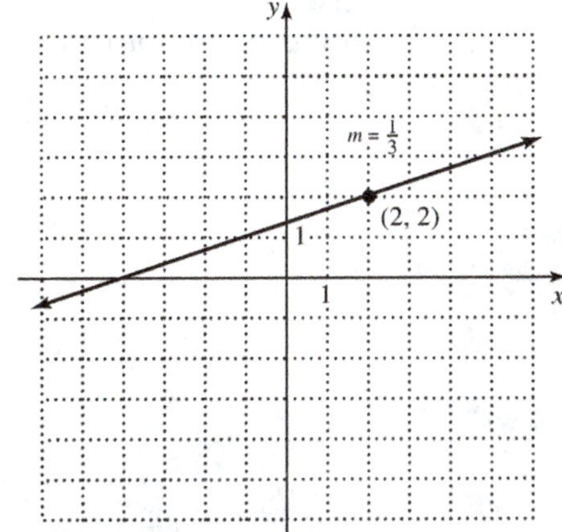

Practice Exercises
7.

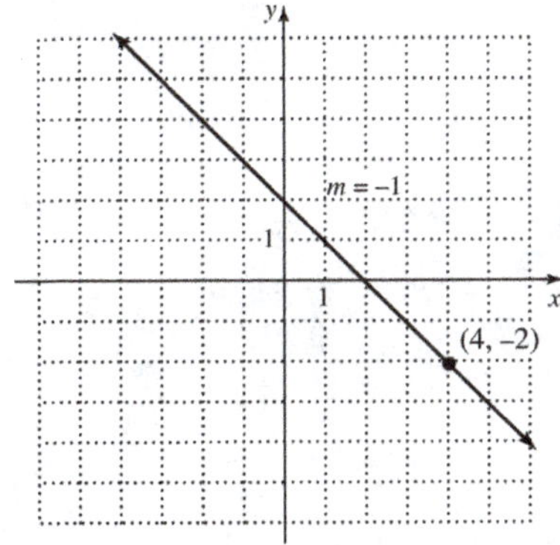

9.

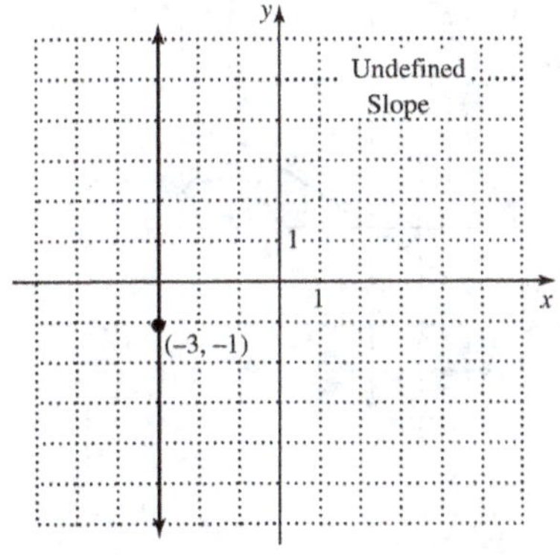

Objective 4
Now Try
6. parallel   7a. perpendicular

Practice Exercises
11. parallel

Objective 5
Now Try
8. 113.5 ft/min   9. an increase of 5 employees/yr

Practice Exercises
13. $125,000/yr   15. $950/yr

Answers

2.3 Writing Equations of Lines

Key Terms

  1. point-slope form   2. standard form   3. slope-intercept form

Objective 1

Now Try

1. $y = \dfrac{7}{9}x + 8$

Practice Exercises

1. $y = \dfrac{3}{2}x - \dfrac{2}{3}$ 3. $y = -\dfrac{6}{5}x + \dfrac{2}{5}$

Objective 2

Now Try

2.

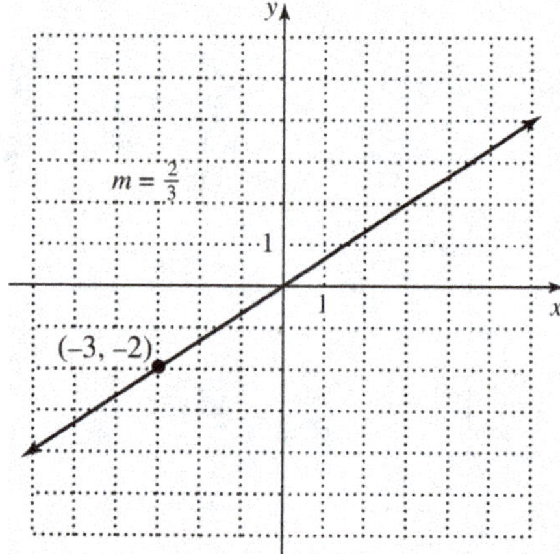

Practice Exercises

5.

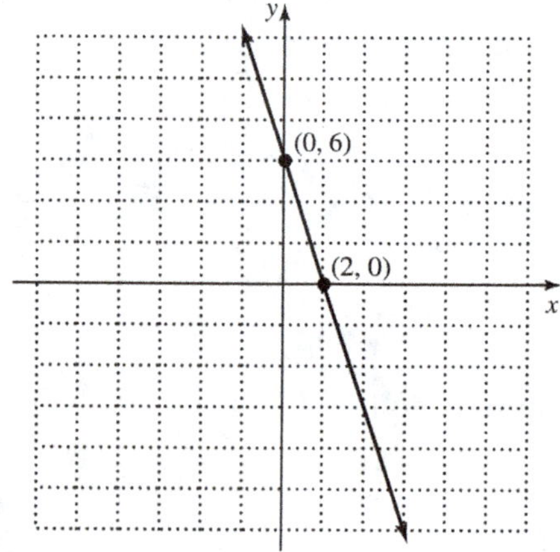

Objective 3

Now Try

3. $y = \dfrac{4}{5}x - \dfrac{44}{5}$

Practice Exercises

7. $3x + 5y = 11$ 9. $2x - 3y = -8$

Objective 4
  Now Try

4. $y = -\dfrac{3}{4}x + \dfrac{81}{4};\ \ 3x + 4y = 81$

  Practice Exercises
11. $x + 3y = -1$

Objective 5
  Now Try
5a. $y = 5$ 5b. $x = -5$

  Practice Exercises
13. $x = -1$ 15. $y = -5$

Objective 6
  Now Try

6a. $y = -\dfrac{3}{4}x + \dfrac{7}{2}$ 6b. $y = \dfrac{4}{3}x + 16$

  Practice Exercises
17. $4x - 3y = -17$

Objective 7
  Now Try
8a. $y = -28.8x + 686$ 8b. $(1,\ 657.2)$

  Practice Exercises
19. (a) $y = 1.25x + 25$; (b) $(0, 25),\ (5, 31.25),\ (10, 37.50)$

2.4   Linear Inequalities in Two Variables

Key Terms
  1. boundary line   2. linear inequality in two variables

Objective 1
  Now Try
1.

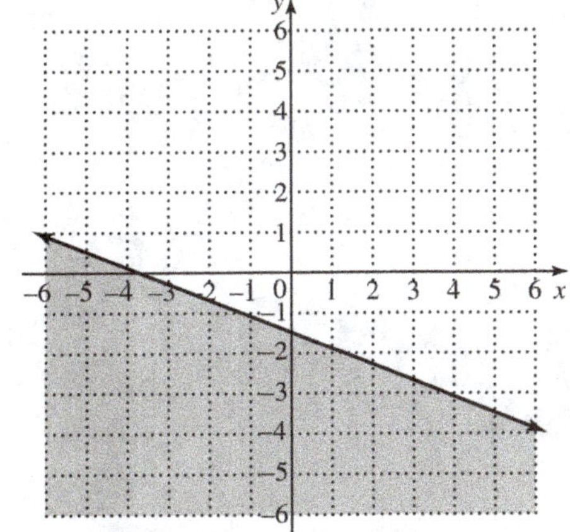

2.

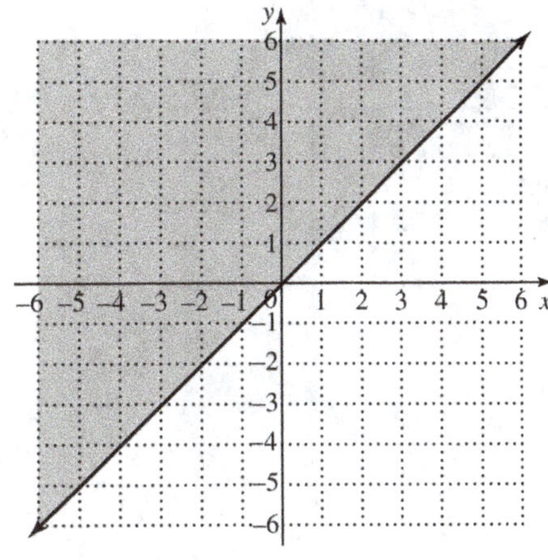

Practice Exercises

1. 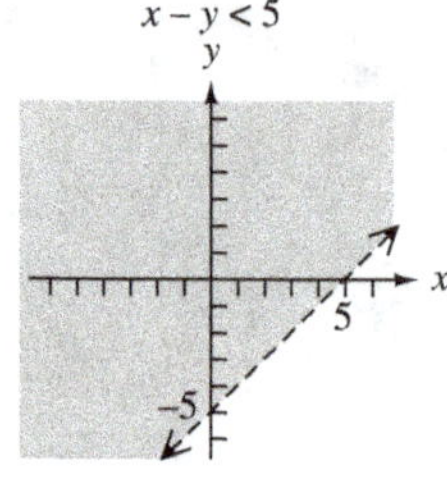

3. 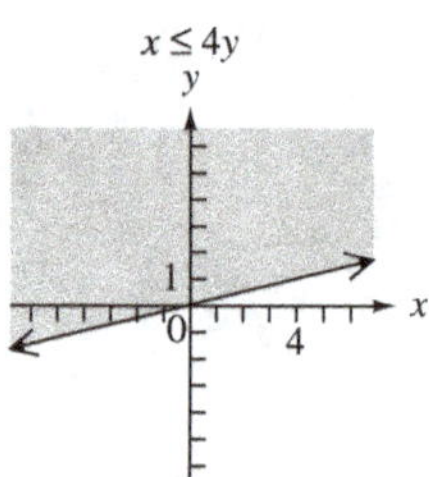

Objective 2
Now Try

4. 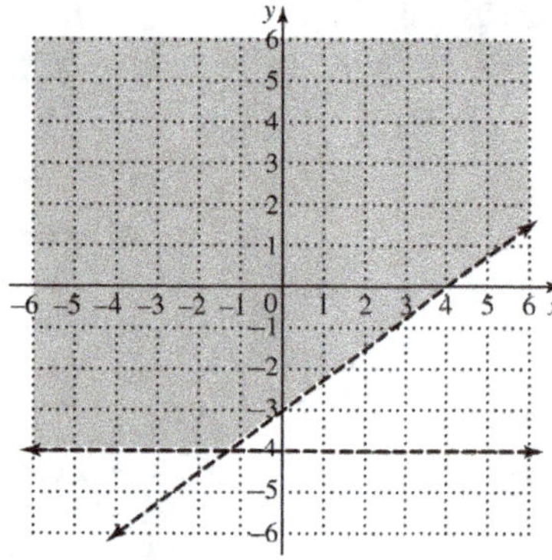

Practice Exercises

5. 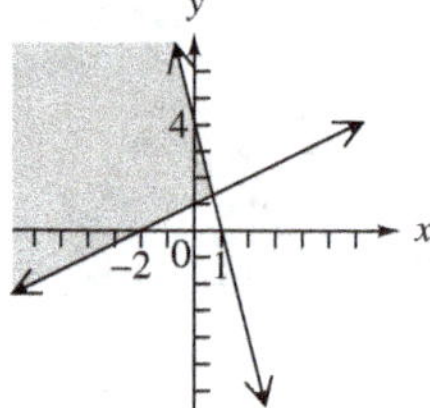

Objective 3
Now Try

5. 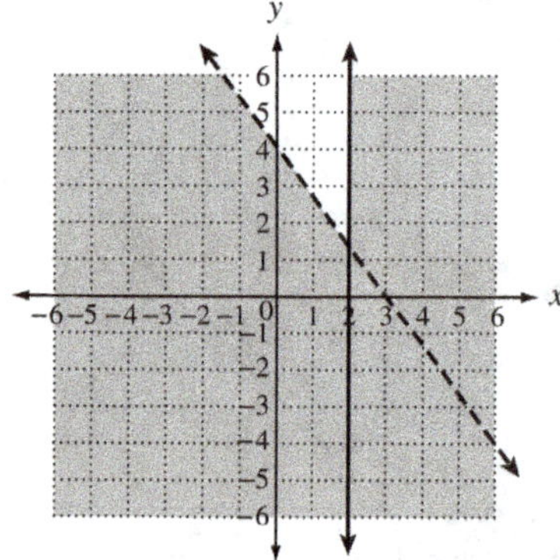

Practice Exercises

7.
$4x - 2y \geq -4$
or $x \geq 1$

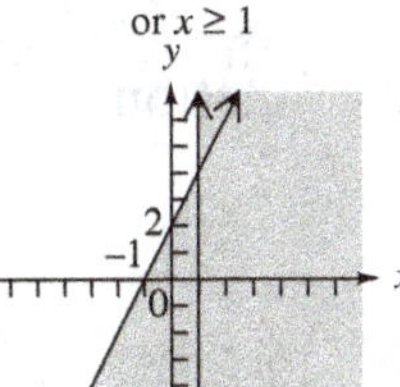

9.
$x + y \geq 0$
or $x - y \geq 0$

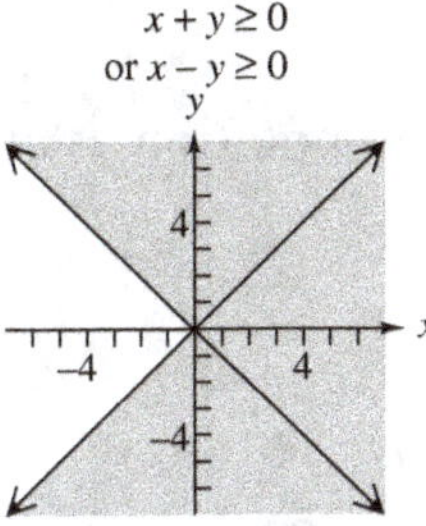

2.5   Introduction to Relations and Functions

Key Terms

1. range

2. relation

3. dependent variable

4. domain

5. function

6. independent variable

Objective 1

Now Try

1. $\{(1,\ 3),\ (2,\ 4),\ (3,\ 6),\ (5,\ 7)\}$

2a. function

2b. not a function

Practice Exercises

1. $\{(1,\ 3),\ (1,\ 4),\ (2,-1),\ (3,\ 7)\}$

3. not a function

Objective 2

Now Try

3. domain: $\{13\}$; range: $\{-2, -1, 4\}$; not a function

4a. domain: $(-\infty,\ \infty)$; range: $(-\infty,\ \infty)$

4b. domain: $[-4,\ \infty)$; range: $(-\infty,\ \infty)$

Practice Exercises

5. function; domain: $\{A, B, C, D, E\}$; range: $\{V, W, X, Z\}$

Objective 3

Now Try

5. function

6. function; domain: $(-\infty,\ \infty)$

Practice Exercises

7. function

9. function; $(-\infty,-6)\cup(-6,\ \infty)$

2.6   Function Notation and Linear Functions

Key Terms
1. linear function   2. function notation   3. constant function

Objective 1
Now Try
1. $f(3)=17$ 2a. $f(-6)=16$ 2b. $f(c)=-c^2-7c+10$

3. $g(a-1)=4a-11$ 4. $f(-6)=21$ 5a. $f(-2)=3$

5b. $f(-3)=2$ 6. $f(x)=-\dfrac{2}{3}x+\dfrac{7}{3},\ f(-3)=\dfrac{13}{3}$

Practice Exercises
1. (a) -13; (b) -7; (c) $-3x-7$ 3. (a) 9; (b) 9; (c) 9

Objective 2
Now Try
7. domain: $(-\infty,\ \infty)$;

    range: $(-\infty,\ \infty)$

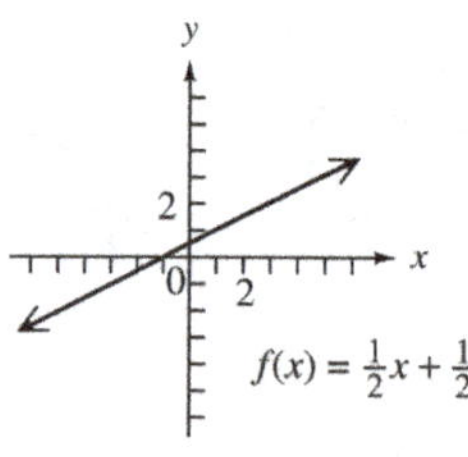

Practice Exercises
5. domain: $(-\infty,\infty)$

    range: $(-\infty,\infty)$

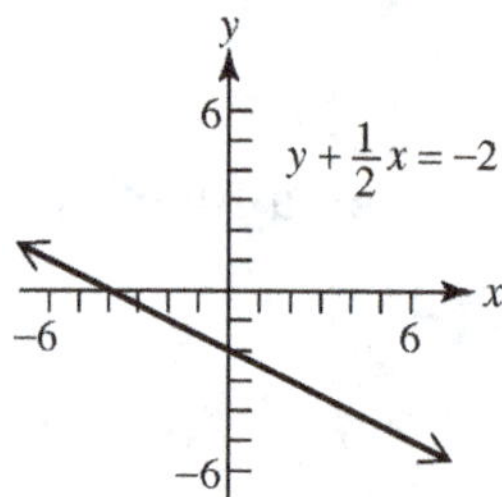

Chapter 3 SYSTEMS OF LINEAR EQUATIONS

3.1   Systems of Linear Equations in Two Variables

Key Terms
1. independent equations   2. consistent system

3. system of equations   4. solution set of a system

5. dependent equations   6. inconsistent system

7. linear system

Objective 1
Now Try
1. yes

Practice Exercises
1. not a solution   3. not a solution

Objective 2
Now Try
2.

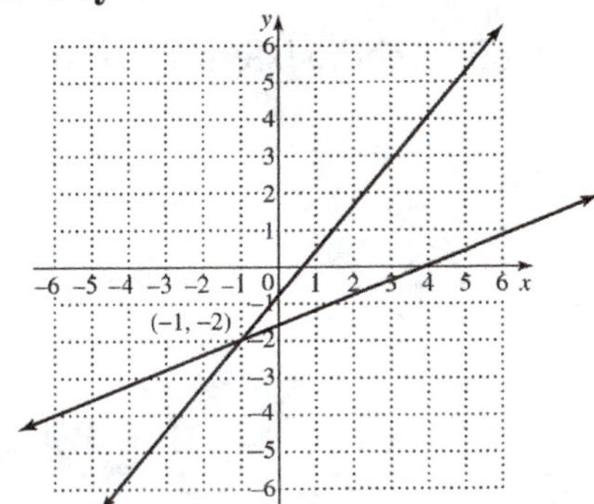

Practice Exercises
5.

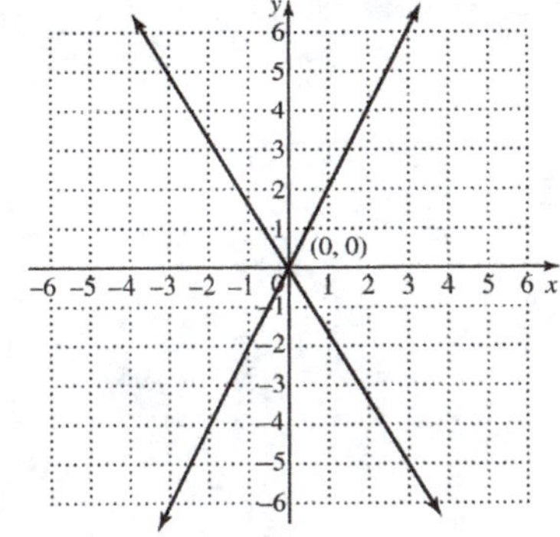

Objective 3
Now Try
4. $\{(5,-1)\}$

Practice Exercises
7. $\{(-3,-4)\}$ 9. $\{(3, 5)\}$

Objective 4
Now Try
7. $\{(-1, 3)\}$

Practice Exercises
11. $\{(6, 8)\}$

Objective 5
Now Try
8. inconsistent; $\varnothing$ 9. dependent; $\{(x, y) \mid 4x + 3y = 12\}$

10. Both are $y = \dfrac{2}{5}x - \dfrac{7}{5}$; infinitely many solutions

Practice Exercises
13. $\varnothing$ 15. $\left\{(x, y) \mid -3x + 2y = 6\right\}$

Answers

3.2   Systems of Linear Equations in Three Variables

Key Terms
1. ordered triple   2. dependent system   3. inconsistent system

Objective 1
  Practice Exercises
  1. The planes intersect in one point.

Objective 2
  Now Try
  1. $\{(-3, 1, 2)\}$
  Practice Exercises
  3. $\{(-1, 2, 1)\}$ 5. $\{(4, -4, 1)\}$

Objective 3
  Now Try
  2. $\{(2, -5, 3)\}$
  Practice Exercises
  7. $\{(4, 2, -1)\}$

Objective 4
  Now Try
  3. $\emptyset$ 4. $\{(x, y, z) \mid x - 5y + 2z = 0\}$ 5. $\emptyset$
  Practice Exercises
  9. $\emptyset$; inconsistent system
  11. $\{(x, y, z) \mid -x + 5y - 2z = 3\}$; dependent equations

3.3   Applications of Systems of Linear Equations

Key Terms
1. elimination method   2. substitution

Objective 1
  Now Try
  1. length: 17 ft; width: 10 ft

  Practice Exercises
  1. square: 8 cm; triangle: 13 cm   3. 21 cm

Objective 2
  Now Try
  2. marigold: $12.29; carnation: $17.60

  Practice Exercises
  5. large: $6; small: $3

Objective 3

  Now Try

3. water: 6 liters; 25% solution: 24 liters

  Practice Exercises

7. $6 coffee: 100 lbs; $12 coffee: 50 lb

9. water: 9 oz; 80% solution: 3 oz

Objective 4

  Now Try

4. Ashley: 5 mph; Taylor: 3 mph

  Practice Exercises

11. distance from school: $1\frac{1}{8}$ mi; jogging time: $\frac{1}{8}$ hr or $7\frac{1}{2}$ min

Objective 5

  Now Try

7. 20 lb of $4 candy; 30 lb of $6 candy; 50 lb of $10 candy

  Practice Exercises

13. $8 coffee: 15 lb; $10 coffee: 12 lb; $15 coffee: 23 lb

Chapter 4 EXPONENTS, POLYNOMIALS, AND POLYNOMIAL FUNCTIONS

4.1   Integer Exponents

Key Terms

1. product rule for exponents

2. quotient rule for exponents

3. power rule for exponents

4. base; exponent

Objective 1

  Now Try

1a. 7^8

1b. 6^{16}

1c. x^{12}

1d. $-24x^7$

1e. $24p^6q^6$

1f. The product rule does not apply

  Practice Exercises

1. 7^7

3. $48k^{18}$

Objective 2

  Now Try

2a. 1

2b. 1

2c. -1

2d. -1

2f. $1, k \neq 0$

2e. 2

Answers

3a. $\dfrac{1}{5^4}$

3b. $\dfrac{1}{9}$

3c. $\dfrac{1}{(6z)^2}$, $z \neq 0$

3d. $\dfrac{12}{z^5}$, $z \neq 0$

3e. $-\dfrac{1}{p^4}$, $p \neq 0$

3f. $\dfrac{1}{(-p)^4}$, $p \neq 0$

4a. $\dfrac{7}{24}$

4b. $\dfrac{2}{35}$

4c. 49

4d. $\dfrac{27}{16}$

Practice Exercises

5. $2r^7$

Objective 3

Now Try

5a. 10^2

5b. q^3

5c. $\dfrac{1}{m^3}$

5d. 8^{10}

5e. $\dfrac{1}{11^{12}}$

5f. 9^4

5g. x^2, $x \neq 0$

5h. The quotient rule does not apply.

Practice Exercises

7. This expression cannot be simplified further.

9. p^8

Objective 4

Now Try

6a. q^{20}

6b. $343x^3$

6c. $\dfrac{8}{125}$

6d. $125r^{15}$

6e. $\dfrac{-27p^{12}}{q^3}$

7a. 625

7b. $\dfrac{81}{49}$

7c. $\dfrac{64}{27}$

Practice Exercises

11. $-\dfrac{128a^7}{b^{14}}$

Objective 5

Now Try

8. $\dfrac{4x}{3y^6}$

Practice Exercises

13. c^{25}

15. $\dfrac{1}{k^4 t^{20}}$

4.2   Scientific Notation

Key Terms

1. scientific notation   2. base; exponent

Objective 1

Now Try

1a. 3.946×10^{6} 1b. 4.8×10^{-4} 1c. -8.49×10^{5}

Practice Exercises

1. 2.3651×10^{4} 2. -4.296×10^{11} 3. -2.208×10^{-4}

Objective 2

Now Try

2a. $9{,}450{,}000$ 2b. 0.0000804 2c. -7.89

Practice Exercises

4. $90{,}000{,}000$ 5. 0.000042 6. $-40{,}020$

Objective 3

Now Try

3. 980 4. 9.5×10^{20} meters

Practice Exercises

7. 9×10^{-1} or 0.9 8. 2.53×10^{2} 9. 4.86×10^{19} atoms

Answers

4.3 Adding and Subtracting Polynomials

Key Terms
1. degree of a term
2. descending powers
3. term
4. trinomial
5. polynomial
6. coefficient (numerical coefficient)
7. monomial
8. degree of a polynomial
9. binomial
10. algebraic expression
11. negative of a polynomial
12. polynomial in x

Objective 1

Now Try
1. $11x^7 - 5x^5 + x^3 - 4$; $11x^7$; 11
2a. trinomial; 3
2b. monomial; 10
2c. binomial; 20
2d. none of these; 6

Practice Exercises
1. $y^3 - 7y^2 + 5y + 8$; y^3; 1
3. binomial; 4

Objective 2

Now Try
3b. $14m + 2n$
3c. $-x^2yz + 13xyz^2$
4. $2x^5 - 14x^3 - x$
5. $10y^3 - 12y^2 + 13y$

Practice Exercises
5. $5a^4 - a^3 - 6a^2 + 28a + 1$

4.4 Polynomial Functions, Graphs, and Composition

Key Terms
1. cubing function
2. composite function
3. polynomial function of degree n
4. squaring function
5. identity function
6. composition

Objective 1

Now Try
1. 1

Practice Exercises
1. (a) -7; (b) -17
3. (a) 28; (b) 198

Objective 2

Now Try
2. $10,278.96

Practice Exercises
5. 236 ft

Objective 3
Now Try

3a. $3x^2 - 6x + 21$ 3b. $9x^2 - 8x + 3$ 4. $7x^2 - x$; 30

Practice Exercises
7. (a) $x^2 + 7x - 13$; (b) $3x^2 + x + 3$ 9. 18

Objective 4
Now Try
5.

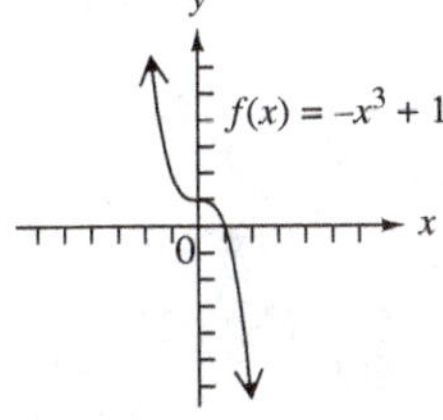

domain: $(-\infty, \infty)$;
range: $(-\infty, \infty)$

Practice Exercises
11.

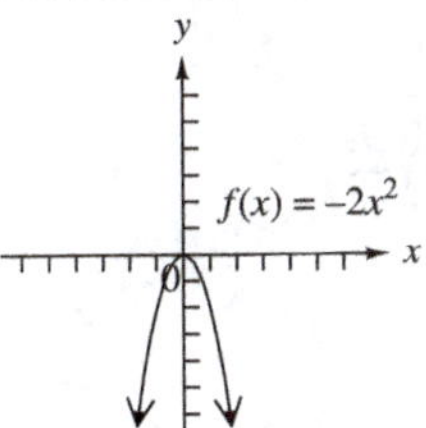

domain: $(-\infty, \infty)$;
range: $(-\infty, 0]$

Objective 5
Now Try
6. −15   8. 76

Practice Exercises
13. a. 25; b. 97; c. $8x^2 - 7$

4.5   Multiplying Polynomials

Key Terms
1. inner product   2. FOIL   3. outer product

Objective 1
Now Try
1. $10x^9 y^3$

Answers

Practice Exercises

1. $16y^7$ 3. $108r^5s^7$

Objective 2

Now Try

2b. $-28x^4 - 56x^3 - 42x^2$ 2d. $3x^4 + 15x^3 - 18x^2$

3a. $32x^2 + 12x - 35$ 3b. $12x^4 - 43x^3 + 35x^2 + 27x - 63$

Practice Exercises

5. $6m^2 + 2m - 20$

Objective 3

Now Try

4. $54z^2 - 66z + 20$

Practice Exercises

7. $6x^2 - 5xy - 6y^2$ 9. $x^2 - 2x - 15$

Objective 4

Now Try

5c. $9m^2 - 64y^2$ 5d. $10x^5 - 250x^3$

Practice Exercises

11. $49x^2 - 9y^2$

Objective 5

Now Try

6b. $p^2 - 6p + 9$ 6c. $16p^2 - 56pq + 49q^2$ 7a. $49p^2 - 42p + 9 - 4q^2$

7d. $625a^4 + 500a^3b + 150a^2b^2 + 20ab^3 + b^4$

Practice Exercises

13. $25y^2 - 30y + 9$ 15. $4x^2 + y^2 - 4xy + 12x - 6y + 9$

Objective 6

Now Try

8. $42x^3 + 47x^2 + 10x;\ -741$

Practice Exercises

17. 25

4.6   Dividing Polynomials

Key Terms
1. dividend
2. quotient
3. divisor

Objective 1

Now Try

1a. $2x^2 + x - 3$

1b. $a - 2 + \dfrac{4}{a}$

1c. $\dfrac{3}{x} - \dfrac{7}{y} + 5$

Practice Exercises

1. $3x^4 + 7x^3 + 5x$

3. $-\dfrac{r}{3s} - \dfrac{2}{3} + \dfrac{s}{r}$

Objective 2

Now Try

2. $5r + 7$

3. $2x^2 + 6x + 3 + \dfrac{2}{x-3}$

4. $6r^2 - 3r + 7 + \dfrac{-35r + 3}{2r^2 - 5}$

5. $p^2 - \dfrac{3}{5}p + 4$

Practice Exercises

5. $9p^3 - 2p + 5$

Objective 3

Now Try

6. $2x + 5 \quad x \neq 4 ; \ 1$

Practice Exercises

7. $4x + 9, \ x \neq 5$

9. $-\dfrac{6}{5}$

Chapter 5 FACTORING

5.1   Greatest Common Factors and Factoring by Grouping

Key Terms
1. factored form
2. greatest common factor

3. factor

Objective 1

Now Try

1a. $6(z - 9)$

1b. $9(4m + 5p)$

1c. There is no common factor other than 1

1d. $14(1 + 2z)$

2a. $11x(x + 4)$

2b. $12x^3(3x^2 - 2x + 4)$

2c. $6k^3(1 - 5k^2 + 7k^4)$

2d. $5mn(9m^3n - 3n^3 + 2m^2)$

2e. $13x^2y(1 + 2xy - 3y^3)$

3a. $(y + 5)(5y - 4)$

3b. $(x - y)(c^2 + d^2)$

3c. $(p + 8q)^2(b - dp - 8dq)$

3d. $-4(x - 9)(x - 1)$

4. $b^2(-7b^2 - 8b + 5); \ -b^2(7b^2 + 8b - 5)$

Answers

Practice Exercises

1. There is no common factor other than one.   3. $(x+2)^2$

Objective 2
Now Try
5. $(9+r)(a-b)$ 6. $(a-5)(y-3)$ 7. $(p+4q)(9x+1)$
8. $(a^2-6)(b^2+8)$

Practice Exercises
5. $(x-4)(1+2y^2)$

5.2   Factoring Trinomials

Key Terms
1. factoring   2. prime polynomial

Objective 1
Now Try
1a. $(x-7)(x+4)$ 1b. $(p+7)(p+3)$ 2. Prime
3. $(x+13a)(x-2a)$ 4. $15y(y+2)(y-4)$

Practice Exercises
1. $(x+9)(x+2)$ 3. $(rs+7)(rs-3)$

Objective 2
Now Try
5. $(3x-1)(6x+5)$

Practice Exercises
5. $(2r-3)(10r+1)$

Objective 3
Now Try
6. $(4x-5)(2x+1)$ 7. $(5x-3y)(3x+2y)$ 9. $9x(4x-5)(x+3)$
8. $-(7x-5)(x+2)$

Practice Exercises
7. $(3p-5)(2p+3)$ 9. $2ab(a-3b)(a-2b)$

Objective 4
Now Try
10. $(3z+7)(4z+19)$ 11. $(4y^2-5)(2y^2+1)$

Practice Exercises
11. $(2z-7)(4z-19)$

5.3   Special Factoring

Key Terms
1. difference of squares   2. perfect square trinomial

Objective 1
Now Try

1a. $(t+13)(t-13)$ 1b. $4(a+5)(a-5)$ 1c. $(5p+8q)(5p-8q)$

1d. $(7r+a-6)(7r-a+6)$ 1e. $(x^2+1)(x+1)(x-1)$

Practice Exercises
1. $(5a-6)(5a+6)$
3. $[q-(2r+3)][q+(2r+3)]$ or $(q-2r-3)(q+2r+3)$

Objective 2
Now Try

2a. $(13x-5)^2$ 2b. prime   2c. $(r+13)^2$
2d. $(m-9+b)(m-9-b)$

Practice Exercises
5. $\left(8p^2+3q^2\right)^2$

Objective 3
Now Try

3a. $(10x-y)(100x^2+10xy+y^2)$ 3b. $(3a-5b)(9a^2+15ab+25b^2)$
3c. $(4k-3n)(16k^2+12kn+9n^2)$

Practice Exercises
7. $(2r-3s)(4r^2+6rs+9s^2)$ 9. $(2a-5b)(4a^2+10ab+25b^2)$

Objective 4
Now Try

4a. $(r+5)(r^2-5r+25)$ 4b. $(6z+1)(36z^2-6z+1)$
4c. $(2t+5r^2)(4t^2-10tr^2+25r^4)$ 4d. $7(x+10)(x^2-10x+100)$
4e. $(a-5+b)(a^2-10a+25-ab+5b+b^2)$

Practice Exercises
11. $(5p+q)(25p^2-5pq+q^2)$

Answers

5.4 A General Approach to Factoring

Key Terms

1. factoring by grouping

2. FOIL

Objective 1

Now Try

1a. $18(x+3)$

1b. $6r^3s(3rs+1)$

1c. $(y+z)(7x-5)$

1d. $(x+1)(2x-3)$

2a. $(11x+5y)(11x-5y)$

2b. $(3t-4w)(9t^2+12tw+16w^2)$

2c. $(2m+1)(4m^2-2m+1)$

2d. prime

3a. $(p-9)^2$

3b. $(6z-7)^2$

3c. $(y-8)(y+1)$

3d. $(4k+1)(k-2)$

3e. $(4x-3y)(x+y)$

3f. $5(3z-2)(2z+1)$

4a. $(c+d)(3c^2-d^2)$

4c. $(6b-1+c)(6b-1-c)$

4d. $(5a-b)(25a^2+5ab+b^2+5a+b)$

Practice Exercises

1. $3ab(4ab+a-3b)$

3. $2(4x-y)(16x^2+4xy+y^2)$

5. $(x-3)(x^2+7)$

5.5 Solving Equations Using the Zero-Factor Property

Key Terms

1. standard form

2. quadratic equation

3. zero-factor property

4. double solution

Objective 1

Now Try

1. $\left\{-\dfrac{5}{4},\dfrac{3}{2}\right\}$

2. $\left\{-4,\dfrac{1}{4}\right\}$

3. $\left\{\dfrac{3}{7}\right\}$

4. $\{0,5\}$

5. $\{-1,1\}$

6. $\left\{-2,\dfrac{5}{3}\right\}$

7. $\{-5,0,7\}$

Practice Exercises

1. $\left\{-\dfrac{5}{2},4\right\}$

3. $\{3\}$

Objective 2

Now Try

8. width: 10ft

9. 1 or 2 seconds

Practice Exercises
5. 18 ft by 30 ft

Objective 3
  Now Try
10. $W = \dfrac{S - 2LH}{2H + L}$

Practice Exercises
7. $b = \dfrac{c}{d - a}$ 9. $w = \dfrac{v}{uv + 1}$

Chapter 6 RATIONAL EXPRESSIONS AND FUNCTIONS

6.1 Rational Expressions and Functions; Multiplying and Dividing

Key Terms
  1. rational function   2. rational expression

Objective 1; Objective 2
  Now Try
1a. 2; $\{x \mid x \neq 2\}$, $(-\infty, 2) \cup (2, \infty)$

1b. $2, 3$; $\{x \mid x \neq 2, 3\}$, $(-\infty, 2) \cup (2, 3) \cup (3, \infty)$

1c. none; $\{x \mid x \text{ is a real number}\}$, $(-\infty, \infty)$

1d. none; $\{x \mid x \text{ is a real number}\}$, $(-\infty, \infty)$

Practice Exercises
1. $\left\{ s \mid s \neq \dfrac{2}{3} \right\}$; $\left(-\infty, \dfrac{2}{3} \right) \cup \left(\dfrac{2}{3}, \infty \right)$

3. $\{q \mid q \neq 1, 2\} \{x \mid x \neq 2, 3\}$; $(-\infty, 2) \cup (2, 3) \cup (3, \infty)$

Objective 3
  Now Try
2. $\dfrac{x - 4}{x - 7}$ 3. -1

Practice Exercises
5. $\dfrac{y + 1}{y - 3}$

Objective 4
  Now Try
4. $\dfrac{36x}{7}$

Practice Exercises

7. $\dfrac{x-3}{x+2}$

9. $\dfrac{m-2}{m-4}$

Objective 5
Practice Exercises

11. $\dfrac{z^2-9}{7z+7}$

Objective 6
 Now Try

5a. $\dfrac{2}{9p}$

5b. $\dfrac{5x}{3}$

5c. $\dfrac{2a+3}{a+3}$

Practice Exercises

13. $-\dfrac{2(m-5)}{m+3}$

15. $\dfrac{4(y-1)}{3(y-3)}$

6.2   Adding and Subtracting Rational Expressions

Key Terms
 1. equivalent expressions

 2. least common denominator (LCD)

Objective 1
 Now Try

1a. $-\dfrac{2}{z^2}$

1b. $\dfrac{1}{c+d}$

1c. $\dfrac{1}{x-1}$

Practice Exercises

1. $\dfrac{4n-7}{m+3}$

3. $\dfrac{1}{k-4}$

Objective 2
 Now Try

2a. $33x^2y^4$

2b. $y(y+6)$

2c. $3(x+1)(x-4)^2$

Practice Exercises
 5. $3(x+2)(x+3)$

Objective 3
 Now Try

3a. $\dfrac{19}{18z}$

3b. $\dfrac{s-30}{s(s-6)}$

4a. 3

4b. $\dfrac{-30}{(x+3)(x-3)}$ 5. $\dfrac{x-12}{x-8}$ 6. $\dfrac{1}{x}$

7. $\dfrac{8x^2+37x+15}{(x+5)(x-5)^2}$

Practice Exercises

7. $\dfrac{-6n^2+27n+12}{(n+4)^2(n-4)}$ 9. $\dfrac{y^2-18y-3}{(y+3)(y+1)(y-2)}$

6.3   Complex Fractions

Key Terms

1. complex fraction   2. LCD

Objective 1

Now Try

1a. $\dfrac{5(x+1)}{3(x-1)}$ 1b. $\dfrac{2y-7}{7y-2}$

Practice Exercises

1. $\dfrac{5(3a+4)}{2a+5}$ 3. $(a+2)^2$

Objective 2

Now Try

2a. $\dfrac{2y-7}{7y-2}$ 2b. $\dfrac{4a^2-8a+1}{3a^2-6a+1}$

Practice Exercises

5. $\dfrac{r^2+3}{5+r^2t}$

Objective 3

Now Try

3a. $\dfrac{1-r}{r(1+r)}$ 3b. $\dfrac{b(b+a^2)}{a(a-b^2)}$

Practice Exercises

7. $\dfrac{7(5k-m)}{4}$ or $\dfrac{35k-7m}{4}$ 9. $\dfrac{3(8-x)}{2(15+x)}$

Objective 4

Now Try

4. $\dfrac{1}{xy-1}$

Practice Exercises

11. $\dfrac{r}{s}$

6.4 Equations with Rational Expressions and Graphs

Key Terms

1. vertical asymptote
2. domain of the variable in a rational equation
3. discontinuous
4. horizontal asymptote

Objective 1

Now Try

1a. $\{x \mid x \neq 0\}$ 1b. $\{x \mid x \neq -8,\ 7\}$

Practice Exercises

1. (a) $0, -1$; (b) $\{x \mid x \neq 0, -1\}$ 3. (a) $-3, 0, 1$; (b) $\{x \mid x \neq -3, 0, 1\}$

Objective 2

Now Try

2. $\{2\}$ 3. $\varnothing$ 4. $\{-2\}$

5. $\left\{\dfrac{4}{3}\right\}$ 6. $\{-1, 6\}$

Practice Exercises

5. $\varnothing$

Objective 3

Now Try

7. 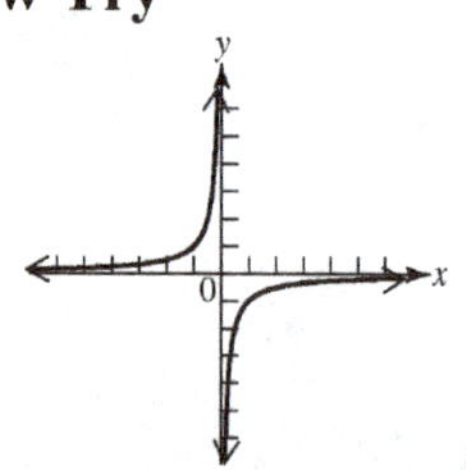

Vertical asymptote: $x = 0$
Horizontal asymptote: $y = 0$

Practice Exercises

7. 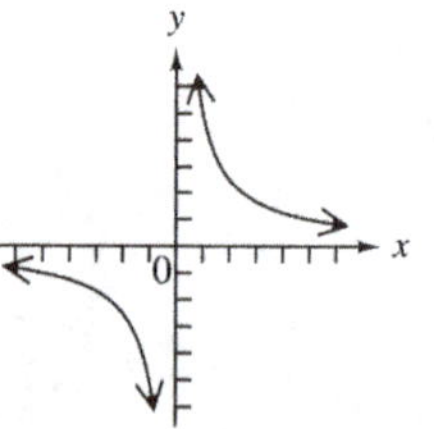

Vertical asymptote: $x = 0$
Horizontal asymptote: $y = 0$

9. 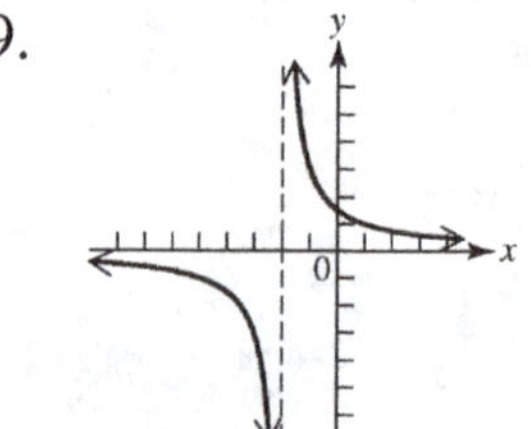

Vertical asymptote: $x = -2$
Horizontal asymptote: $y = 0$

    Copyright © 2020 Pearson Education, Inc.

6.5 Applications of Rational Expressions

Key Terms
  1. ratio   2. proportion

Objective 1
  Now Try
  1. $\dfrac{11}{3}$

  Practice Exercises
  1. $m = \dfrac{75}{8}$ 3. $L = 20$

Objective 2
  Now Try
  2. $R = \dfrac{R_1 R_2}{R_1 + R_2}$ 3. $R_1 = \dfrac{AR_r}{R_2 - A}$

  Practice Exercises
  5. $B = \dfrac{2A - hb}{h}$

Objective 3
  Now Try
  5. 4 gallons

  Practice Exercises
  7. 8 gal   9. 50,000 people

Objective 4
  Now Try
  6. 2 mph   7. 3 mph

  Practice Exercises
  11. Pauline: 60 mph; Pete: 40 mph

Objective 5
  Now Try
  8. $\dfrac{2}{5}$ hour

  Practice Exercises
  13. $2\frac{2}{5}$ hr   15. 2 hr

6.6   Variation

Key Terms
1. constant of variation
2. varies inversely
3. varies directly

Objective 1; Objective 2
Now Try
1. $25; $P = 25x$
2. 125 psi
3. 153.86 sq cm

Practice Exercises
1. $k = 0.25$; $y = 0.25x$
3. 100 newtons

Objective 3
Now Try
4. 640 cycles per second
5. $\dfrac{9}{8}$

Practice Exercises
5. 40 lb

Objective 4
Now Try
6. 1280 psi

Practice Exercises
7. 96
9. 750°

Objective 5
Now Try
7. About 63.5 cm^3

Practice Exercises
11. 9 hr

Chapter 7 ROOTS, RADICALS, AND ROOT FUNCTIONS

7.1   Radical Expressions and Graphs

Key Terms
1. radicand
2. perfect square
3. index (order)
4. square root
5. radical expression
6. principal square root
7. radical
8. cube root

Objective 1
Now Try
1c. 4
1e. $\dfrac{4}{3}$

Practice Exercises
1. 5
3. 9

Objective 2
  Now Try
2b. -11 2e. not a real number   2g. -5

  Practice Exercises
5. 4

Objective 3
  Now Try
3a. 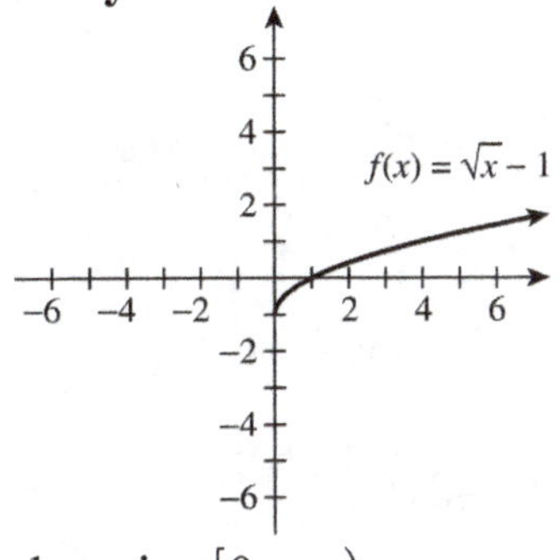

domain: $[0, \infty)$

range : $[-1, \infty)$

3b. 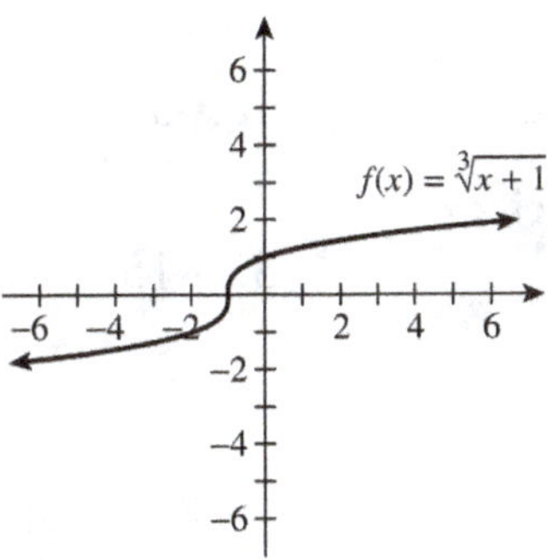

domain: $(-\infty, \infty)$

range: $(-\infty, \infty)$

  Practice Exercises
7. 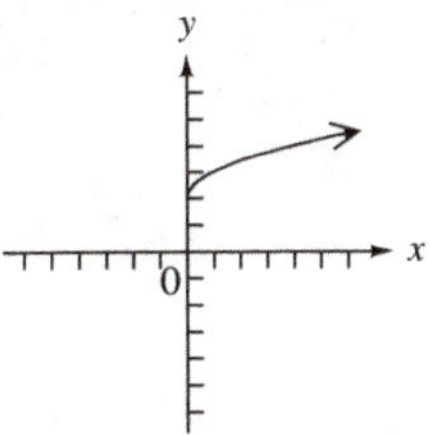

domain: $[0, \infty)$
range: $[2, \infty)$

9. 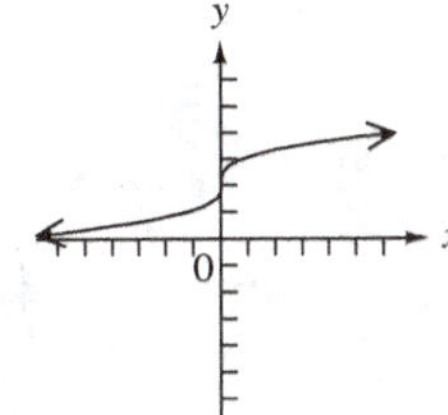

domain: $(-\infty, \infty)$

range: $(-\infty, \infty)$

Objective 4
  Now Try
4a. 73   4d. $|n|$ 5a. 4

5b. -5 5c. -2 5d. $-x^4$

5e. w^{10} 5f. $|x^5|$

  Practice Exercises
11. $-x$

Objective 5
  Now Try
6a. 3.742   6b. -26.038 6c. 7.554
6d. 9.499   7. 23.2 mph

  Practice Exercises
13. 8.883   15. 2.2 sec

Answers

7.2   Rational Exponents

Key Terms
1. power rule for exponents
2. product rule for exponents
3. quotient rule for exponents

Objective 1
Now Try

1a. 6

1b. 11

1c. −2

1d. not a real number

1e. −3

1f. $\dfrac{1}{2}$

Practice Exercises
1. −4

3. −15

Objective 2
Now Try

2a. 9

2b. 8

2c. −216

2d. 9

2e. not a real number

3a. $\dfrac{1}{4}$

3b. $\dfrac{1}{625}$

3c. $\dfrac{16}{9}$

Practice Exercises
5. 7776

Objective 3
Now Try

4a. $\sqrt[3]{23}$

4b. $\left(\sqrt[4]{21}\right)^3$

4c. $5\left(\sqrt[4]{x}\right)^5$

4d. $\left(\sqrt[3]{2x}\right)^4 - 3\left(\sqrt[5]{x}\right)^2$

4e. $\dfrac{1}{\left(\sqrt{x}\right)^3}$

4f. $\sqrt[4]{x^2 - y^2}$

5a. $22^{1/2}$

5b. $10^2 = 100$

5c. $3^2 = 9$

5d. m

Practice Exercises
7. $4\left(\sqrt[5]{y}\right)^2 + \sqrt[5]{5x}$

9. $a^{1/4}$, or $\sqrt[4]{a}$

Objective 4
Now Try

6a. $5^{5/2}$

6b. $a^{2/15}$

6c. $x^2 y^{7/2}$

6d. $\dfrac{x^2}{y^{1/4}}$

6e. $r^{7/6} - r^{19/6}$

7a. $x^{7/6}$

7b. $y^{7/12}$

7c. $x^{1/4}$

Practice Exercises
11. $a^{5/6}$

7.3 Simplifying Radicals, the Distance Formula, and Circles

Key Terms

1. hypotenuse
2. index; radicand
3. legs
4. circle
5. center
6. radius

Objective 1

Now Try

1a. $\sqrt{14}$

1b. $\sqrt{35x}$

1c. $\sqrt{33mn}$

2a. $\sqrt[3]{21}$

2b. $\sqrt[3]{35xy}$

2c. $\sqrt[5]{8w^4}$

2d. cannot be simplified using the product rule

Practice Exercises

1. $\sqrt{42tx}$

3. cannot be simplified using the product rule

Objective 2

Now Try

3a. $\dfrac{6}{7}$

3b. $\dfrac{\sqrt{13}}{9}$

3c. $-\dfrac{7}{5}$

3d. $-\dfrac{a^2}{5}$

3e. $\dfrac{\sqrt[4]{m}}{3}$

Practice Exercises

5. $\dfrac{\sqrt[5]{7x}}{2}$

Objective 3

Now Try

4a. $2\sqrt{21}$

4b. $9\sqrt{2}$

4c. cannot be simplified

4d. $4\sqrt[3]{4}$

4e. $-2\sqrt[5]{16}$

5a. $10y\sqrt{y}$

5b. $4m^2r^4\sqrt{3mr}$

5c. $-2n^2t\sqrt[3]{4nt^2}$

5d. $-3y^2\sqrt[4]{5x^3y}$

6a. $\sqrt[4]{11^3}$, or $\sqrt[4]{1331}$

6b. $\sqrt[5]{z^4}$

6c. $w^3\sqrt{w}$

Practice Exercises

7. $\sqrt[3]{x^2}$

9. $5ab^2\sqrt[3]{10a^2b}$

Objective 4

Now Try

7. $\sqrt[6]{63}$

Practice Exercises

11. $\sqrt[8]{28}$

Objective 5
Now Try

8. $12\sqrt{2}$

Practice Exercises

13. 26   15. $6\sqrt{2}$

Objective 6
Now Try

9. $\sqrt{34}$

Practice Exercises

17. 5

Objective 7
Now Try

10. $x^2 + y^2 = 25$ 11. $(x+5)^2 + (y-4)^2 = 16$

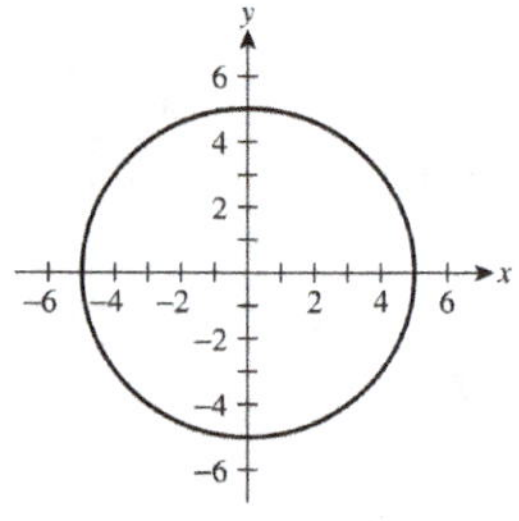
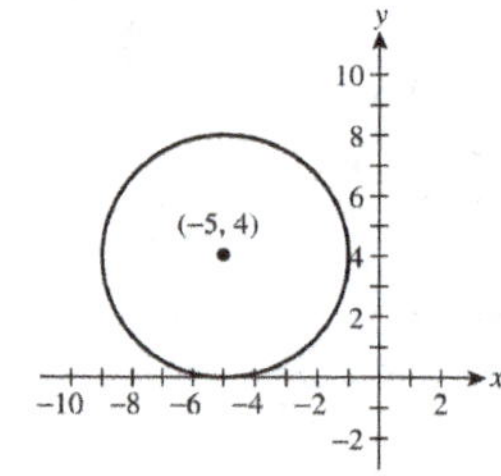

12. $(x-3)^2 + (y+1)^2 = 6$

Practice Exercises

19. $(x-3)^2 + (y+4)^2 = 25$ 21. $x^2 + (y-3)^2 = 2$

7.4 Adding and Subtracting Radical Expressions

Key Terms

1. unlike radicals   2. like radicals

Objective 1
Now Try

1a. $8\sqrt{5}$ 1b. $-4\sqrt{3}$ 1c. $-\sqrt{6}$

1d. $13\sqrt{2z}$ 1e. cannot be simplified

2a. $-3\sqrt[3]{2}$ 2b. $5z\sqrt[4]{2y^2z}$ 2c. $6x^2\sqrt[3]{2x} + 6x^3\sqrt{3x}$

3a. $\dfrac{2\sqrt{5}}{3}$ 3b. $\dfrac{17}{w^2}$

Practice Exercises

1. $12\sqrt{x}$ 3. $\dfrac{8y^2\sqrt[3]{y}}{15}$

7.5 Multiplying and Dividing Radical Expressions

Key Terms

1. conjugate
2. rationalizing the denominator

Objective 1

Now Try

2a. 12

2b. $6 + 3\sqrt{7} - 2\sqrt{2} - \sqrt{14}$

2c. $11 - 6\sqrt{2}$

Practice Exercises

1. $\sqrt{10} - 4\sqrt{5} + 2\sqrt{3} - 4\sqrt{6}$

3. $4 - \sqrt[3]{25}$

Objective 2

Now Try

3. $\dfrac{2\sqrt{15}}{15}$

4. $-\dfrac{3\sqrt{10}}{8}$

5. $\dfrac{\sqrt[3]{10}}{5}$

Practice Exercises

5. $\dfrac{y\sqrt{21b}}{6b}$

Objective 3

Now Try

6. $-2\left(\sqrt{3} + 2\right)$

Practice Exercises

7. $-4\sqrt{3} + 8$

9. $\dfrac{\sqrt{3} + 2\sqrt{6} + \sqrt{2} + 4}{-7}$

Objective 4

Now Try

7. $\dfrac{3 + 2\sqrt{15}}{4}$

Practice Exercises

11. $\dfrac{2 - 9\sqrt{2}}{3}$

7.6 Solving Equations with Radicals

Key Terms

1. extraneous solution
2. radical equation

Objective 1

Now Try

1. $\{10\}$

2. $\varnothing$

Practice Exercises
1. $\{11\}$ 3. $\{13\}$

Objective 2
  Now Try
3. $\{5\}$ 4. $\{3\}$ 5. $\{4\}$

  Practice Exercises
5. $\{5, 6\}$

Objective 3
  Now Try
6. $\{2\}$

  Practice Exercises
7. $\{-31\}$ 9. $\{2\}$

Objective 4
  Now Try
7. $h = \dfrac{3v}{\pi r^2}$

  Practice Exercises
11. $C = \dfrac{1}{4\pi^2 f^2 L}$

7.7   Complex Numbers

Key Terms
  1. complex number   2. complex conjugate   3. imaginary part

  4. real part   5. standard form (of a complex number)

  6. pure imaginary number   7. nonreal complex number

Objective 1
  Now Try
1a. $4i$ 1b. $-12i$ 1c. $i\sqrt{11}$
1d. $8i\sqrt{2}$ 2. $-\sqrt{30}$ 3a. 5
3b. $4i$

  Practice Exercises
1. $-9i\sqrt{2}$ 3. $6i$

Objective 2
  Practice Exercises
5. imaginary

Objective 3

Now Try

4. $10 - 9i$　　　　5. $24 + 4i$

Practice Exercises

7. $-4 + i$　　　　9. $2 - 3i$

Objective 4

Now Try

6a. $-14 + 8i$　　　6b. $17 - 7i$　　　6c. $17 + i$

Practice Exercises

11. $-8 + 6i$

Objective 5

Now Try

7. $\dfrac{18}{29} + \dfrac{13i}{29}$

Practice Exercises

13. $\dfrac{4}{5} - \dfrac{7}{5}i$　　　15. $\dfrac{15}{13} + \dfrac{16}{13}i$

Objective 6

Now Try

8a. 1　　　　8b. i　　　　8c. $-i$

8d. $-i$

Practice Exercises

17. i

Chapter 8　QUADRATIC EQUATIONS, INEQUALITIES, AND FUNCTIONS

8.1　The Square Root Property and Completing the Square

Key Terms

1. square root property　2. completing the square

3. quadratic equation　4. zero-factor property

Objective 1

Now Try

1. $\left\{-\dfrac{7}{3}, 4\right\}$

Practice Exercises

1. $\left\{-\dfrac{1}{3}, \dfrac{2}{5}\right\}$　　　3. $\{-2, 2\}$

Answers

Objective 2
  Now Try
2a. $\left\{\sqrt{13}, -\sqrt{13}\right\}$, or $\left\{\pm\sqrt{13}\right\}$ $\qquad$ 2b. $\left\{3\sqrt{2}, -3\sqrt{2}\right\}$, or $\left\{\pm 3\sqrt{2}\right\}$
  3. About 2.7 seconds

  Practice Exercises
  5. $\left\{-7\sqrt{2},\ 7\sqrt{2}\right\}$

Objective 3
  Now Try

  4. $\{-3, 5\}$ $\qquad$ 5. $\left\{\dfrac{-5-4\sqrt{2}}{2},\ \dfrac{-5+4\sqrt{2}}{2}\right\}$

  Practice Exercises

  7. $\{-6, 2\}$ $\qquad$ 9. $\left\{\dfrac{-4-4\sqrt{2}}{3},\ \dfrac{-4+4\sqrt{2}}{3}\right\}$

Objective 4
  Now Try

  7. $\left\{\dfrac{11-\sqrt{89}}{2},\ \dfrac{11+\sqrt{89}}{2}\right\}$

  Practice Exercises
  11. $\{1, 8\}$

Objective 5
  Now Try
  9. $\left\{-4i\sqrt{2},\ 4i\sqrt{2}\right\}$

  Practice Exercises
  13. $\left\{\dfrac{1}{2}-\dfrac{3}{10}i,\ \dfrac{1}{2}+\dfrac{3}{10}i\right\}$ $\qquad$ 15. $\left\{1+i\sqrt{2},\ 1-i\sqrt{2}\right\}$

8.2 The Quadratic Formula

Key Terms
  1. discriminant $\qquad$ 2. quadratic formula

Objective 1
Objective 2
  Now Try

  1. $\left\{\dfrac{4}{3},\ \dfrac{3}{2}\right\}$ $\qquad$ 2. $\left\{-\dfrac{3}{5}\right\}$ $\qquad$ 3. $\left\{\dfrac{1-\sqrt{7}}{2},\ \dfrac{1+\sqrt{7}}{2}\right\}$
  4. $\{1-2i,\ 1+2i\}$

  Practice Exercises
  1. $\{2, 4\}$ $\qquad$ 3. $\{5-3i,\ 5+3i\}$

Objective 3
Now Try

5a. 81; two rational solutions; factoring

5b. 0; one rational solution; factoring

5c. –48; two nonreal complex solutions; quadratic formula

Practice Exercises

5. C

8.3   Equations That Lead to Quadratic Methods

Key Terms

1. standard form

2. quadratic in form

Objective 1
Now Try

1. $\left\{-7, \frac{5}{4}\right\}$

Practice Exercises

1. $\left\{-\frac{35}{4}, -3\right\}$

3. $\left\{-7, -\frac{7}{2}\right\}$

Objective 2
Now Try

2. 2 mph

3. Tom: 12 hours; Huck: 24 hours

Practice Exercises

5. 550 mph

Objective 3
Now Try

4a. $\{2\}$

4b. $\{8\}$

Practice Exercises

7. $\{2, 5\}$

9. $\{9\}$

Objective 4
Now Try

6. $\{-2, -1, 1, 2\}$

7a. $\{-2, 10\}$

7b. $\{-1, 27\}$

Practice Exercises

11. $\left\{-27, -3\sqrt{3}, 3\sqrt{3}, 27\right\}$

8.4 Formulas and Further Applications

Key Terms
1. quadratic function   2. Pythagorean theorem

Objective 1
Now Try

1a. $t = \pm \dfrac{\sqrt{mxF}}{F}$ 1b. $z = \dfrac{y^2}{6p^2}$ 2. $q = \dfrac{-k \pm k\sqrt{5}}{2p}$

Practice Exercises

1. $d = \dfrac{k^2 l^2}{F^2}$ 3. $a = \dfrac{-c \pm c\sqrt{2}}{b}$

Objective 2
Now Try
3. south: 72 mi; east 54 mi

Practice Exercises
5. 10 ft

Objective 3
Now Try
4. 2.5 in.

Practice Exercises
7. 2.5 ft   9. 4 ft

Objective 4
Now Try
5. 4.1 sec

Practice Exercises
11. 1.2 sec

8.5   Graphs of Quadratic Functions

Key Terms

1. axis   2. vertex   3. quadratic function

4. parabola

Objective 1; Objective 2

Now Try

1. vertex: $(0,-1)$, axis: $x = 0$;
   domain: $(-\infty, \infty)$; range: $[-1, \infty)$

2. Vertex: $(-3, 0)$; axis: $x = -3$
   domain: $(-\infty, \infty)$; range: $[0, \infty)$

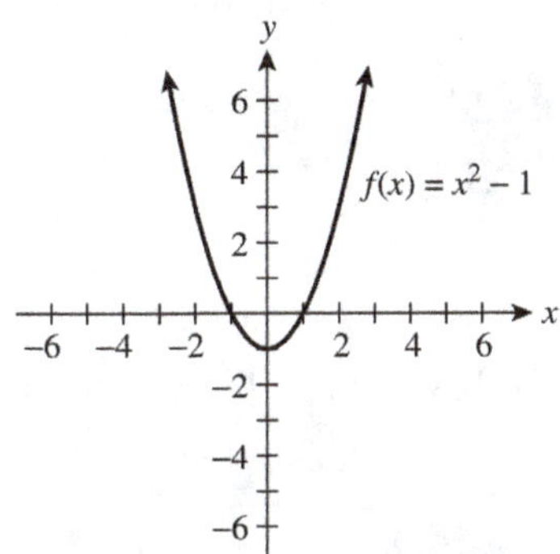
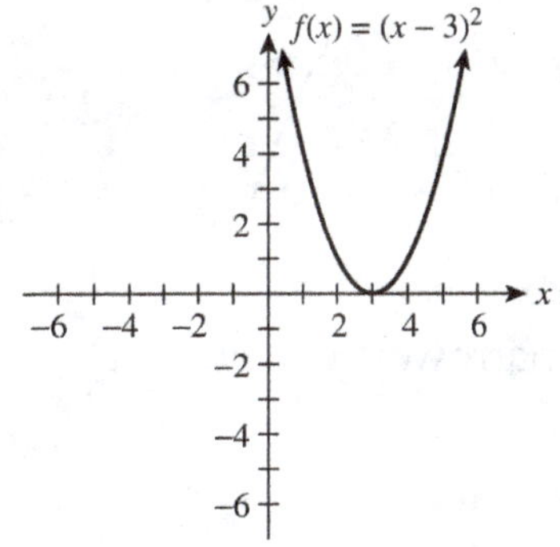

3. vertex: $(-2,-1)$; axis: $x = -2$
   domain: $(-\infty, \infty)$; range: $[-1, \infty)$

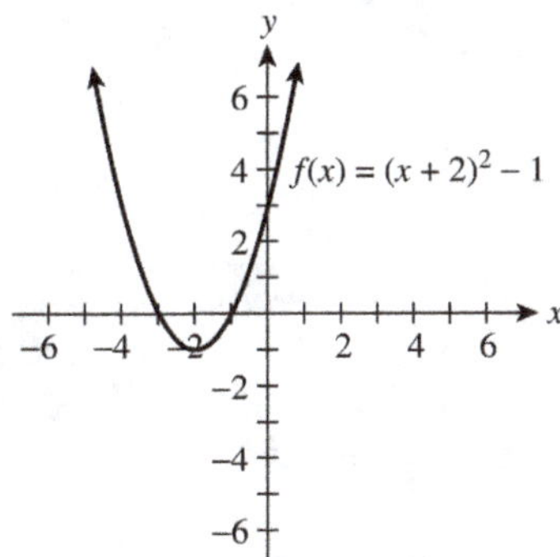

Practice Exercises

1. 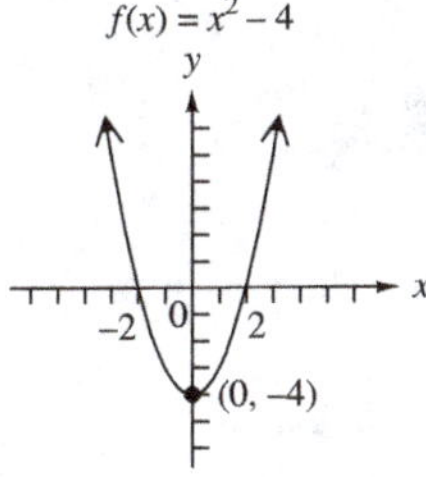

   Vertex: $(0, -4)$
   Axis: $x = 0$
   Domain: $(-\infty, \infty)$
   Range: $[-4, \infty)$

3. 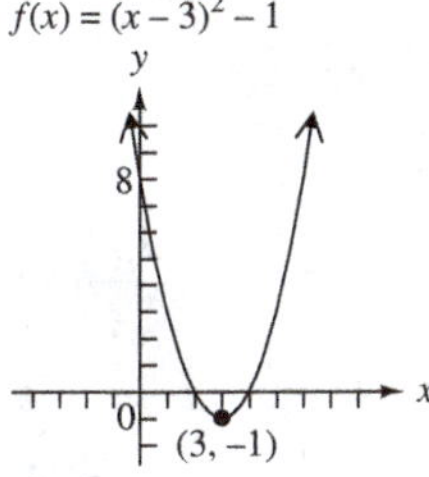

   Vertex: $(3, -1)$
   Axis: $x = 3$
   Domain:
   $(-\infty, \infty)$
   Range: $[-1, \infty)$

Objective 3
Now Try
4. vertex: $(0, 0)$; axis: $x = 0$
   domain: $(-\infty, \infty)$; range: $(-\infty, 0]$

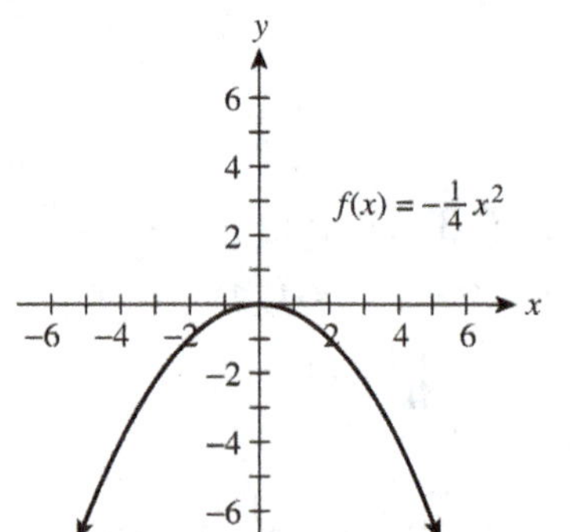

5. vertex: $(1, 1)$; axis: $x = 1$
   domain: $(-\infty, \infty)$; range: $[1, \infty)$

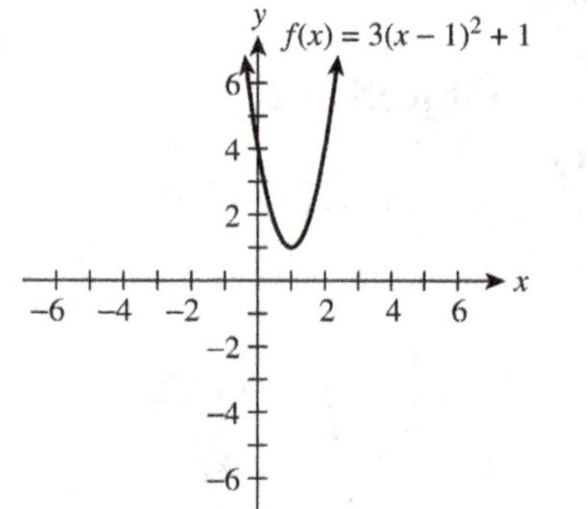

Practice Exercises
5. down; narrower; vertex: $(-1, 0)$; domain: $(-\infty, \infty)$; range: $[0, \infty)$

Objective 4
Now Try
6. $y = 0.01x^2 + 0.77x + 8.36$

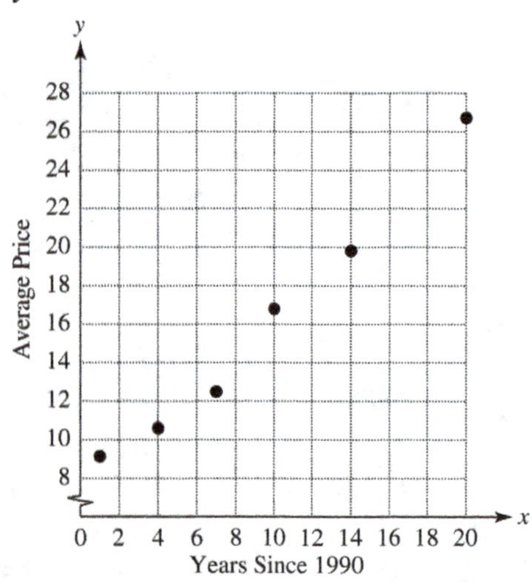

Practice Exercises
7. linear; positive

9. 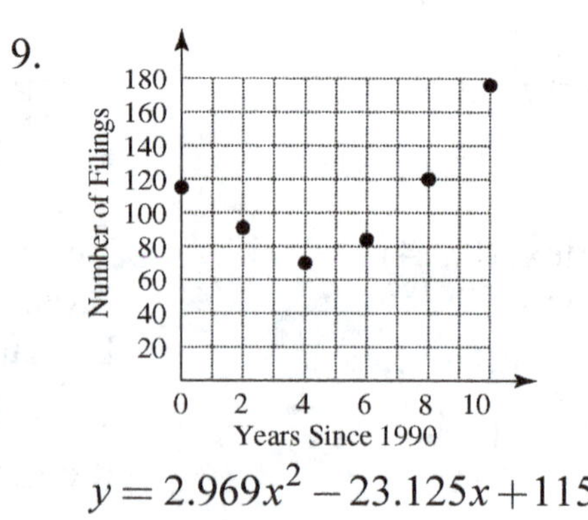

$y = 2.969x^2 - 23.125x + 115$

8.6 More about Parabolas and Their Applications

Key Terms

1. discriminant
2. vertex

Objective 1

Now Try

1. $(3, -5)$
2. $(1, 1)$
3. $\left(\dfrac{3}{2}, \dfrac{1}{2}\right)$

Practice Exercises

1. $(1, 3)$
3. $(-6, 0)$

Objective 2

Now Try

4. vertex: $(-2, 1)$, axis: $x = -2$;
   domain: $(-\infty, \infty)$; range: $[1, \infty)$

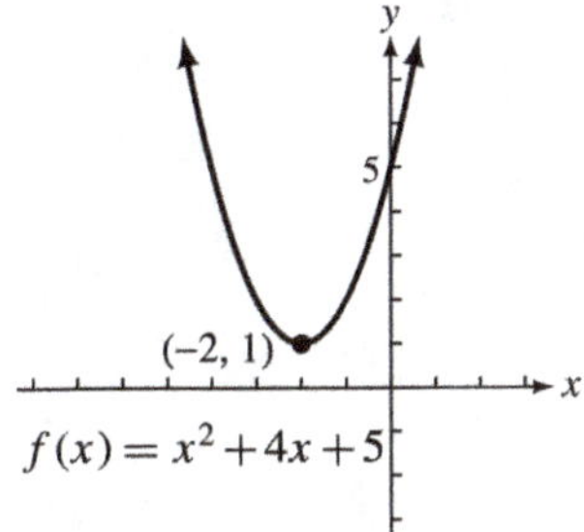

Practice Exercises

5.

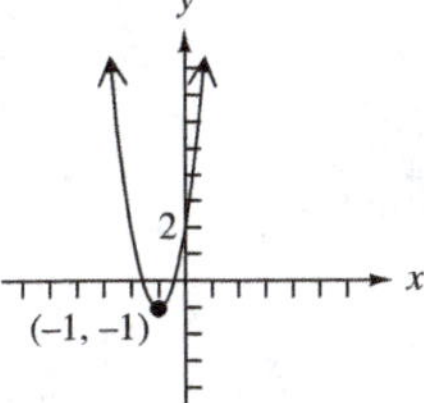

Vertex: $(-1, -1)$
Axis: $x = -1$
Domain: $(-\infty, \infty)$
Range: $[-1, \infty)$

Objective 3

Now Try

5. 1

Practice Exercises

7. 0
9. 1

Answers

Objective 4
Now Try
6. maximum area: 125,000 sq yd; length: 500 yd; width: 250 yd
7. maximum height: 286 feet after 1.5 seconds

Practice Exercises
11. 24 (a square)

Objective 5
Now Try
9. vertex: $(0, 2)$; axis: $y = 2$
   domain: $(-\infty, 0]$; range: $(-\infty, \infty)$

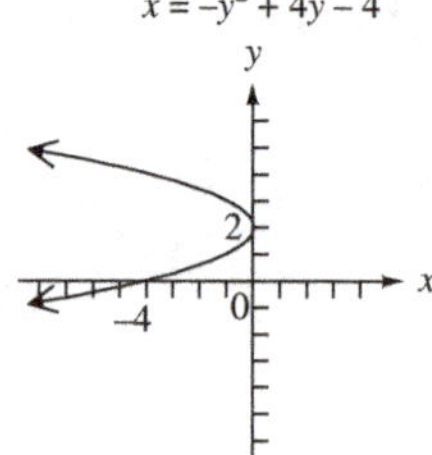

Practice Exercises
13.

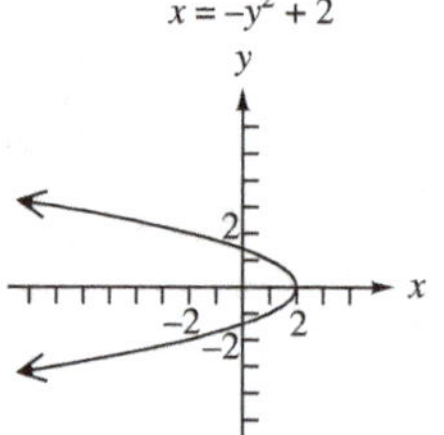

Vertex: $(2, 0)$
Axis: $y = 0$
Domain: $(-\infty, 2]$
Range: $(-\infty, \infty)$

15.

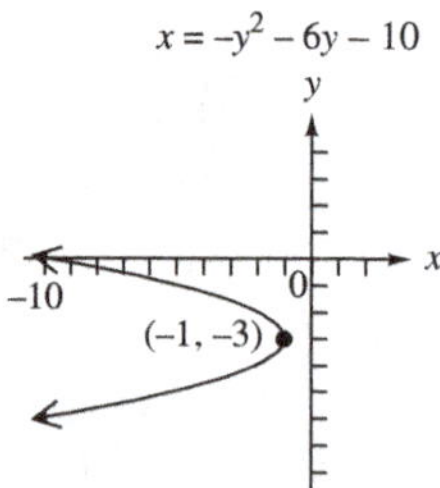

Vertex: $(-1, -3)$
Axis: $y = -3$
Domain:
$(-\infty, -1]$
Range: $(-\infty, \infty)$

8.7 Polynomial and Rational Inequalities

Key Terms
1. rational inequality   2. quadratic inequality

Objective 1
Now Try
1a. $(-\infty, 2)\cup(6, \infty)$ 1b. $(2, 6)$ 2. $(-1, 2)$

4a. $(-\infty, \infty)$ 4b. $\varnothing$

   Copyright © 2020 Pearson Education, Inc.

Practice Exercises

1. $[-1, 2]$ 3. $\varnothing$

Objective 2
Now Try

5. $\left(-\infty, -\dfrac{3}{2}\right] \cup \left[-\dfrac{1}{3}, \dfrac{1}{2}\right]$

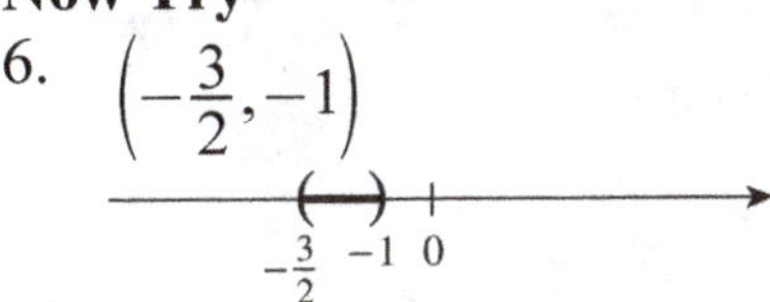

Practice Exercises

5. $(-\infty, -5] \cup [-3, 1]$

Objective 3
Now Try

6. $\left(-\dfrac{3}{2}, -1\right)$

7. $(-\infty, 3) \cup [8, \infty)$

Practice Exercises

7. $(-\infty, 1) \cup [8, \infty)$ 9. $[-2, 3)$

Chapter 9   INVERSE, EXPONENTIAL, AND LOGARITHMIC FUNCTIONS

9.1   Inverse Functions

Key Terms
1. one-to-one function   2. inverse of a function

Objective 1
Now Try

1a. One-to-one; $G^{-1} = \{(2, 3), (-2, -3), (3, 2), (-3, -2)\}$

1b. Not one-to-one   1c. Not one-to-one

Practice Exercises

1. One-to-one; $\{(-1, -3), (2, -2), (3, -1), (4, 0)\}$

3. One-to-one; $\{(0, 0), (1, 1), (-1, -1), (2, 2), (-2, -2)\}$

                    513

Objective 2
Now Try
2a. One-to-one   2b. Not one-to-one

Practice Exercises
5. Not one-to-one

Objective 3
Now Try

3a. $f^{-1}(x) = \dfrac{1}{4}x + \dfrac{1}{4}$ 3b. Not one-to-one   3c. $f^{-1}(x) = \sqrt[3]{\dfrac{x+3}{2}}$

Practice Exercises
7. $f^{-1}(x) = \dfrac{x+5}{2}$ 9. Not one-to-one

Objective 4
Now Try
5.

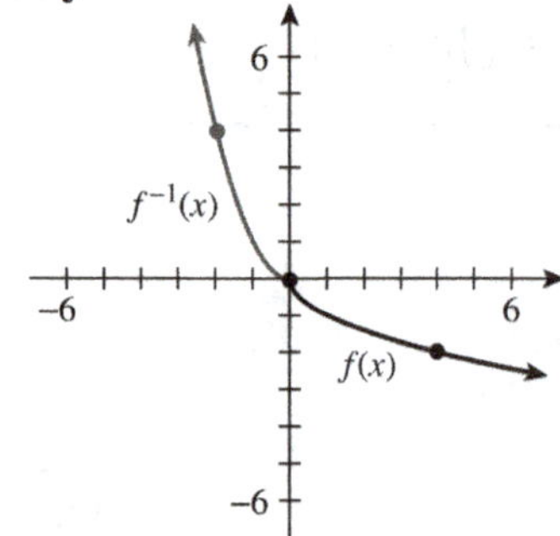

Practice Exercises
11. Not one-to-one

9.2   Exponential Functions

Key Terms
1. inverse   2. exponential equation

Objective 1
Now Try
1a. 8.064   1b. 0.172   1c. 1.246

Practice Exercises
1. 3.737   3. 1.442

Objective 2
Now Try

2. 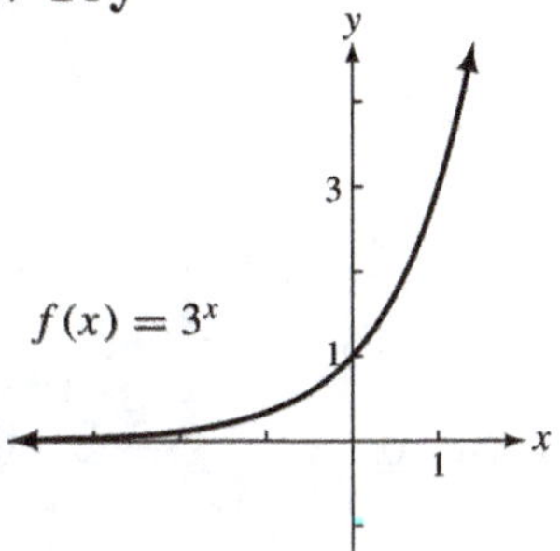

3. 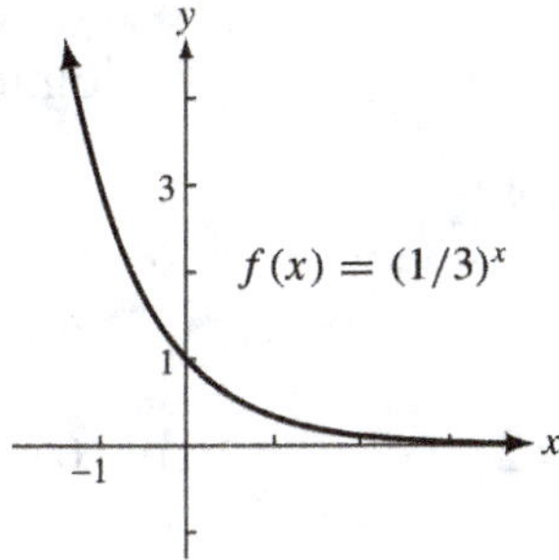

4. 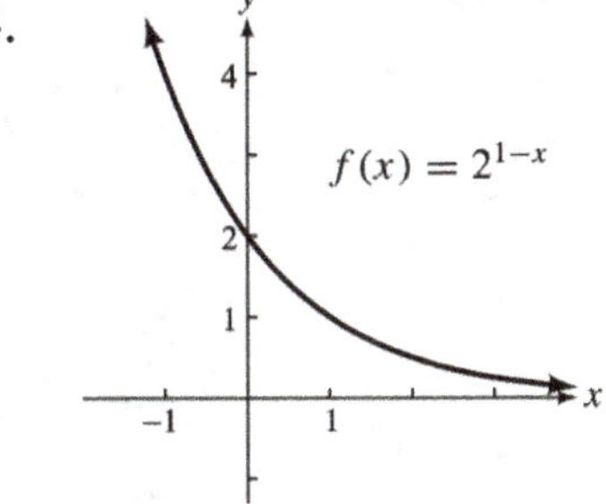

Practice Exercises

5. 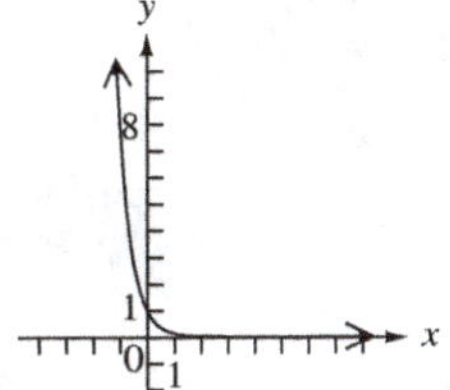

Objective 3
Now Try

5. $\left\{\dfrac{3}{2}\right\}$

6a. $\{-2\}$

6b. $\{-5\}$

6c. $\{-4\}$

Practice Exercises

7. $\left\{\dfrac{1}{2}\right\}$

9. $\{-2\}$

Objective 4
Now Try

7. 256,000

8. about 32,656

Practice Exercises

11. 1 gram

9.3 Logarithmic Functions

Key Terms

1. logarithm
2. logarithmic equation

Objective 1

Objective 2

Now Try

1a. $\log_8 64 = 2$

1b. $16^{-1/2} = \dfrac{1}{4}$

2a. 2.5850

2b. 2.1962

2c. -2.3219

2d. 1.0792

Practice Exercises

1. $10^{-3} = 0.001$

3. 2.9299

Objective 3

Now Try

3a. $\left\{\dfrac{1}{64}\right\}$

3b. $\{40\}$

3c. $2\sqrt{3}$

3d. $\left\{\dfrac{1}{6}\right\}$

Practice Exercises

5. $\{81\}$

Objective 4

Now Try

4a. 1

4c. 0

4d. 0

4e. 3

4g. 9

4i. 3

4j. -1

Practice Exercises

7. 0

9. 5

Objective 5

Now Try

5. 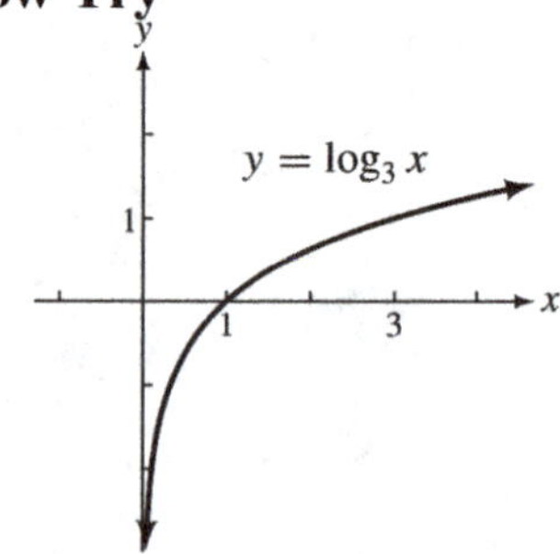

6. 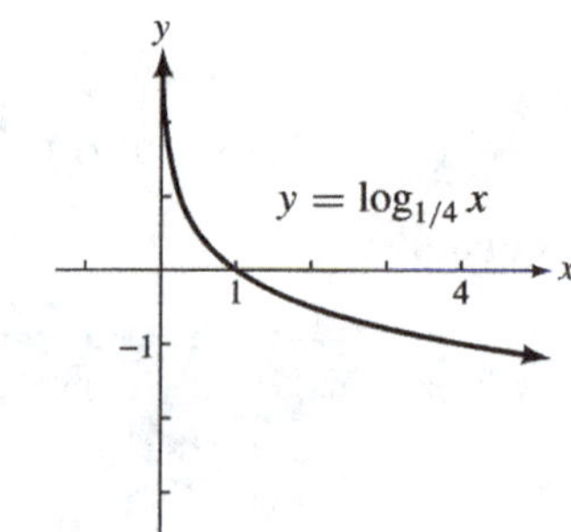

Practice Exercises

11.

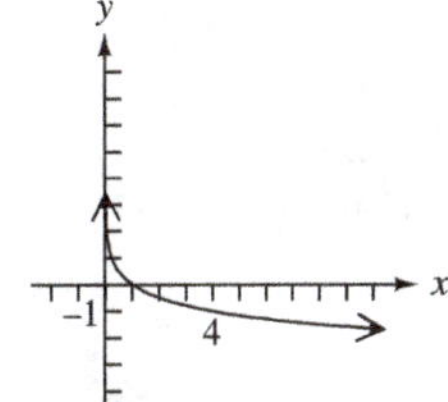

Objective 6
Now Try
7. 16 fish

Practice Exercises
13. 100 mites

9.4   Properties of Logarithms

Key Terms
1. special properties

2. quotient rule for logarithms

3. product rule for logarithms

4. power rule for logarithms

Objective 1
Now Try

1a. $\log_6 5 + \log_6 3$

1b. $\log_5 21$

1c. $1 + \log_4 x$

1d. $4\log_5 x$

Practice Exercises

1. $\log_7 5 + \log_7 m$

3. $\log_4 21$

Objective 2
Now Try

2a. $\log_5 4 - \log_5 9$

2b. $\log_6 \dfrac{x}{3}$

2c. $2 - \log_4 11$

2d. $\left\{ \dfrac{1}{6} \right\}$

Practice Exercises

5. $\log_6 k - \log_6 3$

Objective 3
Now Try

3a. $3\log_6 4$

3c. $\dfrac{1}{2}\log_b 13$

3d. $\dfrac{3}{4}\log_3 x$

3e. $-7\log_3 x$

Practice Exercises

7. $7\log_m 2$

9. $\sqrt[3]{7}$

Objective 4
Now Try

4a. $2 + 3\log_6 x$

4b. $\dfrac{1}{2}(\log_b x - \log_b 3)$

4d. $\log_b \dfrac{xy^4}{z}$

4e. $\log_2 \dfrac{x^2(x-1)}{\sqrt[3]{x^2+1}}$

4f. cannot be rewritten

5a. 7.6293

5b. -3.4594

5c. 9.5097

6a. False

6b. False

Practice Exercises
11. 6.1700

9.5   Common and Natural Logarithms

Key Terms
1. natural logarithm
2. common logarithm

Objective 1
Now Try
1. 2.9928

Practice Exercises
1. 1.7576
3. 4.9401

Objective 2
Now Try
2. $\text{pH} = 7.2076$; rich fen
3. 2.5×10^{-4}
4. 85 dB

Practice Exercises
5. 134 dB

Objective 3
Now Try
5. 4.5850

Practice Exercises
7. 4.3347
9. 3.9120

Objective 4
Now Try
6. 11.9 years

Practice Exercises
11. 3240 years

Objective 5
  Now Try
  7. 3.2266

  Practice Exercises
13. 1.1887   15. -2.3219

9.6   Exponential and Logarithmic Equations; Further Applications

Key Terms
  1. continuous compounding   2. compound interest

Objective 1
  Now Try
  1. 2.465   2. 6.770

  Practice Exercises
  1. $\{1\}$ 3. $\{-4.324\}$

Objective 2
  Now Try
  5. $-1+\sqrt[3]{36}$ 7. $\{1\}$

  Practice Exercises
  5. $\{5\}$

Objective 3
  Now Try
  8. $12,201.90   9. 13.89 years   10a. 5309.18

10b. 34.7 years

  Practice Exercises
  7. $55,200.99   9. 7.7 years

Objective 4
  Now Try
11a. 11.3 mg   11b. 30.0 years

  Practice Exercises
11. 29 years

Chapter 10 NONLINEAR FUNCTIONS, CONIC SECTIONS, AND NONLINEAR SYSTEMS

10.1 Additional Graphs of Functions

Key Terms
1. step function
2. greatest integer function
3. asymptotes

Objective 1
Now Try
1. 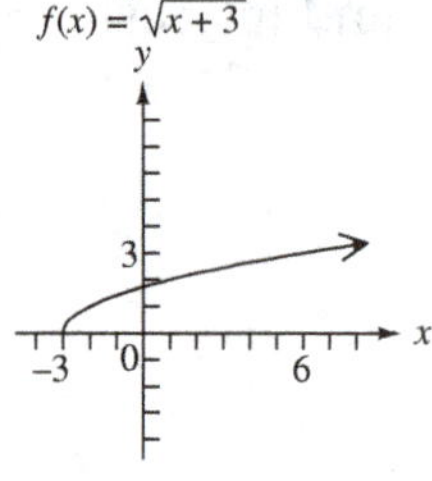

Domain: $[-3, \infty)$

Range: $[0, \infty)$

2. 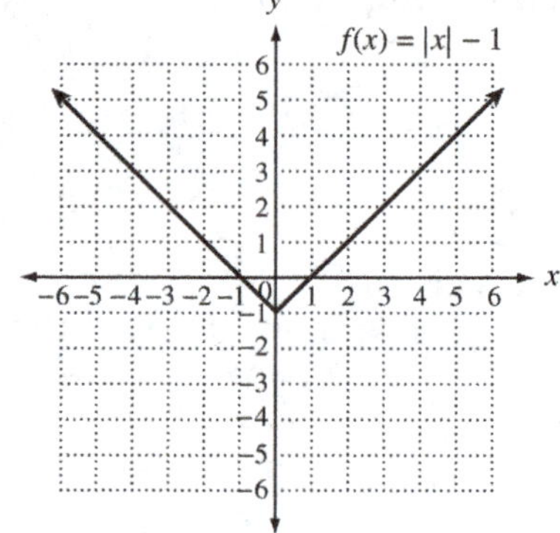

Domain: $(-\infty, \infty)$

Range: $[-1, \infty)$

3. 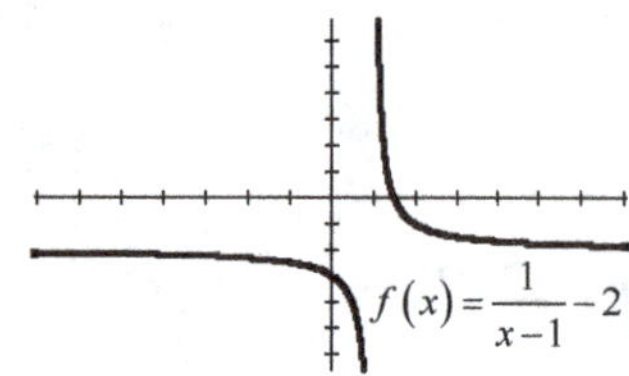

Domain: $(-\infty, 1)\cup(1, \infty)$

Range: $(-\infty, -2)\cup(-2, \infty)$

Practice Exercises
1. 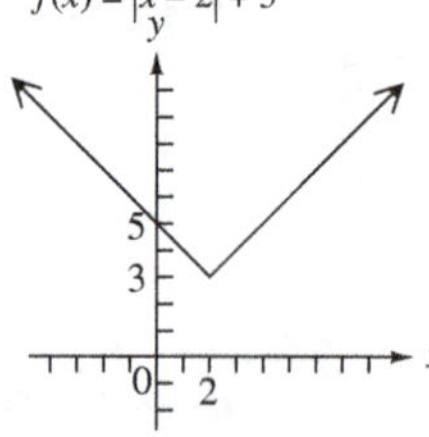

Domain: $(-\infty, \infty)$

Range: $[3, \infty)$

3. 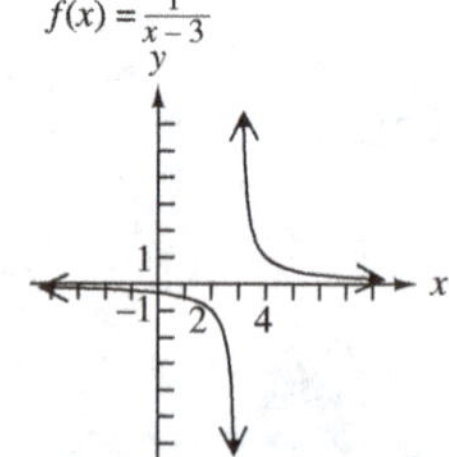

Domain: $(-\infty, 3)\cup(3, \infty)$

Range: $(-\infty, 0)\cup(0, \infty)$

Objective 2
Now Try

4d. 1   4e. −11

5.
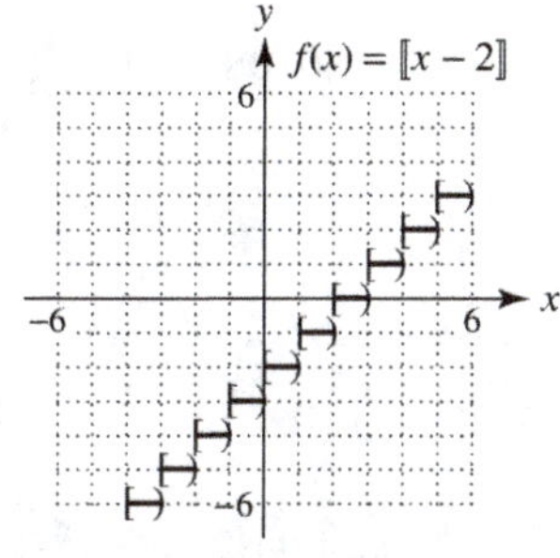

The domain is $(-\infty, \infty)$.

The range is $\{\ldots, -2, -1, 0, 1, 2, \ldots\}$ or the set of integers.

6.
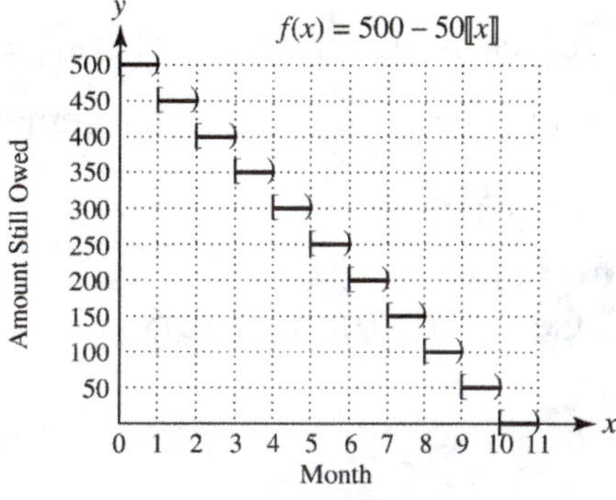

Practice Exercises

5.
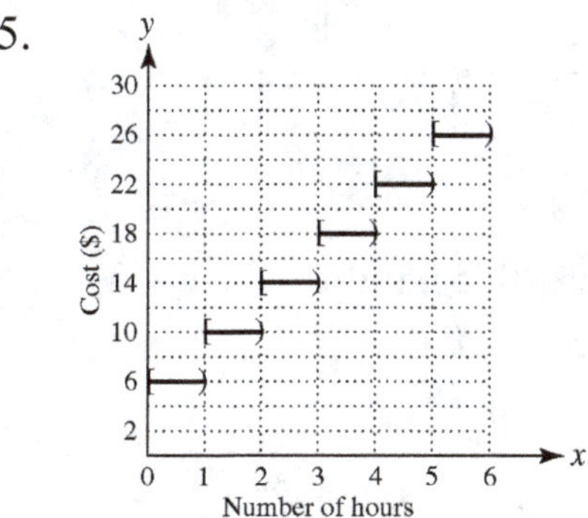

Answers

10.2 Circles Revisited and Ellipses

Key Terms

1. circle
2. ellipse
3. conic sections
4. center of the circle
5. radius
6. center-radius form
7. foci
8. center

Objective 1

Now Try

1a. Center: (0, 0); radius : 5

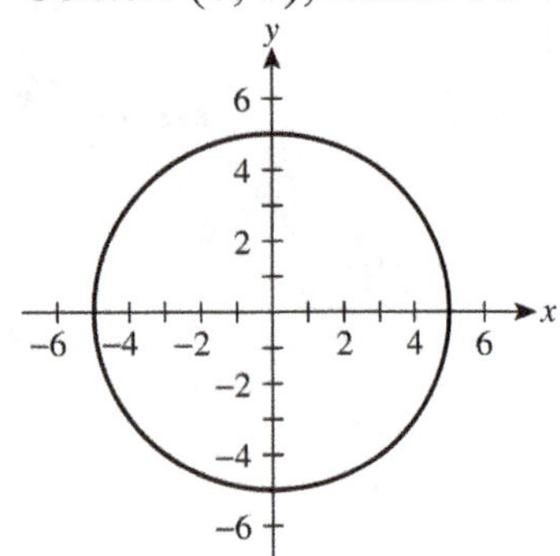

1b. Center: (−5, 4); radius: 4

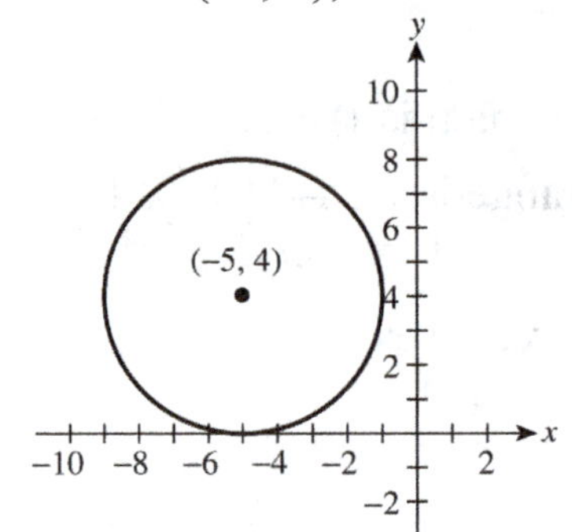

Practice Exercises

1. Center: (1, 4); radius 2

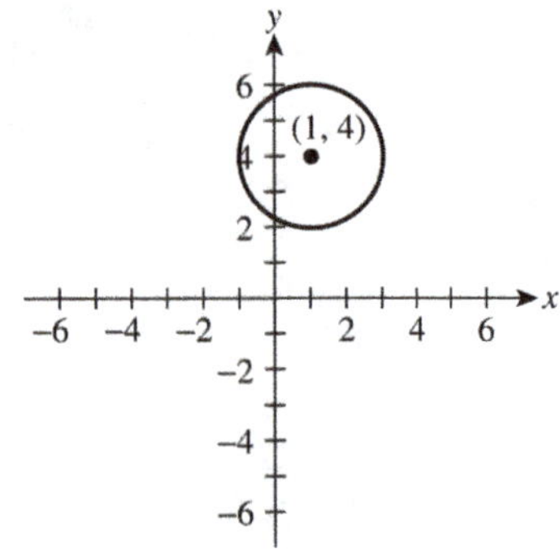

3. Center: (0, 5); radius: 3

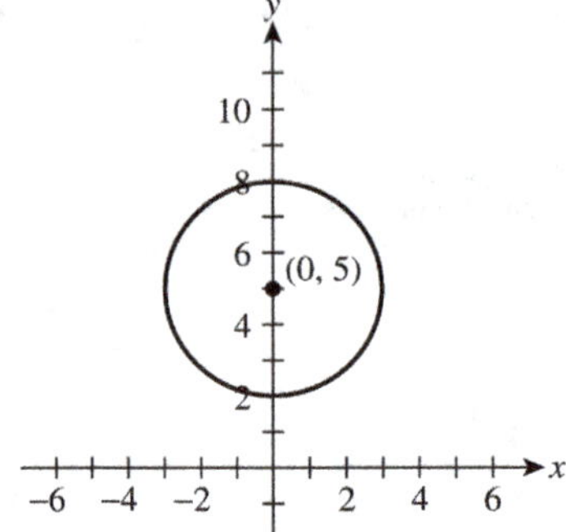

Objective 2

Now Try

2. $x^2 + (y-3)^2 = 2$

Practice Exercises

5. $(x-4)^2 + y^2 = 11$

Objective 3

Now Try

3. Center: (4, 1); radius: $\sqrt{2}$

Practice Exercises

7. Center: (2, −4); radius: 3

9. Center: (−5, −2); radius: 6

Objective 4

Objective 5
  Now Try
  4.

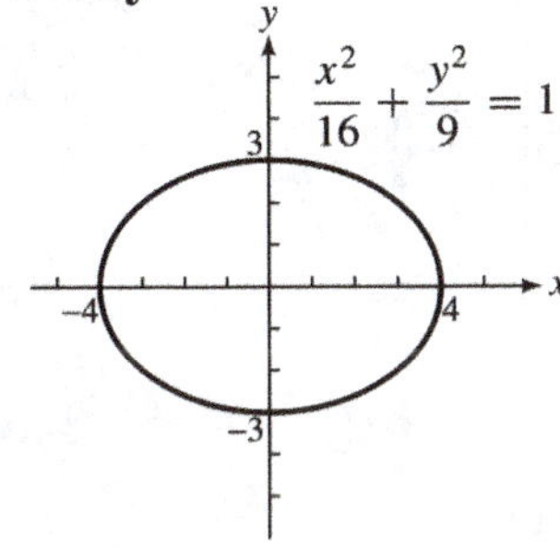

$$\frac{x^2}{16} + \frac{y^2}{9} = 1$$

5.

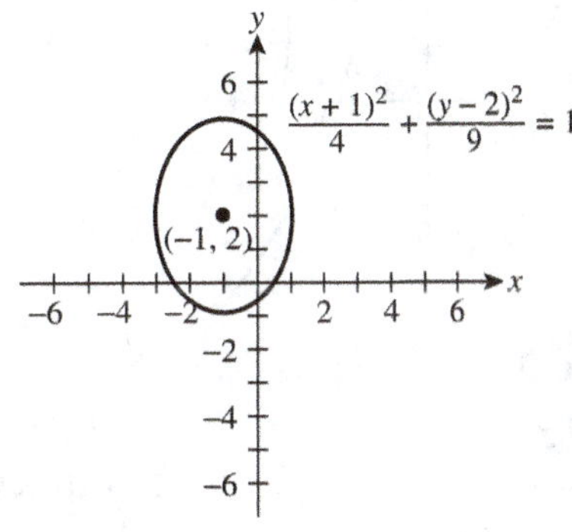

$$\frac{(x+1)^2}{4} + \frac{(y-2)^2}{9} = 1$$

  Practice Exercises
11.

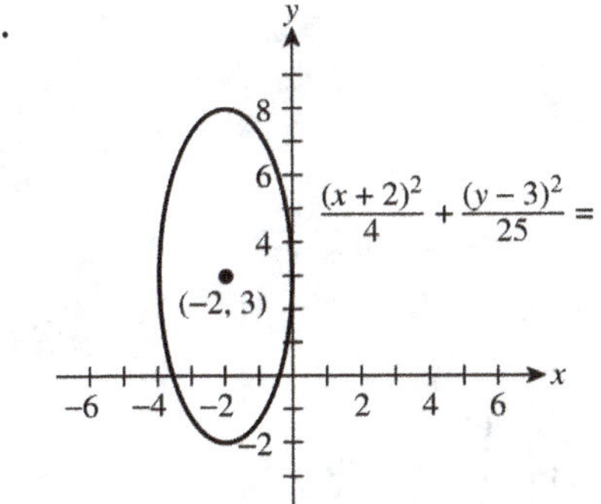

$$\frac{(x+2)^2}{4} + \frac{(y-3)^2}{25} = 1$$

10.3 Hyperbolas and Functions Defined by Radicals

Key Terms
  1. hyperbola
  2. asymptotes
  3. fundamental rectangle
  4. transverse axis
  5. generalized square root function

Objective 1

Objective 2
  Now Try
  1.

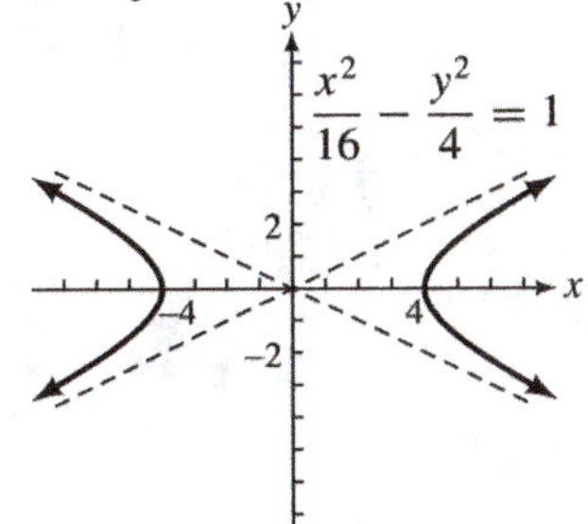

$$\frac{x^2}{16} - \frac{y^2}{4} = 1$$

Practice Exercises

1. 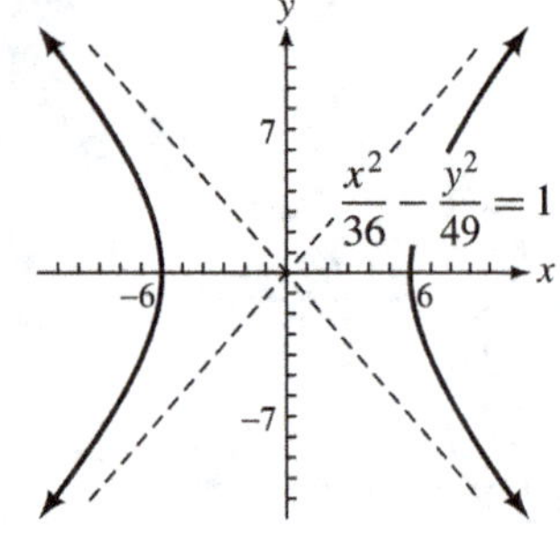

3. 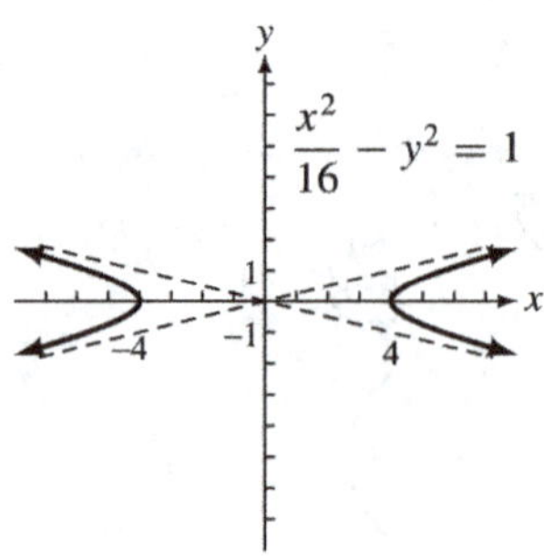

Objective 3
Now Try

3a. hyperbola   3b. ellipse   3c. parabola

Practice Exercises

5. hyperbola

Objective 4
Now Try

4. 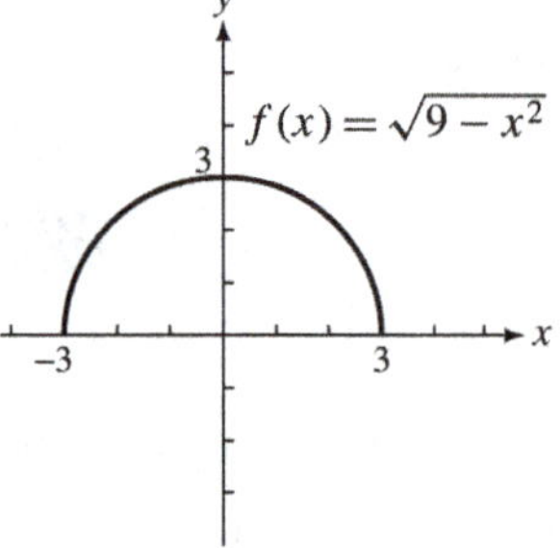

Domain: $[-3, 3]$
Range: $[0, 3]$

5. 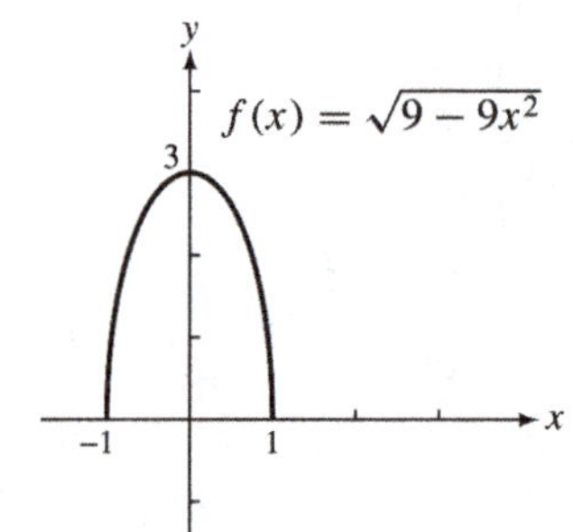

Domain: $[-1, 1]$
Range: $[0, 3]$

Practice Exercises

7. 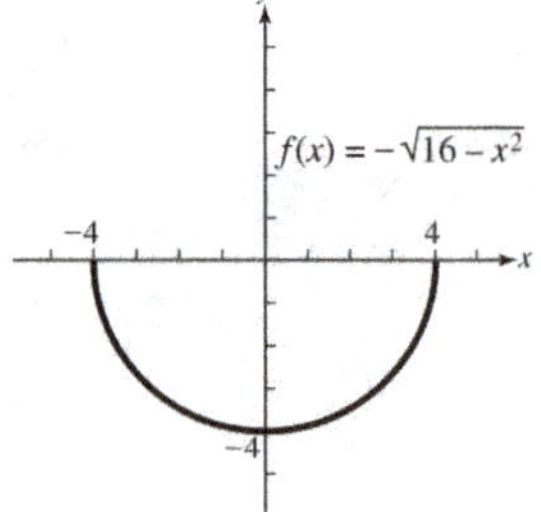

Domain: $[-4, 4]$
Range: $[-4, 0]$

9. 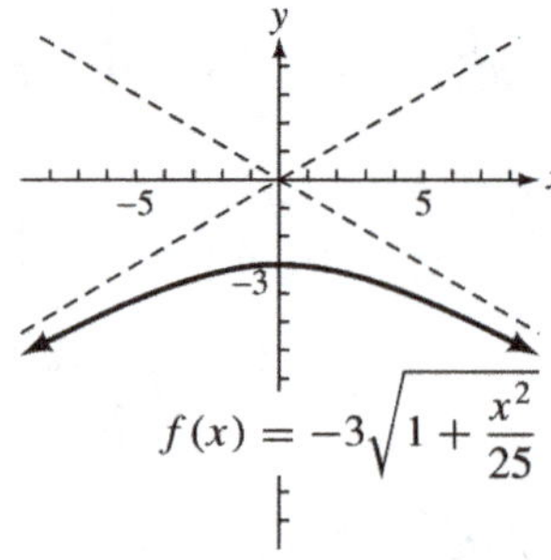

Domain: $(-\infty, \infty)$
Range: $(-\infty, -3]$

10.4 Nonlinear Systems of Equations

Key Terms

1. nonlinear equation   2. nonlinear system of equations

Objective 1

Now Try

1. $\{(-4, 1), (1,-4)\}$ 2. $\{(1, 1)\}$

Practice Exercises

1. $\{(12,-17), (2, 3)\}$ 3. $\{(3, 8), (-4,-6)\}$

Objective 2

Now Try

3. $\left\{(0,-4),\ (2\sqrt{3},\ 2),\ (-2\sqrt{3},\ 2)\right\}$

Practice Exercises

5. $\{(3, 1), (3,-1), (-3, 1), (-3,-1)\}$

Objective 3

Now Try

4. $\{(-4, 1), (-1, 4), (1,-4), (4,-1)\}$

Practice Exercises

7. $\{(2,-3), (-2, 3), (3,-2), (-3, 2)\}$

9. $\{(2, 2), (-2,-2), (2i,-2i), (-2i, 2i)\}$

10.5 Second-Degree Inequalities and Systems of Inequalities

Key Terms

1. system of inequalities   2. second-degree inequality

Objective 1

Now Try

2. 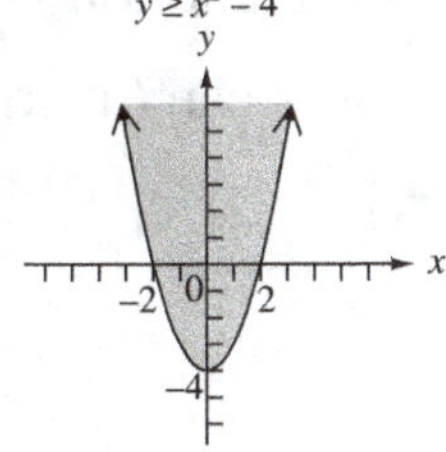

3. 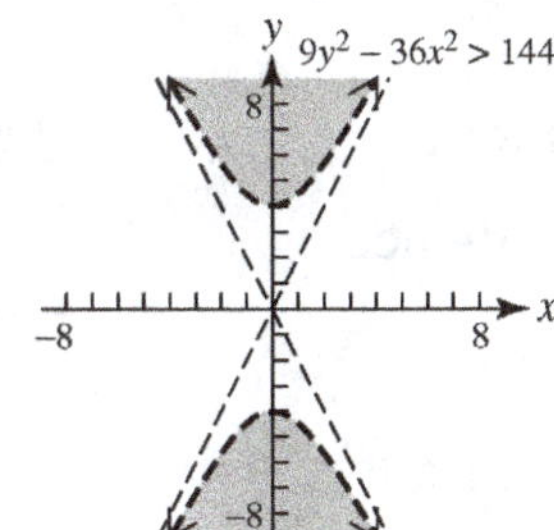

Practice Exercises

1. 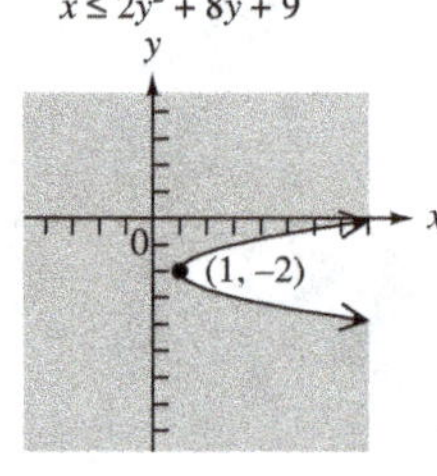

$$x \le 2y^2 + 8y + 9$$

3. 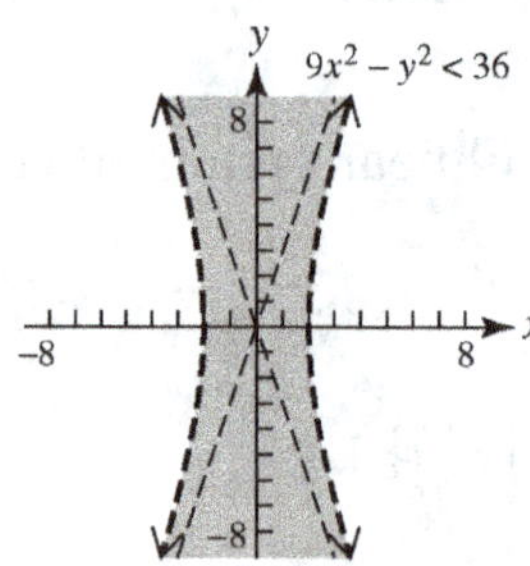

$$9x^2 - y^2 < 36$$

Objective 2
Now Try

5. 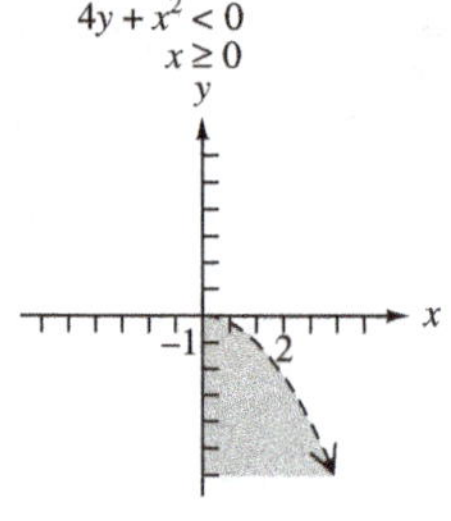

$$4y + x^2 < 0$$
$$x \ge 0$$

7. 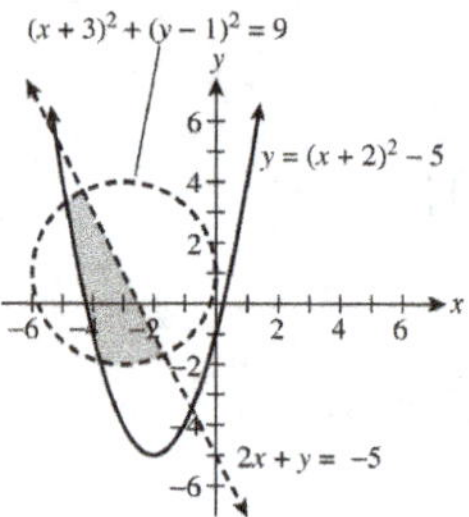

$$(x + 3)^2 + (y - 1)^2 = 9$$
$$y = (x + 2)^2 - 5$$
$$2x + y = -5$$

Practice Exercises

5.

$$x^2 + y^2 \le 25$$
$$3x - 5y > -15$$

Chapter 11 FURTHER TOPICS IN ALGEBRA

11.1   Sequences and Series

Key Terms

1. infinite sequence
2. series
3. summation notation
4. index of summation
5. general term
6. arithmetic mean (average)
7. terms of a sequence
8. finite sequence

Objective 1

Objective 2
Now Try

1. $a_6 = \dfrac{8}{30} = \dfrac{4}{15}$

Practice Exercises

1. $-1, 1, -1, 1, -1$

3. $\dfrac{11}{21}$

Objective 3
Now Try
2. $a_n = (-1)^{n-1} 2^n$

Practice Exercises
5. $a_n = \left(\sqrt{3}\right)^n$

Objective 4
Now Try
3. Month 1: \$620; Month 2: \$607.60; Month 3: \$595.45; Month 4: 583.54
   Remaining balance: \$3593.41

Practice Exercises
7. 16 g

Objective 5
Now Try
4. $1 + 5 + 9 + 13 + 17 + 21 + 25 = 91$

Practice Exercises
9. $5 + 7 + 9 + 11 = 32$ 11. $2 + 5 + 10 + 17 + 26 + 37 = 97$

Objective 6
Now Try
5. $\displaystyle\sum_{i=1}^{6} 4i$

Practice Exercises
13. $\displaystyle\sum_{i=1}^{5} \frac{1}{3i + 2}$

Objective 7
Now Try
6. About 3.38 in.

Practice Exercises
15. 8

Answers

11.2 Arithmetic Sequences

Key Terms
  1. arithmetic sequence (arithmetic progression)   2. common difference

Objective 1
  Now Try
  1. 2   2. 9, 4,–1,–6,–11

  Practice Exercises
  1. 4   3. 6, 3, 0,–3,–6

Objective 2
  Now Try
  3. $a_n = 31 + 4n$

  Practice Exercises
  5. $a_n = \dfrac{1}{3}n + \dfrac{1}{6}$

Objective 3
  Now Try
  4. $41,800

  Practice Exercises
  7. $10,900   9. $11.75

Objective 4
  Now Try
5a. 93   5b. –44   6. 16

  Practice Exercises
  11. $\dfrac{41}{2}$

Objective 5
  Now Try
  7. 39   8. 806   9. 365

  Practice Exercises
  14. 80

11.3   Geometric Sequences

Key Terms

1. common ratio   2. annuity

3. geometric sequence (geometric progression)   4. future value of an annuity

5. ordinary annuity   5. term of the annuity   6. payment period

Objective 1

Now Try

1. $\dfrac{1}{4}$

Practice Exercises

1. $\sqrt{2}$

Objective 2

Now Try

2. $a_n = \left(\sqrt{3}\right)^n$

Practice Exercises

3. $a_n = 6 \cdot 2^{n-1}$ 5. $a_n = -4\left(\dfrac{1}{5}\right)^n$

Objective 3

Now Try

3a. -729 3b. $\dfrac{1}{384}$ 4. $0.8, -4, 20, -100, 500$

Practice Exercises

7. -162

Objective 4

Now Try

5. $\dfrac{728}{3}$ 6. 682

Practice Exercises

9. 2186 11. -258

Objective 5

Now Try

7a. $328,988.05   7b. $132,877.70

Practice Exercises

13. $3464.34

Objective 6
Now Try
8. $\dfrac{25}{6}$ 9. $|r| > 1$, so the sum does not exist.

Practice Exercises
13. $\dfrac{3}{2}$

11.4 The Binomial Theorem

Key Terms
1. binomial theorem (general binomial expansion)   2. Pascal's triangle

Objective 1
Now Try
1. 3,628,800   2a. 1   2b. 36

2c. 9   2d. 126   3. 45

4. $81r^4 - 216r^3 s + 216r^2 s^2 - 96rs^3 + 16s^4$

5. $-\dfrac{x^5}{32} + \dfrac{5}{8}x^4 - 5x^3 + 20x^2 - 40x + 32$

Practice Exercises
1. 35

3. $32y^5 + 240y^4 z + 720y^3 z^2 + 1080y^2 z^3 + 810yz^4 + 243z^5$

Objective 2
Now Try
6. $-5103m^6 r^5$

Practice Exercises
5. $-4320x^2 y^3$